Theoretical Foundations of Programming Methodology

NATO ADVANCED STUDY INSTITUTES SERIES

Proceedings of the Advanced Study Institute Programme, which aims at the dissemination of advanced knowledge and the formation of contacts among scientists from different countries

The series is published by an international board of publishers in conjunction with NATO Scientific Affairs Division

A Life Sciences B Physics	Plenum Publishing Corporation London and New York
C Mathematical and Physical Sciences	D. Reidel Publishing Company Dordrecht, Boston and London
D Behavioural and Social Sciences E Engineering and Materials Sciences	Martinus Nijhoff Publishers The Hague, London and Boston
F Computer and Systems Sciences G Ecological Sciences	Springer Verlag Heidelberg

Series C – Mathematical and Physical Sciences

Volume 91 – Theoretical Foundations of Programming Methodology

Theoretical Foundations of Programming Methodology

Lecture Notes of an International Summer School, directed by F. L. Bauer, E. W. Dijkstra and C. A. R. Hoare

Lecturers: M. Broy, R. M. Burstall, B. Courcelle, E. W. Dijkstra, C. A. R. Hoare, Z. Manna, J. M. Morris, M. Nivat, D. S. Scott, S. Sickel, J. E. Stoy, M. Wirsing

edited by

MANFRED BROY

and

GUNTHER SCHMIDT

Institut für Informatik der Technischen Universität München, B.R.D.

D. Reidel Publishing Company

Dordrecht : Holland / Boston : U.S.A. / London : England

Published in cooperation with NATO Scientific Affairs Division

Library of Congress Cataloging in Publication Data
Main entry under title

Theoretical foundations of programming methodology.

(NATO advanced study institutes series. Series C, Mathematical and physical sciences; v. 91)
Papers presented in 1981 at the Marktoberdorf Summer School on Theoretical Foundations of Programming Methodology, organized under the auspices of the Technical University Münich and sponsored by the NATO Science Committee under the 1981 advanced study institutes program.
'Published in cooperation with NATO Scientific Affairs Division'.
Includes bibliographical references.
1. Electronic digital computers–Programming–Addresses, essays, lectures. I. Broy, M., 1949- II. Schmidt, Gunther, 1939- III. Marktoberdorf Summer School on Theoretical Foundations of Programming Methodology (1981) IV. Technische Universität München. V. NATO Science Committee. VI. Series.
QA76.6.T446 1982 001.64'2 82-12347

ISBN-13: 978-90-277-1462-6 e-ISBN-13: 978-94-009-7893-5
DOI: 10.1007/978-94-009-7893-5

Published by D. Reidel Publishing Company
P.O. Box 17, 3300 AA Dordrecht, Holland

Sold and distributed in the U.S.A. and Canada
by Kluwer Boston Inc.,
190 Old Derby Street, Hingham, MA 02043, U.S.A.

In all other countries, sold and distributed
by Kluwer Academic Publishers Group,
P.O. Box 322, 3300 AH Dordrecht, Holland

D. Reidel Publishing Company is a member of the Kluwer Group

Softcover reprint of the hardcover 1st edition 1982

TABLE OF CONTENTS

PART III: ABSTRACT DATA TYPES

PART IV: INFINITE STRUCTURES

PART V: CONCURRENT PROGRAMS

PREFACE

Long ago, the welfare of a society used to depend heavily on the skill and dedication of its craftsmen - the miller, the blacksmith, the cobbler and the tailor. These craftsmen acquired their skill by a long and poorly paid apprenticeship to some master of their craft. They learned by imitation and experience, and by trial and error. They did not read books or study science, they knew nothing of the theory of their subject, the geometry of their rudimentary drawings, nor the mathematics underlying their primitive calculations. They could not explain how or why they used their methods; yet they worked effectively by themselves or in small teams to complete their tasks at a predicted cost, to a fairly well predicted timescale, and usually to the satisfaction of their clients.

The programmer of today shares many of these attributes of a craftsman. He learns his craft by apprenticeship in an existing team of programmers - but his apprenticeship is highly paid and usually very short. He develops his skill by trial; but mostly by error. He does not study theory, or even read books on Computer Science. He knows nothing of the logical and mathematical foundations of his profession; and he hates to explain or justify, or even to document what he has done. Yet he can often manage to complete his undertaken tasks, sometimes at the predicted time and within the predicted cost, and occasionally even to the satisfaction of his client.

In modern times, we see the emergence of a new class of specialist: the professional engineer. He differs most from the craftsman in

M. Broy and G. Schmidt (eds.), Theoretical Foundations of Programming Methodology, vii–xiii.

that he has spent many years in school and at the university studying the mathematical and scientific foundations of his subject: the differential and integral calculus, the derivation and solution of complex equations, the formulae of mechanics and hydrodynamics and electronics. Throughout his career he will read books and articles to keep abreast of the development of his subject, or to learn newly emerging branches of it.

Above all, he will organise his own work and that of his colleagues in the light of his understanding of relevant theory in order to reduce to a minimum the risk of error leading to catastrophe. As a result, most bridges built by modern engineers do not fall down; most aeroplanes flown by professional pilots do not crash. One day in the future, we hope, one will be able to say the same about programmers: that they can deliver software products - even compilers and operating systems - which never crash, because from the very beginning they contain no errors. But we can realise that hope only if all programmers recognise and fully understand the logical and mathematical foundations of our profession; and on this basis learn to construct mathematical specifications with the same care and completeness as an engineer surveys the site for a new bridge or the route for a new road. Working directly from these specifications the programmer must learn how to construct programs which will meet their specifications with mathematical rigour and certainty. Only then will he justify a claim that the arcane and error-prone craft of computer programming has been transformed into a respectable modern engineering profession.

Let me not underestimate the enormity of the revolution which I advocate. Such revolutions are not easily accepted, because they require us to change the very way we think.

Those who were participants of the Marktoberdorf Summer School on Theoretical Foundations of Programming Methodology will remember the farewell dinner speech of Professor C.A.R. Hoare. With his kind permission, we have presented a freely cited part of it as a Preface. He continued, addressing the participants:

Together we have joined an adventure in search of wisdom. This adventure has only just begun. Tomorrow when you leave the school, you must carry on the good work. I hope you will develop our understanding by your own study and reading and research in the years to come, that you will be able to find useful application for the concepts and methods you have learned here in the design and implementation of your own programs: small ones first to gain experience and confidence, and then larger ones, because that is where these methods are most needed. And I hope that you will be able to inspire and encourage your colleagues, subordinates and even your managers and professors to follow the ideals of mathematical and logical rigour which have been so ably and convincingly expounded by the lecturers at this school. You have to play an important part in the immense task which faces us: to transform a medieval craft of programming to the highest standards of a modern engineering profession.

The papers in this volume represent most of the material presented at the Summer School where the lectures were given by

M. Broy, R.M. Burstall, B. Courcelle,
E.W. Dijkstra, C.A.R. Hoare, Z. Manna,
J.M. Morris, M. Nivat, S. Sickel, J.E. Stoy and M. Wirsing.

We are very grateful to them and the other authors for sending their manuscripts so promptly. We enjoyed the help of the directors of the Summer School and the organising staff.

The Summer School was organized under the auspices of the Technical University Munich and was sponsored by the Nato Science Committee under the 1981 Advanced Study Institutes Programme. Partial support for the conference was provided by the European Research Council and the National Science Foundation.

In order to give at least a glimpse on the atmosphere at the particular lectures, for every lecturer a brief piece of music is given. This music had been composed by Thomas A. Matzner and played by Joe Stoy on an organ in Marktoberdorf during the farewell dinner (The original sequence was I, II, III, IV, VI, IX, XI, X, VIII, VII, XII, V, I.)

Manfred Broy, Gunther Schmidt

Theme I.

(F. L. Bauer)

Gave an appropriate Bavarian flavour
to the entire enterprise

Theme II.

(C.A.R. Hoare)

A very English sound, with clinks and clunks

Theme III.

(E.W. Dijkstra)

An academic presentation, with much formal symmetry

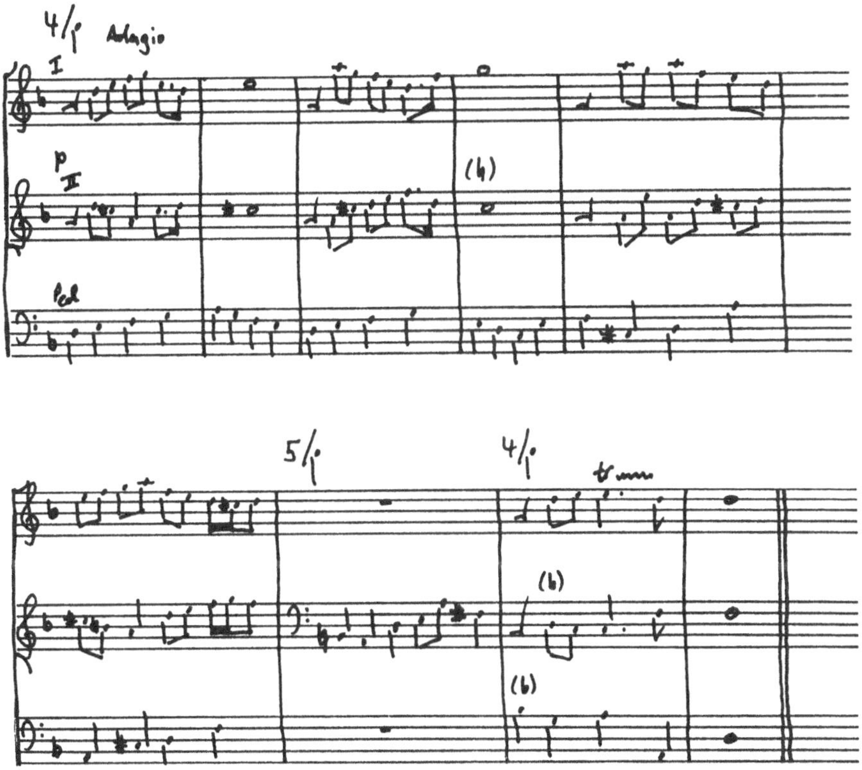

Part I
PROGRAM DEVELOPMENT AND VERIFICATION

In this part, the formal techniques of an orderly development of an imperative sequential program are exemplified with SMOOTHSORT. The LAMBEK-MOSER note presents a derivation of properties of a specified function during the process of program development. Other approaches presented in this part include a Horn clause technique for specification and derivation of programs as well as temporal logic used in order to specify and verify a computer program. These two techniques may be viewed as resembling both the program development and the program verification direction. On the other hand, the investigations of the assignment axiom for linked data structures ("pointers") concentrate on the verification of a given program: in this case, the SCHORR-WAITE algorithm.

SMOOTHSORT, AN ALTERNATIVE FOR SORTING IN SITU

Edsger W.Dijkstra

Burroughs,
Plataanstraat 5,
5671 AL NUENEN,
The Netherlands

Abstract

Heapsort [0] [1] is an efficient algorithm for sorting $m(i\colon 0 \leq i < N)$ in situ; some, however, consider it a disadvantage of heapsort that it absolutely fails to exploit the circumstances in which the sequence is initially nearly sorted. While sharing in general with heapsort its N.log N characteristic, smoothsort does not share this disadvantage: for an initially (nearly) sorted sequence, smoothsort is of order N with a smooth transition between the two. Smoothsort can be viewed as a pure exchange sort that is of order N.log N in the worst case. For brevity's sake we shall describe sorting the integer sequence $m(i\colon 0 \leq i < N)$ in ascending order.

General outline of smoothsort

After a preparation in its first phase, smoothsort builds up the sorted sequence from right to left, i.e. it maintains between q and m

$$P0\colon\ (\underline{A}\ i, j\colon 0 \leq i < j \wedge q \leq j < N\colon m(i) \leq m(j)) \wedge 1 \leq q \leq N \quad ,$$

which is vacuously true for $q = N$ and enjoys the useful property that $P0 \wedge q = 1$ implies that the sequence m is in ascending order. (Since smoothsort modifies m only be swapping its elements, m obiously remains a permutation of the same bag of

M. Broy and G. Schmidt (eds.), Theoretical Foundations of Programming Methodology, 3–17.

values.)

The second relation, built up during smoothsort's first phase and maintained during its second phase, is

P1: the unsorted prefix $m(i: 0 \leq i < q)$ is the postorder traversal of a tree in which no son exceeds its father.

Relation P1 ensures that the rightmost element of the unsorted prefix is its maximum element and that, therefore, q can be decreased by 1 without violating P0 . In order to maintain P1 , however, the decrease $q:= q - 1$ must, in general, be accompanied by a rebuilding of the tree. This clerical obligation has no analogue in heapsort , in which a similar tree is pruned by removing a leaf; in smoothsort the tree is pruned at its root and without precautions it would, in general, fall apart into a forest of subtrees. Smoothsort restores the tree by grafting each subtree of the forest on the root of the subtree to the right of it.

Note that relation P1 has been inspired by the desire to leave the sequence m untouched when initially already in ascending order.

Once the shape of the tree for $q = N$ has been chosen, the grafting procedure sketched above determines the shape of the tree for all smaller values of q . Our desire to construct an algorithm that would be of order N when m is initially (nearly) sorted forced us to derive the shape of the next tree from that of the preceding one. This recurrent computation, which heavily depends on the way in which shapes of trees are represented, is responsible for much of smoothsort's apparent complexity.

The presentation of smoothsort

In our presentation we shall follow the principle of postponing definitions until they are needed and --as a special case-- not introducing variables until they are needed. The latter leads to so-called "program projections". A program is projected on a subset of its variables by omitting the declarations of its other variables and all statements not assigning to any of the variables of the subset projected on; the remaining expressions may only depend on the variables of the subset. Each time we shall give the minimal extension of the subset projected on. In the new statements thereby introduced, the variables introduced earlier are constants.

This way of presentation has the advantage of introducing one complication at a time. It has the disadvantage of hiding the heuristics that led to the algorithm to be presented; the general outline and later remarks have been included to overcome this disadvantage as much as possible. (I think we shall have to learn to live with the fact that presenting the final design in the most disentangled way and giving the heuristics --perhaps even in the form of a possible design history-- are not necessarily compatible goals.) Finally I beg the impatient reader to remember that a program projection --though a legal program-- does not make sense in isolation: its sole purpose is to be extended to something meaningful.

When invariants are given, they precede the repetition of which they are the invariant.

The introduction of q

Projected on the variable q , smoothsort is reduced to

```
|[ q: int; q:= 1 {invariant: 1 ≤ q ≤ N}
 ; do q ≠ N → q:= q + 1 od {invariant: 1 ≤ q ≤ N}
 ; do q ≠ 1 → q:= q - 1 od
]|
```

Variable q denotes the length of the unsorted prefix; the above projection shows that smoothsort as presented here is only defined for $N \geq 1$.

The introduction of r

Projected on the variables (q, r) , smoothsort is reduced to

```
|[ q, r: int; q:= 1; r:= 0 {invariant: q - r = constant}
 ; do q ≠ N → q:= q + 1; r:= r + 1 od {invariant: q - r = const.}
 ; do q ≠ 1 → q:= q - 1; r:= r - 1 od
]|
```

Remark 0. Variable r comes in handy in two ways. Firstly because m(r) is the rightmost element of the unsorted prefix, and secondly because replacing its initialization r:= 0 by r:= X will cause smoothsort to sort the sequence m(i: $X \leq i < X + N$) . Smoothsort accommodates such a shift of origin a little bit more easily than heapsort . (End of Remark 0.)

The introduction of p , b , and c

Invariant P1 states that the unsorted prefix m(i: $0 \leq i < q$) is the postorder traversal of a tree, but does not define the tree. In this section we shall begin to define the tree for the unsorted prefix of length q and how the shape of that tree is recorded using the triple (p, b, c) .

To this purpose we regard the unsorted prefix m(i: $0 \leq i < q$) as a so-called standard concatenation of so-called stretches.

A "stretch" is a subsequence of consecutive elements m(i: $h \leq i < h1$) for some $h < h1$ (which we shall later identify with the postorder traversal of a binary subtree of the tree mentioned in P1). As we shall see later, it is desirable that the number of stretches that concatenated together constitute the unordered prefix is relatively small. Stretches, however, don't come in all possible lengths and when q is not a stretch length we need more stretches to cover m(i: $0 \leq i < q$) . The available stretch lengths are the so-called Leonardo numbers

... 41 25 15 9 5 3 1 1 (-1)

given by

$$LP_0 = LP_1 = 1 \quad \text{and} \quad LP_{n+2} = LP_{n+1} + LP_n + 1$$

(The justification for this choice of available stretch lengths is better postponed.)

The "standard concatenation" of a sequence of length q1 consists of the longest stretch with length $\leq$ q1 , followed by the standard concatenation of the remainder (when not empty).

Remark 1. We leave it as an exercise for the reader to convince himself of the fact that the standard concatenation of a sequence of given length decomposes that sequence into the minimum number of stretches. (End of Remark 1.)

For the sake of the recurrent stretch length computations, we introduce for each stretch length b its "companion", i.e. we maintain

$$(\underline{E}\ n:\ n \geq 0:\ b = LP_n \wedge c = LP_{n-1}) \qquad ;$$

here LP_{-1} is to be taken = -1 . This is achieved by modifying variables b and c using only "up" and "down" , defined by

up: b, c := b + c + 1, b and down: b, c := c, b - c - 1 .

The stretches forming a standard concatenation are given by the triple (p, b, c); more precisely, with a binary representation of p

$$\ldots\ldots \quad p_5\ p_4\ p_3\ p_2\ p_1\ p_0$$

the triple (p, b, c) defines the set of stretches LP_{n+i} for all i such that $p_i = 1$ and n defined by

$$LP_n = b \wedge LP_{n-1} = c$$

Note 0. As a first result, the length of the standard concatenation given by the triple (p, b, c) can --destructively-- be computed by

```
   length:= 0
; do p > 0 →
    if even(p) → p:= p/2; up
    ▯ odd(p)  → length:= length + b; p:=(p - 1)/2; up
    fi
  od                          (End of Note 0.)
```

Note 1. The representation is not unique: the operations "p:= 2*p; down" leave the standard concatenation represented by the triple (p, b, c) unchanged. (End of Note 1.)

The above coding of a standard concatenation is possible because, with the exception of stretch length 1 , which may occur twice in a standard concatenation --e.g. of length 2 or 7 -- , each stretch length occurs at most once, whereas for stretch length 1 we have LP_1 and LP_0 at our disposal. We adopt the additional convention of recording a single stretch of length 1 as LP_1 .

Note 2. We leave it as an exercise for the reader to prove that, as a consequence of the stretch lengths being Leonardo numbers, in the binary representation of p only the two least significant 1's may be adjacent. This fact will be used in out next projection. (End of Note 2.)

We now extend the subset of variables projected on by adding the triple (p, b, c) satisfying the invariant

P2: the length of the standard concatenation represented by the triple (p, b, c) equals q .

```
|[ q, r, p, b, c: int; q:= 1; r:= 0; p, b, c := 1, 1, 1
   {invariant: P2}
 ; do q ≠ N
     → if p mod 8 = 3
         → p:=(p - 1)/2; up; p:=(p - 1)/2; up; p:= p + 1 {b ≥ 3}
       [] p mod 4 = 1
         → down; p:= 2*p
         ; do b ≠ 1 → down; p:= 2*p od; p:= p + 1 {b = 1}
       fi; q:= q + 1; r:= r + 1
   od {invariant: P2}
 ; do q ≠ 1
     → q:= q - 1; r:= r - 1
     ; if b = 1
         → p:= p - 1; do even(p) → p:= p/2; up od {p mod 4 = 1}
       [] b ≥ 3
         → p:= p - 1; down; p:= 2*p + 1; down; p:= 2*p + 1
           {p mod 8 = 3}
       fi
   od
]|
```

Note 3. For the (nonempty!) standard concatenation we have chosen in the above the "normalized" representation with odd(p) . (End of Note 3.)

Note 4. The assertions at the end of each alternative have been given in order to stress that --as it should be!-- the one repeatable statement is the inverse of the other: assertions in the one reappear as guards in the other [2]. (End of Note 4.)

Note 5. The reader may wish to prove that p's property as described in Note 2 is an invariant of both repetitions. (End of Note 5.)

Note 6. The above projection is still of order N . The argument is as follows. In the first repetition the number of "down's" is bounded by the number of "up's" , which is certainly less than 2N . The second repetition is merely the inverse of the first one and the conclusion follows. (End of Note 6.)

The introduction of m

At last the time has come to describe how stretches and the standard concatenation define which order relations between elements of m are maintained by smoothsort . We begin with the stretches, on which the predicates "trusty" and "dubious" will be defined. In accordance with the interpretation of a stretch as the postorder traversal of a binary tree we shall refer to the rightmost element of a stretch as the "root" of that stretch.

Denoting a sequence of length LP_n by $< seq_n >$, we parse for $n \geq 2$

$$< seq_n > = < seq_{n-1} > < seq_{n-2} > < root >$$

where $< root >$ stands for a singleton sequence. Stretch $< seq_n >$ is dubious means that both $< seq_{n-1} >$ and $< seq_{n-2} >$ are trusty. Stretch $< seq_n >$ is trusty means that, in addition, the roots of $< seq_{n-1} >$ and $< seq_{n-2} >$ are at most the root of $< seq_n >$; a stretch of length 1 is by definition both dubious and trusty. As a consequence, the root of a trusty stretch is a maximum element of that stretch.

When stretches thus parsed are viewed as postorder traversals of binary trees, trustiness means that no son exceeds its father. A dubious stretch is made into a trusty one by applying the operation "sift" --a direct inheritance from heapsort -- to its root, where sift is defined as follows: sift applied to an element without larger sons is a skip, sift applied to an element m(r1) that is exceeded by its largest son m(r2) consists of a swap of these two values, followed by an application of sift to m(r2) .

Remark 2. We can now partly justify our choice of the Leonardo numbers as available stretch lengths, i.e. justify why we have not chosen (with the same recurrence relation)

... 33 20 12 7 4 2 1 (0)

The occurrence of lengt 2 would have required a sift able to deal with fathers having one or two sons, like the sift required in heapsort ; thanks to the Leonardo numbers a father has always two sons and, consequently, smoothsort's sift is simpler. (End of Remark 2.)

During the second repetition smoothsort maintains

P3: the stretches of the standard concatenation of the unsorted prefix $m(i: 0 \leq i < q)$ are all trusty.

During the first one it maintains the weaker

P3': of the standard concatenation of the unsorted prefix $m(i: 0 \leq i < q)$ the rightmost stretch is dubious; its other stretches are all trusty.

Remark 3. The weaker P3' has been introduced for reasons of efficiency which cannot be explained now; see, however, Remark 4. (End of Remark 3.)

So much for the order relations captured by the stretches. In addition, smoothsort maintains during the second repetition

P4: the roots of the stretches of the standard concatenation of the unordered prefix $m(i: 0 \leq i < q)$ are ascending from left to right ,

a relation, which is useful since $P3 \wedge P4$ implies that m(r) , the rightmost element of the prefix, is a maximum element of the prefix, and this is the circumstance under which q:= q - 1 maintains P0 . During the first repetition smoothsort maintains the weaker

P4': the roots of the trusty stretches of the standard concatenation of the unordered prefix $m(i: 0 \leq i < q)$ that are also stretches of the standard concatenation of length N are ascending from left to right.

We now have to investigate

1) what to add to the first repetition for the maintenance of P3' $\wedge$ P4'

2) what to insert between the two repetitions in order to transform P3' $\wedge$ P4' into P3 $\wedge$ P4

3) what to add to the second repetition for the maintenance of P3 $\wedge$ P4 .

Investigation 1. In the case p mod 8 = 3 , the standard concatenation ends on a dubious stretch of length b which must be made trusty before it can be combined with the preceding trusty stretch and the following element into a new dubious rightmost stretch. This can be achieved by applying sift to m(r) . Since no new trusty stretch is added to the standard concatenation, P4' is maintained without further measures.

In the case p mod 4 = 1 , the standard concatenation ends on a dubious stretch of length b , which in this step becomes the last but one stretch of the standard concatenation and, hence, must be made trusty. In the case $q + c < N$, it suffices to apply sift to m(r) as before, since this stretch will later disappear from the standard concatenation. In the case $q + c \geq N$, however, just applying sift to m(r) might violate P4' since this stretch of length b also occurs in the standard concatenation of length N . Making a dubious stretch trusty and including its root in the sequence of ascending roots is achieved by applying "trinkle" to m(r) . (As we shall see later, trinkle is like sift , be it for a partly ternary tree.) (End of Investigation 1.)

Investigation 2. The reader may prove that it suffices to apply trinkle to m(r) . (End of Investigation 2.)

Investigation 3. In the case b = 1 , the standard concatenation loses its last stretch, and P3 $\wedge$ P4 is maintained without further measures.

In the case $b \geq 3$, the rightmost stretch of length b is replaced by two trusty ones; hence P3 is maintained. To restore P4 it would suffice to apply trinkle first to the root of the first new stretch and then to the root of the second new stretch, but this would fail to exploit the fact that the new stretches are already trusty to start with. This is exploited by applying "semitrinkle" in order to those roots. (End of Investigation 3.)

Remark 4. From a logical point of view it would be perfectly permissible to replace a call on trinkle by a call on sift , which would make the dubious stretch trusty, followed by a call on semitrinkle , which would include its root in the sequence of ascending roots. After this substitution, each iteration of the first repetition starts with a sift and the whole first repetition is immediately followed by a sift . Since initially the last (and only) stretch is trusty, we can transform the program by removing all calls on sift and inserting a single call on sift at the end of the repeatable statement of the first repetition. This is essentially the program transformation that would be required if we wished to replace P3' by P3 . (The collection of trusty stretches being extended, P4' would require reformulation.)

The version resulting from the above transformation is, however, rejected because a succession of sift and semitrinkle requires in general more comparisons and swaps than trinkle , as will become apparent later. This can be remedied by replacing the single call on sift by guarded calls on either sift or the combination in the form of trinkle (and removal of the calls on semitrinkle from the first repetition, which have now been catered for). P3 would still be valid, P4' would have to be changed. This version, however, is rejected since it would lead to a duplication of the evaluation of the guards p mod 8 = 3 etc. . (End of Remark 4.)

In order to enable the reader to check the code in which the calls on sift , trinkle , and semitrinkle have been inserted, we give their calling conventions. (These conventions are not to be regarded as a recommendation: they have been chosen because in this publication I did not want to make any assumptions about a parameter mechanism.)

Routine sift is applied to the root m(r1) of a stretch of length b1 , of which c1 is the companion. Routine trinkle is applied to the root m(r1) of the last stretch of the standard concatenation represented by the triple (p, b, c) ; this representation need not be normalized. Routine semitrinkle is applied to the root m(r) of a stretch of length c which is preceded by the nonempty standard concatenation represented by the triple (p, b, c) ; again this representation is not necessarily normalized.

Note that "p:=(p - 1)/2; p:=(p - 1)/2; p:= p + 1" has been

simplified to "p:=(p + 1)/4" and that "r:= r - b + c; down; r:= r + c" decreases r by 1 .

```
smoothsort:
|[ q, r, p, b, c, r1, b1, c1: int
 ; q:= 1; r:= 0; p, b, c := 1, 1, 1 {invariant: P3' ∧ P4'}
 ; do q ≠ N
      → r1:= r
      ; if p mod 8 = 3
          → b1, c1 := b, c; sift; p:=(p + 1)/4; up; up
         [] p mod 4 = 1
          → if q + c < N → b1, c1 := b, c; sift
             [] q + c ≥ N → trinkle
            fi; down; p:= 2*p
          ; do b ≠ 1 → down; p:= 2*p od; p:= p + 1
        fi; q:= q + 1; r:= r + 1
   od {P3' ∧ P4'}; r1:= r; trinkle {invariant: P3 ∧ P4}
 ; do q ≠ 1
      → q:= q - 1
      ; if b = 1
          → r:= r - 1; p:= p - 1; do even(p) → p:= p/2; up od
         [] b ≥ 3
          → p:= p - 1; r:= r - b + c
          ; if p = 0 → skip [] p > 0 → semitrinkle fi
          ; down; p:= 2*p + 1; r:= r + c; semitrinkle
          ; down; p:= 2*p + 1
        fi
   od
]|

up1:      b1, c1 := b1 + c1 + 1, b1

down1:    b1, c1 := c1, b1 - c1 - 1
```

```
sift:

 do b1 ≥ 3 →
    |[ r2: int; r2:= r1 - b1 + c1
     ; if m(r2) ≥ m(r1 - 1) → skip
        [] m(r2) ≤ m(r1 - 1) → r2:= r1 - 1; down1
       fi
     ; if m(r1) ≥ m(r2) → b1:= 1
        [] m(r1) < m(r2) → m:swap(r1, r2); r1:= r2; down1
       fi
    ]|
 od


semitrinkle:
   r1:= r - c
 ; if m(r1) ≤ m(r) → skip
    [] m(r1) > m(r) → m:swap(r, r1); trinkle
   fi
```

Trinkle is very similar to sift when we regard each stretch root as the stepson of the root of the stretch to its right. Applied to a root without larger sons, trinkle is a skip; otherwise the root is swapped with its largest son, etc. The trouble with the code is that all sorts of sons may be missing. In the following, trinkle is eventually reduced to a sift , viz. when the stepson relation is no longer of interest.

```
trinkle:
|[ p1: int; p1, b1, c1 := p, b, c
 ; do p1 > 0 →
     |[ r3: int; do even(p1) → p1:= p1/2; up1 od; r3:= r1 - b1
      ; if p1 = 1 cor m(r3) ≤ m(r1) → p1:= 0
         [] p1 > 1 cand m(r3) > m(r1)
          → p1:= p1 - 1
          ; if b1 = 1 → m:swap(r1, r3); r1:= r3
             [] b1 ≥ 3 →
              |[ r2: int; r2:= r1 - b1 + c1
               ; if m(r2) ≥ m(r1 - 1) → skip
                  [] m(r2) ≤ m(r1 - 1)
                   → r2:= r1 - 1; down1; p1:= 2*p1
                 fi
               ; if m(r3) ≥ m(r2)
                   → m:swap(r1, r3); r1:= r3
                  [] m(r3) ≤ m(r2)
                   → m:swap(r1, r2); r1:= r2; down1; p1:= 0
                 fi
              ]|
            fi
        fi
     ]|
  od
]|; sift
```

And this concludes the code, in which I have abstained from implementation dependent optimizations.

In retrospect

While heapsort prunes the tree leaf by leaf, smoothsort prunes the tree at the root, and immediately one of heapsort's charms is lost: while the tree in heapsort remains beautifully balanced, the tree in smoothsort can get very skew indeed. So why bother about smoothsort at all? Well, I wanted to design a sorting algorithm of order N in the best case, of order N.log N in the worst case, and with a smooth transition between the two. (Hence its name.)

This is also the answer to the question why I introduced P4. By dropping P4 one can dispense with trinkle and the code becomes much simpler. The price to be paid is a search for the maximum stretch root in order to establish that m(r) is a maximum element of the unsorted prefix. Though such a simpler sorting algorithm is quite defensible, I rejected the option because it is never of order N .

One can also raise the question why I have not chosen as available stretch lengths: ... 63 31 15 7 3 1 , which seems attractive since each stretch can then be viewed as the postorder traversal of a balanced binary tree. In addition, the recurrence relation would be simpler. But I know why I chose the Leonardo numbers: with balanced binary trees the average number of stretches is $1.2559 \{= \frac{1}{4}(5 + \sqrt{5})({}^{2}\log(1 + \sqrt{5}) - 1)\}$ times the average number of stretches with the Leonardo numbers. (I do not present this ratio as a compelling argument.)

It is possible that others have thought of this algorithm, but have rejected it for valid reasons, as yet unknown to me. I could not find it in the literature and it is not mentioned in [3], a recent article that compares five well-known sorting algorithms when fed with initially nearly sorted sequences. (That article compares Straight Insertion Sort, Shellsort, Straight Merge Sort, Quickersort, and Heapsort.) If it has not been discovered earlier, I would like to know the reason, because all its ingredients are well-known since the discovery of heapsort in 1964.

Besides the possible interest in smoothsort I had another reason for developing it to the degree I did and for writing the above. (It took me three weeks, but I consider them well-spent.) The reason was that I knew beforehand that in trying to present smoothsort in a way as disentangled as possible I would encounter considerable difficulties. I hope they have been surmounted sufficiently well.

Acknowledgements

I am greatly indebted to C.S.Scholten and to all members of the Tuesday Afternoon Club, with whom I had the privilege of discussing the algorithm, its coding, and its presentation. They have helped me clarifying my own thoughts and have suggested several significant simplifications. I am furthermore indebted to D.E.Knuth and W.M.Turski for their comments on the previous version of this text, and to the participants of the Marktoberdorf Summer School, 1981, on whom I could try out my presentation.

References

[0] Williams, J.W.J., Algorithm 232 HEAPSORT C.A.C.M., 7, 6 (June 1964), pp. 347 - 348

[1] Floyd, Robert W., Algorithm 242 TREESORT 3 C.A.C.M., 7, 12 (Dec. 1964), p. 701

[2] Bauer, F.L. and Broy, M. (Ed.), Program Construction, Lecture Notes in Computer Science 69, Berlin, Heidelberg, New York, Springer Verlag, 1979, pp. 54 - 57

[3] Cook, Curtis R. and Kim, Do Jin, Best Sorting Algorithm for Nearly Sorted Lists, C.A.C.M., 23, 11 (Nov. 1980), pp. 620 - 624

LAMBEK AND MOSER REVISITED

Edsger W.Dijkstra

Burroughs,
Plataanstraat 5,
5671 AL NUENEN,
The Netherlands

Abstract

This note shows how we can prove properties of a function by deriving and manipulating a program computing it.

Let f be an ascending sequence of natural numbers, i.e.

$$(\underline{A}\ i, j: 0 \leq i < j: f(i) \leq f(j))$$

that is unbounded, i.e.

$$(\underline{A}\ j: j \geq 0: (\underline{E}\ i: i \geq 0: f(i) > j))$$

The function lambo is defined as follows: lambo(f) is a sequence of natural numbers and g = lambo(f) means that for all $y \geq 0$, $g(y) = x$, where x stands for the minimum value such that $f(x) > y$, or, more formally, where x satisfies

R: $(\underline{A}\ i: 0 \leq i < x: f(i) \leq y) \wedge f(x) > y$

In order to compute g , we design a program with the invariant relation

P: $(\underline{A}\ i: 0 \leq i < x: f(i) \leq y) \wedge f(x) \geq y$

```
x, y := 0, 0; {P}                                              (1)
do f(x) = y → x:= x + 1 {P}
[] f(x) > y → {R} g(y):= x; y:= y + 1 {P} od
```


M. Broy and G. Schmidt (eds.), Theoretical Foundations of Programming Methodology, 19–23.

Note firstly, that the program

$$\underline{\text{do}}\ f(x) = y \rightarrow x:= x + 1\ \underline{\text{od}} \qquad (2)$$

terminates because f is unbounded; note secondly, that the program

$$\underline{\text{do}}\ f(x) > y \rightarrow g(y):= x;\ y:= y + 1\ \underline{\text{od}} \qquad (3)$$

terminates; note, finally, that on account of the last term of P program (1) fails to terminate.

Consider now program (4), in which we assume g to be initialized $g = \text{lambo}(f)$; the same P is again an invariant.

$$\begin{array}{l} x,\ y := 0,\ 0; \{P\} \\ \underline{\text{do}}\ f(x) = y \rightarrow x:= x + 1\ \{P\} \\ \ [\!]\ f(x) > y \rightarrow \{R,\ \text{hence}\ g(y) = x\}\ y:= y + 1\ \{P\} \\ \underline{\text{od}} \quad . \end{array} \qquad (4)$$

Also program (4) fails to terminate; because we are entitled to assert $g(y) = x$ in the second guarded command, the program still fails to terminate when we include that relation in the second guard

$$\begin{array}{l} x,\ y := 0,\ 0;\ \{Q\} \\ \underline{\text{do}}\ f(x) = y \rightarrow x:= x + 1\ \{Q\} \\ \ [\!]\ f(x) > y \wedge g(y) = x \rightarrow y:= y + 1\ \{Q\} \\ \underline{\text{od}} \quad . \end{array} \qquad (5)$$

From the fact that (5) fails to terminate we conclude a further invariant

Q: $g(y) \geq x$.

We conclude this by considering <u>non</u> Q: $g(y) < x$. In that case, (5) reduces to (2), of which <u>non</u> Q is obviously an invariant; because (2) terminates and (5) does not, <u>non</u> Q cannot occur. Having established the invariance of Q , we conclude that the program still fails to terminate when we "strengthen" the first guard with wp("x:= x + 1", Q):

$$\begin{array}{l} x,\ y := 0,\ 0; \\ \underline{\text{do}}\ f(x) = y \wedge g(y) > x \rightarrow x:= x + 1 \\ \ [\!]\ f(x) > y \wedge g(y) = x \rightarrow y:= y + 1\ \underline{\text{od}} \end{array} \qquad (6)$$

But program (6) is symmetric in the pairs (x, f) and (y, g) ; hence

$$(g = \text{lambo}(f)) = (f = \text{lambo}(g)) \quad ,$$

in other words: the function lambo is its own inverse.

Finally, let us consider the program

```
x, y, n := 0, 0, 0;                                              (7)
do f(x) = y ∧ g(y) > x → {x + f(x) = n} x, n := x + 1, n + 1
 [] f(x) > y ∧ g(y) = x → {y + g(y) = n} y, n := y + 1, n + 1
od
```

Program (7) has the obvious invariant $x + y = n$, which justifies the two assertions. They, however, are the weakest preconditions for the invariance of

Q1: the sets $\{i + f(i) \mid 0 \le i < x\}$ and $\{j + g(j) \mid 0 \le j < y\}$ form a partitioning of the first n natural numbers 0 through $n - 1$.

From the fact that x and y are unbounded and from the invariance of Q1 , which is true at initialization, we conclude the second result of Lambek and Moser, viz. that the sets $\{i + f(i) \mid 0 \le i\}$ and $\{j + g(j) \mid 0 \le j\}$ form a partitioning of the natural numbers.

Note. By introducing $n = x + y$ in program (1) we could have derived the second result of Lambek and Moser immediately; it, in turn, implies that the function lambo is its own inverse. But I thought the independent derivation of (6) more fun. (End of Note.)

* * *

I am not quite clear about the moral of the above. We have proved theorems about the function lambo by first deriving a program for it and then massaging the program. For me this is a novel application of semantics preserving program transformations, and this novelty --as all such novelties-- causes some mild excitement. On the other hand we know that a chain of program transformations is so close to a mechanically verifiable proof that it seems vain to hope to prove any "deep" theorems this way. (Here I should add that I get less and less certain about the significance of the supposed difference between "deep" and "shallow" theorems.) It is possibly no more than an occasionally neat way of formulating an otherwise not unusual mathematical argument.

Theme IV.

(J. Morris)

The same theme as Dijkstra's (movement III); more low-key and omitting the formal symmetry, but with Irish accents

A GENERAL AXIOM OF ASSIGNMENT

Joseph M. Morris

Dept. of Computer Science, Trinity College, Dublin 2.

1. INTRODUCTION

The axiomatic method of Floyd [1] and Hoare [2] has become the most popular formal method for reasoning about programs. Axiomatic semantics, more or less complete, exist for various programming languages [3, 4] and the method is widely used in deriving and verifying programs. Some areas of programming, however, have thus far defied a general axiomatic treatment. One such area is that of pointers and linked data structures, which will be the subject of this and the following two papers. The goal is to make manageable the formal verification of list-processing programs, using axiomatic semantics. The present work does not claim to be complete, but is more systematic than previous treatments [5, 6, 7, 8], and is more general. These advantages accrue from the use of Dijkstra's "weakest preconditions" [9] rather than Hoare's "sufficient preconditions".

The path to be followed is this. First, an axiom of assignment is developed that is applicable no matter what the variable on the left hand side of the assignment operator is; in particular it covers assignment to pointers. A formalism is then developed for reasoning about list-processing; it is especially concerned with connectivity relations between the nodes of a data structure. The final paper presents an example of using the formalism to verify a complex list-processing algorithm.

M. Broy and G. Schmidt (eds.), Theoretical Foundations of Programming Methodology, 25–34.

2. A GENERAL AXIOM OF ASSIGNMENT

The axiom of assignment presented in [2] defines assignment to scalar variables only. A more general axiom will now be developed.

Let a "reference" be any term that may appear on the left hand side of the assignment operator in a Pascal-like language. (The word "reference" is chosen over the more common "variable" in order to place emphasis on the name rather than on the object it refers to.) For example, given the following type declarations (using a Pascal-like notation)

```
array1  =  array(1..5) of integer;
array2  =  array(0..20,1..100) of integer;
ptr     =  pointer to rec;
rec     =  record
           year  :  integer;
           days  :  array(0..20) of integer;
           s     :  ptr
           end;
```

and given the following variable declarations

```
i, j    :  integer;
ax, ay  :  array1;
az      :  array2;
p       :  ptr
```

then the following are examples of references

```
i           j
ax(i)       ax(ay(i+j))                   ay(3)
az(i,j*i)   az(j,i)
p.year      p.s.s.days(j)
```

(In Pascal, the preceding line would be written

```
p↑.year     p↑.s↑.s↑.days(j)
```

- the upwards arrow has been omitted for notational brevity. No confusion should arise because only dynamic records are being considered.) The syntax of a reference, expressed in Backus-Naur form, follows; it omits replacement rules for <identifier> and <expression>:

```
<reference>              ::= <basic reference> | <basic reference>
                                              ( <subscript list> )
<basic reference>        ::= <variable identifier> |
                             <pointer expression>.<variable
                                                     identifier>
<pointer expression>     ::= <expression>
<subscript list>         ::= <subscript expression> |
                             <subscript list>, <subscript
                                                   expression>
<subscript expression>   ::= <expression>
<variable identifier>    ::= <identifier>
```

Note that only simple values may be assigned; assignment to entire arrays or entire records is covered in section 4, which deals with concurrent assignment.

A reference refers to a "box" which contains a simple value (once it has been initialized). The machine analogue of a box is a memory word. Each box has a unique identifying name, which will be called its address. If x denotes a reference, then the address of the box referred to by x is denoted by $x\downarrow$. If references x and y satisfy $x\downarrow = y\downarrow$, then x and y are "aliases" of each other. If x denotes a simple reference, then x has no aliases. If x contains pointer expressions or subscript expressions, then x has infinitely many aliases; such aliases are called "dynamic" aliases.

The assignment statement has the form x:= e where x denotes an arbitrary reference and e denotes an expression of the appropriate type. The assignment is assumed to terminate normally, and to be free of side-effects. If x and y denote references, and e denotes an expression, then

$$y_e^{/x} =_{df} \begin{cases} x\downarrow = y\downarrow : e \\ x\downarrow \neq y\downarrow : y \end{cases}$$

(The right hand side of the above definition will be linearized to

<u>if</u> $x\downarrow = y\downarrow$ <u>then</u> e <u>else</u> y.)

Letting R denote an arbitrary postcondition, the proposed axiom of assignment is

$$wp(x:= e,R) =_{df} R_e^{/x}$$

where $R_e^{/x}$ stands for R with every reference y in R replaced by $y_e^{/x}$ (the usual assumption re free and bound variables applies - see [2]). Less formally, $R_e^{/x}$ stands for R with each alias of x in R replaced by e. The justification of the axiom follows immediately from the justification of the original axiom [2]. Note that if x denotes a simple reference, then $R_e^{/x}$ is equivalent to R_e^x, and so the old and new axioms are identical in this special case.

In order to evaluate $R_e^{/x}$, it is necessary to be able to determine the truth or falsity of x↓ = y↓ for arbitrary references x and y. Fortunately, this is easily determined. The most general form of a reference is

<pointer expression>.<variable identifier>(<subscript list>)

The <variable identifier> is always present, but one or both of <pointer expression> and <subscript list> - call these the "qualifying expressions of the reference" - may be empty. When a qualifying expression is empty, its associated separator(s) - '.', '(', ')' - are omitted. The following table gives examples of the compositions of references in the above terms. The variables used are as previously declared; 'E' denotes the empty expression.

reference	pointer expression	variable identifier	subscript list
i	E	i	E
ax(i)	E	ax	i
ax(ay(i+j))	E	ax	ay(i+j)
az(i,j*i)	E	az	i,j*i
p.year	p	year	E
p.s.s	p.s	s	E
p.s.s.days(j)	p.s.s	days	j

Now informally, two references are aliases of one another if and only if

(i) their variable identifiers are textually identical, and
(ii) corresponding qualifying expressions are equal in value.

In applying rule (ii), note that two empty expressions are deemed equal, and that two subscript lists are equal in value if they have equal lengths and are element-wise equal in value.

More formally, let reference x have the form px.X(<u>ex</u>), and reference y have the form py.Y(ey) where px and py denote pointer expressions; X and Y denote variable identifiers; and <u>ex</u> (=$ex_1,\ldots,ex_n$) and <u>ey</u> (=$ey_1,\ldots,ey_n$) denote subscript lists. Furthermore, let 'X' and 'Y' denote the textual value of X and Y, respectively, and let the predicate ($\forall i(1\leq i\leq n)$:$ex_i = ey_i$) be denoted by (<u>ex</u> = <u>ey</u>). Then

$$x\downarrow = y\downarrow =_{df} \text{'X'} = \text{'Y'} \ \underline{\text{and}}\ px = py\ \underline{\text{and}}\ \underline{ex} = \underline{ey}.$$

The following table gives examples, using variables previously declared; to distinguish between variable identifiers and qualifying expressions, the former are typed in upper case, and the latter in lower case:

Reference x	Reference y	x↓ = y↓
I	I	<u>true</u>
I	AX(i)	<u>false</u>
AX(i)	AY(3)	<u>false</u>
AX(i)	AX(ax(i+j))	i = ax(i+j)
AZ(i,j+i)	AZ(j,i)	i = j <u>and</u> j + i = i (≡ i = j = 0)
p.s.s.DAYS(j)	p.s.DAYS(i)	p.s.s = p.s. <u>and</u> j = i
p	p.S	<u>false</u>

3. EXAMPLES

Some examples will help to clarify the new axiom. The examples will use variables declared as follows:

i,j	:	integer
b	:	array1
c	:	array2
p,q,u,v	:	ptr

where the types are as previously declared. In evaluating $R^{/x}_{e}$, no substitution will be made for references y in R satisfying x↓ ≠ y↓ where this can be determined by inspection.

<u>Example 1</u>

wp(b(j):= j, b(j) = j)
≡ $[b(j) = j]^{/b(j)}_{j}$
≡ $b(j)^{/b(j)}_{j} = j_{j}$
≡ j = j
≡ <u>true</u> □

<u>Example 2</u>

wp(b(i):= 1, b(i) = b(j))
≡ $[b(i) = b(j)]^{/b(i)}_{1}$
≡ $b(i)^{/b(i)}_{1} = b(j)^{/b(i)}_{1}$
≡ 1 = (<u>if</u> i = j <u>then</u> 1 <u>else</u> b(j))
≡ i = j <u>or</u> b(j) = 1 □

Example 3

$$
\begin{aligned}
& wp(b(j) := j, b(b(j)) = j) \\
\equiv\ & [b(b(j)) = j]_{j}^{/b(j)} \\
\equiv\ & [b(b0) = j \ \underline{\text{and}}\ b(j) = b0]_{j}^{/b(j)} \\
& \text{(where b0 denotes some constant)} \\
\equiv\ & b(b0)_{j}^{/b(j)} = j \ \underline{\text{and}}\ b(j)_{j}^{/b(j)} = b0 \\
\equiv\ & (\underline{\text{if}}\ j = b0\ \underline{\text{then}}\ j\ \underline{\text{else}}\ b(b0)) = j\ \underline{\text{and}}\ j = b0 \\
\equiv\ & (\underline{\text{if}}\ j = j\ \underline{\text{then}}\ j\ \underline{\text{else}}\ b(j)) = j \\
\equiv\ & \underline{\text{true}}
\end{aligned}
$$

One may avoid introducing the constant b0 by nesting substitutions:

$$
\begin{aligned}
& [b(b(j)) = j]_{j}^{/b(j)} \\
\equiv\ & b(b(j)_{j}^{/b(j)})_{j}^{/b(j)} = j \\
\equiv\ & b(j)_{j}^{/b(j)} = j \\
\equiv\ & j = j \\
\equiv\ & \underline{\text{true}}
\end{aligned}
$$

□

Example 4

$$
\begin{aligned}
& wp(b(b(j)) := j, b(j) = j) \\
\equiv\ & [b(j) = j]_{j}^{/b(b(j))} \\
\equiv\ & b(j)_{j}^{/b(b(j))} = j \\
\equiv\ & (\underline{\text{if}}\ b(j) = j\ \underline{\text{then}}\ j\ \underline{\text{else}}\ b(j)) = j \\
\equiv\ & b(j) = j
\end{aligned}
$$

□

It is clear from the preceding examples that in the special case of assignment to array elements, the new axiom is identical with the earlier axiomatic definition of assignment to array elements, given in [3].

Example 5

$$wp(u := v, u.s = v) \equiv v.s = v$$

□

Example 6

$$
\begin{aligned}
& wp(u.s := v, p.s = v) \\
\equiv\ & [p.s = v]_{v}^{/u.s} \\
\equiv\ & p.s_{v}^{/u.s} = v \\
\equiv\ & (\underline{\text{if}}\ p = u\ \underline{\text{then}}\ v\ \underline{\text{else}}\ p.s) = v \\
\equiv\ & p = u\ \underline{\text{or}}\ p.s = v
\end{aligned}
$$

□

Example 7

$$
\begin{aligned}
&\mathrm{wp}(u.s.s := v,\ p.s = u)\\
\equiv\ &[p.s = u]^{/u.s.s}_{v}\\
\equiv\ &p.s^{/u.s.s}_{v} = u\\
\equiv\ &(\underline{\text{if}}\ p = u.s\ \underline{\text{then}}\ v\ \underline{\text{else}}\ p.s) = u\\
\equiv\ &(p = u.s\ \underline{\text{and}}\ v = u)\ \underline{\text{or}}\ (p \neq u.s\ \underline{\text{and}}\ p.s = u) \qquad \square
\end{aligned}
$$

Example 8

$$
\begin{aligned}
&\mathrm{wp}(u.s := v,\ p.s.s = v)\\
\equiv\ &[p.s.s = v]^{/u.s}_{v}\\
\equiv\ &(p.s^{/u.s}_{v}).s^{/u.s}_{v} = v\\
\equiv\ &(p = u\ \underline{\text{and}}\ v.s^{/u.s}_{v} = v)\ \underline{\text{or}}\\
&(p \neq u\ \underline{\text{and}}\ p.s.s^{/u.s}_{v} = v)\\
\equiv\ &(p = u\ \underline{\text{and}}\ v = u)\ \underline{\text{or}}\\
&(p = u\ \underline{\text{and}}\ v \neq u\ \underline{\text{and}}\ v.s = v)\ \underline{\text{or}}\\
&(p \neq u\ \underline{\text{and}}\ p.s = u)\ \underline{\text{or}}\\
&(p \neq u\ \underline{\text{and}}\ p.s \neq u\ \underline{\text{and}}\ p.s.s = v) \qquad \square
\end{aligned}
$$

Example 9

$$
\begin{aligned}
&\mathrm{wp}(p.days(j) := i,\ p.s.s.days(i) = j)\\
\equiv\ &[p.s.s.days(i) = j]^{/p.days(j)}_{i}\\
\equiv\ &p.s.s.days(i)^{/p.days(j)}_{i} = j\\
\equiv\ &(\underline{\text{if}}\ p = p.s.s\ \underline{\text{and}}\ j = i\ \underline{\text{then}}\ i\\
&\underline{\text{else}}\ p.s.s.days(i)) = j\\
\equiv\ &(p = p.s.s\ \underline{\text{and}}\ j = i)\ \underline{\text{or}}\ ((p \neq p.s.s\ \underline{\text{or}}\ j \neq i)\\
&\qquad \underline{\text{and}}\ p.s.s.days(i) = j) \qquad \square
\end{aligned}
$$

It is clear from the preceding examples that the new axiom is as easily applied to pointer-based records as to array elements.

4. AN AXIOM OF CONCURRENT ASSIGNMENT

The conventional axiom of concurrent assignment, stated in [9], holds good only when the variables being assigned to are simple and different. This section presents a new axiom of concurrent assignment which removes those restrictions. The axiom is an extension of the axiom of section 2.

The concurrent assignment statement has the form $\underline{x} := \underline{e}$ where $\underline{x}$ denotes an arbitrary list of references $x_1, \ldots x_n$ and $\underline{e}$ denotes a list of expressions $e_1, \ldots, e_n$ of appropriate types. The assignment is executed in three stages:

(i) evaluate $x_i\downarrow$, for each i, and call the result id_i;
(ii) evaluate e_i, for each i, and call the result v_i;
(iii) assign v_i to box id_i, for each i.

To avoid any conflict that might arise when the id_i's are not all different, the assignment in (ii) is carried out in left-to-right order; this follows a convention established in [10]. The axiom will assume that the assignment terminates normally.

Define for an arbitrary reference y,

$$y^{/\underline{x}}_{\underline{e}} =_{df}$$
$$\begin{cases} y\downarrow = x_n\downarrow : e_n \\ y\downarrow \neq x_n\downarrow \ \underline{and}\ y\downarrow = x_{n-1}\downarrow : e_{n-1} \\ \vdots \\ y\downarrow \neq x_n\downarrow \ \underline{and} \ldots \underline{and}\ y\downarrow \neq x_2\downarrow \ \underline{and}\ y\downarrow = x_1\downarrow : e_1 \\ y\downarrow \neq x_n\downarrow \ \underline{and} \ldots \underline{and}\ y\downarrow \neq x_1\downarrow : y \end{cases}$$

(The right hand side of the above definition will be linearized to $\underline{if}\ y\downarrow = x_n\downarrow\ \underline{then}\ e_n\ \underline{else}\ \underline{if}\ y\downarrow = e_{n-1}\downarrow\ \underline{then}\ e_{n-1}\ \underline{else} \ldots \underline{else}\ y$.) Intuitively, $y^{/\underline{x}}_{\underline{e}}$ is equal to $e_{i'}$, where $x_{i'}$ is the rightmost x_i satisfying $x_i\downarrow = y\downarrow$; if no such x_i exists, then $y^{/\underline{x}}_{\underline{e}}$ is equal to y. With an arbitrary postcondition R, the new axiom of concurrent assignment is

$$wp(\underline{x} := \underline{e}, R) =_{df} R^{/\underline{x}}_{\underline{e}}$$

where $R^{/\underline{x}}_{\underline{e}}$ stands for R with every eference y in R replaced by $y^{/\underline{x}}_{\underline{e}}$ (the usual assumption re free and bound variables applies.) The reader may easily convince himself (or herself) that the axiom accords with the operational description given above. When evaluating $R^{/\underline{x}}_{\underline{e}}$ in practice, it is not necessary to substitute for references y in R satisfying $x_i\downarrow \neq y\downarrow$ for $1 \leq i \leq n$, where this this can be determined by inspection. Otherwise, it is only necessary to replace reference y by $y^{/\underline{x}'}_{\underline{e}'}$, where $\underline{x}'$ and e' are x and $\underline{e}$, respectively, with each x_i and e_i such that $x_i\downarrow \neq y\downarrow$, where this can be determined by inspection, removed. For example, $a(i)^{/j,a(j)}_{1,2}$ is equivalent to $a(i)^{/a(j)}_{2}$. Some examples will clarify the use of the axiom. The variables used in the examples are as declared in section 2.

Example 10

$$\begin{aligned} & wp(\text{"}i,b(i) := i+1, 5\text{"}, b(i) = i \\ \equiv\ & [b(i) = i]^{/i,\ b(i)}_{i+1,\ 5} \\ \equiv\ & b(i^{/i}_{i+1})^{/b(i)}_{5} = i^{/i}_{i+1} \\ \equiv\ & b(i+1)^{/b(i)}_{5} = i+1 \\ \equiv\ & b(i+1) = i+1 \end{aligned}$$

□

<u>Example 11</u>

$$\begin{aligned} & wp(\text{"p.s, q.s:= u, v"}, p.s = u) \\ \equiv\ & [p.s = u]^{p.s,q.s}_{u,v} \\ \equiv\ & p.s^{p.s,q.s}_{u,v} = u \\ \equiv\ & (\underline{\text{if}}\ p = q\ \underline{\text{then}}\ v\ \underline{\text{else}}\ u) = u \\ \equiv\ & p \neq q\ \underline{\text{or}}\ v = u \end{aligned}$$

□

Assignment to entire arrays or records - call it "compound assignment" - may be defined by regarding the compound assignment as a concurrent assignment to the elements of the array or record in some agreed order. The axiom of concurrent assignment then applies. For example, given the declarations

```
b:  array(1..5) of integer;
r:  pointer to record
    f:  integer;
    g:  array(1..3) of integer
    end
```

then b:= r↑, 10 (where r↑ denotes the entire record pointed to by r) may be regarded as equivalent to

```
b(1), b(2), b(3), b(4), b(5) := r.f, r.g(1), r.g (2),
                                r.g (3), 10
```

REFERENCES

[1] Floyd, R.W.: *Assigning Meanings to Programs*, Proc. Symp. in Applied Mathematics, Vol. 19, J.T. Schwartz (ed.), American Mathematical Society, 1967, pp. 19-32.

[2] Hoare, C.A.R.: *An Axiomatic Basis for Computer Programming*, Comm. ACM, Vol. 12, No. 10, October 1969, pp. 576-580, 583.

[3] Hoare, C.A.R., and Wirth, N.: *An Axiomatic Definition of the Programming Language Pascal*, Acta Informatica, Vol. 2, No. 4, 1973, pp. 335-355.

[4] London, R.L., Guttag, J.V., Horning, J.J., Lampson, B.W., Mitchell, J.G., and Popek, G.J.: *Proof Rules for the Programming Language Euclid*, Acta Informatica, Vol. 10, 1978, pp. 1-26.

[5] Burstall, R.M.: *Some Techniques for Proving Correctness of Programs which Alter Data Structures*, Machine Intelligence 7, D. Michie (ed.)., American Elsevier, New York, 1972, pp. 23-50.

[6] Kowaltowski, T.: *Data Structures and Correctness of Programs*, Jrnl. ACM, Vol. 26, No. 2, 1979, pp. 283-301.

[7] Laventhal, M.S.: *Verifying Programs which Operate on Data Structures*, Proc. Int. Conf. on Reliable Software, 1975, pp. 420-426.

[8] Luckham, D.C., and Suzuki, N.: *Verification of Array, Record, and Pointer Operations in Pascal*, ACM Trans. Programming Languages and Systems, Vol. 1, No. 2, 1979, pp. 226-244.

[9] Dijkstra, E.W.: *A Discipline of Programming*, Prentice-Hall, Englewood Cliffs, N.J., 1976.

[10] Gries, D.: *The Multiple Assignment Statement*, IEEE Trans. Software Engineering, Vol. SE-4, No. 2, 1978, pp. 89-93.

ASSIGNMENT AND LINKED DATA STRUCTURES

Joseph M. Morris

Dept. of Computer Science, Trinity College, Dublin 2.

1. INTRODUCTION

The formal treatment of linked data structures is significantly more complex than that of simpler structures because it is necessary to handle not just properties of individual nodes, but also relationships between the nodes. The most fundamental of these relationships is that of connectivity. If the verification of programs manipulating linked data structures is to be manageable, then it is essential that connectivity relations can be comfortably manoeuvred through changes in the data structure. A change in the data structure is effected by assignment to a pointer field of a node. This paper discusses the axiom of assignment when the assignment is to a pointer field of a node, and the postcondition is a connectivity relation.

2. RESTRICTED CONNECTIVITY RELATIONS

A data structure is a finite collection of uniform elements called "nodes". Each node has a unique name, the empty node having the distinguished name <u>nil</u>. In this paper a "pointer variable" (sometimes abbreviated to "pointer") is to be regarded as a variable of type "node name". A node is an aggregate of elements called "fields", which in implementations are simply variables. The fields of a particular node will be denoted by the familiar "dot" notation - e.g. x.s, x.key etc. A data structure is called "linked" if its nodes have at least one pointer field. The presence of pointer fields in nodes introduces the possibility of "connectivity" or "reachability" between nodes. Connectivity relations are important in the specification and

M. Broy and G. Schmidt (eds.), Theoretical Foundations of Programming Methodology, 35–41.

hence in the construction and verification of programs that manipulate linked data structures.

The simplest connectivity between nodes x and y (which have pointer field s) is denoted by $x \xrightarrow{s} y$, defined: there exists an ordered list of nodes $x_0, x_1, \ldots, x_n$ $(n \geq 0)$ such that

(i) $x_0 = x$
(ii) $x_n = y$
(iii) $x_{i+1} = x_i.s$ for $0 \leq i < n$

Such a list of nodes constitutes a "path" from node x to node y. If there exists a path from node x to node y, then there exists a non-cyclic path. Consequently, it will be assumed that the path also satisfies

(iv) $x_i \neq x_j$ for $i \neq j$ and $0 \leq i,j \leq n$

If there exists a path from node x to node y that is independent of the value of z.s for some node z, then nodes x, y and z satisfy $x \xrightarrow{s} y|\{z\}$. More generally, if K denotes a set of nodes, then $x \xrightarrow{s} y|K$ is defined: there exists a set of nodes $x_0, x_1, \ldots, x_n$ satisfying (i) to (iv) above and

(v) $x_i \notin K$ for $0 \leq i < n$

(Note that $x_n \in K$ is not excluded.) Clearly $x \xrightarrow{s} y$ is equivalent to $x \xrightarrow{s} y|\phi$ where ϕ denotes the empty set. The following equivalences are useful and easily proved:

$$x \xrightarrow{s} y \equiv x = y \ \underline{\text{or}}\ x.s \xrightarrow{s} y \quad (1)$$
$$x \xrightarrow{s} y|K \equiv x = y \ \underline{\text{or}}\ (x \notin K \ \underline{\text{and}}\ x.s \xrightarrow{s} y|K) \quad (2)$$

In this section, connectivity is restricted to connectivity via a given pointer field - 's'. More general connectivity relations will be considered in section 3.

Some useful properties of these connectivity relations are presented below. First, however, consider the following problem which shows how these relations may be used in specifying programs that manipulate linked data structures. In this example, the reader is assumed to be familiar with the terminology of list processing. Suppose that a non-cyclic list is composed of nodes having, say, an integer field h and a pointer field s. Let the value of pointer p be the first node in the list. Now an example of a list processing problem is this: given that at least one node of the list has its h field equal to a given constant h0, determine the first such node. That p is the head of a non-cyclic list containing h0 in one of its h fields may be formulated as

$$p \xrightarrow{s} \underline{\text{nil}} \ \underline{\text{and}}\ \exists x (p \xrightarrow{s} x): x.h = h0$$

Locating the first node q in the given list satisfying q.h = h0 is equivalent to establishing the truth of

$$p \xrightarrow{s} q \ \underline{\text{and}}\ q.s \xrightarrow{s} nil \ \underline{\text{and}}\ q.h = h0 \ \underline{\text{and}}$$
$$\forall x (p \xrightarrow{s} x | \{q\}) : x = q \ \underline{\text{or}}\ x.h \neq h0$$

Constructing programs from such specifications, and verifying them, will require the evaluation of weakest preconditions such as wp(u.s:= v, $p \xrightarrow{s} q$) where u, v, p, and q are pointers. The following lemmas allow such weakest preconditions to be easily evaluated. In what follows, x, y, and z denote nodes of a data structure; K and L denote sets of nodes; e, e1, and e2 denote expressions (of type "node name"); and s denotes an arbitrary pointer field of a node.

Lemma 1 $\quad x \xrightarrow{s} y \equiv x \xrightarrow{s} y | \{y\}$

Proof. The implication to the left is obvious. To prove the implication to the right, let the path from node x to node y be $x_0, \ldots, x_n$ ($n \geq 0$) and infer $x_i \neq y$ for $0 \leq i < n$ from conditions (ii) and (iv) above. $x \xrightarrow{s} y | \{y\}$ follows immediately. □

Lemma 2 $\quad x.s \xrightarrow{s} y \equiv x.s \xrightarrow{s} y | \{x\}$

Proof. The implication to the left is obvious. To prove the implication to the right, let the path from node x.s to node y be $x_0, \ldots, x_n$ ($n \geq 0$) and infer $x_i \neq x$ for $0 \leq i < n$ as follows. Condition (iv) guarantees $x.s (= x_0) \neq x_i$ for $0 < i \leq n$, and hence, by condition (iii), $x \neq x_i$ for $0 \leq i < n$. $x \xrightarrow{s} y | \{x\}$ follows immediately. □

Lemma 3 $\quad x \xrightarrow{s} y | K \ \underline{\text{and}}\ x \xrightarrow{s} y | L \equiv x \xrightarrow{s} y | K \cup L$

Proof. Consider the implication to the right. Let the path from node x to node y be $x_0, \ldots, x_n$ ($n \geq 0$). Now $x \xrightarrow{s} y | K$ guarantees $x_i \notin K$ for $0 \leq i < n$ and $x \xrightarrow{s} y | L$ guarantees $x_i \notin L$ for $0 \leq i < n$. $x \xrightarrow{s} y | K \cup L$ follows immediately. The implication to the left is similarly proved. □

Lemma 4 $\quad$ If $z \notin K$, then
$$x \xrightarrow{s} y | K \equiv x \xrightarrow{s} y | K \cup \{z\} \ \underline{\text{or}}\ (x \xrightarrow{s} z | K \ \underline{\text{and}}\ z.s \xrightarrow{s} y | K)$$

Proof. Consider first the implication to the right. Let the path from node x to node y be $x_0, \ldots, x_n$ ($n \geq 0$). Either z occurs in the path or it does not. If $z \neq x_i$ for $0 \leq i < n$ then $x \xrightarrow{s} y | K \cup \{z\}$ follows from lemma 3, while if $z = x_j$ for some j satisfying $0 \leq j < n$ then ($x \xrightarrow{s} z | K$ and $z.s \xrightarrow{s} y | K$) is easily inferred. Consider now the implication to the left. If $x \xrightarrow{s} y | K \cup \{z\}$ holds then $x \xrightarrow{s} y | K$ follows from lemma 3. Otherwise, let the path from node x to node z be $x_0, \ldots, x_j$ ($j \geq 0$) and the path from node z.s to node y be $x_{j+1}, \ldots, x_n$ ($n \geq j+1$). Now $x \xrightarrow{s} z | K$ guarantees $x_i \notin K$ for $0 \leq i < j$;

$z.s \xrightarrow{s} y|K$ guarantees $x_i \notin K$ for $j+1 \leq i < n$; and the assumption $z \notin K$ is $x_j \notin K$. Hence $x_i \notin K$ for $0 \leq i < n$ and $x \xrightarrow{s} y|K$ follows immediately. □

Lemma 5 If $z \in K$, then (in the notation of [1])
$$[x \xrightarrow{s} y|K]_e^{/z.s} = x \xrightarrow{s} y|K$$

Proof. Suppose there is a path $x_0, \ldots, x_n$ $(n \geq 0)$ from node x to node y. The substitution of e for z.s or its aliases can only affect condition (iii). But the assumption $z \in K$ and condition (v) guarantees $x_i \neq z$ for $0 \leq i < n$, and hence condition (iii) is invariant with respect to the substitution. □

Lemma 6 If $z \in K$, then
$$[e1 \xrightarrow{s} e2|K]_e^{/z.s} = (e1_e^{/z.s} \xrightarrow{s} e2_e^{/z.s}|K)$$

Proof. $[e1 \xrightarrow{s} e2|K]_e^{/z.s}$

$\equiv [e1 = x0 \text{ and } e2 = y0 \text{ and } x0 \xrightarrow{s} y0|K]_e^{/z.s}$ where x0 and y0 denote some constants

$\equiv e1_e^{/z.s} = x0 \text{ and } e2_e^{/z.s} = y0 \text{ and } x0 \xrightarrow{s} y0|K$

$\equiv e1_e^{/z.s} \xrightarrow{s} e2_e^{/z.s}|K$ □

The following examples will demonstrate how weakest preconditions can be evaluated, when the postcondition is a connectivity relation, and the assignment is to the pointer field of a node of the data structure. The assignment axiom of [1] is used.

In the examples p, q, u, v and t are pointers; e denotes an expression (of type "node name"); and s is a pointer field of a node.

Example 1

$$\begin{aligned}
& wp(u.s := e, p \xrightarrow{s} q) \\
\equiv\ & [p \xrightarrow{s} q]_e^{/u.s} \\
\equiv\ & [p \xrightarrow{s} q|\{u\} \text{ or } (p \xrightarrow{s} u \text{ and } u.s \xrightarrow{s} q)]_e^{/u.s} \\
\equiv\ & [p \xrightarrow{s} q|\{u\} \text{ or } (p \xrightarrow{s} u|\{u\} \text{ and } u.s \xrightarrow{s} q|\{u\})]_e^{/u.s} \\
\equiv\ & p \xrightarrow{s} q|\{u\} \text{ or } (p \xrightarrow{s} u|\{u\} \text{ and } e \xrightarrow{s} q|\{u\}) \\
\equiv\ & p \xrightarrow{s} q|\{u\} \text{ or } (p \xrightarrow{s} u \text{ and } e \xrightarrow{s} q|\{u\})
\end{aligned}$$
□

Example 2

$$\begin{aligned}
& wp(p.s := q, p.s \xrightarrow{s} p) \\
\equiv\ & [p.s \xrightarrow{s} p]_q^{/p.s} \\
\equiv\ & [p.s \xrightarrow{s} p|\{p\}]_q^{/p.s} \\
\equiv\ & q \xrightarrow{s} p|\{p\} \\
\equiv\ & q \xrightarrow{s} p
\end{aligned}$$
□

Example 3

$$
\begin{aligned}
& wp(u.s:=v,\ u.s.s \xrightarrow{s} u.s) \\
\equiv\ & [u.s.s \xrightarrow{s} u.s]^{/u.s}_{v} \\
\equiv\ & [u.s.s \xrightarrow{s} u.s \,|\, \{u\}\ \underline{or}\ (u.s.s \xrightarrow{s} u\ \underline{and}\ u.s \xrightarrow{s} u.s)]^{/u.s}_{v} \\
\equiv\ & [u.s.s \xrightarrow{s} u.s \,|\, \{u\}\ \underline{or}\ u.s.s \xrightarrow{s} u \,|\, \{u\}]^{/u.s}_{v} \\
\equiv\ & (u.s^{/u.s}_{v}).s^{/u.s}_{v} \xrightarrow{s} u.s^{/u.s}_{v} \,|\, \{u\}\ \underline{or}\ (u.s^{/u.s}_{v}.s^{/u.s}_{v} \xrightarrow{s} u \,|\, \{u\} \\
\equiv\ & v.s^{/u.s}_{v} \xrightarrow{s} v \,|\, \{u\}\ \underline{or}\ v.s^{/u.s}_{v} \xrightarrow{s} u \\
\equiv\ & (u = v\ \underline{and}\ v \xrightarrow{s} v \,|\, \{u\})\ \underline{or}\ (u \neq v\ \underline{and}\ v.s \xrightarrow{s} v \,|\, \{u\}) \\
& \underline{or}\ (u = v\ \underline{and}\ v \xrightarrow{s} u)\ \underline{or}\ (u \neq v\ \underline{and}\ v.s \xrightarrow{s} u) \\
\equiv\ & u = v\ \underline{or}\ v.s \xrightarrow{s} v \,|\, \{u\}\ \underline{or}\ v.s \xrightarrow{s} u \\
\equiv\ & v \xrightarrow{s} u\ \underline{or}\ v.s \xrightarrow{s} v \qquad \square
\end{aligned}
$$

Example 4

$$
\begin{aligned}
& wp(\text{"}u.s:=v;\ t:=u.s\text{"}, p \xrightarrow{s} q \,|\, \{t\}) \\
\equiv\ & wp(u.s:=v. p \xrightarrow{s} q \,|\, \{u.s\}) \\
\equiv\ & [p \xrightarrow{s} q \,|\, \{u.s\}]^{/u.s}_{v} \\
\equiv\ & [p \xrightarrow{s} q \,|\, \{u.s,u\}\ \underline{or}\ (p \xrightarrow{s} u \,|\, \{u.s,\ u\}\ \underline{and}\ u.s \xrightarrow{s} q \,|\, \{u.s,u\})]^{/u.s}_{v} \\
\equiv\ & p \xrightarrow{s} q \,|\, \{v,u\}\ \underline{or}\ (p \xrightarrow{s} u \,|\, \{v\}\ \underline{and}\ v \xrightarrow{s} q \,|\, \{v,u\}) \\
\equiv\ & p \xrightarrow{s} q \,|\, \{u,v\}\ \underline{or}\ (p \xrightarrow{s} u \,|\, \{v\}\ \underline{and}\ v = q) \qquad \square
\end{aligned}
$$

3. UNRESTRICTED CONNECTIVITY

If node y is reachable from node x by sequencing via any pointer field at each step then $x \longrightarrow y$, defined: there exists a sequence of nodes $x_0, x_1, \ldots, x_n$ $(n \geqslant 0)$ such that

(ia) $x_0 = x$

(iia) $x_p = y$

(iiia) $\exists f : x_{i+1} = x_i.f$ for $0 \leqslant i < n$, where f is quantified over the pointer field of node x_i.

As before, the shortest path will be chosen and hence

(iva) $x_i \neq x_j$ for $i \neq j$ and $0 \leqslant i,j \leqslant n$.

If $x \longrightarrow y$ and the path from x to y is independent of the value of field z.f for some node z then $x \longrightarrow y \,|\, \{(z,f)\}$. More generally, if K denotes a set of ordered pairs (z,f) where z denotes a node and f a pointer field of the node then $x \longrightarrow y \,|\, K$ is defined to be $x \longrightarrow y$ with condition (iiia) strengthened to

(iiib) $\exists f : (x_i, f) \notin K$ and $x_{i+1} = x_i.f$ for $0 \leqslant i < n$, where f is quantified over the pointer fields of node x_i.

The ordered pair (z,f) will also be written as z_f. Note that $x \longrightarrow y$ is equivalent to $x \longrightarrow y \,|\, \phi$. Analagous to (1) and (2) of section 2, it can be shown:

$$x \longrightarrow y \equiv x = y \ \underline{\text{or}}\ \exists f\colon x.f \longrightarrow y$$
$$x \longrightarrow y|K \equiv x = y \ \underline{\text{or}}\ \exists f\colon (x,f) \notin K \ \underline{\text{and}}\ x.f \longrightarrow y|K)$$

where the notation is retained from above.

The following six lemmas, analagous to lemmas 1 to 6 of section 2, give useful properties of the preceding two relations. Their proofs, which are similar to the proofs of the earlier lemmas, are left as an exercise. In what follows, x, y, and z denote nodes of a data structure; K and L denote sets of ordered pairs (z,f) as explained above; e,e1, and e2 denote expressions (of type "node name"); and f denotes an arbitrary pointer field of a node.

Lemma 1a $\quad x \longrightarrow y \equiv x \longrightarrow y|\{y_f\}$

Lemma 2a $\quad x.f \longrightarrow y \equiv x.f \longrightarrow y|\{x_f\}$

Lemma 3a $\quad x \longrightarrow y|K \ \underline{\text{and}}\ x \longrightarrow y|L \equiv x \longrightarrow y|K \cup L$

Lemma 4a $\quad$ If $z_f \notin K$, then

$$x \longrightarrow y|K \equiv x \longrightarrow y|K \cup \{z_f\} \ \underline{\text{or}}\ (x \longrightarrow z|K \ \underline{\text{and}}\ z.f \longrightarrow y|K)$$

Lemma 5a $\quad$ If $z_f \in K$, then

$$\left[x \longrightarrow y|K\right]^{/z.f}_{e} = x \longrightarrow y|K$$

Lemma 6a $\quad$ If $z_f \in K$, then

$$\left[e1 \longrightarrow e2|K\right]^{/z.f}_{e} = (e1^{/z.f}_{e} \longrightarrow e2^{/z.f}_{e}|K)$$

Example 5

If u, p, and q are pointers, e denotes an expression, and f is a pointer field, then

$$\begin{aligned}
& wp(u.f := e,\ p \longrightarrow q) \\
\equiv\ & [p \longrightarrow q]^{/u.f}_{e} \\
\equiv\ & [p \longrightarrow q|\{u_f\} \ \underline{\text{or}}\ (p \longrightarrow u|\{u_f\} \ \underline{\text{and}}\ u.f \longrightarrow q|\{u_f\})]^{/u.f}_{e} \\
\equiv\ & p \longrightarrow q|\{u_f\} \ \underline{\text{or}}\ (p \longrightarrow u \ \underline{\text{and}}\ e \longrightarrow q|\{u_f\})
\end{aligned}$$

4. CONNECTIVITY RELATIONS AND CONCURRENT ASSIGNMENT

The following example illustrates the use of the foregoing lemmas when several nodes of a data structure are concurrently altered. The example is from Gries [2].

Example 6

Given that $p \neq q$

$$wp(\text{"}p,\ p.s,\ q.s := p.s,\ q.s,\ p\text{"},\ p \xrightarrow{s} \underline{nil})$$
$$\equiv \left[p \xrightarrow{s} \underline{nil}\right]^{p\ ,p.s,q.s}_{p.s,q.s,p}$$
$$\equiv \left[p = p0\ \underline{and}\ p0 \xrightarrow{s} \underline{nil}\right]^{p\ ,p.s,q.s}_{p.s,q.s,p} \quad \text{where p0 denotes some constant}$$
$$\equiv\ p.s = p0\ \underline{and}\ \left[p0 \xrightarrow{s} \underline{nil}\right]^{p.s,q.s}_{q.s,p}$$

Denoting the set $\{p,q\}$ by K, it is not difficult to argue that $p0 \xrightarrow{s} \underline{nil}$ is equivalent to

$$\begin{aligned}
&p0 \xrightarrow{s} \underline{nil}\,|K\ \underline{or}\\
&(p0 \xrightarrow{s} q\,|K\ \underline{and}\ q.s \xrightarrow{s} \underline{nil}\,|K)\ \underline{or}\\
&(p0 \xrightarrow{s} p\,|K\ \underline{and}\ p.s \xrightarrow{s} \underline{nil}\,|K)\ \underline{or}\\
&(p0 \xrightarrow{s} q\,|K\ \underline{and}\ q.s \xrightarrow{s} p\,|K\ \underline{and}\ p.s \xrightarrow{s} \underline{nil}\,|K)\ \underline{or}\\
&(p0 \xrightarrow{s} p\,|K\ \underline{and}\ p.s \xrightarrow{s} q\,|K\ \underline{and}\ q.s \xrightarrow{s} \underline{nil}\,|K)
\end{aligned}$$

Therefore $\left[p0 \xrightarrow{s} \underline{nil}\right]^{p.s,q.s}_{q.s,p}$

$$\begin{aligned}
\equiv\ &p0 \xrightarrow{s} \underline{nil}\,|K\ \underline{or}\\
&(p0 \xrightarrow{s} q\,|K\ \underline{and}\ p \xrightarrow{s} \underline{nil}\,|K)\ \underline{or}\\
&(p0 \xrightarrow{s} p\,|K\ \underline{and}\ q.s \xrightarrow{s} \underline{nil}\,|K)\ \underline{or}\\
&(p0 \xrightarrow{s} q\,|K\ \underline{and}\ p \xrightarrow{s} p\,|K\ \underline{and}\ q.s \xrightarrow{s} \underline{nil}\,|K)\ \underline{or}\\
&(p0 \xrightarrow{s} p\,|K\ \underline{and}\ q.s \xrightarrow{s} q\,|K\ \underline{and}\ p \xrightarrow{s} \underline{nil}\,|K)
\end{aligned}$$

Substituting p.s. for p0 in the above, simplifying, and observing that the second and final terms are false, the result is

$$\begin{aligned}
&p.s \xrightarrow{s} \underline{nil}\,|\{q\}\ \underline{or}\\
&(p.s \xrightarrow{s} p\,|\{q\}\ \underline{and}\ q.s \xrightarrow{s} \underline{nil}\,|\{p\})\ \underline{or}\\
&(p.s \xrightarrow{s} q\ \underline{and}\ q.s \xrightarrow{s} \underline{nil}\,|\{p\})
\end{aligned}$$

which, by combining the first and last terms and further simplifying, reduces to

$$p.s \xrightarrow{s} \underline{nil}\ \underline{or}\ (p.s \xrightarrow{s} p\ \underline{and}\ q.s \xrightarrow{s} \underline{nil})$$

REFERENCES

[1] Morris, J.M.: *A General Axiom of Assignment*, this volume.

[2] Gries, D.: *The Multiple Assignment Statement*, IEEE Trans. Software Engineering, Vol. SE-4, No. 2, 1978, pp. 89-93.

A PROOF OF THE SCHORR-WAITE ALGORITHM

Joseph M. Morris

Dept. of Computer Science, Trinity College, Dublin 2.

1. INTRODUCTION

The preceding two papers laid a theoretical foundation for verifying list-processing programs. The present paper exercises the formalism in a verification of that most recalcitrant algorithm, the Schorr-Waite graph-marking algorithm.

The Schorr-Waite algorithm [1] which traverses a graph and "marks" its nodes, has acquired the status of a standard testbed for proof techniques concerned with complex data structures [2, 3, 4, 5, 6]. According to Topor [2]: "The algorithm is of interest because of the clever way it avoids using any extra storage by manipulating pointers within the structure and restoring them all at the end. It thus presents a challenge to any proof procedure". The intention is to develop a rigorous and formal argument for the correctness of the algorithm.

2. THE PROBLEM

A graph is composed of nodes. Each node has a unique name, and a variable of type "node name" is called a "pointer". Each node x has four fields of interest: two boolean fields denoted by x.m ("the mark bit") and x.c ("the control bit"), respectively, and two pointer fields denoted by x.l and x.r, respectively. A graph is a finite collection of nodes.

It is given that at the outset each mark bit is <u>false</u>, and that there is a distinguished node called "the root". The purpose of the Schorr-Waite algorithm is to set the mark bits of all nodes reachable from the root, and only those nodes. The control bits

M. Broy and G. Schmidt (eds.), Theoretical Foundations of Programming Methodology, 43–51.

are for housekeeping purposes, and their value before or after execution of the algorithm is not of interest.

3. THE ALGORITHM

Let "root" be a pointer to the given root. For simplicity of coding, it is assumed that no pointer fields in the nodes are equal to nil. This standard trick imposes no serious restriction because, if necessary, the empty node can be represented in the graph by some other distinguished node. For convenience in the proof, a virtual root V is introduced; V.l = V.r = root initially. With pointers p, t, and q the program is:

```
SETUP:   p:= V; t:= root;
         do p ≠ V or not t.m →
            if not t.m →
PUSH:           q:= p; p:= t; t:= t.l; p.l:= q;
                p.m:= true; p.c:= false
            [] t.m and not p.c →
SWING:          q:= t; t:= p.r; p.r:= p.l; p.l:= q;
                p.c:= true
            [] t.m and p.c →
POP:            q:= t; t:= p; p:= p.r; t.r:= q
            fi
         od.
```

The program has four actions, each consisting of a sequence of assignments; the actions have been labelled for reference in the text.

In the sequel, all quantifications are over the nodes of the given graph (V is excluded). The original values of pointer fields will be denoted by zero subscripting. Thus the initial conditions are

$$\forall x: x.l = x.l_0 \text{ and } x.r = x.r_0 \text{ and not } x.m$$
$$V.l_0 = V.r_0 = \text{root and } V.m$$

The proof obligation is to show that on termination of the program R0 and R1 holds, as follows:

R0: $\forall x: x.l = x.l_0 \text{ and } x.r = x.r_0$

R1: $\forall x: x.m \equiv \text{root} \longrightarrow x$

The proof of R0 is called the structural proof, and the proof of R1 is called the functional proof.

4. THE INVARIANTS

The following informal description assumes a familiarity with marking algorithms.

The essential and distinctive feature of the Schorr-Waite algorithm is its use of an implicit stack encoded within the graph as it is being traversed. The top of the stack is node p, and the tail is node V. If x ($\neq$V) is a node "on the stack", then its following node is either x.l or x.r, and the control bit x.c is used to indicate which one. To be more precise, first define

$$x.s =_{df} \underline{\text{if}}\ x.c\ \underline{\text{then}}\ x.r\ \underline{\text{else}}\ x.l$$
$$x.\bar{s} =_{df} \underline{\text{if}}\ x.c\ \underline{\text{then}}\ x.l\ \underline{\text{else}}\ x.r$$

The stack satisfies invariants (i) and (ii):

(i) Node p is the first node on the stack, and node V is the last.
(ii) If x($\neq$V) is a node on the stack then x.s is its successor.

The relation between the values of pointer fields and their original values is given by invariants (iii) to (vi):

(iii) Each node x not on the stack has $x.l = x.l_0$ and $x.r = x.r_0$.
(iv) If x is a node on the stack then $x.\bar{s} = x.\bar{s}_0$.
(v) If y is the successor of x on the stack then $y.s_0 = x$; in other words, $x.s.s_0 = x$.
(vi) $p.s_0 = t$.

The final two invariants, (vii) and (viii), keep track of the marking of nodes:

(vii) All stack nodes are marked.
(viii) Every marked node not on the stack has both its successor nodes marked. More precisely, if node x is not on the stack, and x.m, then x.l.m and x.r.m.

The above eight facts describe the behaviour of the Schorr-Waite algorithm. Facts (i) to (vii) are used in the structural proof; they are gathered into formal invariants P0, P1 and P2 below (the notation of the preceding paper is used).

$$\text{P0:}\quad \forall x{:}\ (p \xrightarrow{s} x\ \underline{\text{and}}\ x.\bar{s} = x.\bar{s}_0\ \underline{\text{and}}\ x.s.s_0 = x\ \underline{\text{and}}\ x.m)$$
$$\quad\quad \underline{\text{or}}\ (\underline{\text{not}}\ p \xrightarrow{s} x\ \underline{\text{and}}\ x.l = x.l_0\ \underline{\text{and}}\ x.r = x.r_0)$$
$$\text{P1:}\quad p.s_0 = t$$
$$\text{P2:}\quad p \xrightarrow{s} V$$

It will be postulated that V.s = nil; p2 then becomes $p \xrightarrow{s} \underline{\text{nil}}$.

R0 follows from P0, V.s = nil, and p = V, which holds on termination, as then all nodes other than V must satisfy the second term of P0.

The functional proof hangs on the invariance of fact (viii) above, more precisely stated as

P3: ∀x: not x.m or $p \xrightarrow{s} x$ or (x.l.m and x.r.m)
The proof of P3 will require the auxiliary invariant
P4: ∀x: not x.c or not x.m or x.l.m

To infer the truth of R1 from P3 and the termination of the algorithm (which will be proved later), R1 is split in two, thus:

R1a: ∀x: x.m or not root ⟶ x
R1b: ∀x: not x.m or root ⟶ x
R1 states that all nodes reachable from root are marked; R1b says that all nodes not reachable from root are unmarked.

Consider R1a. Now on termination of the program, p = V and t.m holds; therefore, from P1, t = root and hence root.m. Moreover, p = V guarantees not $p \xrightarrow{s}$ root. Therefore, by P3, both of the roots' successors are marked, and their successors in turn are marked, etc. Hence all nodes x satisfying root ⟶ x also satisfy x.m. Therefore R1a is established.

Consider R1b. As nodes not reachable from root cannot enter the game, and as initially ∀x: not x.m holds, R1b is an invariant of the program. (The reader so inclined may easily formalize this argument.)

In the proofs which follow P => wp(S,Q) will be abbreviated to {P}S{Q} where P and Q denote arbitrary predicates and S denotes an arbitrary statement or program

5. P0 IS INVARIANT

P0 is established by SETUP as

∀x: not $p \xrightarrow{s} x$ and $x.l = x.l_0$ and $x.r = x.r_0$ holds initially.

The proof of the invariance of P0 is carried out by considering wp(X,P0) where X is PUSH, SWING, or POP, and applying the basic theorem for the alternative construct [7]. The strategy in each of the three cases is this. First wp(X, $p \xrightarrow{s} x$) is evaluated to determine the set of nodes for which the value of $p \xrightarrow{s} x$ changes. It turns out that each action changes the value of $p \xrightarrow{s} x$ for at most one node - call it n1. Next it is observed that X assigns to the fields of precisely one node - call it n2 - and therefore the terms of P0 other than $p \xrightarrow{s} x$ are changed for n2 only. It turns out, moreover, that n1, if it exists, equals n2. Given this happy circumstance and the truth of P0 before execution of X, it suffices to evaluate wp(X, P0') where P0' is the body of P0 with n2 substituted for x. The details follow.

Consider PUSH. It is easy to show

wp(PUSH, $p \xrightarrow{s} x$) ≡ (t = x or $p \xrightarrow{s} x \mid \{t\}$);

this is done in Fig. 1. Now PO and $\underline{\text{not}}$ t.m in the guard of PUSH together imply $\underline{\text{not}}$ $p \xrightarrow{s} t$, from which it is easy to infer:

$$p \xrightarrow{s} x|\{t\} = p \xrightarrow{s} x.$$

$\{t = x \ \underline{\text{or}}\ p \xrightarrow{s} x|\{t\}\}$
q:= p;
$\{t = x \ \underline{\text{or}}\ q \xrightarrow{s} x|\{t\}\}$
p:= t;
$\{p = x \ \underline{\text{or}}\ q \xrightarrow{s} x|\{p\}\}$
t := t.l;
$\{p = x \ \underline{\text{or}}\ q \xrightarrow{s} x|\{p\}\}$
p.l:= q;
$\{p = x \ \underline{\text{or}}\ p.l \xrightarrow{s} x|\{p\}\}$
p.m:= $\underline{\text{true}}$;
$\{p = x \ \underline{\text{or}}\ p.l \xrightarrow{s} x|\{p\}\}$
p.c:= $\underline{\text{false}}$
$\{p = x \ \underline{\text{or}}\ (p.l \xrightarrow{s} x|\{p\} \ \underline{\text{and}}\ \underline{\text{not}}\ p.c)\}$
$\{p = x \ \underline{\text{or}}\ p.s \xrightarrow{s} x|\{p\}\}$
$\{p \xrightarrow{s} x\}$

Fig. 1 ($p \xrightarrow{s} x$ w.r.t. PUSH)

$\{t.r = t.r_0 \ \underline{\text{and}}\ p.s_0 = t\}$
q:= p;
$\{t.r = t.r_0 \ \underline{\text{and}}\ q.s_0 = t\}$
p:= t;
$\{p.r = p.r_0 \ \underline{\text{and}}\ q.s_0 = p\}$
t:= t.l;
$\{p.r = p.r_0 \ \underline{\text{and}}\ q.s_0 = p\}$
p.l:= q;
$\{p.r = p.r_0 \ \underline{\text{and}}\ p.l.s_0 = p\}$
p.m:= $\underline{\text{true}}$;
$\{p.r = p.r_0 \ \underline{\text{and}}\ p.l.s_0 = p \ \underline{\text{and}}\ p.m\}$
p.c:= $\underline{\text{false}}$
$\{p.r = p.r_0 \ \text{and}\ p.l.s_0 = p \ \underline{\text{and}}\ p.m \ \underline{\text{and}}\ \underline{\text{not}}\ p.c\}$
$\{p.\bar{s} = p.\bar{s}_0 \ \underline{\text{and}}\ p.s.s_0 = p \ \underline{\text{and}}\ p.m\}$

Fig. 2 (PO w.r.t. PUSH)

Hence PUSH changes the value of $p \xrightarrow{s} x$ for x = (the old) t only, i.e. for x = (the new) p as it may be shown, trivially, that PUSH establishes p = (the old) t.

Moreover, PUSH assigns to fields of node p only. Therefore it only remains to verify that on termination of PUSH, node p complies with PO. This is done in Fig. 2, which proves

$\{t.r = t.r_0 \ \underline{\text{and}}\ p.s_0 = t\}$
PUSH
$\{p.\bar{s} = p.\bar{s}_0 \ \underline{\text{and}}\ p.s.s_0 = p \ \underline{\text{and}}\ p.m\}$

The first term of the foregoing precondition is guaranteed by P0 and not t.m in the guard of PUSH; the second term is P1. Hence PUSH leaves P0 invariant.

Consider SWING. It is easy to prove

$$wp(SWING,\ p \overset{s}{\longrightarrow} x) \equiv (p = x \ \underline{or}\ p.l \overset{s}{\longrightarrow} x)$$

Now not p.c in the guard of SWING implies p.l is p.s, and hence the right hand side of the above equivalence becomes $p \overset{s}{\longrightarrow} x$. Therefore SWING does not change the value of $p \overset{s}{\longrightarrow} x$ for any node x.

Moreover, SWING assigns to fields of node p only. Therefore it only remains to verify that on termination of SWING, node p complies with P0. Now it is easy to show:

$\{t = p.l_0$ and $p.l.s_0 = p$ and $p.m\}$
SWING
$\{p.\bar{s} = p.\bar{s}_0$ and $p.s.s_0 = p$ and $p.m\}$

Given not p.c in the guard of SWING, p.r is p.s and so the foregoing precondition may be written as

$t = p.s_0$ and $p.s.s_0 = p$ and $p.m$.

The first term is P1, and the other terms are implied by P0 and $p \neq V$. $p \neq V$, in turn, is guaranteed by the outermost guard and t.m in the guard of SWING.

The invariance of P0 with respect to POP is similarly proven. POP changes the value of $p \overset{s}{\longrightarrow} x$ for x = (the old) p only, i.e. for x = (the new) t as it may be shown, trivially, that POP establishes t = (the old) p. Furthermore, POP assigns to fields of node t only, so it only remains to verify that on termination of POP, node t complies with P0, and this is easily shown.

Therefore P0 is invariant.

6. P1 IS INVARIANT

P1 is established by SETUP as $V.s_0 = root$. It is an easy task, done in Fig. 3, to show

$\{t.l = t.l_0\}$ PUSH $\{p.s_0 = t\}$;

the precondition is implied by P0 and not t.m in the guard of PUSH.

```
{t.l0 = t.l}
q:= p;
{t.l0 = t.l}
p:= t;
{p.l0 = t.l}
t:= t.l;
{p.l0 = t}
p.l:= q;
{p.l0 = t}
p.m:= true;
{p.l0 = t}
p.c:= false
{p.l0 = t and not p.c}
{p.s0 = t}
```

Fig. 3 (P1 w.r.t. PUSH)

```
{p.r0 = p.r}
q:= t;
{p.r0 = p.r}
t:= p.r;
{p.r0 = t}
p.r:= p.l;
{p.r0 = t}
p.l:= q;
{p.r0 = t}
p.c:= true
{p.r0 = t and p.c}
{p.s0 = t}
```

Fig. 4 (P1 w.r.t. SWING)

Fig. 4 shows

$$\{p.r = p.r_0\} \text{ SWING } \{p.s_0 = t\};$$

given not p.c in the guard of SWING, the precondition is $p.\bar{s} = p.\bar{s}_0$, which is guaranteed by P0. Finally, it may be shown:

$$\{p.r.s_0 = p\} \text{ POP } \{p.s_0 = t\};$$

given p.c in the guard of POP, the precondition is $p.s.s_0 = p$, which is implied by P0 and $p \neq V$. $p \neq V$, in turn, is guaranteed by the outermost guard and t.m in the guard of POP.

Therefore P1 is invariant.

7. P2 IS INVARIANT

P2 is trivially established by SETUP. The proof of the invariance of P2 is similar to the proof of P0. The proof of P0 required the evaluation of wp(X, $p \xrightarrow{s} x$) for X equal to PUSH, SWING or POP. In each case replace x by V. The resulting preconditions will be guaranteed by P2 and $p \neq V$ in the outer guard.

Therefore P2 is invariant, and this concludes the structural proof.

8. P3 IS INVARIANT

P3 is established by SETUP as $\forall x$: not x.m holds initially. PUSH establishes p.m, but P3 is safe as $p \xrightarrow{s} p$ is true. SWING affects nothing referenced by P3. POP, it was seen in Section 5, only affects node t; and it is easy to show {p.l.m and t.m} POP {t.l.m and t.r.m}

The second term of the foregoing precondition is assured by the guard of POP. The guard also guarantees p.c; P0 guarantees p.m; consequently P4 guarantees p.l.m, the first term of the precondition.

Therefore P3 is invariant.

9. P4 IS INVARIANT

P4 is established by SETUP as initially ∀x; not x.m holds. PUSH assigns to node p but establishes not p.c. It was shown in Section 5 that SWING only affects node p. It is trivial to show wp(SWING, p.l.m) = t.m, and t.m is guaranteed by the guard. POP affects nothing referenced by P4.

Therefore P4 is invariant.

This completes the functional proof.

10. TERMINATION

Let N be the number of nodes in the given graph. Let f1 be the number of nodes satisfying x.m, f2 be the number of nodes satisfying x.m and x.c, and f3 be the number of nodes satisfying x.m and not $p \xrightarrow{s} x$, where x is a node of the given graph. Then the function f = f1 + f2 + f3 is bounded from above by 3 * N and is incremented by 1 on each execution of the loop; PUSH increments f1 only, SWING increments f2 only, and POP increments f3 only. Hence the program terminates.

This completes the proof.

REFERENCES

[1] Schorr, H., and Waite, W.M.: *"An Efficient Machine-Independent Procedure for Garbage Collection in Various List Structures,"* Comm. ACM, Vol. 10, No. 8, 1967, pp. 501-506.

[2] Topor, R.W.: *"The Correctness of the Schorr-Waite List Marking Algorithm,"* Acta Informatica, Vol. 11, 1979, pp. 211-221.

[3] Gries, D.: *"The Schorr-Waite Graph Marking Algorithm,"* Acta Informatica, Vol. 11, 1979, pp. 223-232.

[4] de Roever, W.P.: *"On Backtracking and Greatest Fixpoints,"* Formal Descriptions of Programming Concepts, E.J. Neuhold (ed.), North-Holland, 1978, pp. 621-636.

[5] Kowaltowski, T.: *"Data Structures and Correctness of Programs,"* Jrnl. ACM, Vol. 26, No. 2, 1979, pp. 283-301.

[6] Yelowitz, L., and Duncan, A.G.: *Abstractions, Instantiations, and Proofs of Marking Algorithms,"* Proc. Symp. Artificial Intelligence and Programming Languages, Sigplan, Vol. 12, No. 8, 1977, pp. 13-21.

[7] Dijkstra, E.W.: *"A Discipline of Programming,"* Prentice-Hall, Englewood Cliffs, N.J., 1976.

VERIFICATION OF SEQUENTIAL PROGRAMS: TEMPORAL AXIOMATIZATION

ZOHAR MANNA

Computer Science Dept.
Stanford University
Stanford, CA

Applied Mathematics Dept.
The Weizmann Institute
Rehovot, Israel

Abstract

This is one in a series of reports describing the application of temporal logic to the specification and verification of computer programs.

In earlier reports, we introduced temporal logic as a tool for reasoning about concurrent programs and specifying their properties [MP1] and presented proof principles for establishing these properties ([MP2]). Here, we restrict ourselves to deterministic, sequential programs. We present a proof system in which properties of such programs, expressed as temporal formulas, can be proved formally.

Our proof system consists of three parts: a *general part* elaborating the properties of temporal logic, a *domain part* giving an axiomatic description of the data domain, and a *program part* giving an axiomatic description of the program under consideration.

We illustrate the use of the proof system by giving two alternative formal proofs of the total correctness of a simple program.

This research was supported in part by the National Science Foundation under grants MCS80-06930, in part by the Office of Naval Research under Contract N00014-76-C-0687, and in part by the United States Air Force Office of Scientific Research under Contract AFOSR-81-0014.

M. Broy and G. Schmidt (eds.), Theoretical Foundations of Programming Methodology, 53–102.

1. INTRODUCTION

Temporal logic is a modal logic in which we impose special restrictions on the models of interpretation ([PRI], [RU],[PNU],[GPSS], [MP1]). A ***universe*** for temporal logic consists of a collection of ***states*** (***worlds***). A state s' is accessible from a state s if through development in time, s can change into s'. We concentrate on histories of development which are linear and discrete. Thus, the models of temporal logic consist of ω-sequences, *i.e.*, infinite sequences of the form $\sigma = s_0, s_1, \ldots$. In such a sequence, s_j is accessible from s_i if $i \leq j$. On these states we define an ***immediate accessibility relation*** ρ which is required to be a function. That means that every state s has exactly one other state s' such that $\rho(s, s')$. This corresponds to our intuition that in a discrete time model each instant has exactly one immediate successor. the transitive reflexive closure of ρ, $R = \rho^*$, is the ***accessibility relation***; intuitively, $R(s, s')$ holds when s' is either identical to s or lies in the future of s.

We first describe the temporal language we are going to use. This language is designed specially for the application we have in mind, namely reasoning about programs, and is not necessarily the most general temporal language possible.

The language uses a set of basic symbols consisting of individual variables and constants, and proposition, function and predicate symbols. The set is partitioned into two subsets: global and local symbols. The ***global symbols*** have a uniform interpretation over the complete universe and do not change their values or meanings from one state to another. The ***local symbols***, on the other hand, may assume different meanings and values in different states of the universe. For our purpose, the only local symbols that interest us are local individual variables. We will have global symbols of all types.

We use the regular set of boolean connectives: $\wedge$, $\vee$, $\supset$, $\equiv$, and $\sim$ together with the equality operator $=$ and the first-order quantifiers $\forall$ and $\exists$. This set is referred to as the ***classical operators***. The quantifiers $\forall$ and $\exists$ are applied only to global individual variables.

The ***modal operators*** used are: $\Box$, $\Diamond$, $\bigcirc$, and $\mathcal{U}$, which are called respectively the ***always***, ***sometime***, ***next*** and ***until*** operators. The first three operators are unary while the $\mathcal{U}$ operator is binary. We use the *next* operator $\bigcirc$ in two different ways - as a temporal operator applied to formulas and as a temporal operator applied to terms.

A ***model*** (I, α, σ) for our language consists of a (global) interpretation I, a (global) assignment α and a sequence of states σ.

- The ***interpretation*** I specifies a nonempty domain D and assigns concrete elements, functions and predicates to the (global) individual constants, function and predicate symbols.

- The *assignment* α assigns a value over the appropriate domain to each of the global free individual variables.

- The *sequence* $\sigma = s_0, s_1, \ldots$ is an infinite sequence of states. Each *state* s_i assigns values to the local free individual variables and propositions.

  For a sequence

$$\sigma = s_0, s_1, \ldots$$

we denote by

$$\sigma^{(i)} = s_i, s_{i+1}, \ldots$$

the i-truncated suffix of σ.

Given a temporal formula w, we present below an inductive definition of the truth value of w in a model (I, α, σ). The value of a subformula or term τ under (I, α, σ) is denoted by $\tau|_\sigma^\alpha$, I being implicitly assumed.

Consider first the evaluation of terms:

- For a local individual variable or local proposition y:

$$y|_\sigma^\alpha = y_{s_0},$$

  i.e., the value assigned to y in s_0, the first state of σ.

- For a global individual variable or global proposition u:

$$u|_\sigma^\alpha = \alpha[u],$$

  i.e., the value assigned to u by α.

- For an individual constant the evaluation is given by I:

$$c|_\sigma^\alpha = I[c].$$

- For a k-ary function f:

$$f(t_1, \ldots, t_k)|_\sigma^\alpha = I[f](t_1|_\sigma^\alpha, \ldots, t_k|_\sigma^\alpha),$$

  i.e., the value is given by the application of the interpreted function $I[f]$ to the values of $t_1, \ldots, t_k$ evaluated in the environment (I, α, σ).

- For a term t:

$$\bigcirc t|_{\sigma}^{\alpha} = t|_{\sigma^{(1)}}^{\alpha},$$

i.e., the value of $\bigcirc t$ in $\sigma = s_0, s_1, \ldots$ is given by the value of t in the shifted sequence $\sigma^{(1)} = s_1, s_2, \ldots$.

Consider now the evaluation of formulas:

- For a k-ary predicate p (including equality):

$$p(t_1, \ldots, t_k)|_{\sigma}^{\alpha} = I[p](t_1|_{\sigma}^{\alpha}, \ldots, t_k|_{\sigma}^{\alpha}).$$

Here again, we evaluate the arguments in the environment and then test $I[p]$ on them.

- For a disjunction:

$$(w_1 \vee w_2)|_{\sigma}^{\alpha} = \textit{true} \quad \textit{iff} \quad w_1|_{\sigma}^{\alpha} = \textit{true} \quad \text{or} \quad w_2|_{\sigma}^{\alpha} = \textit{true}.$$

- For a negation:

$$(\sim w)|_{\sigma}^{\alpha} = \textit{true} \quad \textit{iff} \quad w|_{\sigma}^{\alpha} = \textit{false}.$$

- For a next-time application:

$$\bigcirc w|_{\sigma}^{\alpha} = w|_{\sigma^{(1)}}^{\alpha}.$$

Thus $\bigcirc w$ means: w will be true in the *next* instant – read "next w".

- For an all-times application:

$$\Box w|_{\sigma}^{\alpha} = \textit{true} \quad \textit{iff} \quad \text{for every } k \geq 0, w|_{\sigma^{(k)}}^{\alpha} = \textit{true},$$

i.e., w is true for all suffix sequences of σ. Thus $\Box w$ means: w is true for *all* future instants (including the present) – read "always w" or "henceforth w".

- For a some-time application:

$$\Diamond w|_{\sigma}^{\alpha} = \textit{true} \quad \textit{iff}$$
$$\text{there exists a } k \geq 0 \text{ such that } w|_{\sigma^{(k)}}^{\alpha} = \textit{true},$$

i.e., w is *true* on at least one suffix of σ. Thus $\Diamond w$ means: w will be true for *some* future instant (possibly the present) – read "sometimes w" or "eventually w".

- For an until application:

$$w_1 \mathcal{U} w_2|_{\sigma}^{\alpha} = true \quad \textit{iff} \quad \text{for some } k \geq 0,\ w_2|_{\sigma^{(k)}}^{\alpha} = true \text{ and } \text{for all } i,\ 0 \leq i < k,\ w_1|_{\sigma^{(i)}}^{\alpha} = true.$$

Thus $w_1 \mathcal{U} w_2$ means: there is a future instant in which w_2 holds, and such that *until* that instant w_1 continuously holds – read "w_1 until w_2"([KAM], [GPSS]).

- For a universal quantification:

$$(\forall u.w)|_{\sigma}^{\alpha} = true \quad \textit{iff} \quad \text{for every } d \in D,\ w|_{\sigma}^{\alpha'} = true,$$

where $\alpha' = \alpha \circ [u \leftarrow d]$ is the assignment obtained from α by assigning d to u.

- For an existential quantification:

$$(\exists u.w)|_{\sigma}^{\alpha} = true \quad \textit{iff} \quad \text{for some } d \in D,\ w|_{\sigma}^{\alpha'} = true,$$

where $\alpha' = \alpha \circ [u \leftarrow d]$.

A formula w is *valid* if it is true in every model (I, α, σ).

Having defined valid formulas, we naturally look for a deductive system. In such a system we take some of the valid formulas as basic axioms and provide a set of sound inference rules by which we hope to be able to prove the other valid formulas as theorems. In order to denote the fact that a formula w is a theorem derivable in our deductive system we will write $\vdash w$. This will be the case if w is an axiom or is derivable from the axioms by a *proof* using the inference rules of the system.

We partition our deductive system into a *general part* dealing with the general temporal properties of discrete linear sequences, a *domain part* which gives an axiomatic description of the necessary knowledge about the domain, and a *program part* which gives an axiomatic description of a particular program.

We start with the general part, describing first the axiomatic system for propositional temporal logic in which we do not admit predicates or quantification. We treat first the "classical" modal operators $\Box$ and $\Diamond$ (the *modal system*), and later add the special operators $\bigcirc$ and $\mathcal{U}$ (the *temporal system*).

2. THE $\Box$ ("ALWAYS") AND $\Diamond$ ("SOMETIME") OPERATORS

Axioms:

*A*1.	$\vdash \ \sim\Diamond w \equiv \Box\sim w$
*A*2.	$\vdash \ \Box(w_1 \supset w_2) \ \supset \ (\Box w_1 \supset \Box w_2)$
*A*3.	$\vdash \ \Box w \supset w$
*A*4.	$\vdash \ \Box w \supset \Box\Box w$

Axiom *A*1 defines $\Diamond$ as the dual of $\Box$; it states that at all times w is false *iff* it is not the case that sometime w holds. Axiom *A*2 states that if universally w_1 implies w_2 then if at all times w_1 is true then so is w_2. Axiom *A*3 establishes the present as part of the future by stating that if w is true at all future times it must be true of the present. Axiom *A*4 states that if w holds in the future, it holds in the future of the future.

Inference rules:

*R*1.	If w is an instance of a propositional tautology then $\vdash w$ (*Propositional Tautology – PT*)
*R*2.	If $\vdash w_1 \supset w_2$ and $\vdash w_1$ then $\vdash w_2$ (*Modus Ponens – MP*)
*R*3.	If $\vdash w$ then $\vdash \Box w$ ($\Box$ *Insertion* – $\Box I$)

All these rules are sound. The soundness of *R*1 and *R*2 is obvious. Note that in *R*1 we also include modal instances of tautologies; we may substitute an arbitrary modal formula for a proposition letter in obtaining an instance. For example $\Box w \supset \Box w$ is a modal instance of the tautology $p \supset p$. To justify *R*3, we recall that validity of w means that w is true in *all* models, hence $\Box w$ is also valid.

This system provides a logical basis for "propositional" modal reasoning. In Modal Logic circles, this system is known as *S*4 (see, *e.g.*, [HC]). This system constrains *R* to be reflexive (*A*3) and transitive (*A*4).

Before demonstrating some theorems that can be proved in this system, we develop several useful derived rules:

Propositional Reasoning — *PR*

$$\frac{\begin{array}{l}\vdash (w_1 \wedge w_2 \wedge \ldots \wedge w_n) \supset w \\ \vdash w_1, \ \vdash w_2, \ \ldots, \text{ and } \vdash w_n\end{array}}{\vdash w}$$

The notation above is used to describe inference rules. It has the general form

$$\frac{\vdash \varphi_1, \ \vdash \varphi_2 \ \ldots, \ \vdash \varphi_m}{\vdash \psi}$$

and means that if we have already proved $\varphi_1, \ldots, \varphi_m$ (the *assumptions* of the rule), we are allowed by this rule to infer ψ (the *conclusion* of the rule).

proof:

The rule follows from the propositional tautology (Rule $R1$)

$$\vdash \ [(w_1 \wedge w_2 \wedge \ldots \wedge w_n) \supset w] \ \supset \ [w_1 \supset (w_2 \supset (\ldots (w_n \supset w) \ldots))]$$

by applying MP (Rule $R2$) $n+1$ times. ∎

Whenever we apply this derived rule without indicating the antecedent

$$\vdash \ (w_1 \wedge w_2 \ldots \wedge w_n) \supset w,$$

it means that this formula is simply an instance of a propositional tautology.

□□ *Rules*

(a) $$\frac{\vdash w_1 \supset w_2}{\vdash \Box w_1 \supset \Box w_2}$$

(b) $$\frac{\vdash w_1 \equiv w_2}{\vdash \Box w_1 \equiv \Box w_2}$$

proof of (a):

1.	$\vdash \ w_1 \supset w_2$	given
2.	$\vdash \ \Box(w_1 \supset w_2)$	by $\Box I$
3.	$\vdash \ \Box(w_1 \supset w_2) \supset (\Box w_1 \supset \Box w_2)$	by $A2$
4.	$\vdash \ \Box w_1 \supset \Box w_2$	by 2, 3, and MP

Rule (b) then follows by propositional reasoning, since

$$[(w_1 \supset w_2) \wedge (w_2 \supset w_1)] \equiv (w_1 \equiv w_2)$$

is a tautology. ∎

$\Diamond\Diamond$ *Rules*

(a) $\dfrac{\vdash w_1 \supset w_2}{\vdash \Diamond w_1 \supset \Diamond w_2}$ (b) $\dfrac{\vdash w_1 \equiv w_2}{\vdash \Diamond w_1 \equiv \Diamond w_2}$

---

proof of (a):

1.	$\vdash w_1 \supset w_2$	given
2.	$\vdash \sim w_2 \supset \sim w_1$	by *PR*
3.	$\vdash \Box \sim w_2 \supset \Box \sim w_1$	by $\Box\Box$
4.	$\vdash \sim \Diamond w_2 \supset \sim \Diamond w_1$	by *A1* and *PR*
5.	$\vdash \Diamond w_1 \supset \Diamond w_2$	by *PR*

Rule (b) then follows by propositional reasoning. ∎

Equivalence Rule - ER

Let w' be the result of replacing an occurrence of a subformula v_1 in w by v_2. Then

$$\frac{\vdash v_1 \equiv v_2}{\vdash w \equiv w'}$$

---

proof:

By induction on the structure of w.

Case: w is v_1. Then w' is v_2 and $\vdash v_1 \equiv v_2$ implies $\vdash w \equiv w'$

Case: w is of the form $\sim u$. We assume that $\vdash v_1 \equiv v_2$ implies $\vdash u \equiv u'$. Then by propositional reasoning $\vdash \sim u \equiv \sim u'$, *i.e.*, $\vdash w \equiv w'$.

Case: w is of the form $u_1 \vee u_2$. We assume that if $\vdash v_1 \equiv v_2$, then $\vdash u_1 \equiv u_1'$ and $\vdash u_2 \equiv u_2'$. Then by propositional reasoning $\vdash (u_1 \vee u_2) \equiv (u_1' \vee u_2')$, *i.e.*, $\vdash w \equiv w'$.

The cases where w is of form $u_1 \wedge u_2$, $u_1 \supset u_2$, *etc.* are similar.

Case: w is of the form $\Box u$. We assume that if $\vdash v_1 \equiv v_2$, then $\vdash u \equiv u'$. By the $\Box\Box$-rule, $\vdash \Box u \equiv \Box u'$, *i.e.*, $\vdash w \equiv w'$.

The case in which w is of the form $\Diamond u$ is treated similarly, using the $\Diamond\Diamond$-rule. ∎

Some theorems that can be derived in the system are:

*T*1. $\vdash w \supset \Diamond w$

proof:

1.	$\vdash (\Box\sim w) \supset \sim w$	by *A*3
2.	$\vdash w \supset (\sim\Box\sim w)$	by *PR*
3.	$\vdash w \supset \Diamond w$	by *A*1 and *PR*

The theorem implies (by *MP*)

$\Diamond$ *Insertion* – $\Diamond I$

$$\frac{\vdash w}{\vdash \Diamond w}$$

We can derive the converse of axiom *A*4 as stated in the modal system, and thus prove:

*T*2. $\vdash \Box w \equiv \Box\Box w$

proof:

1.	$\vdash \Box w \supset \Box\Box w$	by *A*4
2.	$\vdash \Box w \supset w$	by *A*3
3.	$\vdash \Box\Box w \supset \Box w$	by $\Box\Box$
4.	$\vdash \Box w \equiv \Box\Box w$	by 1, 3, and *PR*

*T*3. $\vdash \Diamond w \equiv \Diamond\Diamond w$

proof:

1.	$\vdash \Box\sim w \equiv \Box\Box\sim w$	by *T*2
2.	$\vdash \sim\Box\sim w \equiv \sim\Box\Box\sim w$	by *PR*
3.	$\vdash \Diamond w \equiv \sim\Box\sim\Diamond w$	by *A*1 and *ER*
4.	$\vdash \Diamond w \equiv \Diamond\Diamond w$	by *A*1 and *PR*

Because of these last two theorems we can collapse any string of consecutive identical modalities such as $\Box \cdots \Box$ or $\Diamond \cdots \Diamond$ into a single modality of the same type.

Note that to derive line 3 from line 2 we could not use propositional reasoning (PR), but we had to use the equivalence rule (ER). The subformula $\Box \sim w$ in

$$2. \vdash \ \ldots \ \equiv \ \sim \Box\Box \sim w$$

was replaced by the equivalent subformula $\sim \Diamond w$ to obtain

$$3. \vdash \ \ldots \ \equiv \ \sim \Box \sim \Diamond w.$$

But this replacement is inside $\Box$ and thus cannot be justified by propositional reasoning. The replacement done on the left-hand side of the equivalence can be justified by propositional reasoning.

$T4.$ $\vdash \ (\Diamond \sim w) \ \equiv \ (\sim \Box w)$

proof:

1.	$\vdash \ (\sim\sim w) \ \equiv \ w$	by PT
2.	$\vdash \ (\Box \sim\sim w) \ \equiv \ \Box w$	by $\Box\Box$
3.	$\vdash \ (\sim \Diamond \sim w) \ \equiv \ \Box w$	by $A1$ and PR
4.	$\vdash \ (\Diamond \sim w) \ \equiv \ (\sim \Box w)$	by PR

$T5.$ $\vdash \ \Box(w_1 \supset w_2) \ \supset \ (\Diamond w_1 \supset \Diamond w_2)$

proof:

1.	$\vdash \ (w_1 \supset w_2) \ \equiv \ (\sim w_2 \supset \sim w_1)$	by PT
2.	$\vdash \ \Box(w_1 \supset w_2) \ \equiv \ \Box(\sim w_2 \supset \sim w_1)$	by $\Box\Box$
3.	$\vdash \ \Box(\sim w_2 \supset \sim w_1) \ \supset \ (\Box \sim w_2 \supset \Box \sim w_1)$	by $A2$
4.	$\vdash \ (\Box \sim w_2 \supset \Box \sim w_1) \ \equiv \ (\sim \Diamond w_2 \supset \sim \Diamond w_1)$	by $A1$ and PR
5.	$\vdash \ (\sim \Diamond w_2 \supset \sim \Diamond w_1) \ \equiv \ (\Diamond w_1 \supset \Diamond w_2)$	by PT
6.	$\vdash \ \Box(w_1 \supset w_2) \ \supset \ (\Diamond w_1 \supset \Diamond w_2)$	by 2, 3, 4, 5, and PR

$T6.$ $\vdash \ \Box(w_1 \wedge w_2) \ \equiv \ (\Box w_1 \wedge \Box w_2)$

proof:

1.	$\vdash \ (w_1 \wedge w_2) \ \supset \ w_1$	by PT
2.	$\vdash \ \Box(w_1 \wedge w_2) \ \supset \ \Box w_1$	by $\Box\Box$
3.	$\vdash \ (w_1 \wedge w_2) \ \supset \ w_2$	by PT
4.	$\vdash \ \Box(w_1 \wedge w_2) \ \supset \ \Box w_2$	by $\Box\Box$
5.	$\vdash \ \Box(w_1 \wedge w_2) \ \supset \ (\Box w_1 \wedge \Box w_2)$	by 2, 4, and PR
6.	$\vdash \ w_1 \ \supset \ (w_2 \supset w_1 \wedge w_2)$	by PT
7.	$\vdash \ \Box w_1 \ \supset \ \Box(w_2 \supset (w_1 \wedge w_2))$	by $\Box\Box$

8. $\vdash \Box(w_2 \supset (w_1 \wedge w_2)) \supset (\Box w_2 \supset \Box(w_1 \wedge w_2))$ by *A2*
9. $\vdash \Box w_1 \supset (\Box w_2 \supset \Box(w_1 \wedge w_2))$ by 7, 8, and *PR*
10. $\vdash (\Box w_1 \wedge \Box w_2) \supset \Box(w_1 \wedge w_2)$ by *PR*

11. $\vdash \Box(w_1 \wedge w_2) \equiv (\Box w_1 \wedge \Box w_2)$ by 5, 10, and *PR*

T7. $\vdash \Diamond(w_1 \vee w_2) \equiv (\Diamond w_1 \vee \Diamond w_2)$

proof:

1. $\vdash \Box(\sim w_1 \wedge \sim w_2) \equiv (\Box\sim w_1 \wedge \Box\sim w_2)$ by *T6*
2. $\vdash \Box\sim(w_1 \vee w_2) \equiv \sim(\sim\Box\sim w_1 \vee \sim\Box\sim w_2)$ by *ER*
3. $\vdash \sim\Diamond(w_1 \vee w_2) \equiv \sim(\Diamond w_1 \vee \Diamond w_2)$ by *A1* and *PR*
4. $\vdash \Diamond(w_1 \vee w_2) \equiv (\Diamond w_1 \vee \Diamond w_2)$ by *PR*

Note that because of the universal character of $\Box$ it can be distributed over $\wedge$ (Theorem *T6*), while $\Diamond$, which is of existential character can be distributed over $\vee$ (Theorem *T7*).

T8. $\vdash \Diamond(w_1 \wedge w_2) \supset (\Diamond w_1 \wedge \Diamond w_2)$

proof:

1. $\vdash \Diamond(w_1 \wedge w_2) \supset \Diamond w_1$ by *PT* and $\Diamond\Diamond$
2. $\vdash \Diamond(w_1 \wedge w_2) \supset \Diamond w_2$ by *PT* and $\Diamond\Diamond$
3. $\vdash \Diamond(w_1 \wedge w_2) \supset (\Diamond w_1 \wedge \Diamond w_2)$ by 1, 2, and *PR*

T9. $\vdash (\Box w_1 \vee \Box w_2) \supset \Box(w_1 \vee w_2)$

proof:

1. $\vdash \Box w_1 \supset \Box(w_1 \vee w_2)$ by *PT* and $\Box\Box$
2. $\vdash \Box w_2 \supset \Box(w_1 \vee w_2)$ by *PT* and $\Box\Box$
3. $\vdash (\Box w_1 \vee \Box w_2) \supset \Box(w_1 \vee w_2)$ by 1, 2, and *PR*

T10. $\vdash (\Box w_1 \wedge \Diamond w_2) \supset \Diamond(w_1 \wedge w_2)$

proof:

1. $\vdash \Box(w_1 \supset \sim w_2) \supset (\Box w_1 \supset \Box\sim w_2)$ by *A2*
2. $\vdash \Box\sim(w_1 \wedge w_2) \supset \sim(\Box w_1 \wedge \sim\Box\sim w_2)$ by *ER*
3. $\vdash \sim\Diamond(w_1 \wedge w_2) \supset \sim(\Box w_1 \wedge \Diamond w_2)$ by *A1* and *PR*
4. $\vdash (\Box w_1 \wedge \Diamond w_2) \supset \Diamond(w_1 \wedge w_2)$ by *PR*

another proof (without using ER):

1. $\vdash w_1 \supset (w_2 \supset (w_1 \wedge w_2))$ by *PT*

2. $\vdash \Box w_1 \supset \Box(w_2 \supset (w_1 \wedge w_2))$ by $\Box\Box$
3. $\vdash \Box(w_2 \supset (w_1 \wedge w_2)) \supset (\Diamond w_2 \supset \Diamond(w_1 \wedge w_2))$ by $T5$
4. $\vdash \Box w_1 \supset (\Diamond w_2 \supset \Diamond(w_1 \wedge w_2))$ by 2, 3, and PR
5. $\vdash (\Box w_1 \wedge \Diamond w_2) \supset \Diamond(w_1 \wedge w_2)$ by PR

The following derived rules correspond to proof rules existing in most axiomatic verification systems:

Consequence Rules – $\Diamond Q$ *and* $\Box Q$

$$\frac{\vdash w_1 \supset w_2 \quad \vdash w_2 \supset \Diamond w_3 \quad \vdash w_3 \supset w_4}{\vdash w_1 \supset \Diamond w_4} \qquad \frac{\vdash w_1 \supset w_2 \quad \vdash w_2 \supset \Box w_3 \quad \vdash w_3 \supset w_4}{\vdash w_1 \supset \Box w_4}$$

proof of $\Diamond Q$:

1. $\vdash w_1 \supset w_2$ given
2. $\vdash w_2 \supset \Diamond w_3$ given
3. $\vdash w_3 \supset w_4$ given
4. $\vdash \Diamond w_3 \supset \Diamond w_4$ by 3 and $\Diamond\Diamond$
5. $\vdash w_1 \supset \Diamond w_4$ by 1, 2, 4, and PR

The $\Box Q$ rule is proved similarly by the $\Box\Box$-rule.

Concatenation Rule – $\Diamond C$ *and* $\Box C$

$$\frac{\vdash w_1 \supset \Diamond w_2 \quad \vdash w_2 \supset \Diamond w_3}{\vdash w_1 \supset \Diamond w_3} \qquad \frac{\vdash w_1 \supset \Box w_2 \quad \vdash w_2 \supset \Box w_3}{\vdash w_1 \supset \Box w_3}$$

proof of $\Diamond C$:

1. $\vdash w_1 \supset \Diamond w_2$ given
2. $\vdash w_2 \supset \Diamond w_3$ given
3. $\vdash \Diamond w_2 \supset \Diamond\Diamond w_3$ by 2 and $\Diamond\Diamond$
4. $\vdash \Diamond w_2 \supset \Diamond w_3$ by $T3$ and PR
5. $\vdash w_1 \supset \Diamond w_3$ by 1, 4, and PR

The $\Box C$ rule is proved similarly by the $\Box\Box$-rule.

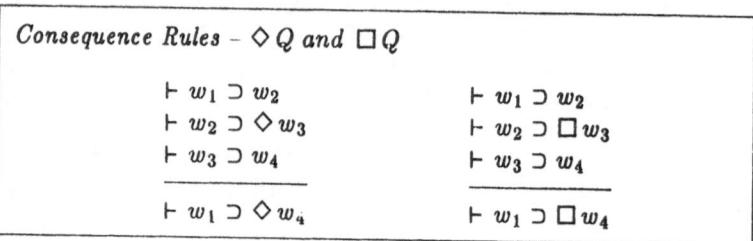

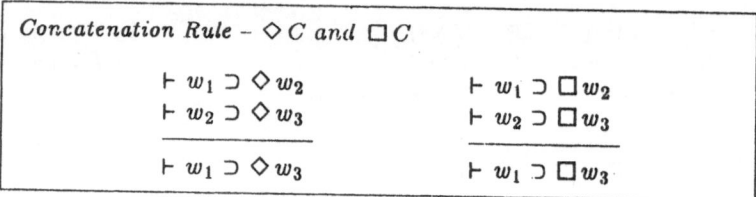

3. THE $\bigcirc$ ("NEXT") AND $\mathcal{U}$ ("UNTIL") OPERATORS

Axioms:

$C1.$	$\vdash \ \sim\Diamond w \equiv \Box{\sim}w$
$C2.$	$\vdash \ \Box(w_1 \supset w_2) \supset (\Box w_1 \supset \Box w_2)$
$C3.$	$\vdash \ \Box w \supset w$
$C4.$	$\vdash \ \bigcirc{\sim}w \equiv {\sim}\bigcirc w$
$C5.$	$\vdash \ \bigcirc(w_1 \supset w_2) \supset (\bigcirc w_1 \supset \bigcirc w_2)$
$C6.$	$\vdash \ \Box w \supset \bigcirc w$
$C7.$	$\vdash \ \Box w \supset \bigcirc\Box w$
$C8.$	$\vdash \ \Box(w \supset \bigcirc w) \supset (w \supset \Box w)$
$C9.$	$\vdash \ w_1 \,\mathcal{U}\, w_2 \equiv [w_2 \vee (w_1 \wedge \bigcirc(w_1 \,\mathcal{U}\, w_2))]$
$C10.$	$\vdash \ w_1 \,\mathcal{U}\, w_2 \supset \Diamond w_2.$

Axioms $C1 - C3$ are the same as $A1 - A3$ in the modal system.

Axiom $C4$ establishes $\bigcirc$ as self-dual. Consequently it implies that the next instant exists and is unique, and restricts our models to linear sequences (no branching).

Axiom $C5$ is the analogue of $C2$ for the $\bigcirc$ operator. Axiom $C6$ states that the next instant is one of the reachable states, *i.e.*, it is also part of the future. Axiom $C7$ is a weaker version of $A4$, $\vdash \Box w \supset \Box\Box w$, and can be used together with $C8$ to prove $A4$ as a theorem in this system. Axiom $C8$ is the "computational induction" axiom; it states that if a property is inherited over one step transitions, it is invariant over any suffix sequence whose first state satisfies w. Axiom $C9$ defines the ***until*** operator by distributing its effect into what is implied for the present and what is implied for the next instant. Axiom $C10$ simply states that "w_1 until w_2" implies that w_2 will eventually happen.

Inference rules:

$R1$. If w is an instance of a propositional tautology then $\vdash w$ (*Propositional Tautology – PT*)
$R2$. If $\vdash w_1 \supset w_2$ and $\vdash w_1$ then $\vdash w_2$ (*Modus Ponens – MP*)
$R3$. If $\vdash w$ then $\vdash \Box w$ ($\Box$ *Insertion* – $\Box I$)

These rules are identical to $R1 - R3$ of the modal system. Since axioms $C1$, $C2$ and $C3$ are identical to axioms $A1$, $A2$ and $A3$ and we will show later that axiom $A4$ is derivable in this system, it follows that all the derived rules of inference and the theorems in the modal system are also derivable in this system. Here are several additional derived rules:

$\bigcirc$ *Insertion* – $\bigcirc I$:

$$\frac{\vdash w}{\vdash \bigcirc w}$$

proof:

1.	$\vdash w$	given
2.	$\vdash \Box w$	by $\Box I$
3.	$\vdash \bigcirc w$	by $C6$ and MP

$\bigcirc\bigcirc$ *Rules*

(a) $$\frac{\vdash w_1 \supset w_2}{\vdash \bigcirc w_1 \supset \bigcirc w_2}$$ (b) $$\frac{\vdash w_1 \equiv w_2}{\vdash \bigcirc w_1 \equiv \bigcirc w_2}$$

proof of (a):

1.	$\vdash w_1 \supset w_2$	given
2.	$\vdash \bigcirc(w_1 \supset w_2)$	by $\bigcirc I$
3.	$\vdash \bigcirc w_1 \supset \bigcirc w_2$	by $C5$ and MP

Rule (b) follows by propositional reasoning.

Computational Induction Rule – *CI*

$$\frac{\vdash\ w \supset \bigcirc w}{\vdash\ w \supset \Box w}$$

proof:

1.	$\vdash\ w \supset \bigcirc w$	given
2.	$\vdash\ \Box(w \supset \bigcirc w)$	by $\Box I$
3.	$\vdash\ \Box(w \supset \bigcirc w) \supset (w \supset \Box w)$	by $C8$
4.	$\vdash\ w \supset \Box w$	by 2, 3, and MP

Backward Induction Rule – *BI*

$$\frac{\vdash\ \bigcirc w \supset w}{\vdash\ \Diamond w \supset w}$$

proof:

1.	$\vdash\ \bigcirc w \supset w$	given
2.	$\vdash\ \sim w \supset \sim\bigcirc w$	by PR
3.	$\vdash\ \sim w \supset \bigcirc\sim w$	by $C4$ and PR
4.	$\vdash\ \sim w \supset \Box\sim w$	by CI
5.	$\vdash\ \sim w \supset \sim\Diamond w$	by $C1$ and PR
6.	$\vdash\ \Diamond w \supset w$	by PR

$\bigcirc$ *Consequence* Rule – $\bigcirc Q$

$$\frac{\begin{array}{l}\vdash\ w_1 \supset w_2\\ \vdash\ w_2 \supset \bigcirc w_3\\ \vdash\ w_3 \supset w_4\end{array}}{\vdash\ w_1 \supset \bigcirc w_4}$$

proof:

1.	$\vdash\ w_1 \supset w_2$	given
2.	$\vdash\ w_2 \supset \bigcirc w_3$	given
3.	$\vdash\ w_3 \supset w_4$	given
4.	$\vdash\ \bigcirc w_3 \supset \bigcirc w_4$	by $\bigcirc\bigcirc$
5.	$\vdash\ w_1 \supset \bigcirc w_4$	by 1, 2, 4, and PR

Note that we do not have a $\bigcirc$ concatenation rule.

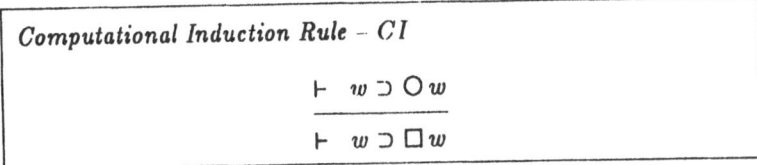

A simple theorem of this system is:

T11. $\vdash \ \bigcirc w \supset \Diamond w$

proof:

1.	$\vdash (\Box \sim w) \supset (\bigcirc \sim w)$	by *C6*
2.	$\vdash (\sim \bigcirc \sim w) \supset (\sim \Box \sim w)$	by *PR*
3.	$\vdash \bigcirc w \supset \Diamond w$	by *C1*, *C4*, and *PR*

T12. $\vdash \ \Box w \supset \Box\Box w$

proof:

1.	$\vdash \Box w \supset \bigcirc \Box w$	by *C7*
2.	$\vdash \Box w \supset \Box\Box w$	by *CI*

This is the "missing" axiom *A4*. We have all axioms and rules of the previous system, therefore we can deduce all theorems and derived rules of the modal system.

The following special rule is very useful in proving *until* theorems:

Next to Present Rule – NP

$$\frac{\begin{array}{l}\vdash (\bigcirc w_1 \equiv \bigcirc w_2) \supset (w_1 \equiv w_2) \\ \vdash w_1 \supset \Diamond(w_1 \wedge w_2) \\ \vdash w_2 \supset \Diamond(w_1 \wedge w_2)\end{array}}{\vdash w_1 \equiv w_2}$$

proof:

1.	$\vdash w_1 \supset \Diamond(w_1 \wedge w_2)$	given
2.	$\vdash w_2 \supset \Diamond(w_1 \wedge w_2)$	given
3.	$\vdash (w_1 \vee w_2) \supset \Diamond(w_1 \wedge w_2)$	by 1, 2, and *PR*
4.	$\vdash (w_1 \wedge w_2) \supset (w_1 \equiv w_2)$	by *PT*
5.	$\vdash \Diamond(w_1 \wedge w_2) \supset \Diamond(w_1 \equiv w_2)$	by $\Diamond\Diamond$
6.	$\vdash \bigcirc(w_1 \equiv w_2) \supset (w_1 \equiv w_2)$	given
7.	$\vdash \Diamond(w_1 \equiv w_2) \supset (w_1 \equiv w_2)$	by *BI*
8.	$\vdash (w_1 \vee w_2) \supset (w_1 \equiv w_2)$	by 3, 5, 7, and *PR*
9.	$\vdash w_1 \equiv w_2$	by *PR*

We extend now the Equivalence Rule (*ER*) to handle the $\bigcirc$ and $\mathcal{U}$ operators.

> *Equivalence Rule – ER*
>
> Let w' be the result of replacing an occurrence of a subformula v_1 in w by v_2. Then
>
> $$\frac{\vdash v_1 \equiv v_2}{\vdash w \equiv w'}$$

proof:

As before, the proof is by induction on the structure of w. The cases where w is w_1 or of form $\sim u$, $u_1 \vee u_2$, $u_1 \supset u_2$, etc. are treated as in the ER derived rule above.

Case: w is of form $\bigcirc u$. We assume that if $\vdash v_1 \equiv v_2$, then $\vdash u \equiv u'$. Then by the $\bigcirc\bigcirc$-rule $\vdash \bigcirc u \equiv \bigcirc u'$, *i.e.* $\vdash w \equiv w'$.

The cases where w is of form $\Box u$ and $\Diamond u$ are proved similarly by the $\Box\Box$-rule and $\Diamond\Diamond$-rule, respectively. The case that w is of form $u_1 \,\mathcal{U}\, u_2$ needs a more detailed proof.

Case: w is of form $u_1 \,\mathcal{U}\, u_2$. We assume that if $\vdash v_1 \equiv v_2$, then $\vdash u_1 \equiv u_1'$ and $\vdash u_2 \equiv u_2'$. We attempt to use the Next to Present derived rule (NP) taking w_1 to be $u_1 \,\mathcal{U}\, u_2$ and w_2 to be $u_1' \,\mathcal{U}\, u_2'$.

1.	$\vdash u_1 \equiv u_1'$	induction hypothesis
2.	$\vdash u_2 \equiv u_2'$	induction hypothesis
3.	$\vdash u_1 \,\mathcal{U}\, u_2 \equiv [u_2 \vee (u_1 \wedge \bigcirc(u_1 \,\mathcal{U}\, u_2)])$	by $C9$
4.	$\vdash u_1' \,\mathcal{U}\, u_2' \equiv [u_2' \vee (u_1' \wedge \bigcirc(u_1' \,\mathcal{U}\, u_2'))]$	by $C9$
5.	$\vdash u_1' \,\mathcal{U}\, u_2' \equiv [u_2 \vee (u_1 \wedge \bigcirc(u_1' \,\mathcal{U}\, u_2'))]$	by 1, 2, 4, and PR
6.	$\vdash [\bigcirc(u_1 \,\mathcal{U}\, u_2) \equiv \bigcirc(u_1' \,\mathcal{U}\, u_2')] \supset [(u_1 \,\mathcal{U}\, u_2) \equiv (u_1' \,\mathcal{U}\, u_2')]$	by 3, 5, and PR
7.	$\vdash u_1 \,\mathcal{U}\, u_2 \supset \Diamond u_2$	by $C10$
8.	$\vdash u_2 \supset [(u_1 \,\mathcal{U}\, u_2) \wedge (u_1' \,\mathcal{U}\, u_2')]$	by 3, 5, and PR
9.	$\vdash u_1 \,\mathcal{U}\, u_2 \supset \Diamond[(u_1 \,\mathcal{U}\, u_2) \wedge (u_1' \,\mathcal{U}\, u_2')]$	by 7, 8, and $\Diamond Q$
10.	$\vdash u_1' \,\mathcal{U}\, u_2' \supset \Diamond u_2'$	by $C10$
11.	$\vdash \Diamond u_2 \equiv \Diamond u_2'$	by 2 and $\Diamond\Diamond$
12.	$\vdash u_1' \,\mathcal{U}\, u_2' \supset \Diamond u_2$	by 10, 11, and PR
13.	$\vdash u_1' \,\mathcal{U}\, u_2' \supset \Diamond[(u_1 \,\mathcal{U}\, u_2) \wedge (u_1' \,\mathcal{U}\, u_2')]$	by 8, 12, and $\Diamond Q$
14.	$\vdash (u_1 \,\mathcal{U}\, u_2) \equiv (u_1' \,\mathcal{U}\, u_2')$	by 6, 9, 13, and NP

This concludes the proof. ∎

"next" theorems

*T*13. $\vdash \bigcirc(w_1 \wedge w_2) \equiv (\bigcirc w_1 \wedge \bigcirc w_2)$

proof:

1.	$\vdash \bigcirc(w_1 \supset \sim w_2) \supset (\bigcirc w_1 \supset \bigcirc \sim w_2)$	by $C5$
2.	$\vdash \sim(\bigcirc w_1 \supset \bigcirc \sim w_2) \supset \sim \bigcirc(w_1 \supset \sim w_2)$	by PR
3.	$\vdash \sim(\bigcirc w_1 \supset \sim \bigcirc w_2) \supset \bigcirc \sim(w_1 \supset \sim w_2)$	by $C4$ and PR
4.	$\vdash (\bigcirc w_1 \wedge \bigcirc w_2) \supset \bigcirc(w_1 \wedge w_2)$	by ER
5.	$\vdash (w_1 \wedge w_2) \supset w_1$	by PT
6.	$\vdash \bigcirc(w_1 \wedge w_2) \supset \bigcirc w_1$	by $\bigcirc\bigcirc$
7.	$\vdash (w_1 \wedge w_2) \supset w_2$	by PT
8.	$\vdash \bigcirc(w_1 \wedge w_2) \supset \bigcirc w_2$	by $\bigcirc\bigcirc$
9.	$\vdash \bigcirc(w_1 \wedge w_2) \supset (\bigcirc w_1 \wedge \bigcirc w_2)$	by 6, 8, and PR
10.	$\vdash \bigcirc(w_1 \wedge w_2) \equiv (\bigcirc w_1 \wedge \bigcirc w_2)$	by 4, 9, and PR

*T*14. $\vdash \bigcirc(w_1 \vee w_2) \equiv (\bigcirc w_1 \vee \bigcirc w_2)$

proof:

1.	$\vdash \bigcirc(\sim w_1 \wedge \sim w_2) \equiv (\bigcirc \sim w_1) \wedge (\bigcirc \sim w_2)$	by $T13$
2.	$\vdash \bigcirc(\sim w_1 \wedge \sim w_2) \equiv (\sim \bigcirc w_1) \wedge (\sim \bigcirc w_2)$	by $C4$ and PR
3.	$\vdash \bigcirc \sim(w_1 \vee w_2) \equiv (\sim \bigcirc w_1) \wedge (\sim \bigcirc w_2)$	by ER and PR
4.	$\vdash \sim \bigcirc(w_1 \vee w_2) \equiv \sim(\bigcirc w_1 \vee \bigcirc w_2)$	by $C4$ and PR
5.	$\vdash \bigcirc(w_1 \vee w_2) \equiv (\bigcirc w_1 \vee \bigcirc w_2)$	by PR

*T*15. $\vdash \bigcirc(w_1 \supset w_2) \equiv (\bigcirc w_1 \supset \bigcirc w_2)$

proof:

1.	$\vdash \bigcirc(\sim w_1 \vee w_2) \equiv (\bigcirc \sim w_1) \vee (\bigcirc w_2)$	by $T14$
2.	$\vdash \bigcirc(\sim w_1 \vee w_2) \equiv (\sim \bigcirc w_1) \vee (\bigcirc w_2)$	by $C4$ and PR
3.	$\vdash \bigcirc(w_1 \supset w_2) \equiv (\bigcirc w_1 \supset \bigcirc w_2)$	by ER and PR

*T*16. $\vdash \bigcirc(w_1 \equiv w_2) \equiv (\bigcirc w_1 \equiv \bigcirc w_2)$

proof:

1.	$\vdash [\bigcirc(w_1 \supset w_2) \wedge \bigcirc(w_2 \supset w_1)]$ $\equiv [(\bigcirc w_1 \supset \bigcirc w_2) \wedge (\bigcirc w_2 \supset \bigcirc w_1)]$	by $T15$ and PR

2. $\vdash$ $\bigcirc[(w_1 \supset w_2) \wedge (w_2 \supset w_1)]$
$\equiv [(\bigcirc w_1 \supset \bigcirc w_2) \wedge (\bigcirc w_2 \supset \bigcirc w_1)]$ by $T13$ and PR
3. $\vdash$ $\bigcirc(w_1 \equiv w_2) \equiv (\bigcirc w_1 \equiv \bigcirc w_2)$
by ER and PR

$T17.$ $\vdash$ $\bigcirc \Box w \equiv \Box \bigcirc w$

proof:

1. $\vdash$ $\bigcirc w \supset (w \supset \bigcirc w)$ by PT
2. $\vdash$ $\Box \bigcirc w \supset \Box(w \supset \bigcirc w)$ by $\Box\Box$
3. $\vdash$ $\Box(w \supset \bigcirc w) \supset \bigcirc \Box(w \supset \bigcirc w)$ by $C7$
4. $\vdash$ $\bigcirc \Box(w \supset \bigcirc w) \supset \bigcirc(w \supset \Box w)$ by $C8$ and $\bigcirc\bigcirc$
5. $\vdash$ $\bigcirc(w \supset \Box w) \supset (\bigcirc w \supset \bigcirc \Box w)$ by $C5$
6. $\vdash$ $\Box \bigcirc w \supset (\bigcirc w \supset \bigcirc \Box w)$ by 2, 3, 4, 5, and PR
7. $\vdash$ $\Box \bigcirc w \supset \bigcirc w$ by $C3$
8. $\vdash$ $\Box \bigcirc w \supset \bigcirc \Box w$ by 6, 7, and PR

9. $\vdash$ $\bigcirc \Box w \supset \bigcirc \bigcirc \Box w$ by $C7$ and $\bigcirc\bigcirc$
10. $\vdash$ $\bigcirc \Box w \supset \Box \bigcirc \Box w$ by CI
11. $\vdash$ $\bigcirc \Box w \supset \bigcirc w$ by $C3$ and $\bigcirc\bigcirc$
12. $\vdash$ $\Box \bigcirc \Box w \supset \Box \bigcirc w$ by $\Box\Box$
13. $\vdash$ $\bigcirc \Box w \supset \Box \bigcirc w$ by 10, 12, and PR

14. $\vdash$ $\bigcirc \Box w \equiv \Box \bigcirc w$ by 8, 13, and PR

$T18.$ $\vdash$ $\bigcirc \Diamond w \equiv \Diamond \bigcirc w$

proof:

1. $\vdash$ $\bigcirc \Box {\sim} w \equiv \Box \bigcirc {\sim} w$ by $T17$
2. $\vdash$ ${\sim} \bigcirc \Diamond w \equiv {\sim} \Diamond \bigcirc w$ by $C1$, $C4$, and ER
3. $\vdash$ $\bigcirc \Diamond w \equiv \Diamond \bigcirc w$ by PR

$T19.$ $\vdash$ $\Box w \equiv (w \wedge \bigcirc \Box w)$

proof:

1. $\vdash$ $\Box w \supset w$ by $C3$
2. $\vdash$ $\Box w \supset \bigcirc \Box w$ by $C7$
3. $\vdash$ $\Box w \supset (w \wedge \bigcirc \Box w)$ by 1, 2, and PR

4. $\vdash$ $\bigcirc \Box w \supset \bigcirc(w \wedge \bigcirc \Box w)$ by $\bigcirc\bigcirc$
5. $\vdash$ $(w \wedge \bigcirc \Box w) \supset \bigcirc(w \wedge \bigcirc \Box w)$ by PR
6. $\vdash$ $(w \wedge \bigcirc \Box w) \supset \Box(w \wedge \bigcirc \Box w)$ by CI
7. $\vdash$ $\Box(w \wedge \bigcirc \Box w) \supset (\Box w \wedge \Box \bigcirc \Box w)$ by $T6$

8. $\vdash \Box(w \wedge \bigcirc\Box w) \supset \Box w$ by PR
9. $\vdash (w \wedge \bigcirc\Box w) \supset \Box w$ by 6, 8, and PR

10. $\vdash \Box w \equiv (w \wedge \bigcirc\Box w)$ by 3, 9, and PR

T20. $\vdash \Diamond w \equiv (w \vee \bigcirc\Diamond w)$

proof:

1. $\vdash \Box{\sim}w \equiv ({\sim}w \wedge \bigcirc\Box{\sim}w)$ by $T19$
2. $\vdash {\sim}\Diamond w \equiv {\sim}(w \vee {\sim}\bigcirc\Box{\sim}w)$ by $C1$ and PR
3. $\vdash {\sim}\Diamond w \equiv {\sim}(w \vee \bigcirc\Diamond w)$ by $C4$, $C1$, and ER
4. $\vdash \Diamond w \equiv (w \vee \bigcirc\Diamond w)$ by PR

T21. $\vdash (w \wedge \Diamond{\sim}w) \supset \Diamond(w \wedge \bigcirc{\sim}w)$.

This is the dual of the "computational induction" axiom $C8$. It states that if w is true now and is false in the future, then there exists some instant such that w is true at that instant and false at the next.

proof:

1. $\vdash \Box(w \supset \bigcirc w) \supset (w \supset \Box w)$ by $C8$
2. $\vdash {\sim}(w \supset \Box w) \supset {\sim}\Box(w \supset \bigcirc w)$ by PR
3. $\vdash (w \wedge {\sim}\Box w) \supset \Diamond(w \wedge {\sim}\bigcirc w)$ by $T4$ and ER
4. $\vdash (w \wedge \Diamond{\sim}w) \supset \Diamond(w \wedge \bigcirc{\sim}w)$ by $T4$, $C4$, and ER

"until" theorems

T22. $\vdash (\bigcirc w_1)\mathcal{U}(\bigcirc w_2) \equiv \bigcirc(w_1 \mathcal{U} w_2)$

Denoting

$$w_1^* : \quad (\bigcirc w_1)\mathcal{U}(\bigcirc w_2)$$

$$w_2^* : \quad \bigcirc(w_1 \mathcal{U} w_2)$$

we have to show $\vdash w_1^* \equiv w_2^*$. We will use the Next to Present derived rule (NP).

proof:

1. $\vdash w_1^* \equiv \bigcirc w_2 \vee (\bigcirc w_1 \wedge \bigcirc w_1^*)$ by $C9$

2. $\vdash \bigcirc(w_1 \mathcal{U} w_2) \equiv \bigcirc(w_2 \vee (w_1 \wedge \bigcirc(w_1 \mathcal{U} w_2)))$ by $C9$ and $\bigcirc\bigcirc$

3. $\vdash \overset{*}{w}_2 \equiv \bigcirc w_2 \vee (\bigcirc w_1 \wedge \bigcirc \overset{*}{w}_2)$ by 2, *T*13, *T*14, and *PR*

4. $\vdash (\bigcirc \overset{*}{w}_1 \equiv \bigcirc \overset{*}{w}_2) \supset (\overset{*}{w}_1 \equiv \overset{*}{w}_2)$ by 1, 3 and *PR*

5. $\vdash \bigcirc w_2 \supset (\overset{*}{w}_1 \wedge \overset{*}{w}_2)$ by 1, 3 and *PR*

6. $\vdash \Diamond \bigcirc w_2 \supset \Diamond(\overset{*}{w}_1 \wedge \overset{*}{w}_2)$ by $\Diamond\Diamond$

7. $\vdash (\bigcirc w_1 \mathcal{U} \bigcirc w_2) \supset \Diamond \bigcirc w_2$ by *C*10

8. $\vdash \overset{*}{w}_1 \supset \Diamond(\overset{*}{w}_1 \wedge \overset{*}{w}_2)$ by 6, 7 and *PR*

9. $\vdash w_1 \mathcal{U} w_2 \supset \Diamond w_2$ by *C*10

10. $\vdash \bigcirc(w_1 \mathcal{U} w_2) \supset \Diamond \bigcirc w_2$ by 9, $\bigcirc\bigcirc$, and *T*18

11. $\vdash \overset{*}{w}_2 \supset \Diamond(\overset{*}{w}_1 \wedge \overset{*}{w}_2)$ by 6, 10, and *PR*

12. $\vdash \overset{*}{w}_1 \equiv \overset{*}{w}_2$ by 4, 8, 11 and *NP*

*T*23. $\vdash (w_1 \wedge w_2)\mathcal{U} w_3 \equiv [(w_1 \mathcal{U} w_2) \wedge (w_2 \mathcal{U} w_3)]$

Denoting

$$\overset{*}{w}_1 : \quad (w_1 \wedge w_2)\mathcal{U} w_3$$

$$\overset{*}{w}_2 : \quad (w_1 \mathcal{U} w_3) \wedge (w_2 \mathcal{U} w_3)$$

we have to show $\vdash \overset{*}{w}_1 \equiv \overset{*}{w}_2$. We will again use the derived rule *NP*.

proof:

1. $\vdash \overset{*}{w}_1 \equiv w_3 \vee ((w_1 \wedge w_2) \wedge \bigcirc \overset{*}{w}_1)$ by *C*9

2. $\vdash w_1 \mathcal{U} w_3 \equiv w_3 \vee (w_1 \wedge \bigcirc(w_1 \mathcal{U} w_3))$ by *C*9

3. $\vdash w_2 \mathcal{U} w_3 \equiv w_3 \vee (w_2 \wedge \bigcirc(w_2 \mathcal{U} w_3))$ by *C*9

4. $\vdash (w_1 \mathcal{U} w_3) \wedge (w_2 \mathcal{U} w_3) \equiv$
$w_3 \vee ((w_1 \wedge w_2) \wedge \bigcirc(w_1 \mathcal{U} w_3) \wedge \bigcirc(w_2 \mathcal{U} w_3))$
by 2, 3, and *PR*

5. $\vdash \overset{*}{w}_2 \equiv w_3 \vee ((w_1 \wedge w_2) \wedge \bigcirc \overset{*}{w}_2)$ by 4, *T*13, and *PR*

6. $\vdash (\bigcirc \overset{*}{w}_1 \equiv \bigcirc \overset{*}{w}_2) \supset (\overset{*}{w}_1 \equiv \overset{*}{w}_2)$ by 1, 5, and *PR*

7. $\vdash w_3 \supset (\overset{*}{w}_1 \wedge \overset{*}{w}_2)$ by 1, 5, and *PR*

8. $\vdash \Diamond w_3 \supset \Diamond(\overset{*}{w}_1 \wedge \overset{*}{w}_2)$ by $\Diamond\Diamond$

9. $\vdash (w_1 \wedge w_2)\mathcal{U}w_3 \supset \Diamond w_3$ by $C10$

10. $\vdash \overset{*}{w}_1 \supset \Diamond(\overset{*}{w}_1 \wedge \overset{*}{w}_2)$ by 7, 9, and PR

11. $\vdash w_1\mathcal{U}w_3 \supset \Diamond w_3$ by $C10$

12. $\vdash (w_1\mathcal{U}w_3) \wedge (w_2\mathcal{U}w_3) \supset \Diamond w_3$ by PR

13. $\vdash \overset{*}{w}_2 \supset \Diamond(\overset{*}{w}_1 \wedge \overset{*}{w}_2)$ by 8, 12, and PR

14. $\vdash \overset{*}{w}_1 \equiv \overset{*}{w}_2$ by 6, 10, 13, and NP

4. QUANTIFIERS

Since we intend to use terms and predicates in our reasoning we have to extend our system to admit individual variables, terms and quantification. Let us consider additional axioms involving quantifiers and their interaction with modalities.

Axioms:

$D1.$ $\vdash \sim\exists x.w \equiv \forall x.\sim w$

$D2.$ $\vdash (\forall x.w(x)) \supset w(t)$
where t is any term globally free for x in w

$D3.$ $\vdash (\forall x.\Box w) \supset (\Box\forall x.w)$

$D4.$ $\vdash (\forall x.\bigcirc w) \supset (\bigcirc\forall x.w)$

In these axioms x is any global individual variable. Axioms $D1$ and $D2$ are the usual predicate calculus axioms: $D1$ defines $\exists$ as the dual of $\forall$ and $D2$ is the *instantiation axiom*. Axiom $D3$ is known as the Barcan formula connecting the two universal operators $\forall$ and $\Box$. Axiom $D4$ is the Barcan formula for the $\bigcirc$ operator. The axioms state that since both operators have universal characteristics they commute.

A term t is said to be ***globally free for*** x in w if substitution of t for all free occurrences of x in w: (a) does not create new bound occurrences of (global) variables, and (b) does not create new occurrences of local variables in the scope of a modal operator. A trivial case: if t is x itself, then t is free for x. Condition (b) in this definition is essential. For, otherwise, we could derive the formula

$$(\forall x.\Diamond(x < y)) \supset \Diamond(y < y),$$

which is not valid for a local variable y.

An additional rule of inference is:

Inference rule:

R4. $\forall$ *Insertion* – $\forall I$

$$\frac{\vdash w_1 \supset w_2}{\vdash w_1 \supset \forall x.w_2}$$

where x is not free in w_1.

We have the derived rule

Instantiation Rule – *INST*

$$\frac{\vdash w(x)}{\vdash w(t)}$$

where t is any term globally free for x in w.

proof:

1.	$\vdash w(x)$	given
2.	$\vdash \forall x.w(x)$	by $\forall I$ (taking w_1 to be *true*)
3.	$\vdash (\forall x.w(x)) \supset w(t)$	by *D2*
4.	$\vdash w(t)$	by 2, 3, and *MP*

The following are the duals of *D2* and *R4* for the existential quantifier $\exists$:

T24. $\vdash w(t) \supset \exists x.w(x)$

where t is any term globally free for x in w.

proof:

1.	$\vdash (\forall x. \sim w(x)) \supset \sim w(t)$	by *D2*
2.	$\vdash (\sim \exists x.w(x)) \supset \sim w(t)$	by *D1* and *PR*
3.	$\vdash w(t) \supset \exists x.w(x)$	by *PR*

Note that we need here again the additional condition (b) that the substitution of t for x in w does not create new occurrences of local variables in the scope of a modal operator. For otherwise, we could deduce from $T24$

$$\Box(y \leq y) \supset \exists u. \Box(y \leq u),$$

which is not valid for a local variable y.

$\exists$ *Insertion* – $\exists I$

$$\frac{\vdash w_1 \supset w_2}{\vdash \exists x.w_1 \supset w_2}$$

where x is not free in w_2.

proof:

1.	$\vdash w_1 \supset w_2$	given
2.	$\vdash \sim w_2 \supset \sim w_1$	by PR
3.	$\vdash \sim w_2 \supset \forall x. \sim w_1$	by $\forall I$ ($R4$)
4.	$\vdash \sim w_2 \supset \sim \exists x.w_1$	by $D1$ and PR
5.	$\vdash \exists x.w_1 \supset w_2$	by PR

$\forall\forall$ *Rules*

(a) $$\frac{\vdash w_1 \supset w_2}{\vdash \forall x.w_1 \supset \forall x.w_2}$$ (b) $$\frac{\vdash w_1 \equiv w_2}{\vdash \forall x.w_1 \equiv \forall x.w_2}$$

proof of (a):

1.	$\vdash \forall x.w_1 \supset w_1$	by $D2$
2.	$\vdash w_1 \supset w_2$	given
3.	$\vdash \forall x.w_1 \supset w_2$	by PR
4.	$\vdash \forall x.w_1 \supset \forall x.w_2$	by $\forall I$

Rule (b) then follows by propositional reasoning.

$\exists\exists$ *Rules* :

(a) $$\frac{\vdash w_1 \supset w_2}{\vdash \exists x.w_1 \supset \exists x.w_2}$$ (b) $$\frac{\vdash w_1 \equiv w_2}{\vdash \exists x.w_1 \equiv \exists x.w_2}$$

proof of (a):

1.	$\vdash w_1 \supset w_2$	given

2.	$\vdash (\sim w_2) \supset (\sim w_1)$	by PR
3.	$\vdash (\forall x. \sim w_2) \supset (\forall x. \sim w_1)$	by $\forall\forall$
4.	$\vdash (\sim\exists x.w_2) \supset (\sim\exists x.w_1)$	by $D1$ and PR
5.	$\vdash \exists x.w_1 \supset \exists x.w_2$	by PR

Rule (b) then follows by propositional reasoning.

The last two rules are, of course, classical rules of the predicate calculus, and are brought here only for the sake of completeness and later reference.

We extend now the Equivalence Rule (ER), given above for propositional formulas, to handle predicate formulas as well.

Equivalence Rule – ER

Let w' be the result of replacing an occurrence of a subformula v_1 in w by v_2. Then

$$\frac{\vdash v_1 \equiv v_2}{\vdash w \equiv w'}$$

proof:

The proof is by induction on the structure of w. The cases where w is w_1 or of form $\sim u$, $u_1 \vee u_2$, $u_1 \supset u_2$, $\Box u$, $\Diamond u$, $\bigcirc u$ and $u_1 \mathcal{U} u_2$, are treated as before.

Case: w is of form $\forall x.u$. We assume that if $\vdash v_1 \equiv v_2$, then $\vdash u \equiv u'$. Then by the $\forall\forall$-rule $\vdash \forall x.u \equiv \forall x.u'$, *i.e.* $\vdash w \equiv w'$.

The case where w is of form $\exists x.u$, is proved similarly by the $\exists\exists$-rule. ∎

Deduction Rule – DED

$$\frac{w_1 \vdash w_2}{\vdash (\Box w_1) \supset w_2}$$

where the $\forall I$ rule (Rule $R4$) is never applied to a free variable of w_1 in the derivation of $w_1 \vdash w_2$.

That is, if under the assumption w_1 we can derive $\vdash w_2$, where rule $R4$ is never applied to a free variable of w_1, then there exists a proof establishing $\vdash (\Box w_1) \supset w_2$. We clearly must also be careful in using any theorem or derived rule such that the $\forall I$ rule was used in its proof.

The additional $\Box$ operator in the conclusion is obviously necessary since in general $w_1 \vdash w_2$ does not imply $\vdash w_1 \supset w_2$. For example, obviously $w \vdash \Box w$

is true (an immediate application of Rule $R3$: $\vdash w$ by assumption and therefore $\vdash \Box w$ by $\Box I$); but $\vdash w \supset \Box w$ is false.

proof:

The proof of the modal Deduction Rule follows the same arguments used in the proof of the classical Deduction Rule of Predicate Calculus. We replace each line $\vdash u_i$ in the proof of $w_1 \vdash w_2$ by the line $\vdash \Box w_1 \supset u_i$, and show that this transformation preserves soundness. That is

given	show
$\vdash u_1$	$\vdash (\Box w_1) \supset u_1$
$\vdash u_2$	$\vdash (\Box w_1) \supset u_2$
$\vdots$	$\vdots$
$\vdash u_i$	$\vdash (\Box w_1) \supset u_i$
$\vdots$	$\vdots$
$\vdash u_m$	$\vdash (\Box w_1) \supset u_m$.
i.e. $\vdash w_2$	*i.e.* $\vdash (\Box w_1) \supset w_2$

where u_i is either the assumption w_1, an axiom, or derived from previous u_j's by some rule of inference.

The proof is by a complete induction on i. We assume that for all $k < i$, $\vdash (\Box w_1) \supset u_k$, and prove that $\vdash (\Box w_1) \supset u_i$.

Case: u_i is an axiom.

1.	$\vdash u_i$	axiom
2.	$\vdash (\Box w_1) \supset u_i$	by PR

Note that $\vdash w'$ implies $\vdash w \supset w'$ for any w, by propositional reasoning.

Case: u_i is w_1.

1.	$\vdash (\Box w_1) \supset w_1$	by $C3$

Case: u_i is obtained by Rule $R1$, *i.e.*, u_i is an instance of a tautology.

1.	$\vdash u_i$	by PT
2.	$\vdash (\Box w_1) \supset u_i$	by PR

Case: u_i is obtained by Rule $R2$ (using previous $\vdash u_k$ and $\vdash u_k \supset u_i$).

1.	$\vdash (\Box w_1) \supset u_k$	induction hypothesis

2. $\vdash (\Box w_1) \supset (u_k \supset u_i)$ — induction hypothesis
3. $\vdash (\Box w_1) \supset u_i$ — by 1, 2, and PR

Case: u_i is obtained by Rule $R3$ (using previous $\vdash u_k$), *i.e.*, u_i is $\Box u_k$.

1. $\vdash (\Box w_1) \supset u_k$ — induction hypothesis
2. $\vdash (\Box\Box w_1) \supset \Box u_k$ — by $\Box\Box$
3. $\vdash (\Box w_1) \supset (\Box\Box w_1)$ — by $T12$
4. $\vdash (\Box w_1) \supset \Box u_k$ — by 2, 3, and PR

Case: u_i is obtained by Rule $R4$ (using previous $\vdash u \supset v$, *i.e.* u_k, to get $\vdash u \supset \forall x.v$, *i.e.* u_i, where x is not free in u).

By our deduction rule assumption, we know also that x is not free in w_1.

1. $\vdash (\Box w_1) \supset (u \supset v)$ — induction hypothesis
2. $\vdash ((\Box w_1) \wedge u) \supset v$ — by PR
3. $\vdash ((\Box w_1) \wedge u) \supset \forall x.v$ — by $R4$ (since x is not free in u or w_1)
4. $\vdash (\Box w_1) \supset (u \supset \forall x.v)$ — by PR ∎

A different approach to coping with the application of $\Box$ insertion rule (Rule $R3$) is to forbid it altogether. We then get the following restricted deduction rule:

Restricted Deduction Rule — RDED

$$\frac{w_1 \vdash w_2}{\vdash w_1 \supset w_2}$$

Provided $\Box I$ (Rule $R3$) is never applied and $\forall I$ (Rule $R4$) is never applied to a free variable of w_1 in the derivation of $w_1 \vdash w_2$.

Here, we are not allowed to use rule $\Box I$ or any theorem or derived rule that $\Box I$ was used in its proof.

The proof of $RDED$ follows exactly that of DED except that the case in which Rule $R3$ is applied does not arise.

Predicate Theorems

$T25.$ $\vdash (\sim\forall x.w) \equiv (\exists x. \sim w)$

proof:

1. $\vdash (\sim\sim w) \equiv w$ — by PT

2.	$\vdash (\forall x. \sim\sim w) \equiv \forall x.w$	by $\forall\forall$
3.	$\vdash (\sim\exists x. \sim w) \equiv \forall x.w$	by *D*1 and *PR*
4.	$\vdash \sim\forall x.w \equiv \exists x. \sim w$	by *PR*

*T*26. $\vdash \forall x.(w_1 \wedge w_2) \equiv (\forall x.w_1 \wedge \forall x.w_2)$

proof:

1.	$\vdash \forall x.w_1 \supset w_1$	by *D*2
2.	$\vdash \forall x.w_2 \supset w_2$	by *D*2
3.	$\vdash (\forall x.w_1 \wedge \forall x.w_2) \supset (w_1 \wedge w_2)$	by 1, 2, and *PR*
4.	$\vdash (\forall x.w_1 \wedge \forall x.w_2) \supset \forall x.(w_1 \wedge w_2)$	by $\forall I$
5.	$\vdash (w_1 \wedge w_2) \supset w_1$	by *PT*
6.	$\vdash \forall x.(w_1 \wedge w_2) \supset \forall x.w_1$	by $\forall\forall$
7.	$\vdash (w_1 \wedge w_2) \supset w_2$	by *PT*
8.	$\vdash \forall x.(w_1 \wedge w_2) \supset \forall x.w_2$	by $\forall\forall$
9.	$\vdash \forall x.(w_1 \wedge w_2) \supset (\forall x.w_1 \wedge \forall x.w_2)$	by 6, 8, and *PR*
10.	$\vdash \forall x.(w_1 \wedge w_2) \equiv (\forall x.w_1 \wedge \forall x.w_2)$	by 4, 9, and *PR*

*T*27. $\vdash \exists x.(w_1 \vee w_2) \equiv (\exists x.w_1 \vee \exists x.w_2)$

proof:

1.	$\vdash \forall x.(\sim w_1 \wedge \sim w_2) \equiv (\forall x. \sim w_1 \wedge \forall x. \sim w_2)$	by *T*26
2.	$\vdash \forall x. \sim (w_1 \vee w_2) \equiv (\forall x. \sim w_1 \wedge \forall x. \sim w_2)$	by *ER*
3.	$\vdash \sim\exists x.(w_1 \vee w_2) \equiv (\sim\exists x.w_1 \wedge \sim\exists x.w_2)$	by *D*1 and *PR*
4.	$\vdash \exists x.(w_1 \vee w_2) \equiv (\exists x.w_1 \vee \exists x.w_2)$	by *PR*

*T*28. $\vdash (\forall x. \Box w) \equiv (\Box \forall x.w)$

proof:

1.	$\vdash (\forall x.w) \supset w$	by *D*2
2.	$\vdash (\Box \forall x.w) \supset \Box w$	by $\Box\Box$
3.	$\vdash (\Box \forall x.w) \supset (\forall x. \Box w)$	by $\forall I$
4.	$\vdash (\forall x. \Box w) \supset (\Box \forall x.w)$	by *D*3
5.	$\vdash (\forall x. \Box w) \equiv (\Box \forall x.w)$	by 3, 4, and *PR*

alternative proof of $\vdash (\Box \forall x.w) \supset (\forall x. \Box w)$

1.	$\vdash \forall x.w$	assumption
2.	$\vdash w$	by *D*2 and *MP*
3.	$\vdash \Box w$	by $\Box I$

4. $\vdash\ \forall x.\,\Box\, w$ — by $\forall I$

Thus, $\forall x.w \vdash \forall x.\,\Box\, w$ and by the deduction rule

5. $\vdash\ (\Box\,\forall x.w) \supset (\forall x.\,\Box\, w)$

$T29.\ \vdash\ (\exists x.\,\Diamond\, w) \equiv (\Diamond\,\exists x.w)$

proof:

1. $\vdash\ (\forall x.\,\Box \sim w) \equiv (\Box\,\forall x.\sim w)$ — by $T28$
2. $\vdash\ (\forall x.\sim \Diamond\, w) \equiv (\Box \sim \exists x.w)$ — by $C1$, $D1$, and ER (twice)
3. $\vdash\ (\sim \exists x.\,\Diamond\, w) \equiv (\sim \Diamond\,\exists x.w)$ — by $C1$, $D1$ and PR
4. $\vdash\ (\exists x.\,\Diamond\, w) \equiv (\Diamond\,\exists x.w)$ — by PR

$T30.\ \vdash\ (\bigcirc\,\forall x.w) \equiv (\forall x.\bigcirc\, w)$

proof:

1. $\vdash\ (\forall x.\bigcirc\, w) \supset (\bigcirc\,\forall x.w)$ — by $D4$
2. $\vdash\ \forall x.w \supset w$ — by $D2$
3. $\vdash\ (\bigcirc\,\forall x.w) \supset \bigcirc\, w$ — by $\bigcirc\bigcirc$
4. $\vdash\ (\bigcirc\,\forall x.w) \supset (\forall x.\bigcirc\, w)$ — by $\forall I$
5. $\vdash\ (\forall x.\bigcirc\, w) \equiv (\bigcirc\,\forall x.w)$ — by 1, 4, and PR

$T31.\ \vdash\ (\bigcirc\,\exists x.w) \equiv (\exists x.\bigcirc\, w)$

proof:

1. $\vdash\ (\forall x.\bigcirc \sim w) \equiv (\bigcirc\,\forall x.\sim w)$ — by $T30$
2. $\vdash\ (\forall x.\sim \bigcirc\, w) \equiv (\bigcirc \sim \exists x.w)$ — by $C4$, $D1$, and ER
3. $\vdash\ (\sim \exists x.\bigcirc\, w) \equiv (\sim \bigcirc\,\exists x.w)$ — by $C4$, $D1$, and PR
4. $\vdash\ (\exists x.\bigcirc\, w) \equiv (\bigcirc\,\exists x.w)$ — by PR

Theorem $T28$ implies the commutativity of $\forall$ with $\Box$: Both have a universal character, with one quantifying over individuals and the other quantifying over states. Similarly, Theorem $T29$ implies the commutativity of $\exists$ with $\Diamond$. The last two theorems ($T30$ and $T31$) imply the commutativity of $\forall$ and $\exists$ with $\bigcirc$.

5. EQUALITY

Equality is handled by the following axioms:

Axioms:

$E1.\ \vdash\ t = t$ for any term t

$E2.\ \vdash (t_1 = t_2) \supset [w(t_1, t_1) \equiv w(t_1, t_2)]$
and t_2 is any term globally free for t_1 in w.

Axiom $E1$ states the ***reflexivity*** of equality. Axiom $E2$ states the ***substitutivity*** property of equality. We use $w(t_1, t_2)$ to indicate that t_2 replaces ***some*** of the occurrences of t_1 in w.

Recall that a term t_2 is said to be ***globally free for*** t_1 in w if substitution of t_2 for all free occurrences of t_1 in w : (a) does not create new bound occurrences of (global) variables, and (b) does not create new occurrences of local variables in the scope of a modal operator.

Note that the classical axiom for substitutivity of equality $E2$

$$\vdash\ (t_1 = t_2) \supset [w(t_1, t_1) \equiv w(t_1, t_2)]$$

(where t_2 is free for t_1 in w) is not correct if w contains modal operators. We could take $w(t_1, t_2)$ to be $\Box(t_1 = t_2)$ and deduce from $E2$

$$\vdash\ (t_1 = t_2) \supset [\Box(t_1 = t_1) \equiv \Box(t_1 = t_2)],$$

i.e.,

$$\vdash\ (t_1 = t_2) \supset \Box(t_1 = t_2),$$

which is not a valid statement (since $t_1 = t_2$ may contain local variables). But we have the following theorem for arbitrary formulas.

$T32$. *Substitutivity of Equality*

$$\vdash\ \Box(t_1 = t_2) \supset [w(t_1, t_1) \equiv w(t_1, t_2)]$$

where t_2 is free for t_1 in w.

proof:

By induction on the structure of w.

Case: w contains no modal operators. Then

1.	$\vdash$	$(t_1 = t_2) \supset [w(t_1, t_1) \equiv w(t_1, t_2)]$	by $E2$
2.	$\vdash$	$\Box(t_1 = t_2) \supset (t_1 = t_2)$	by $C3$
3.	$\vdash$	$\Box(t_1 = t_2) \supset [w(t_1, t_1) \equiv w(t_1, t_2)]$	by MP

Case: w is of the form $\Box u$. Then

1.	$\vdash$	$\Box(t_1 = t_2) \supset [u(t_1, t_1) \equiv u(t_1, t_2)]$	induction hypothesis
2.	$\vdash$	$\Box(t_1 = t_2)$	assumption
3.	$\vdash$	$u(t_1, t_1) \equiv u(t_1, t_2)$	by MP
4.	$\vdash$	$\Box u(t_1, t_1) \equiv \Box u(t_1, t_2)$	by $\Box\Box$

Thus, $\Box(t_1 = t_2) \vdash \Box u(t_1, t_1) \equiv \Box u(t_1, t_2)$

4.	$\vdash$	$\Box\Box(t_1 = t_2) \supset [\Box u(t_1, t_1) \equiv \Box u(t_1, t_2)]$	by DED
5.	$\vdash$	$\Box(t_1 = t_2) \supset [\Box u(t_1, t_1) \equiv \Box u(t_1, t_2)]$	by $T2$ and PR

The cases in which w is of the form $\Diamond u$, $\bigcirc u$, $\forall x.u$, and $\exists x.u$ are treated similarly, using the $\Diamond\,\Diamond$-rule, the $\bigcirc\,\bigcirc$-rule, the $\forall\forall$-rule, and the $\exists\exists$-rule, respectively.

Case: w is of the form $u \,\mathcal{U}\, v$.

1.	$\vdash$	$\Box(t_1 = t_2) \supset [u(t_1, t_1) \equiv u(t_1, t_2)]$	induction hypothesis
2.	$\vdash$	$\Box(t_1 = t_2) \supset [v(t_1, t_1) \equiv v(t_1, t_2)]$	induction hypothesis
3.	$\vdash$	$\Box(t_1 = t_2)$	assumption
4.	$\vdash$	$u(t_1, t_1) \equiv u(t_1, t_2)$	by 1, 3, and MP
5.	$\vdash$	$v(t_1, t_1) \equiv v(t_1, t_2)$	by 2, 3, and MP
6.	$\vdash$	$[u(t_1, t_1) \,\mathcal{U}\, v(t_1, t_1)] \equiv [u(t_1, t_2) \,\mathcal{U}\, v(t_1, t_2)]$	by 4, 5, and ER

Thus, $\Box(t_1 = t_2) \vdash (u(t_1, t_1) \,\mathcal{U}\, v(t_1, t_1)) \equiv (u(t_1, t_2) \,\mathcal{U}\, v(t_1, t_2))$

7. $\vdash$ $\Box\Box(t_1 = t_2) \supset [(u(t_1, t_1) \,\mathcal{U}\, v(t_1, t_1)) \equiv (u(t_1, t_2) \,\mathcal{U}\, v(t_1, t_2))]$
by DED

8. $\vdash$ $\Box(t_1 = t_2) \supset [(u(t_1, t_1) \,\mathcal{U}\, v(t_1, t_1)) \equiv (u(t_1, t_2) \,\mathcal{U}\, v(t_1, t_2))]$
by $T2$ and PR ∎

T33. Commutativity of Equality

$$\vdash (t_1 = t_2) \supset (t_2 = t_1)$$

proof:

1. $\vdash \ (t_1 = t_2) \supset [(t_1 = t_1) \equiv (t_2 = t_1)]$ — by $E2$
2. $\vdash \ t_1 = t_1$ — by $E1$
3. $\vdash \ (t_1 = t_2) \supset (t_2 = t_1)$ — by 1, 2, and PR

T34. Transitivity of Equality

$$\vdash \ [(t_1 = t_2) \land (t_2 = t_3)] \supset (t_1 = t_3)$$

proof:

1. $\vdash \ (t_1 = t_2) \supset [(t_1 = t_3) \equiv (t_2 = t_3)]$ — by $E2$
2. $\vdash \ [(t_1 = t_2) \land (t_2 = t_3)] \supset (t_1 = t_3)$ — by PR

T35. Term Equality

(a) $\vdash \ \Box(t_1 = t_2) \supset (\tau(t_1) = \tau(t_2))$ for any term τ

(b) $\vdash \ (t_1 = t_2) \supset (\tau(t_1) = \tau(t_2))$
where τ does not contain the next operator.

Here, $\tau(t_2)$ is the result of replacing an occurrence of t_1 in τ by t_2.

proof of (a):

1. $\vdash \ \Box(t_1 = t_2) \supset [(\tau(t_1) = \tau(t_2)) \equiv (\tau(t_2) = \tau(t_2))]$ — by $T32$
2. $\vdash \ \tau(t_2) = \tau(t_2)$ — by $E1$
3. $\vdash \ \Box(t_1 = t_2) \supset (\tau(t_1) = \tau(t_2))$ — by 1, 2, and PR

proof of (b):

1. $\vdash \ (t_1 = t_2) \supset [(\tau(t_1) = \tau(t_2)) \equiv (\tau(t_2) = \tau(t_2))]$ — by $E2$ (no $\bigcirc$ in τ)
2. $\vdash \ \tau(t_2) = \tau(t_2)$ — by $E1$
3. $\vdash \ (t_1 = t_2) \supset (\tau(t_1) = \tau(t_2))$ — by 1, 2, and PR

6. FRAME AXIOMS AND RULES

The use of the next operator $\bigcirc$ applied to terms is governed by the axioms:

Axioms:

$N1.$	$\vdash\ \bigcirc f(t_1, \ldots, t_n) = f(\bigcirc t_1, \ldots, \bigcirc t_n)$ for any function f and terms $t_1, \ldots, t_n$
$N2.$	$\vdash\ \bigcirc p(t_1, \ldots, t_n) \equiv p(\bigcirc t_1, \ldots, \bigcirc t_n)$ for any predicate p and terms $t_1, \ldots, t_n$
$N3.$	$\vdash\ \bigcirc(t_1 = t_2) \equiv (\bigcirc t_1 = \bigcirc t_2)$

Axiom $N3$ is a special case of $N2$ where p is the equality predicate.

These axioms are consistent with the evaluation rules that we gave which stated that to evaluate an expression $\bigcirc\, \mathcal{E}(t_1, \ldots t_n)$, we can evaluate $\mathcal{E}(\bigcirc t_1, \ldots \bigcirc t_n)$ regardless of whether $\mathcal{E}$ is a term or a logical expression.

Recall that we split the set of our symbols into two subsets: global and local symbols. The logical consequence of this convention is the following frame axiom:

$FA.$	*Frame Axiom* $\vdash x = \bigcirc x$ for every global variable x

We can therefore prove by induction on the structure of the term t and the formula w the following *frame theorems*:

$T36.$ For a term t and formula w

(a) $\vdash t = \bigcirc t$ provided t does not contain local symbols

(b) $\vdash w \equiv \Box w$ provided w does not contain local symbols

(c) $\vdash w(\bigcirc y_1, \ldots, \bigcirc y_n) \equiv \bigcirc w(y_1, \ldots, y_n)$
provided $y_1, \ldots, y_n$ are all the local variables in w.

A derived frame rule that we will be using is

Frame Rule - FR

$$\frac{\vdash\ w_1 \supset \Diamond w_2}{\vdash\ (w \wedge w_1) \supset \Diamond(w \wedge w_2)}$$

provided w does not contain local symbols.

proof:

1. $\vdash\ w \supset \Box w$ — by $T36$
2. $\vdash\ w_1 \supset \Diamond w_2$ — given
3. $\vdash\ (w \wedge w_1) \supset (\Box w \wedge \Diamond w_2)$ — by 1, 2, and PR
4. $\vdash\ (\Box w \wedge \Diamond w_2) \supset \Diamond(w \wedge w_2)$ — by $T10$
5. $\vdash\ (w \wedge w_1) \supset \Diamond(w \wedge w_2)$ — by 3, 4, and PR

7. DOMAIN PART

The next part of the system contains domain axioms that specify the necessary properties of the domain of interest. Thus, to reason about programs manipulating natural numbers, we need the set of Peano Axioms, and to reason about trees we need a set of axioms giving the basic properties of trees and the basic operations defined on them.

An essential axiom schema for many domains is the *induction axiom schema.* This (and all other schemas) should be formulated to admit modal instances as subformulas. Thus the induction principle for natural numbers can be stated as follows:

Induction Axiom

$$\vdash\ [R(0) \wedge \forall n(R(n) \supset R(n+1))] \supset R(k)$$

for any statement R.

One instance of this principle, which will be used later, is obtained by taking $R(n)$ to be $\Box(Q(n) \supset \Diamond \psi)$:

Induction Theorem

$$\begin{aligned}\vdash\ & \{\Box(Q(0) \supset \Diamond \psi) \\ & \wedge \forall n[\Box(Q(n) \supset \Diamond \psi) \supset \Box(Q(n+1) \supset \Diamond \psi)]\} \\ & \supset \Box(Q(k) \supset \Diamond \psi).\end{aligned}$$

Similar induction theorems exist for other domains and depend on well-founded orderings existing in those domains.

Using this induction theorem we can derive the following useful induction rule:

Induction Rule – IND

$$\frac{\vdash\ Q(0) \supset \Diamond\psi \qquad \vdash\ Q(n+1) \supset (\Diamond\psi \vee \Diamond Q(n))}{\vdash\ Q(k) \supset \Diamond\psi}$$

IND is useful for proving convergence of a loop: Show that $Q(0)$ guarantees $\Diamond\psi$ and that for each n, either $Q(n+1)$ implies $Q(n)$ across the loop or it already establishes $\Diamond\psi$ and no further execution is necessary. Then $Q(k)$ ensures that the loop is executed *at most* k times and that $\Diamond\psi$ is established on the last iteration or earlier.

proof:

1.	$\vdash\ Q(0) \supset \Diamond\psi$	given
2.	$\vdash\ \Box(Q(0) \supset \Diamond\psi)$	by $\Box I$
3.	$\vdash\ Q(n+1) \supset (\Diamond\psi \vee \Diamond Q(n))$	given
4.	$\vdash\ \Box(Q(n) \supset \Diamond\psi) \supset (\Diamond Q(n) \supset \Diamond\psi)$	by $T5$, $T3$ and PR
5.	$\vdash\ [(\Diamond Q(n) \supset \Diamond\psi) \wedge (\Diamond\psi \vee \Diamond Q(n))] \supset \Diamond\psi$	by PT
6.	$\vdash\ [Q(n+1) \wedge \Box(Q(n) \supset \Diamond\psi)] \supset \Diamond\psi$	by 3, 4, 5 and PR
7.	$\vdash\ \Box(Q(n) \supset \Diamond\psi) \supset (Q(n+1) \supset \Diamond\psi)$	by PR
8.	$\vdash\ \Box\Box(Q(n) \supset \Diamond\psi) \supset \Box(Q(n+1) \supset \Diamond\psi)$	by $\Box\Box$
9.	$\vdash\ \Box(Q(n) \supset \Diamond\psi) \supset \Box(Q(n+1) \supset \Diamond\psi)$	by $T2$ and PR
10.	$\vdash\ \forall n[\Box(Q(n) \supset \Diamond\psi) \supset \Box(Q(n+1) \supset \Diamond\psi)]$	by $\forall I$
11.	$\vdash\ \Box(Q(k) \supset \Diamond\psi)$	by 2, 10, and Induction Theorem
12.	$\vdash\ Q(k) \supset \Diamond\psi$	by $C3$ and MP ∎

8. PROGRAM PART

Our proof system must be augmented by additional axioms that reflect the structure of the program under consideration. These additional axioms constrain the state sequences to be exactly the set of execution sequences of the program under study. This releases us from the need to express program text syntactically in the system; all necessary information is captured by constraints on the accessibility relation that are expressed by the additional axioms.

For simplicity, we assume that the program is represented by a directed graph whose nodes are the program locations or labels and whose edges represent transitions between the labels. A transition is an instruction of the general form

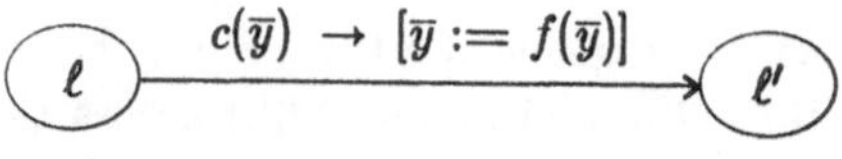

Here, $c(\overline{y})$ is a condition (possibly the trivial condition ***true***) under which the transition replacing $\overline{y}$ by $f(\overline{y})$ should be taken, where $\overline{y} = (y_1, \ldots, y_n)$ is the vector of program variables.

We assume that the programs are sequential and deterministic; in other words, all the conditions $c_1, \ldots, c_k$ on transitions departing from any node are ***exhaustive***, *i.e.*, $\bigvee_{i=1}^{k} c_i(\overline{y}) = \textit{true}$, and ***mutually exclusive***. In order to uniformly satisfy this requirement we add "***true*** $\rightarrow$ [ ]" self-transitions to all the exit nodes.

A first generic axiom states that in every state s, $at\,\ell$ is true for exactly one label ℓ. Let L denote the set of all labels in the program; we have

Location Axiom – LA
$\vdash \sum_{\ell \in L} at\,\ell = 1.$

We use here the abbreviation $\sum p_i = 1$ or $p_1 + \cdots + p_n = 1$ to mean that ***exactly*** one of the p_i's is true; $p_i = 1$ if p_i is true and $p_i = 0$ if p_i is false.

The role of the other axioms, called the ***transition axioms***, is to introduce our knowledge about the program into the system. Since the system does not provide direct tools for speaking about programs (such as mentioning program text in Hoare's formalism or Dynamic Logic), the transition axioms represent the program by characterizing the possible state transitions under the execution of the program. For any transition:

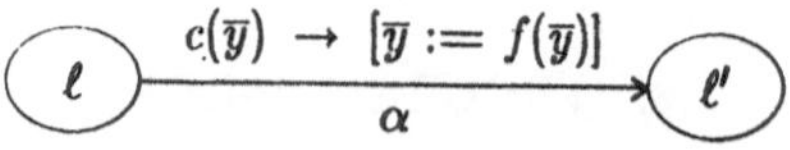

we generate a transition axiom F_α. This axiom corresponds to a "forward" propagation (***symbolic execution***) across the transition α:

Forward transition axiom
$F_\alpha: \quad \vdash \ [at\,\ell \wedge c(\overline{y}) \wedge \overline{y} = \overline{u}] \ \supset \ \bigcirc[at\,\ell' \wedge \overline{y} = f(\overline{u})],$
where $\overline{u}$ are auxiliary global variables.

This axiom states: If at any state, execution is at ℓ, $c(\overline{y})$ holds, and the current values of $\overline{y}$ are $\overline{u}$, then at the next state we will be at ℓ' with $\overline{y} = f(\overline{u})$.

A different approach that suggests an alternative axiom schema is obtained by "backward" substitution (derivation of the *weakest precondition*)

Backward transition axiom
$B_\alpha: \quad \vdash \ [at\,\ell \wedge c(\overline{y}) \wedge P(f(\overline{y}))] \ \supset \ \bigcirc[at\,\ell' \wedge P(\overline{y})],$
where P is any state predicate (*i.e.*, without modalities).

Here $P(f(\overline{y}))$ denotes the substitution of $f(\overline{y})$ for all free occurrences of $\overline{y}$ in $P(\overline{y})$. This form of the axiom expresses the effect of the transition on an arbitrary "state" predicate P; *i.e.*, a predicate P that does not contain any modal operators. It says that if $at\,\ell \wedge c(\overline{y})$ and $P(f(\overline{y}))$ hold, then we are guaranteed to reach ℓ' with $P(\overline{y})$ on the next step.

The predicate P may not contain modalities. As a counterexample, consider the program segment

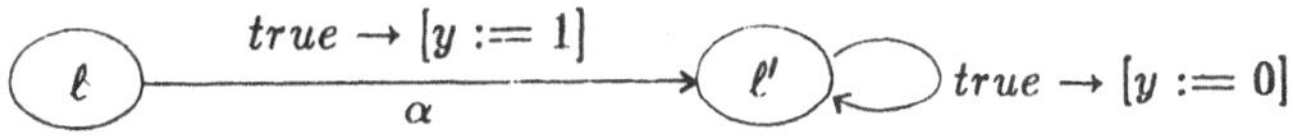

with

$$P(y): \quad \Box(y = 1).$$

The appropriate instance of the backward axiom for α is

$$B_\alpha: \quad \vdash \ [at\,\ell \wedge true \wedge \Box(1 = 1)] \ \supset \ \bigcirc[at\,\ell' \wedge \Box(y = 1)],$$

which clearly does not correctly reflect the computation of the program.

F_α and B_α are equivalent and can be derived from each other. That is

For every transition α:
B_α holds for every P if and only if F_α holds

proof: B_α for every $P \Rightarrow F_\alpha$.

1. $\vdash [at\,\ell \land c(\overline{y}) \land P(f(\overline{y}))] \supset \bigcirc[at\,\ell' \land P(\overline{y})]$ by B_α, given
2. $\vdash [at\,\ell \land c(\overline{y}) \land f(\overline{y}) = f(\overline{u})] \supset \bigcirc[at\,\ell' \land \overline{y} = f(\overline{u})]$
   taking $P(\overline{y})$ to be $\overline{y} = f(\overline{u})$,
   where $\overline{u}$ are auxiliary global variables
3. $\vdash [at\,\ell \land c(\overline{y}) \land \overline{y} = \overline{u}] \supset [at\,\ell \land c(\overline{y}) \land f(\overline{y}) = f(\overline{u})]$
   by $T35(b)$ and PR
4. $\vdash [at\,\ell \land c(\overline{y}) \land \overline{y} = \overline{u}] \supset \bigcirc[at\,\ell' \land \overline{y} = f(\overline{u})]$
   by 2, 3 and PR

which is the desired F_α. ∎

proof: $F_\alpha \Rightarrow B_\alpha$ for every P.

Let P be an arbitrary state predicate and $\overline{u}$ auxiliary global variables not in P. Then

1. $\vdash [at\,\ell \land c(\overline{y}) \land \overline{y} = \overline{u}] \supset \bigcirc[at\,\ell' \land \overline{y} = f(\overline{u})]$ F_α, given
2. $\vdash \bigcirc[at\,\ell' \land \overline{y} = f(\overline{u})] \supset [\bigcirc\, at\,\ell' \land \bigcirc(\overline{y} = f(\overline{u}))]$ by $T13$
3. $\vdash \bigcirc(\overline{y} = f(\overline{u})) \supset ((\bigcirc\,\overline{y}) = f(\bigcirc\,\overline{u}))$ by $N3$ and $N1$
4. $\vdash \overline{u} = \bigcirc\,\overline{u}$ by FA, since $\overline{u}$ is global
5. $\vdash f(\overline{u}) = f(\bigcirc\,\overline{u})$ by $T35(b)$
6. $\vdash \bigcirc(\overline{y} = f(\overline{u})) \supset ((\bigcirc\,\overline{y}) = f(\overline{u}))$ by 3, 5, $E2$, and PR
7. $\vdash [at\,\ell \land c(\overline{y}) \land \overline{y} = \overline{u}] \supset [\bigcirc\, at\,\ell' \land (\bigcirc\,\overline{y}) = f(\overline{u})]$
   by 1, 2, 6, and PR
8. $\vdash [\overline{y} = \overline{u} \land P(f(\overline{y}))] \supset P(f(\overline{u}))$
   by $E2$ (no modal operators in P) and PR
9. $\vdash [at\,\ell \land c(\overline{y}) \land \overline{y} = \overline{u} \land P(f(\overline{y}))]$
   $\supset [\bigcirc\, at\,\ell' \land (\bigcirc\,\overline{y}) = f(\overline{u}) \land P(f(\overline{u}))]$ by 7, 8, and PR
10. $\vdash ((\bigcirc\,\overline{y}) = f(\overline{u})) \supset (P(\bigcirc\,\overline{y}) \equiv P(f(\overline{u})))$ by $E2$ and PR
11. $\vdash P(\bigcirc\,\overline{y}) \equiv \bigcirc\,P(\overline{y})$ by $T36(c)$
12. $\vdash [(\bigcirc\,\overline{y}) = f(\overline{u}) \land P(f(\overline{u}))] \supset \bigcirc\,P(\overline{y})$ by 10, 11, and PR
13. $\vdash [at\,\ell \land c(\overline{y}) \land \overline{y} = \overline{u} \land P(f(\overline{y}))] \supset [\bigcirc\, at\,\ell' \land \bigcirc\,P(\overline{y})]$
    by 9, 12, and PR

14. $\vdash \; [at\,\ell \wedge c(\overline{y}) \wedge \overline{y} = \overline{y} \wedge P(f(\overline{y}))] \supset [\bigcirc\, at\,\ell' \wedge \bigcirc P(\overline{y})]$ by *INST*

15. $\vdash \; [at\,\ell \wedge c(\overline{y}) \wedge P(f(\overline{y}))] \supset [\bigcirc\, at\,\ell' \wedge \bigcirc P(\overline{y})]$ by $E1$ and PR

16. $\vdash \; [at\,\ell \wedge c(\overline{y}) \wedge P(f(\overline{y}))] \supset \bigcirc[at\,\ell' \wedge P(\overline{y})]$ by $T13$ and PR

which is the desired B_α. ∎

We often use a weaker form of the transition axioms:

$F'_\alpha: \quad \vdash \; [at\,\ell \wedge c(\overline{y}) \wedge \overline{y} = \overline{u}] \supset \Diamond[at\,\ell' \wedge \overline{y} = f(\overline{u})]$

and

$B'_\alpha: \quad \vdash \; [at\,\ell \wedge c(\overline{y}) \wedge P(f(\overline{y}))] \supset \Diamond[at\,\ell' \wedge P(\overline{y})]$

obtained from F_α and B_α, respectively, by replacing $\bigcirc$ with $\Diamond$. The weaker forms follow by $T11$, *i.e.* $\vdash \bigcirc w \supset \Diamond w$.

9. THE INVARIANCE PRINCIPLE

We now present a general method for proving invariance properties of programs, *i.e.*, properties that hold continuously throughout the execution. Such properties are expressible by formulas of form

$$\vdash \; [at\,\ell_0 \wedge \phi(\overline{x})] \supset \Box Q(\overline{y}).$$

That is, $Q(\overline{y})$ is invariantly true for every computation starting at ℓ_0 with input $\overline{x}$ satisfying the precondition $\phi(\overline{x})$.

Let ℓ be any label in the program under consideration and let its outgoing transitions be of the form

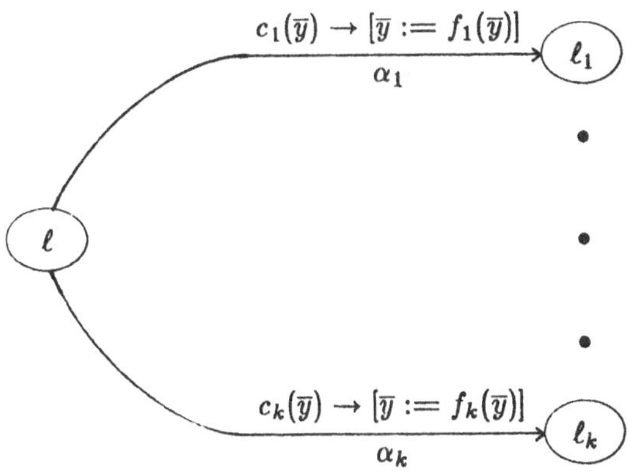

Recall that we assume that $c_1(\bar{y}), \ldots, c_k(\bar{y})$ are exhaustive, *i.e.* $\bigvee_{i=1}^{k} c_i(\bar{y}) = \mathit{true}$, and mutually exclusive. We denote by L the set of all labels in P. We have

Invariance Principle:

Let $Q(\bar{y})$ be a state predicate (with no modalities) and labels describing a property of program P with input condition $\phi(\bar{x})$.

If

(a) Q is true initially, *i.e.*,
$$\vdash \ [at\,\ell_0 \wedge \phi(\bar{x})] \ \supset \ Q(\bar{y})$$

(b) Q is maintained along any transition α in P, *i.e.*,
$$\vdash \ [at\,\ell \wedge c_\alpha(\bar{y}) \wedge Q(\bar{y})] \ \supset \ Q(f_\alpha(\bar{y})),$$

then Q is invariantly true, *i.e.*,
$$\vdash \ [at\,\ell_0 \wedge \phi(\bar{x})] \ \supset \ \Box Q(\bar{y}).$$

proof:

Consider an arbitrary label ℓ and an arbitrary transition α_i, $1 \leq i \leq k$, from ℓ to ℓ_i.

1. $\vdash \ [at\,\ell \wedge c_i(\bar{y}) \wedge Q(\bar{y})] \ \supset \ [at\,\ell \wedge c_i(\bar{y}) \wedge Q(f_i(\bar{y}))]$ — by (b) and PR
2. $\vdash \ [at\,\ell \wedge c_i(\bar{y}) \wedge Q(f_i(\bar{y}))] \ \supset \ \bigcirc[at\,\ell_i \wedge Q(\bar{y})]$ — by B_{α_i}
3. $\vdash \ [at\,\ell \wedge c_i(\bar{y}) \wedge Q(\bar{y})] \ \supset \ \bigcirc[at\,\ell_i \wedge Q(\bar{y})]$ — by 1, 2 and PR
4. $\vdash \ [at\,\ell \wedge c_i(\bar{y}) \wedge Q(\bar{y})] \ \supset \ \bigcirc Q(\bar{y})$ — by $T13$ and PR

5. $\vdash \ \bigvee_{i=1}^{k}[at\,\ell \wedge c_i(\bar{y}) \wedge Q(\bar{y})] \ \supset \ \bigcirc Q(\bar{y})$ — by PR (taking the disjunction over all transitions from ℓ)
6. $\vdash \ [at\,\ell \wedge \bigvee_{i=1}^{k} c_i(\bar{y}) \wedge Q(\bar{y})] \ \supset \ \bigcirc Q(\bar{y})$ — by PR
7. $\vdash \ \bigvee_{i=1}^{k} c_i(\bar{y}) \ = \ \mathit{true}$ — assumption
8. $\vdash \ [at\,\ell \wedge Q(\bar{y})] \ \supset \ \bigcirc Q(\bar{y})$ — by PR

9. $\vdash \ \bigvee_{\ell \in L}[at\,\ell \wedge Q(\bar{y})] \ \supset \ \bigcirc Q(\bar{y})$ — by PR (taking the disjunction over all labels of P)

10. $\vdash \; [(\bigvee_{\ell \in L} at\,\ell) \wedge Q(\bar{y})] \;\supset\; \bigcirc Q(\bar{y})$ by PR

11. $\vdash \; \bigvee_{\ell \in L} at\,\ell \;=\; true$ by Location Axiom and PR

12. $\vdash \; Q(\bar{y}) \;\supset\; \bigcirc Q(\bar{y})$ by 10, 11 and PR

13. $\vdash \; Q(\bar{y}) \;\supset\; \Box Q(\bar{y})$ by CI

14. $\vdash \; [at\,\ell_0 \wedge \phi(\bar{x})] \;\supset\; Q(\bar{y})$ by (a)

15. $\vdash \; [at\,\ell_0 \wedge \phi(\bar{x})] \;\supset\; \Box Q(\bar{y})$ by 13, 14 and PR

10. EXAMPLE: INTEGER EXPONENTIATION PROGRAM

Consider for example the following program IE over the integers, which raises a real number x_1 to an integer x_2, *i.e.* $x_1{}^{x_2}$, where $x_2 \geq 0$. We assume that $0^0 = 1$.

Program IE (Integer Exponentiation):

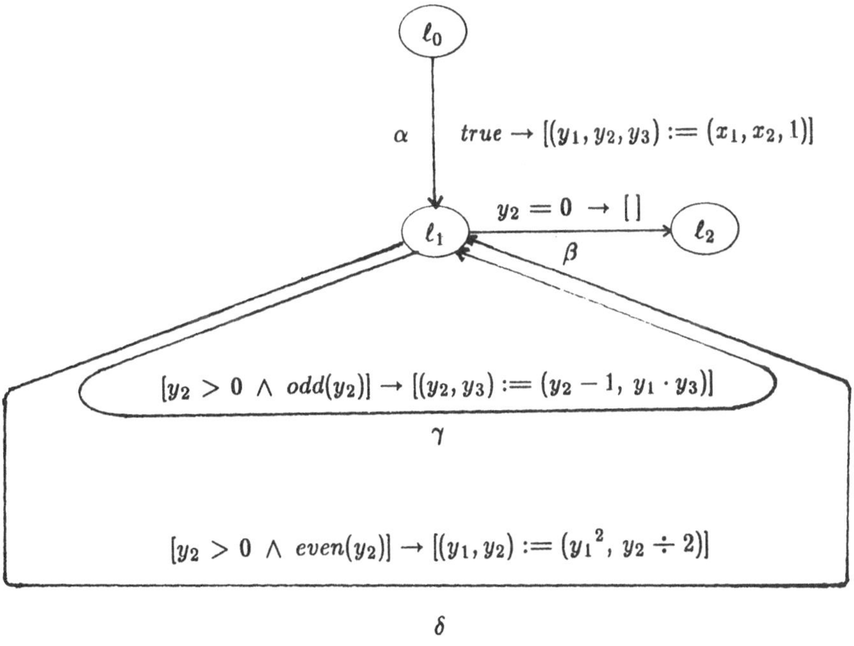

Let

$$\phi: \; at\,\ell_0 \;\wedge\; x_2 \geq 0$$

$$\psi : \; at\,\ell_2 \; \wedge \; y_3 = x_1{}^{x_2}.$$

We would like to use our proof system to establish the total correctness of program *IE* with respect to ϕ and ψ; we will show

$$\vdash \; \phi \; \supset \; \Diamond \psi.$$

In the proof we ignore type considerations such as $real(x_1)$ and $integer(x_2)$. (See [BUR], [MW]).

PROOF 1: Using Backward Transition Axioms

The backward transition axiom schemata corresponding to this program (taking the weaker form, with $\Diamond$ rather than $\bigcirc$) are:

$$B'_\alpha : \quad \vdash \; [at\,\ell_0 \; \wedge \; P(x_1, x_2, 1)] \; \supset \; \Diamond[at\,\ell_1 \; \wedge \; P(y_1, y_2, y_3)]$$

$$B'_\beta : \quad \vdash \; [at\,\ell_1 \; \wedge \; y_2 = 0 \; \wedge \; P(y_1, y_2, y_3)] \qquad \supset \; \Diamond[at\,\ell_2 \; \wedge \; P(y_1, y_2, y_3)]$$

$$B'_\gamma : \quad \vdash \; [at\,\ell_1 \; \wedge \; y_2 > 0 \; \wedge \; odd(y_2) \; \wedge \; P(y_1,\; y_2 - 1,\; y_1 \cdot y_3)] \qquad \supset \; \Diamond[at\,\ell_1 \; \wedge \; P(y_1, y_2, y_3)]$$

$$B'_\delta : \quad \vdash \; [at\,\ell_1 \; \wedge \; y_2 > 0 \; \wedge \; even(y_2) \; \wedge \; P(y_1{}^2,\; y_2 \div 2,\; y_3)] \qquad \supset \; \Diamond[at\,\ell_1 \; \wedge \; P(y_1, y_2, y_3)].$$

We prove

(a) $\vdash \phi \supset \Diamond \exists k.Q(k, \overline{y})$

(b) $\vdash (\exists k.Q(k, \overline{y})) \supset \Diamond \psi$, or equivalently $\vdash Q(k, \overline{y}) \supset \Diamond \psi$,

where

$$Q(n, \overline{y}) : \; at\,\ell_1 \; \wedge \; (0 \le y_2 \le n) \; \wedge \; y_3 \cdot y_1{}^{y_2} = x_1{}^{x_2}.$$

Here, $0 \le y_2 \le n$ is used to establish the termination, and $y_3 \cdot y_1{}^{y_2} = x_1{}^{x_2}$ is the invariant used to establish the correctness.

Clearly, by rule $\Diamond C$, parts (a) and (b) imply the desired result $\vdash \phi \supset \Diamond \psi$.

proof of (a):

1. $\vdash \; 1 \cdot x_1{}^{x_2} = x_1{}^{x_2}$ by domain
2. $\vdash \; \phi \; \supset \; [at\;\ell_0 \wedge x_2 \ge 0 \wedge 1 \cdot x_1{}^{x_2} = x_1{}^{x_2}]$ by *PR*

3. $\vdash$ $[at\, \ell_0 \wedge x_2 \geq 0 \wedge 1 \cdot x_1^{x_2} = x_1^{x_2}]$
$\supset \Diamond[at\, \ell_1 \wedge y_2 \geq 0 \wedge y_3 \cdot y_1^{y_2} = x_1^{x_2}]$ by B'_α
where P is $y_2 \geq 0 \wedge y_3 \cdot y_1^{y_2} = x_1^{x_2}$

4. $\vdash$ $(y_2 \geq 0) \supset (0 \leq y_2 \leq y_2)$ by domain

5. $\vdash$ $[at\, \ell_1 \wedge y_2 \geq 0 \wedge y_3 \cdot y_1^{y_2} = x_1^{x_2}]$
$\supset [at\, \ell_1 \wedge (0 \leq y_2 \leq y_2) \wedge y_3 \cdot y_1^{y_2} = x_1^{x_2}]$ by 4 and PR

6. $\vdash$ $[at\, \ell_1 \wedge y_2 \geq 0 \wedge y_3 \cdot y_1^{y_2} = x_1^{x_2}]$
$\supset \exists k[at\, \ell_1 \wedge (0 \leq y_2 \leq k) \wedge y_3 \cdot y_1^{y_2} = x_1^{x_2}]$ by $T24$

7. $\vdash$ $\phi \supset \Diamond \exists k.Q(k, \overline{y})$ by 2, 3, 6 and $\Diamond Q$

proof of (b): We use the induction rule *IND*:

$$\frac{\begin{array}{ll}(b_1) & \vdash Q(0,\overline{y}) \supset \Diamond\psi \\ (b_2) & \vdash Q(n+1,\overline{y}) \supset [\Diamond\psi \vee \Diamond Q(n,\overline{y})]\end{array}}{\vdash Q(k,\overline{y}) \supset \Diamond\psi}$$

proof of (b_1):

8. $\vdash$ $[(0 \leq y_2 \leq 0) \wedge y_3 \cdot y_1^{y_2} = x_1^{x_2}] \supset [y_2 = 0 \wedge y_3 = x_1^{x_2}]$
by domain

9. $\vdash$ $Q(0, \overline{y}) \supset [at\, \ell_1 \wedge y_2 = 0 \wedge y_3 = x_1^{x_2}]$
by PR

10. $\vdash$ $[at\, \ell_1 \wedge y_2 = 0 \wedge y_3 = x_1^{x_2}] \supset \Diamond[at\, \ell_2 \wedge y_3 = x_1^{x_2}]$
by B'_β, where P is $y_3 = x_1^{x_2}$

11. $\vdash$ $Q(0, \overline{y}) \supset \Diamond\psi$ by 9, 10 and PR

proof of (b_2):

case 1: $y_2 = 0$.

12. $\vdash$ $[y_2 = 0 \wedge y_3 \cdot y_1^{y_2} = x_1^{x_2}] \supset [y_2 = 0 \wedge y_3 = x_1^{x_2}]$
by domain

13. $\vdash$ $[Q(n+1, \overline{y}) \wedge y_2 = 0] \supset [at\, \ell_1 \wedge y_2 = 0 \wedge y_3 = x_1^{x_2}]$
by PR

14. $\vdash\ [at\,\ell_1 \wedge y_2 = 0 \wedge y_3 = x_1{}^{x_2}] \supset \Diamond[at\,\ell_2 \wedge y_3 = x_1{}^{x_2}]$
by B'_β, where P is $y_3 = x_1{}^{x_2}$

15. $\vdash\ [Q(n+1,\bar{y}) \wedge y_2 = 0] \supset \Diamond\psi$ by 13, 14 and *PR*

case 2: $y_2 > 0 \wedge odd(y_2)$.

16. $\vdash\ [y_2 > 0 \wedge (0 \leq y_2 \leq n+1) \wedge y_3 \cdot y_1{}^{y_2} = x_1{}^{x_2}]$
$\supset [(0 \leq y_2 - 1 \leq n) \wedge (y_1 \cdot y_3) \cdot y_1{}^{y_2 - 1} = x_1{}^{x_2}]$
by domain

17. $\vdash\ [Q(n+1,\bar{y}) \wedge y_2 > 0 \wedge odd(y_2)] \supset [at\,\ell_1 \wedge y_2 > 0 \wedge$
$odd(y_2) \wedge (0 \leq y_2 - 1 \leq n) \wedge (y_1 \cdot y_3) \cdot y_1{}^{y_2-1} = x_1{}^{x_2}]$
by *PR*

18. $\vdash\ [at\,\ell_1 \wedge y_2 > 0 \wedge odd(y_2) \wedge (0 \leq y_2 - 1 \leq n)$
$\wedge (y_1 \cdot y_3) \cdot y_1{}^{y_2-1} = x_1{}^{x_2}]$
$\supset \Diamond[at\,\ell_1 \wedge (0 \leq y_2 \leq n) \wedge y_3 \cdot y_1{}^{y_2} = x_1{}^{x_2}]$
by B'_γ, where P is $(0 \leq y_2 \leq n) \wedge y_3 \cdot y_1{}^{y_2} = x_1{}^{x_2}$

19. $\vdash\ [Q(n+1,\bar{y}) \wedge y_2 > 0 \wedge odd(y_2)] \supset \Diamond Q(n,\bar{y})$
by 17, 18, and *PR*

case 3: $y_2 > 0 \wedge even(y_2)$.

20. $\vdash\ [even(y_2) \wedge (0 \leq y_2 \leq n+1) \wedge y_3 \cdot y_1{}^{y_2} = x_1{}^{x_2}]$
$\supset [(0 \leq y_2 \div 2 \leq n) \wedge y_3 \cdot (y_1{}^2)^{y_2 \div 2} = x_1{}^{x_2}]$
by domain

21. $\vdash\ [Q(n+1,\bar{y}) \wedge y_2 > 0 \wedge even(y_2)] \supset [at\,\ell_1 \wedge y_2 > 0$
$\wedge\ even(y_2) \wedge (0 \leq y_2 \div 2 \leq n) \wedge y_3 \cdot (y_1{}^2)^{y_2 \div 2} = x_1{}^{x_2}]$
by *PR*

22. $\vdash\ [at\,\ell_1 \wedge y_2 > 0 \wedge even(y_2) \wedge (0 \leq y_2 \div 2 \leq n)$
$\wedge\ y_3 \cdot (y_1{}^2)^{y_2 \div 2} = x_1{}^{x_2}]$
$\supset \Diamond[at\,\ell_1 \wedge (0 \leq y_2 \leq n) \wedge y_3 \cdot y_1{}^{y_2} = x_1{}^{x_2}]$
by B'_δ, where P is $(0 \leq y_2 \leq n) \wedge (y_3 \cdot y_1{}^{y_2} = x_1{}^{x_2})$

23. $\vdash\ [Q(n+1,\bar{y}) \wedge y_2 > 0 \wedge even(y_2)] \supset \Diamond Q(n,\bar{y})$
by 21, 22, and *PR*

To summarize, we showed

15. $\vdash\ [Q(n+1,\bar{y}) \wedge y_2 = 0] \supset \Diamond\psi$ case 1

$$19. \quad \vdash \ [Q(n+1,\overline{y}) \ \wedge \ y_2 > 0 \ \wedge \ odd(y_2)] \ \supset \ \Diamond Q(n,\overline{y}) \qquad \text{case 2}$$

$$23. \quad \vdash \ [Q(n+1,\overline{y}) \ \wedge \ y_2 > 0 \ \wedge \ even(y_2)] \ \supset \ \Diamond Q(n,\overline{y}) \qquad \text{case 3}$$

Then since

$$24. \quad \vdash \ Q(n+1,\overline{y}) \ \supset \ [y_2 = 0 \ \vee \ (y_2 > 0 \wedge odd(y_2)) \vee (y_2 > 0 \wedge even(y_2))] \qquad \text{by domain}$$

it follows that

$$25. \quad \vdash \ Q(n+1,\overline{y}) \ \supset \ [\Diamond \psi \ \vee \ \Diamond Q(n,\overline{y})] \qquad \text{by 15, 19, 23, 24 and } PR$$

This concludes the first proof of the total correctness of our example. ∎

PROOF 2: Using Forward Transition Axioms

For comparison, let us now prove the total correctness of program *IE* using the forward transition axioms. The proof turns out to be longer than the previous one using the backward axioms.

The forward transition axiom schemas corresponding to the program (taking again the weaker form, with $\Diamond$ rather than $\bigcirc$) are:

$$F'_\alpha : \ \vdash \ at\,\ell_0 \ \supset \ \Diamond[at\,\ell_1 \ \wedge \ \overline{y} = (x_1, x_2, 1)]$$

$$F'_\beta : \ \vdash \ [at\,\ell_1 \ \wedge \ y_2 = 0 \ \wedge \ \overline{y} = \overline{u}] \ \supset \ \Diamond[at\,\ell_2 \ \wedge \ \overline{y} = \overline{u}]$$

$$F'_\gamma : \ \vdash \ [at\,\ell_1 \ \wedge \ y_2 > 0 \ \wedge \ odd(y_2) \ \wedge \ \overline{y} = \overline{u}] \ \supset \ \Diamond[at\,\ell_1 \ \wedge \ \overline{y} = (u_1, u_1 - 1, u_1 \cdot u_3)]$$

$$F'_\delta : \ \vdash \ [at\,\ell_1 \ \wedge \ y_2 > 0 \ \wedge \ even(u_2) \ \wedge \ \overline{y} = \overline{u}] \ \supset \ \Diamond[at\,\ell_1 \ \wedge \ \overline{y} = (u_1{}^2, u_2 \div 2, u_3)]$$

Again, let

$$\phi : \ at\,\ell_0 \ \wedge \ x_2 \geq 0$$

$$\psi : \ at\,\ell_2 \ \wedge \ y_3 = x_1{}^{x_2}.$$

we would like to establish the total correctness of the program, *i.e.*,

$$\vdash \phi \supset \Diamond \psi.$$

As before, we prove

$(a) \quad \vdash \phi \supset \Diamond \exists k. Q(k, \overline{y})$

$(b) \quad \vdash (\exists k. Q(k, \overline{y})) \supset \Diamond \psi$, or equivalently, $\vdash Q(k, \overline{y}) \supset \Diamond \psi$,

where

$$Q(n, \overline{y}): \quad at\,\ell_1 \wedge (0 \leq y_2 \leq n) \wedge y_3 \cdot y_1{}^{y_2} = x_1{}^{x_2}.$$

Parts (a) and (b) implies the desired result $\vdash \phi \supset \Diamond \psi$ by rule $\Diamond C$. We proceed to prove (a) and (b).

proof of (a):

1. $\vdash at\,\ell_0 \supset \Diamond[at\,\ell_1 \wedge \overline{y} = (x_1, x_2, 1)]$ by F'_α
2. $\vdash [at\,\ell_0 \wedge x_2 \geq 0] \supset \Diamond[at\,\ell_1 \wedge \overline{y} = (x_1, x_2, 1) \wedge x_2 \geq 0]$ by FR
3. $\vdash x_2 \geq 0 \supset [1 \cdot x_1{}^{x_2} = x_1{}^{x_2} \wedge (0 \leq x_2 \leq x_2)]$ by domain
4. $\vdash [\overline{y} = (x_1, x_2, 1) \wedge 1 \cdot x_1{}^{x_2} = x_1{}^{x_2} \wedge (0 \leq x_2 \leq x_2)]$ $\supset [y_3 \cdot y_1{}^{y_2} = x_1{}^{x_2} \wedge (0 \leq y_2 \leq y_2)]$ by $E2$ and PR
5. $\vdash [at\,\ell_1 \wedge \overline{y} = (x_1, x_2, 1) \wedge x_2 \geq 0]$ $\supset [at\,\ell_1 \wedge y_3 \cdot y_1{}^{y_2} = x_1{}^{x_2} \wedge (0 \leq y_2 \leq y_2)]$ by 3, 4, and PR
6. $\vdash [at\,\ell_1 \wedge y_3 \cdot y_1{}^{y_2} = x_1{}^{x_2} \wedge (0 \leq y_2 \leq y_2)]$ $\supset \exists k[at\,\ell_1 \wedge y_3 \cdot y_1{}^{y_2} = x_1{}^{x_2} \wedge (0 \leq y_2 \leq k)]$ by $T24$
7. $\vdash [at\,\ell_0 \wedge x_2 \geq 0]$ $\supset \Diamond \exists k[at\,\ell_1 \wedge y_3 \cdot y_1{}^{y_2} = x_1{}^{x_2} \wedge (0 \leq y_2 \leq k)]$ by 2, 5, 6, $\Diamond Q$ and PR

i.e.,

7′. $\vdash \phi \supset \Diamond \exists k. Q(k, \overline{y})$.

proof of (b): We use the induction rule IND:

$$\frac{\begin{array}{l}(b_1) \quad \vdash Q(0, \overline{y}) \supset \Diamond \psi \\ (b_2) \quad \vdash Q(n+1, \overline{y}) \supset [\Diamond \psi \vee \Diamond Q(n, \overline{y})]\end{array}}{\vdash Q(k, \overline{y}) \supset \Diamond \psi}$$

In our proof we use the special consequence rule

> *Consequence* $\exists\Diamond$ *rule* – $\exists\Diamond Q$
>
> $$\begin{array}{l} \vdash w_1 \supset \exists u.w_2 \\ \vdash w_2 \supset \Diamond w_3 \\ \vdash w_3 \supset w_4 \\ \hline \vdash w_1 \supset \Diamond w_4 \end{array}$$
>
> where u is not free in w_4.

proof of rule:

(1)	$\vdash\ w_1 \supset \exists u.w_2$	given
(2)	$\vdash\ w_2 \supset \Diamond w_3$	given
(3)	$\vdash\ \exists u.w_2 \supset \exists u.\Diamond w_3$	by $\exists\exists$
(4)	$\vdash\ \exists u.w_2 \supset \Diamond \exists u.w_3$	by $T29$ and PR
(5)	$\vdash\ w_3 \supset w_4$	given
(6)	$\vdash\ \exists u.w_3 \supset w_4$	by $\exists I$, since u not free in w_4
(7)	$\vdash\ w_1 \supset \Diamond w_4$	by (1), (4), (6), and $\Diamond Q$

proof of (b_1):

8. $\vdash\ (0 \le y_2 \le 0) \supset (y_2 = 0)$ by domain

9. $\vdash\ Q(0,\overline{y}) \supset [at\,\ell_1\ \wedge\ \overline{y} = \overline{y}\ \wedge\ y_2 = 0\ \wedge\ y_3 \cdot y_1{}^{y_2} = x_1{}^{x_2}]$
by $E1$ and PR

10. $\vdash\ Q(0,\overline{y})$
$\supset\ \exists\overline{u}.[at\,\ell_1\ \wedge\ \overline{y} = \overline{u}\ \wedge\ u_2 = 0\ \wedge\ u_3 \cdot u_1{}^{u_2} = x_1{}^{x_2}]$
by $T24$ and PR

11. $\vdash\ [at\,\ell_1\ \wedge\ u_2 = 0\ \wedge\ \overline{y} = \overline{u}] \supset \Diamond[at\,\ell_2\ \wedge\ \overline{y} = \overline{u}]$
by F'_β, $E2$, and PR

12. $\vdash\ [at\,\ell_1\ \wedge\ \overline{y} = \overline{u}\ \wedge\ u_2 = 0\ \wedge\ u_3 \cdot u_1{}^{u_2} = x_1{}^{x_2}]$
$\supset\ \Diamond[at\,\ell_2 \wedge \overline{y} = \overline{u} \wedge u_2 = 0 \wedge u_3 \cdot u_1{}^{u_2} = x_1{}^{x_2}]$ by FR

13. $\vdash\ [u_2 = 0\ \wedge\ u_3 \cdot u_1{}^{u_2} = x_1{}^{x_2}] \supset u_3 = x_1{}^{x_2}$ by domain

14. $\vdash\ [at\,\ell_2\ \wedge\ u_2 = 0\ \wedge\ u_3 \cdot u_1{}^{u_2} = x_1{}^{x_2}]$
$\supset\ [at\,\ell_2\ \wedge\ u_3 = x_1{}^{x_2}]$ by PR

15. $\vdash\ [at\,\ell_2\ \wedge\ \overline{y} = \overline{u}\ \wedge\ u_2 = 0\ \wedge\ u_3 \cdot u_1{}^{u_2} = x_1{}^{x_2}]$
$\supset\ [at\,\ell_2\ \wedge\ y_3 = x_1{}^{x_2}]$ by $E2$ and PR

16. $\vdash\ Q(0,\overline{y}) \supset \Diamond\psi$ by 10, 12, 15 and $\exists\Diamond Q$

proof of (b_2): We have to consider three cases: $y_2 = 0$, $y_2 > 0 \wedge odd(y_2)$, and $y_2 > 0 \wedge even(y_2)$. Let us only prove the last case.

Case 3: $y_2 > 0 \wedge even(y_2)$.

17. $\vdash\ [Q(n+1,\bar{y}) \wedge y_2 > 0 \wedge even(y_2)]$
$\supset\ [at\,\ell_1 \wedge \bar{y} = \bar{y} \wedge y_2 > 0 \wedge even(y_2) \wedge (0 \le y_2 \le n+1) \wedge y_3 \cdot {y_1}^{y_2} = {x_1}^{x_2}]$ by $E1$ and PR

18. $\vdash\ [Q(n+1),\bar{y}) \wedge y_2 > 0 \wedge even(y_2)]$
$\supset\ \exists \bar{u}.[at\,\ell_1 \wedge \bar{y} = \bar{u} \wedge u_2 > 0 \wedge even(u_2) \wedge (0 \le u_2 \le n+1) \wedge u_3 \cdot {u_1}^{u_2} = {x_1}^{x_2}]$ by $T24$ and PR

19. $\vdash\ [at\,\ell_1 \wedge \bar{y} = \bar{u} \wedge u_2 > 0 \wedge even(u_2)]$
$\supset\ \Diamond[at\,\ell_1 \wedge \bar{y} = ({u_1}^2, u_2 \div 2, u_3)]$ by F'_6, $E2$, and PR

20. $\vdash\ [at\,\ell_1 \wedge \bar{y} = \bar{u} \wedge u_2 > 0 \wedge even(u_2) \wedge (0 \le u_2 \le n+1) \wedge u_3 \cdot {u_1}^{u_2} = {x_1}^{x_2}]$
$\supset\ \Diamond[at\,\ell_1 \wedge \bar{y} = ({u_1}^2, u_2 \div 2, u_3) \wedge even(u_2) \wedge (0 \le u_2 \le n+1) \wedge u_3 \cdot {u_1}^{u_2} = {x_1}^{x_2}]$
by FR

21. $\vdash\ [even(u_2) \wedge (0 \le u_2 \le n+1) \wedge u_3 \cdot {u_1}^{u_2} = {x_1}^{x_2}]$
$\supset\ [(0 \le u_2 \div 2 \le n) \wedge u_3 \cdot ({u_1}^2)^{u_2 \div 2} = {x_1}^{x_2}]$ by domain

22. $\vdash\ [at\,\ell_1 \wedge \bar{y} = ({u_1}^2, u_2 \div 2, u_3) \wedge even(u_2) \wedge (0 \le u_2 \le n+1) \wedge u_3 \cdot {u_1}^{u_2} = {x_1}^{x_2}]$
$\supset\ [at\,\ell_1 \wedge (0 \le y_2 \le n) \wedge y_3 \cdot {y_1}^{y_2} = {x_1}^{x_2}]$ by $E2$ and PR

23. $\vdash\ [Q(n+1,\bar{y}) \wedge y_2 > 0 \wedge even(y_2)] \supset \Diamond Q(n,\bar{y})$
by 18, 20, 22, and $\exists \Diamond Q$

To summarize, we can show

$\vdash\ [Q(n+1,\bar{y}) \wedge y_2 = 0] \supset \Diamond \psi$ case 1

$\vdash\ [Q(n+1,\bar{y}) \wedge y_2 > 0 \wedge odd(y_2)] \supset \Diamond Q(n,\bar{y})$ case 2

$\vdash\ [Q(n+1,\bar{y}) \wedge y_2 > 0 \wedge even(y_2)] \supset \Diamond Q(n,\bar{y})$ case 3

Then since

$\vdash\ Q(n+1,\bar{y})$
$\supset\ [y_2 = 0 \vee (y_2 > 0 \wedge odd(y_2)) \wedge (y_2 > 0 \wedge even(y_2))]$
by domain

it follows that

$\vdash\ Q(n+1,\bar{y}) \supset [\Diamond \psi \vee \Diamond Q(n,\bar{y})]$ by PR

This concludes the alternative proof of the total correctness of our example.

Acknowledgement

This exposition of temporal axiomatization and its application to the verification of deterministic, sequential programs emerged from work done in collaboration with Amir Pnueli on the temporal verification of concurrent programs. I wish to thank Amir for a most stimulating and enjoyable collaboration.

I am indebted to Jacques Hagelstein, Yoni Malachi, Ben Moszkowski, Pierre Wolper and Frank Yellin for their careful reading of the manuscript and for providing insightful comments. Special thanks are due to Evelyn Eldridge-Diaz for TEXing the manuscript.

REFERENCES

[BMP] Ben-Ari, M., Z. Manna and A. Pnueli, "The temporal logic of branching time," Proceedings of the Eighth ACM Symposium on Principles of Programming Languages, Williamsburg, VA, Jan. 1981, pp. 169-176.

[BUR] Burstall, R.M., "Program Proving as Hand Simulation with a Little Induction," Proc. IFIP Congress, Amsterdam, The Netherlands (1974), North Holland, pp. 308-312.

[GPSS] Gabbay D., A. Pnueli, S. Shelah, and J. Stavi, "The Temporal Analysis of Fairness," Proc. 7th POPL, Las Vegas, NV (January 1980), pp. 163-173.

[HC] Hughes, G.E. and M.J. Cresswell, *An Introduction to Modal Logic*, Methuen & Co., London, 1968.

[MP1] Manna, Z. and A. Pnueli, "Verification of concurrent programs: The temporal framework," in *The Correctness Problem in Computer Science* (R.S. Boyer and J S. Moore, eds.), International Lecture Series in Computer Science, Academic Press, London, 1981. Also, Computer Science Report, Stanford University, Stanford, CA (June 1981).

[MP2] Manna, Z. and A. Pnueli, "Verification of concurrent programs: temporal proof principles," Proc. of the Workshop on Logics of Programs (Yorktown-Heights, NY), Springer-Verlag Lecture Notes in Computer Science, 1981.

[MW] Manna, Z. and R. Waldinger, "Is 'Sometime' Sometimes Better Than 'Always'?: Intermittent Assertions in Proving Program Correctness," CACM, Vol. 21, No. 2, pp. 159-172 (February 1978), pp. 159-172.

[PNU] Pnueli, A., "The Temporal Logic of Program," Proc. 18th FOCS, Providence, RI (November 1977), pp. 46-57.

[PRI] Prior, A., *Past, Present and Future*, Oxford University Press, 1967.

[RU] Rescher and Urquhart, *Temporal Logic*, Springer Verlag, 1971.

Theme V.

(Z. Manna)

In an exuberant style, with occasional gentle interruptions

SPECIFICATION AND DERIVATION OF PROGRAMS

Sharon Sickel

President, Logical Paradox, Inc.
26, Moreno Drive
Santa Cruz, Calif. 95060, U.S.A.

SECTION 1: A BRIEF LOGIC REVIEW

A formal theory is a four-tuple:

1) A countable set of symbols; sequences of symbols are called expressions.
2) A subset of the expressions, called the well-formed formulas (WFFs).
3) A subset of the WFFs, known as the set of axioms.
4) A finite set {R1,...,Rn} of mappings between WFFs, called rules of inference. If rule R maps WFFs w1 and w2 onto w3, we say that w3 is derived from w1 and w2 by rule R.

A term is:

1) a constant, denoted by the name of a person or thing.

or

2) a variable, denoted by a name distinct from all constants.

or

3) an n-ary function applied to n terms.

An atomic formula is:

1) an n-ary predicate symbol applied to n terms.

or

2) 'if' atomic formula 'then' atomic formula 'else' atomic formula.

M. Broy and G. Schmidt (eds.), Theoretical Foundations of Programming Methodology, 103–133.

A *conjunct* is:

1) an atomic formula.

or

2) A & C where A is an atomic formula and C is a conjunct.

A *Horn clause* is:

1) an atomic formula.

or

2) A `->' B, where A is a conjunct and B is an atomic formula (frequently written B `<-' A).

Some examples of Horn clauses are:

```
1)  Ackerman (0,n) = n+1
2)  Ackerman (n+1,0) = x      <- Ackerman (n,1) = x
3)  Ackerman (m+1,n+1) = y  <- Ackerman (m+1,n) = x
                                  & Ackerman (m,x) = y
```

We denote that a well-formed formula w is provable by $\vdash$ w. A1,A2,...,An $\vdash$ B denotes that WFF B follows from hypotheses A1,A2,...,An.

Example: Theory COMP:

Symbols: predicate, variable, function and constant names, (,), &, ->

WFFs:

1) an atomic formula.

or

2) A -> B, where A is a conjunct and B is an atomic formula.

Axioms:

A1: A(x) -> A(t) if x is a variable and t is a term (This denotes the replacement of all copies of variable x in WFF A by the term t.)

Rules of Inference:

R1(Modus Ponens): A, A->B |- B

R2: B1&...&Bn -> A, Bj
|- B1&..&B(j-1) & B(j+1)&...&Bn -> A

R3: B1&...&Bn -> A, C -> Bj
|- B1&..&B(j-1)& C &B(j+1)&...&Bn -> A

Quantifiers are not included in the above definitions. However, all WFFs are assumed to be universally quantified for all variables. Within a WFF, any variables named alike are assumed to be the same and must be bound identically.

A proof is a finite sequence w1,...,wn of WFFs such that for each i, either wi is an axiom, or wi is derived from some of the preceding WFFs by one of the rules of inference. A theorem is a WFF w such that there exists a proof in which w is the last WFF.

For example, consider the following primitive recursive definition of addition:

H1:Add(0,x)=x
H2:Add(x,y)=z -> Add(s(x),y)=s(z)

Numerals are abbreviations for s(..s(0)..), e.g. "3" represents s(s(s(0))). Using the clauses of that definition as hypotheses and Theory COMP above, the derivation of Add(3,4) = 7 follows.

Step	Reason
1. Add(0,4)=4	H1
2. Add(0,4)=4 -> Add(1,4)=5	H2
3. Add(1,4)=5 -> Add(2,4)=6	H2
4. Add(2,4)=6 -> Add(3,4)=7	H2
5. Add(3,4)=7	3 applications of Modus Ponens to 1,2,3,4

For a second example of a proof, assume the following hypothesis concerning the postorder traversal of a binary tree. A full discussion follows in a later section. The operator ':' is used as a generalized append for both items and lists. It is associative. The empty string, ( ), is both a left and right identity under this operator. The constant, 'et', denotes the empty tree.

H1: Postorder(et,nil)
H2: Postorder(g(et,x,et),x)
H3: Postorder(t1,x) & Postorder(t2,y) -> Postorder(g(t1,r,t2),x:y:r)

Now derive Postorder(g(g(et,4,et),6,g(g(et,3,et),2,g(et,1,et))), 4:3:1:2:6):

1.	Postorder(g(et,3,et),3)	H2
2.	Postorder(g(et,1,et),1)	H2
3.	Postorder(g(et,3,et),3) & Postorder(g(et,1,et),1) -> Postorder(g(g(et,3,et),2,g(et,1,et)),3:1:2)	H3
4.	Postorder(g(et,1,et),1) -> Postorder(g(g(et,3,et),2,g(et,1,et)),3:1:2)	1,3,R2
5.	Postorder(g(g(et,3,et),2,g(et,1,et)),3:1:2)	2,4,R1
6.	Postorder(g(et,4,et),4)	H2
7.	Postorder(g(et,4,et),4)) & Postorder(g(g(et,3,et),2,g(et,1,et)),3:1:2) -> Postorder(g(4,6,g(3,2,1)),4:3:1:2:6)	H3
8.	Postorder(g(g(et,3,et),2,g(et,1,et)),3:1:2) -> Postorder(g(4,6,g(3,2,1)),4:3:1:2:6)	6,7,R2
9.	Postorder(g(4,6,g(3,2,1)),4:3:1:2:6)	5,8,R1

Def'n.: An <u>interpretation</u> consists of:

a) a non-empty domain D over which variables may range

b) association of each predicate symbol with a relation among elements in D

c) association of each function symbol with a function that is fully defined for elements of D and is closed over D.

d) association of each constant with some individual element of D

A WFF is said to be <u>logically valid</u> iff it is true for every interpretation. A WFF is <u>satisfiable</u> iff there is an interpretation for which A is true for at least one replacement of all free variables by elements of the domain.

A tautology is a WFF that is true in all interpretations.

Soundness implies that only tautologies are derivable. A theory S is sound if and only if it has the property that, if A is provable in S, then A is a tautology. In general, there is no requirement that axioms be true or that rules of inference preserve truth. A & not A is an acceptable axiom under the definition of a formal theory since the only requirement on the choice of axioms is that they be WFFs. A theory including that axiom, however, derives some false WFFs. To have confidence in derived results requires the use of sound theories.

SECTION 2: DECLARATIVE DEFINITIONS

Logic programs are a subset of well-formed-formulas (WFFs) of first order predicate calculus that define a relation and that can be used to drive the computation of one or more missing values of that relation. Specifically, we restrict WFFs to be Horn Clauses , i.e. they have the form B <- A1 & A2 & ... & An where B and the Ai's are atomic formulae, n >_ 0 and B may or may not be empty. It has been shown that all WFFs of first-order predicate logic have equivalent Horn clause counterparts. Any names for predicates, functions, variables, and constants are allowed if they form four mutually disjoint sets. We will, in general, try to use names for these objects that reflect the semantics of the object, relationship or function being represented. We will also allow some familiar function symbols, such as +, -, *.

In this section we use a Prolog-like notation to construct some definitions. First, we define the ubiquitous data-type, stacks (Guttag,1975). There is a single constant, es, the empty stack and a single constructor function for stacks, push. The semantics of the function push are, that given any stack we can create a new stack by pushing an object on the top. We use the predicate STACK to be a unary predicate that is true exactly when its argument is an object of the data type stack. We axiomatically describe this data type as follows:

STACK(es)
STACK(S) -> STACK(push(n,S))

In this case we have not denoted the type of the stacked objects. To do so simply requires another predicate in the hypothesis of the recursive rule. For example, a stack of natural numbers could be denoted:

STACK(es)
STACK(S) & NATNUM(n) -> STACK(push(n,S))

Definition by recursion is well-suited for defining functions over an inductively defined set. The definition consists of:

1) Basis Cases - define the value of the function on the primitive objects, in terms of previously-defined functions. A basis case is also referred to as a termination case.
2) Recursive Cases - describe the value of the function in terms of itself applied to simpler objects.

Functions on Stacks

POP(stack_1)=stack_2: stack_1 with the top element removed is stack_2.

POP(es)=es

POP(push(n,S))=S

TOP(stack)=value: Value is the top item on the stack.

TOP(es)=error

TOP(push(n,S))=n

where error is a unique constant indicating that the function is undefined at this point.

ADD(stack_1)=stack_2: stack_1 with the top two elements of the stack replaced by their sum is stack_2. For example:

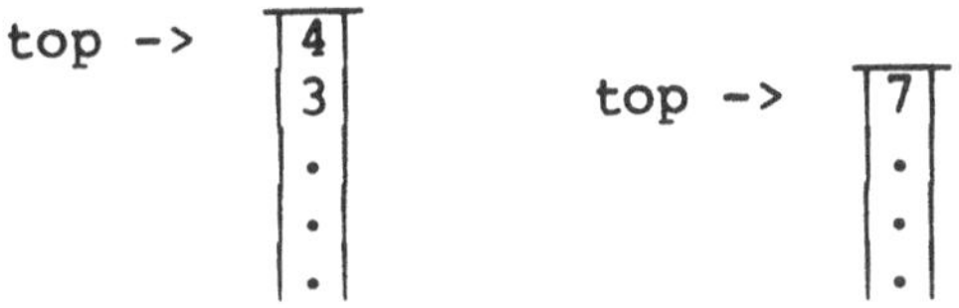

ADD(es)=error

ADD(push(x,es))=error

ADD(push(x,push(y,S)))=push(x+y,S)

MULT(stack_1)=stack_2: stack_1 with the top two elements of the stack replaced by their product is stack_2.

MULT(es)=error

MULT(push(x,es))=error

MULT(push(x,push(y,S)))=push(x*y,S)

Text Editing

A book is a sequence of pages, each of which is a sequence of lines, each of which is a sequence of characters. The following are some appropriate type definitions.

```
BOOK(´( )´)

BOOK(´(´B´)´) & PAGE(´(´P´)´) -> BOOK(´(´B:P´)´)


PAGE(´( )´)

PAGE(´(´P´)´) & LINE(´(´L´)´) -> PAGE(´(´P:L´)´)


LINE(´( )´)

LINE(´(´L´)´) & CHAR(C) -> LINE(´(´L:C´)´)
```

Notice that we have used a constructor function that adds new text on the end instead of the front. That is because we normally create text left-to-right and from the top of a page to the bottom, and we want our editing program to relate to the natural structure of the problem.

Now consider an example of a book, Bookx = (((This is) (an example)) ((of a) (two page) (book))).

```
        Page 1:                          Page 2:

   ______________                   ______________
   |            |                   |            |
   |  This is   |                   |    of a    |
   |            |                   |  two page  |
   | an example |                   |    book    |
   |            |                   |            |
   --------------                   --------------
```

A <u>cursor</u> is a pointer into the text. There are two kinds of cursors that we shall deal with, character cursors and line cursors. We represent the positioning of the line cursor as an ordered pair (li,lf) that divides the line into two sublines, li and lf, where li contains the list of characters preceding the cursor and lf contains the rest of the line, beginning with the character currently being pointed to by the cursor. It is possible for the cursor to be off either end of the line.

ADDCURSOR initializes the cursor at the beginning of the indicated line. Constant ! is added to the left and right to indicate the margins.

```
ADDCURSOR(´(´L´)´) = ( !,L:! )
```

DELETECURSOR removes the cursor from a line.

```
DELETECURSOR(( !:L1, L2:! )=´(´L1:L2´)´
```

The cursor in the following line points to the "b" in "brown".

```
    The quick brown fox
              ^
              ^
represented
    (( !The quick ),(brown fox! ))
```

We can move the cursor to the right (SPACE):

```
The quick brown fox
           ^
           ^
represented
    (( !The quick b),(rown fox! ))
```

or move it to the left (BACKSPACE):

```
The quick brown fox
         ^
         ^
represented
    (( !The quick),( brown fox! )).
```

The parentheses are added to clarify the existence of spaces.

Now we can define SPACE and BACKSPACE.

SPACE((li,lf),(li´,lf´)): If cursor is (li,lf) and we hit space, the cursor becomes (li´,lf´).

SPACE((li,!))=(li,!)

Char(c) -> SPACE((li,c:lf))=(li:c,lf)

BACKSPACE((li,lf)),(li´,lf´)): If cursor is (li,lf) and we hit backspace then the cursor becomes (li´,lf´).

BACKSPACE((!,lf))=(!,lf)

Char(c) -> BACKSPACE((li:c,lf))=(li,c:lf)

More function on lines:

INSERT(x,(li,lf))=(li´,lf): If cursor is at (li,lf) and we insert x to the left of the cursor, the line is now (li´,lf).

INSERT(x,(li,lf))=(li:x,lf)

In this case, a single axiom suffices to define the function.

DELETE(cl)=c2: If we delete the character pointed to by cursor cl, then we result in a line described by cursor c2. Expressing cl as (li,lf), DELETE causes the first character of lf to be removed and the cursor moves to the front of the rest of lf.

DELETE((li,!))=(li,!)

Char(c) -> DELETE((li,c:lf))=(li,lf)

Page cursors do the similar actions to pages that line cursors do to lines; the definitions are constructed accordingly.

Binary Trees

So far, all of the data structures that we have been using are linear, in that each constructed object results from the application of a constructor function to a single, simpler object of the type. We will now look at a data structure that has an added dimension. A binary tree is a set of nodes that are connected in such a way that any given node can have zero to two nodes descended from it. Nodes that have only empty trees as subtrees are leaves. Nodes that have non-empty subtrees are internal nodes. The single tree that is the subtree of no other is the root of the tree. A slightly different notation is used here, one that is more commonly found in the logic programming literature.

´A <-´ is equivalent to ´A´.
´A <- B´ is equivalent to ´B -> A´.

The following is a logic specification of the data type, binary tree.

```
B_TREE(et) <-

B_TREE(graft(t1,r,t2))
        <- B_TREE(t1)
         & B_TREE(t2)
```

The empty tree (et) is the primitive object of the data type. The constructed case of a binary tree is the composition of two binary trees and a node. We call the constructor function in this case "graft". For example, consider the tree in Figure 1.

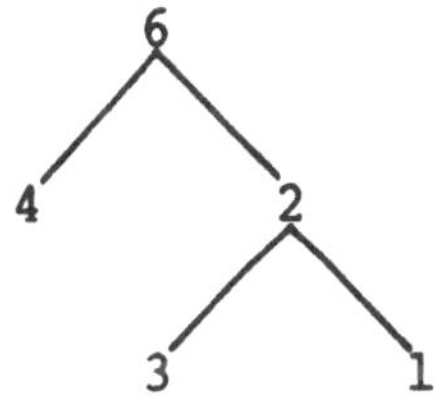

Figure 1

Using the above definition for binary tree, we can represent the tree in Figure 1 by the expression

graft(graft(et,4,et),6,graft(graft(et,3,et),2,graft(et,1,et)))

When displaying trees, we do not show the empty-tree components. We can restrict this binary tree to be a tree of natural numbers by putting a constraint on r in the definition of B_TREE, namely that r be a natural number:

```
B_TREE(et) <-

B_TREE(graft(t1,r,t2))
       <- B_TREE(t1)
        & B_TREE(t2)
        & NATNUM(r)
```

A similar modification could be made to the definition to construct binary trees of any type of object.

We now define some functions that access components of the trees.

ROOT(t,x): x is the root node of binary tree t.

```
ROOT(et, error) <-

ROOT(graft(t1,r,t2),x) <-
```

LEFT(t1,t2): t2 is the left subtree of t1.

```
LEFT(et, error) <-

LEFT(graft(t1,r,t2),t1) <-
```

RIGHT(t1,t2): t2 is the right subtree of t1.

```
RIGHT(et, error) <-

RIGHT(graft(t1,r,t2),t2) <-
```

INORDER(x,y): Given binary tree x, the flattening y of x is the list of all the nodes of x such that every node n follows all nodes in n's left subtree and n precedes every node in n's right subtree. For example,INORDER applied to the tree in Figure 1 is 4:6:3:2:1.

```
INORDER(et,()) <-

INORDER(graft(t1,r,t2),x:r:y)
     <- INORDER(t1,x)
      & INORDER(t2,y)
```

POSTORDER(x,y): List y is a flattening of tree x, in which for every node n of x the nodes of the left subtree of n precede the nodes of the right subtree of n, and n follows all the nodes in its subtrees. For example, Postorder applied to the tree in Figure 1 is 4:3:1:2:6.

```
POSTORDER(et,()) <-

POSTORDER(graft(t1,r,t2),x:y:r)
     <- POSTORDER(t1,x)
      & POSTORDER(t2,y)
```

SUMTREE(t,s): s is the sum of the nodes in binary tree t of natural numbers.

```
SUMTREE(et,0) <-

SUMTREE(graft(t1,val,t2),n1+n2+val)
     <- SUMTREE(t1,n1)
      & SUMTREE(t2,n2)
```

MPL(t,n): n is the maximum path length in tree t.

```
MPL(et,-1) <-

MPL(graft(t1,x,t2),n+1)
     <- MPL(t1,n1)
      & MPL(t2,n2)
      & MAX(n1,n2,n)
```

LEFTMOST(t,n): n is the leftmost node of tree t.

```
LEFTMOST(et, error) <-

LEFTMOST(graft(et,n,t),n) <-

LEFTMOST(graft(t1,r,t2),n)
     <- t1=/=et
      & LEFTMOST(t1,n)
```

IS_IN(n,t): n is a node in binary tree t.

```
IS_IN(r,graft(t1,r,t2)) <-

IS_IN(n,graft(t1,r,t2))
     <-IS_IN(n,t1)

IS_IN(n,graft(t1,r,t2)
     <-IS_IN(n,t2)
```

SUBTREE(t1,t2): t1 is a subtree of t2.

```
SUBTREE(t,t) <-

SUBTREE(t3,graft(t1,r,t2))
     <-SUBTREE(t3,t1)

SUBTREE(t3,graft(t1,r,t2))
     <-SUBTREE(t3,t2)
```

ORDERED(t): t is an ordered tree, i.e. for all nodes n in t, all nodes in the left subtree of n are less than n, and all nodes in the right subtree are greater than n.

```
ORDERED(et) <-

ORDERED(graft(t1,r,t2))
     <- ORDERED(t1)
      & ORDERED(t2)
      & MAXNODE(t1) = m & m<r
      & MINNODE(t2) = n & r<n
```

where MAX and MIN find the largest and smallest nodes, respectively, of a tree and LT(x,y) is true of x is less than y.

INSERT(n,ot1,ot2): Value n inserted into ordered tree ot1 is ordered tree ot2. Duplicate nodes are discarded.

```
INSERT(n,et,graft(et,n,et)) <-

INSERT(n,graft(t1,n,t2),graft(t1,n,t2)) <-

INSERT(n,graft(t1,m,t2),graft(t3,m,t2))
     <- LT(n,m)
      & INSERT(n,t1,t3)

INSERT(n,graft(t1,m,t2),graft(t1,m,t3))
     <- LT(m,n)
      & INSERT(n,t2,t3)
```

SECTION 3: PROCEDURAL INTERPRETATION OF HORN CLAUSES

In the previous section, we have seen logic used as a specification medium for expressing static input/output relationships. It is possible to give these static definitions a procedural interpretation. In Section 1 two theorems were derived under Theory COMP:

1) the derivation of Add(3,4)=7 from addition axioms

and

2) the derivation of the postorder traversal of a binary tree from the axioms defining postorder.

Any theory for first order predicate calculus that had similar WFF syntax could be used to make those derivations, and in fact any automatic theorem prover for first order logic could be used as a Prolog interpreter. It is possible to do derivations more efficiently for the Horn clause form of Prolog than for arbitrary logic WFFs. Theory COMP is the theoretical foundation of every Prolog interpreter; every execution of a Prolog program models a derivation under Theory COMP. If Theory COMP is sound, then we may proceed with confidence that derived results follow logically from the definitions, and at the results are correct if the original defintions were. The soundness of Theory COMP has been proved (Kowalski,1979). The details of a Prolog interpreter and the proof of soundness are not repeated here.

It may be useful to view Prolog definitions from a procedural perspective and relate them to conventional programming constructs. The procedural view of Horn clauses is:

<u>Type of Horn Clause</u> <u>Interpretation</u>

B <- A1 & ... & An

This is treated as a procedure in which B is the procedure head, including the name and parameter list. Each Ai, i = 1..n, is a statement of the procedure. To call procedure B, we match the call with the name and parameter list of B. This determines the values being passed in as input and provides the connections needed to return the output values. To execute procedure B, we call procedures A1,...,An, in any order. In the process of executing the Ai's we bind some variables in the parameters of B.

B <-

This asserts that B is true. In the procedural sense, this is a procedure having no body. The effect of calling it is to do the match with the parameter list, and then return. 'B<-' frequently is denoted simply 'B'.

<- A1 & ... & An

The procedural interpretation of this clause is that of a main program, i.e. it is an unnamed procedure neither receiving nor returning parameters in which the Ai´s are the statements, or subgoals to be accomplished. Usually there is only one clause of this form in the logic program, and that is the top-level call to the relation that defines the program´s input/output relation. The program proves a relationship between the input and output by constructing output having the proper relationship to the input.

Binding Variables

A call P(x1,...,xk) to a procedure has the effect of finding a Horn clause of the form

P(y1,...,yk) <- Q1 & Q2 & ... & Qn

and then unifying (xi,yi), i = 1..k. (See the section on unification, below.) This has the effect of passing input values in the xi´s to the yi´s, and may also cause values to be returned from the yi´s to the xi´s on the return from the procedure.

Example. Consider the logic program:

Fact(x,y): is true iff factorial(x)=y.
Times(x,y)=z: is true iff x*y=z.

1) FACT(0,1) <-
2) FACT(n+1,z) <- FACT(n,x) & TIMES(n+1,x)=z
3) <- FACT(2,ANS)

Clauses 1) and 2) are two separate procedures that together define the factorial function. Clause 1 is the procedure that terminates the computation by assigning the output to be 1 if the input is 0. Clause 2 is the procedure that reduces a factorial computation to a recursive call and a multiplication. We assume here that TIMES has been appropriately defined by other axioms or is considered a primitive function. Clause 3 is the main procedure and can serve the role of forcing a computation of factorial(2) and putting the answer into variable ANS.

Computation Trace

Let's follow the computation. FACT(2,ANS) is a call to a procedure. It cannot call clause 1 because 2 =/= 0. So a call to clause 2 is made. FACT(2,ANS) must be made to match FACT(n+1,z). This can happen if we bind n to 1 and ANS to z. So the procedure with input values substituted looks like:

```
FACT(2,ANS) <- FACT(1,x) & TIMES(2,x)=ANS
```

i.e. ANS = 2*factorial(1). This gives us two new calls:

```
FACT(1,x)
                      and
TIMES(2,x)=ANS.
```

Let us consider only the calls to factorial. FACT(1,x) cannot call clause 1 because 0 =/= 1. So clause 2 is called again. The matching produces procedure:

```
FACT(1,x) <- FACT(0,x') & TIMES(1,x')=x
```

This produces two new calls. Considering again only the factorial call: FACT(0,x') calls clause 1. (Clause 2 will not be a legal call if we have properly defined the natural numbers, i.e. <u>not</u>(0 = n+1). After the binding, the procedure produced from clause 1 is:

```
FACT(0,1) <-
```

which is unchanged, but x' is bound to 1. Since this last procedure has no body, we simply return. Assuming we have also succeeded in doing the multiplications, we have computed ANS through the bindings, i.e.

```
ANS bound to z.
  z bound to 2*x.
    x bound to 1*x'.
      x' bound to 1.
```

Therefore ANS = 2*(1*1) = 2.

The above activity is analagous to a derivation of Fact(2,2) in COMP assuming the subproofs for TIMES are made available.

Unification Algorithm

There are two key components of such an interpreter that we will elucidate, the Unification Algorithm and the Backtracker. A discussion and derivation of the backtracker is included in the next section.

The Unification Algorithm selects an appropriate substitution of terms for variables in WFF´s. The first version of unification is a more conventional one as might be found in many references on resolution, e.g.(Robinson,1971). Substitutions and their applications are denoted as usual.

Unification, Version 1

```
let B = the non-empty set of expressions to be matched

Initialize: sub(0) := empty, k:=0, fail:=false.
while: B  sub(k) is not a singleton & not fail do:

Find a difference set D = {t1,t2} of two elements of B.
If t1 is a variable & doesn´t occur in t2,then
   sub(k+1) := sub(k)  (t2/t1)
   k := k+1;
elseif t2 is a variable & doesn´t occur in t1, then
   sub(k+1) := sub(k)  (t1/t2)
   k := k+1;
else fail := true.
If not fail, then return sub(k) else error.
```

The next version of the algorithm is written in Prolog.

Unification, Version 2

```
Unify(e,e,empty) <-

Unify(e1,e2,sub)
      <- Difference Set(e1,e2) = {t1,t2}
      & t1 is a variable
      & t1 does not occur in t2
      & Unify(e1 (t2/t1),e2 (t2/t1),sub´)
      & sub = (t2/t1) sub´
```

There are two functional differences between Versions 1 and 2. Version 2 works on a pair of expressions to be unified rather than a set as in Version 1, and Version 2 does not return a result in case of failure. The reason for including these two versions is both to help in understanding the mechanism that chooses the substitution to be made to the variables of a called procedure and to contrast the notational style of Prolog and an Algol-like program.

SECTION 4: PROGRAM EVOLUTION

"Programming is the process of transforming the abstract to the executable." (Guttag,1975)

It could be argued that the Prolog programs we have seen here are both abstract and executable. However, some are more efficient than others. Further, what is considered to be an abstract specification by some is considered an executable program by others. For example, Guttag and Horning published an algebraic definition of the Alto display (Guttag/Horning,1980). They claimed it was a specification only. Davis viewed it as a logic program and with minor modifications, executed it (Davis,1980). It is a matter of opinion what the criteria are for a program being finished or even for the model of the target execution engine. For simplicity in this section, we assume as targets a conventional Algol-like language and a conventional von Neumann machine.

Two partial derivations are given below. The first is included in an overview of a derivation of a logic program version of Hoare's program FIND and the proof that it is logically implied by the original specification. The proof is simplified by the fact that both the original and final versions are defined in first-order logic. The second is a derivation of a backtracker for a logic interpreter. It is included to show how to develop complicated programs incrementally, using simplified but related programs in the development.

The Program FIND

Hoare proved the correctness of the program FIND (Hoare,1971). The problem solved by FIND is as follows: Given an array A of N comparable elements, and a natural number, f, permute the array such that all elements smaller than A[f] appear earlier in the array and all larger elements later. This is similar to one iteration in the Quicksort algorithm. Sickel and McKeeman reproved that result by deriving from the original specification a logic program computationally equivalent to FIND (Sickel/McKeeman,1980). The proof is a formal one done in predicate logic. The proof style is highly structured to allow the use of automated deduction tools in doing such proofs. The Sickel/McKeeman derivation is composed of three steps:

1) the original abstract specification:

```
Find of list L and f is list A:r:B
        if L is a permutation of A:r:B
          & A <_ r <_ B
          & | A:r | = f
```

2) an intermediate version that was computationally similar to the final except that lists were used instead of arrays

3) the final version, paraphrased:

```
Find of array A(m)..A(n) and f is array A'(m)..A'(n)
      if m>_n & A=A'

    or

      if m<n
        & splitting A(m)..A(n) based on A(f) is A'(i)..A'(j)
        & if f <_ j

              then Find of A'(m)..A'(j) and f is A''(m)..A''(j)
                   & A''(j+1)..A''(n) = A'(j+1)..A'(n)

              else if i <_ f

                   then Find of A'(i)..A'(n) and f is A''(i)..A''(n)
                        & A''(m)..A''(i-1) = A'(m)..A'(i-1)

                   else Find of A'(f) and f is A''(f)
                        & A''(m)..A''(n) = A'(m)..A'(n)
```

To more closely model the Hoare's version, recursive calls could be replaced by branches, and creation of new arrays replaced by assignment into array A. Hoare's program FIND is included for comparison:

```
begin
 comment  This program operates on an array A[1:N],and a
   value of f(1 ≤ f ≤ N).Its effect is to rearrange
   the elements of A in such a way that:
     ∀ p,q(1 ≤ p ≤ f ≤ q ≤ N  A[p] ≤ A[f] ≤ A[q]);
 integer m,n;  comment
     m ≤ f & ∀ p,q(1 ≤ p < m ≤ q ≤ N  A[p] ≤ A[q]),
     f ≤ n & ∀ p,q(1 ≤ p ≤ n < q ≤ N  A[p] ≤ A[q]):
   m:=1; n:=N;
 while m < n do
 begin integer r,i,j,w;
   comment
      m ≤ i & ∀ p(1 ≤ p < i  A[p] ≤ r),
      j ≤ n & ∀ q(j < q ≤ N  r ≤ A[q]);
    r:=A[f]; i:=m; j:=n;
  while i ≤ j do
  begin while A[i] < r do i := i+1;
    while r < A[j] do j:= j-1;
    comment A[j] ≤ r ≤ A[i];
    if i ≤ j then
    begin w:= A[i]; A[i]:=A[j]; A[j]:=w;
      comment A[i] ≤ r ≤ A[j];
      i := i+1; j := j-1
    end
  end increase i and decrease j;
  if f ≤ j then n := j
else if i ≤ f then m := i
  else go to L
end reduce middle part;
L:
end Find
```

Backtracker

The second Prolog interpreter component we consider is the Backtracker. The Backtracker organizes the selection and application of hypotheses. If a search for a derivation is unable to progress, the derivation is backed-up, and an earlier choice is changed. It is the responsibility of the Backtracker to admit the possibility that any applicable definition be chosen at any decision point, but precludes the repetition of a previous failed choice.

A partial development of a logic interpreter backtracker follows. Figure 2 shows the overall developmental dependencies. Versions 1-4 are given.

```
Ground Case        Version 1:           Version 2:          Version 3:
(no variable   :     Simplistic ------> Complete  ------>  Pascal
 bindings)           Logic                 Logic
                       |                     |                 |
                       |                     |                 |
                       V                     V                 V
General Case|      Version 4:           Version 5:         Version 6:
(including     :     Simplistic ------> Complete  ------>  Pascal
 variable             Logic                 Logic
 bindings)
```

Figure 2.

Proofs of correctness are neither provided nor claimed. The point being made here is that there is added clarity from seeing the progression of the final program through some more easily understood definitions. The six versions do not satisfy the same input/output relations. Instead, they solve six related problems. Version 6, like Version 3 but worse, would be hard to understand in isolation, but is clearer in the context of the progression.

<u>Version 1.</u> This version gives only a declarative definition of the solution of a logic program goal. The necessary actions are not specified in the program but are left to the backtracking interpreter required to execute it. In versions 1-3 variable bindings are ignored.

SG(A) iff goal A is solved.

```
   SG(A) <- (A´<-´Rhs) in Clauselist
          & SGS(Rhs)
```

SGS(Rhs) iff list of subgoals, Rhs, is solved.

```
   SGS(nil) <-

   SGS(x:Rhs) <- SG(x)
               & SGS(Rhs)
```

<u>Version 2.</u> The second version is defined in logic, but the backtracking logic is included in the program. It is much more detailed and contains no implicit global variables (like ´Clauselist´ above).

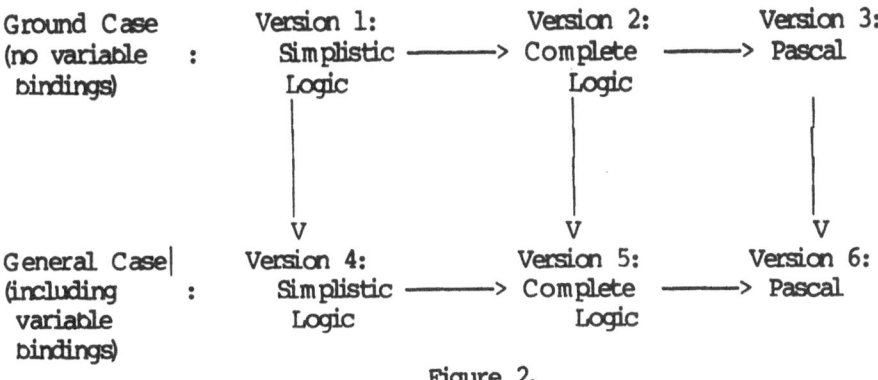

Data Type Definitions:

```
Dictionary(nil) <-

Dictionary( (Pred,Def):D )
   <- Definition( (Pred,Def) )
    & Dictionary(D)
    & NotOccur(Pred,D)
```

```
Definition(nil) <-

Definition(Pred , (Pred:´<-´:Rhs):Def)
   <- Clause(Pred:´<-:Rhs)
    & Definition( (Pred,Def) )
```

Other Definitions:

NotOccur(Pred,D) iff no definition for predicate Pred is in dictionary D.

```
NotOccur(Pred, nil) <-

NotOccur(Pred, (Def,Pred2):D)
   <- Pred1=/=Pred2
    & NotOccur(Pred,D)
```

SG(Webster,Goal,GoalDef,GoodClause,Bool) iff

1) Goal is achievable using GoalDef as its definition and dictionary Webster for other definitions and GoodClause is the first clause of GoalDef that solves Goal and Bool is true,

OR

2) Goal is not achievable under the above conditions and Bool is false.

```
SG(Webster,Goal,nil,x,false) <-

SG(Webster,Goal,(Goal:´<-´:nil):OtherClauses,(Goal:´<-´:nil),true) <

SG(Webster,Goal,(Goal:´<-´:Rhs):OtherClauses,(Goal:´<-´:Rhs),true)
     <- LookUp(Webster,Rhs,DefFirst)
      & SGS(Webster,Rhs,DefFirst,true)

SG(Webster,Goal,(Goal:´<-´:Rhs):OtherClauses, y,bool)
     <- LookUp(Webster,Rhs,DefFirst)
      & SGS(Webster,Rhs,DefFirst,false)
      & SG(Webster,Goal, OtherClauses, y,bool)
```

```
SGS(Webster,PredList,DefFirstPred,Bool) iff
   1) the entire list of subgoals PredList is solvable using
      DefFirstPred as the definition of the first subgoal and
      dictionary Webster for other definitions and Bool is true,
                   OR
   2) PredList is not solvable under those conditions and Bool
      is false.

     SGS(Webster,nil,y,true) <-

     SGS(Webster,FirstPred:RestRhs,DefFirstPred,false)
        <- SG(Webster,FirstPred,DEfFirstPred,y,false)

     SGS(Webster,FirstPred:RestRhs,DefFirstPred,true)
        <- SG(Webster,FirstPred,DefFirstPred,y,true)
         & LookUp(Webster,RestRhs,DefFirst)
         & SGS(Webster,RestRhs,DefFirst,true)

     SGS(Webster,FirstPred:RestRhs,DefFirstPred,Bool)
        <- SG(Webster,FirstPred,DefFirstPred,GoodClause,true)
         & LookUp(Webster,RestRhs,DefFirst)
         & SGS(Webster,RestRhs,DefFirst,false)
         & Successor(DefFirstPred,GoodClause,RemainderDef)
         & SGS(Webster,FirstPred:RestRhs,RemainderDef,Bool)
```

```
Successor(DefFirstPred,GoodClause,RemainderDef) iff RemainderDef
is the list of clauses of DefFirstPred that follow GoodClause.

     Successor(nil,GoodClause,nil) <-

     Successor(GoodClause:RestDef,GoodClause,RestDef) <-

     Successor(x:RestDef,GoodClause,RemainderDef)
        <- x =/= GoodClause
         & Successor(RestDef,GoodClause,RemainderDef)
```

```
LookUp(Webster,Rhs,DefFirstRhs) iff the defintion of the first
   subgoal of Rhs that appears in dictionary Webster is DefFirstRhs.

     LookUp(x,nil,nil) <-

     LookUp(nil,x,nil) <-

     LookUp(FirstPred,Def):Webster, FirstPred:RestRhs,Def) <-

     LookUp(x:Webster,FirstPred:RestRhs,Def)
        <- x =/= FirstPred
         & LookUp(Webster,FirstPred:RestRhs,Def)
```

Version 3. Version 3 is a Pascal program corresponding to the preceding logic program.

```
{Data Type Definitions}

Type
  dictionary      =  ^definition;
  clausept        =  ^clause;
  subgoalpt       =  ^subgoal;

  definition      =  record
                        pred         : string;
                        firstclause  : clausept;
                        nextdef      : dictionary;
                       end;

  clause          =  record
                        predname     : string;
                        rhs          : subgoalpt;
                        nextclause   : clausept;
                       end;
  subgoal         =  record
                        preddef      : dictionary;
                        termlist     : termpt;
                        nextgoal     : subgoalpt;
                       end;

{ We assume that the NotOccur check is done in the
  dictionary builder rather than dynamically checking
  it each time that it is used.

}
```

```
{Procedure Definitions}


function  SG ( Webster        : dictionary;
               Goal           : subgoalpt;
               DefGoal        : clausept;
               VAR GoodClause:  clausept   ) : boolean;
  var  Body      : subgoalpt;
       DefFirst  : clausept;


  procedure  Lookup ( Webster           : dictionary;
                      Rhs               : subgoalpt;
                      VAR DefFirstRhs   : clausept );
                {* Assume that the parser builds the subgoal
                   structures such that the preddef field of a
                   subgoal record points to the definitions of
                   the predicate of that subgoal.              *}
     BEGIN
     DefFirstRhs  :=  nil;
     if       Rhs  <>  nil
       then  begin
             if       Rhs^.preddef  <>  nil
               then  DefFirstRhs  :=  Rhs^.preddef^.firstclause;
             end;
     END;
                {____________________________}


        {** Note that Successor was made into a function **}

  function  Successor ( DefFirstPred  :  clausept;
                        GoodClause    :  clausept  ) : clausept;
     BEGIN
     if       DefFirstPred  =  nil
       then  Successor  :=  nil
       else  begin
             if       DefFirstPred  =  GoodClause
               then  Successor  :=  DefFirstPred^.nextclause
               else  Successor  :=  Successor(
                          DefFirstPred^.nextclause,GoodClause );
             end;
     END;
                          {____________________________}
```

```
function  SGS ( Webster       :  dictionary;
                PredList      :  subgoalpt;
                DefFirstPred  :  clausept ) : boolean;
   var  DefRest     :  clausept;
        GoodClause  :  clausept;
   BEGIN
   if       PredList  =  nil
     then  SGS  :=  true
     else  begin
           if       SG( Webster,PredList,DefFirstPred,GoodClause )
             then  begin
                   Lookup( Webster,PredList^.nextgoal,DefRest );
                   if       SGS( Webster,PredList^.nextgoal,DefRest )
                     then  SGS  :=  true
                     else  SGS  :=  SGS( Webster,PredList,
                                Successor( DefFirstPred,GoodClause ) );
                   end
             else  SGS  :=  false;
           end;
   END;
                          {____________________}


BEGIN      {** Main body of SG *}
if       DefGoal  =  nil
  then  SG         :=  false
  else  begin
        Body       :=  DefGoal^.Rhs;
        if       Body  =  nil
          then  begin
                GoodClause  :=  DefGoal;
                SG          :=  true;
                end
          else  begin
                Lookup( Webster,Body,DefFirst );
                if       SGS( Webster,Body,DefFirst )
                  then  begin
                        GoodClause  :=  DefGoal;
                        SG          :=  true;
                        end
                  else  SG          :=  SG( Webster,Goal,
                                DefGoal^.nextclause,GoodClause );
                end;
        end;
END;
```

Version 4. Version 4 is comparable to Verion 1 with the additional consideration of variable bindings.

Simple General Case - includes variables as parameters

SG(A(x),Gamma) iff goal A(x) Gamma is solved.

```
SG(A(x),Gamma)
   <- (A(y)´<-´Rhs) in Clauselist
    & Unify(x,y,Alpha)
    & Apply(Alpha,Rhs,Rhs2)
    & SGS(Rhs2,Beta)
    & Compose(Alpha,Beta,Gamma´)
    & Restrict(Gamma´,x,Gamma)
```

SGS(Rhs,Gamma) iff Rhs Gamma is solved.

```
SGS(nil,x) <-

SGS(FirstGoal:RestGoals,Gamma)
   <- SG(FirstGoal,Alpha)
    & Apply(Alpha,RestGoals,RestGoals2)
    & SGS(RestGoals2,Beta)
    & Compose(Alpha,Beta,Gamma)
```

where

Unify(Term1,Term2,Thete) iff Theta is the result of unifying Term1 and Term2.

Apply(Theta,GoalList,GoalList2) iff GoalList2 is GoalList with Theta applied to each of it goals.

Compose(Alpha,Beta,Gamma) iff Gamma is the substitution resulting from composing substitutions Alpha and Beta.

Restrict(Gamma´,x,Gamma) iff Gamma is the subset of substitutions of Gamma´ that replace variables occurring in expression x.

Versions 5 and 6 could be constructed analagously to Versions 2 and 3 respectively, adding unification, variable bindings, and restoring bindings as backtracking occurs.

References

[Davis,1980] Ruth Davis, Runnable Specification As a Design Tool, Proceedings of the Logic Programming Workshop, Debrecen, Hungary, July 14-16, 1980.

[Guttag, 1975] John Guttag, The Specification and Application to Programming of Abstract Data Types, Ph.D. Thesis, University of Toronto, 1975.

[Guttag/Horning,1980] John Guttag and J. Horning, "Formal Specification As a Design Tool, Proceedings of the ACM Symposium on Principles of Programming Languages,1980.

[Hoare,1971] C.A.R. Hoare, Proof of a Program: FIND. CACM 14 January, 1971.

[Kowalski,1979] Robert Kowalski, Logic for Problem Solving, North Holland Publishing, New York, 1979.

[Robinson,1971] J.A. Robinson, Computational Logic: The Unification Computation, Machine Intelligence 6, Edinburgh University Press, New York, 1971.

[Sickel/McKeeman,1980] Sharon Sickel and W.M. McKeeman, Hoare's Program FIND Revisited, Proceedings of the Logic Programming Workshop, Debrecen, Hungary, July 14-16, 1980.

Theme VI.

(S. Sickel)

Quotations of Horn clauses, and a general tendency towards concrete syntax in the reverse of the expected order

REPAYING OUR DEBTS

Edsger W.Dijkstra

Burroughs,
Plataanstraat 5,
5671 AL NUENEN,
The Netherlands

Abstract

This text surveys a number of the ways in which computing science could (or should) influence the rest of our mathematical culture.

To Whom it may concern!

There is a wide-spread belief that, whenever a new area of human endeavour has been shown to be amenable to mathematical treatment, the rest of mathematics got eventually enriched by that development. It seems likely, and today I don't feel inclined to challenge the belief. But it raises a question.

Over the last 25 years, and at an increasing pace during the last 10 or so, programming has become "an area of human endeavour amenable to mathematical treatment": the art --or black magic!-- of programming has evolved into a scientific discipline. In its relation to the rest of mathematics, computing science has so far been largely at the receiving end. The question I would like to raise is: "Has the time come to repay our debts?".

Is programming as a mathematical activity sufficiently novel? In other words, do we have any currency in which to repay our debts? Though aware of the danger of being accused of myopia, I

M. Broy and G. Schmidt (eds.), Theoretical Foundations of Programming Methodology, 135–141.

venture an affirmative answer. I could give that affirmative answer more or less ex cathedra: the programming challenge being like nothing mathematicians have ever faced before, the mathematics that grew to meet it must be sufficiently unusual. But such an answer is, of course, only convincing when we preach to the already converted. Slightly more convincing is perhaps a public confession: when I saw Hoare's first correctness proof of a program, I declared most emphatically that such a ballet of symbols was <u>not</u> my cup of tea! I have rarely been so wrong; several years later I was dancing the ballet myself, and liked it.

If I needed a catch-phrase to capture what seems novel, I would come up with something like "problems of scale in a formal environment". Coping mathematically with the programming problem obviously implies that we regard the programming language we use as some sort of formal system, and each program we consider as some sort of formal object. As long as we ignore problems of scale, there is nothing novel in that approach, for it would be rather similar to what all sorts of logicians do. But programs are often big! Admittedly, most formal experiments have been carried out with rather small programs, but sizes are increasing, and how to push the barrier still further is becoming an explicitly stated research topic.

I am perfectly willing to endorse the opinion that most vigorous mathematics results from a subtle balance between applications of formal techniques and of common sense. It is that view of mathematics that may explain some of programming's mathematical "novelty". In the previous paragraph I have explained the intrinsically formal nature of the programming activity; on the other hand, the programmer's environment --the modern computer-- truly deserves the predicate "general purpose". The tension between the two might explain my feeling that the balance between formal techniques and common sense is more critical in programming than in many other areas of mathematics.

Over the last ten years, the predicate calculus became our tool for daily reasoning, its use gradually replacing more and more of the usual verbal arguments. This development was quite clearly forced upon us by the nature of the subject matter we had to deal with; the limited use the usual mathematician makes of the predicate calculus seems an indication of the "novelty" of the programmer's task.

A more wide-spread adoption of the predicate calculus in the rest of mathematics would be a very natural candidate for computing science's influence. I would consider that influence very wholesome. Firstly, it seems a very effective way of cleaning up the presentation of many an argument. (We did experiments with a few examples, more or less chosen by accident, and each time the improvement was considerable.) Secondly, the discipline is purifying: formalization works as an "early-warning system" when things are getting contorted. Thirdly, it clarifies, since it reveals all sorts of mathematical distinctions --such as "proof by infinite regress"-- as artefacts of an in retrospect unfortunate formulation of the argument.

In the long run its influence could be very profound indeed. With sufficient familiarity and sufficient manipulative agility the formal argument becomes the most concise way of capturing our "understanding": it becomes the way in which we prefer to "understand". Eventually it may invalidate the widely-held dogma of the unteachability of thinking.

In the above we need not be deterred by the predictable objection from mathematicians of the more conservative persuasion, who will point out that in most of existing mathematics the logical structure of the arguments is not intricate enough to warrant their formalization, since, even if their observation is correct, there remains the question of the chicken and the egg.

* * *

Another channel through which computing science could exert a potentially profound influence on the rest of mathematics is the computing scientist's attitude towards notational conventions. In the early days, when large sections of the computing community considered the constraints of the limited character set of the punched card sacrosanct, we have committed the most terrible sins and --I am sorry to say-- limited printing facilities of modern terminals seem nowadays to invite similar sinful behaviour.

That physical austerity has not been totally unwholesome. With B a predicate on the natural numbers, I denote quantification by

$$(\underline{E}\ x\colon 0 \le x < X\colon B\ x) \qquad \text{and} \qquad (\underline{A}\ x\colon 0 \le x < X\colon B\ x)$$

Historically, the use of underlining for the creation of new characters has been inspired by the constraints of the typewriter (and of the Flexowriter in particular). But even in handwriting I prefer the $\underline{E}$ and the $\underline{A}$ over the $\exists$ and the $\forall$ respectively. It can immediately be generalized to

$$(\underline{N}\ x\colon 0 \le x < X\colon B\ x) \qquad (1)$$

for the number of distinct values x in the range $0 \le x < X$, such that $B\ x$ holds. (This is just to illustrate that the "upsinde-down convention" is an unfortunate one.)

The recursive definition of (1) is left as an exercise for the reader. The underlined "$\underline{N}$" in (1) is, however, only a minor issue: it has been included because
a) I could not work without it myself, and
b) I never encountered it in the mathematical literature.

I did encounter in the literature --even several times-- something else, viz. the remark that it is very clumsy to express formally that the integer sequence $A(i\colon 0 \le i < N)$ is a permutation of the integer sequence $B(i\colon 0 \le i < N)$. But, using (1), one can do so very easily! I could not escape the impression that, thanks to computing science, a somewhat different tradition of using symbolism has emerged.

* * *

Here is another observation. I think the average mathematician is not as keen as he could be to avoid avoidable case analyses. Let me give you one simple example from Ross Honsberger's "Mathematical Morsels". The example is interesting, since it has been taken from a book in a series explicitly devoted to "the purpose of furthering the ideal of excellence in mathematical exposition".

(Begin of quotation.)

COLORING THE PLANE

Suppose each point of the plane is colored red or blue. Show that some rectangle has its vertices all the same color.

<u>Solution</u>:

Any set of 7 points must contain at least 4 which are the same color. Of 7 points on a line, then, we must have 4 collinear points P_1, P_2, P_3, P_4 which are all the same color, say red. If these points are projected onto two other lines parallel to the first, two collinear quadruples of points (Q_1, Q_2, Q_3, Q_4), (R_1, R_2, R_3, R_4) are obtained which determine several rectangles among themselves and others with the P_i. Now, if any 2 of the

P_1	Q_1	R_1
P_2	Q_2	R_2
P_3	Q_3	R_3
P_4	Q_4	R_4

Q-s are red, an all red rectangle $P_iP_jQ_jQ_i$ results. Similarly for 2 red R's. If neither of these cases hold, then some 3 (or more) of the Q's and some 3 (or more) of the R's must be blue. But these trios of blue points cannot avoid being lined up so that a pair of each trio face each other to yield an all blue rectangle. The conclusion follows. (End of quotation.)

Compare the alternative exposition. Consider a rectangular array of 7 rows of 3 points each. In each row one of the colours dominates. Select <u>four</u> rows in which the same colour dominates and select in each of those two points of the dominant colour. At least two of such latter selections occurred in the same pair of columns, since a pair of columns can only be selected in <u>three</u> different ways. The conclusion follows.

Honsberger's "say red" and his final three-case analysis have completely disappeared. (So have his P's, Q's, and R's; and all the subscripts.) In a comparable degree of verbosity the alternative exposition is only half as long. (The example has been chosen by coincidence: I encountered the problem when I had already started on this text.)

People with a programming background are more alert at trying to avoid case analyses than other mathematicians, who don't seem to mind so much. I made this observation in the mid-seventies --during my, at the time, numerous travels-- and it has never amazed me: as a programmer you know only too well that you have to "pay" for each case analysis as reflected by an alternative construct "<u>if</u> ... [] ... <u>fi</u>" in your program, and the analogy between program structure and proof structure is too obvious to be missed. The tendency to avoid avoidable nomenclature (such as Honsberger's P's, Q's, and R's) is in all probability spin-off of the same influence.

* * *

The discrete nature of much of computing's subject matter has quite naturally increased the role of mathematical induction. First an observation. In order to prove for some predicate P

$$(\underline{A}\ n\colon n \geq 0\colon P\ n) \tag{2}$$

a proof by mathematical induction proves

$$(\underline{A}\ n\colon n \geq 0\colon P'\ n) \tag{3}$$

with P' given by

$$(P'\ n) = (P\ n) \vee (\underline{E}\ i\colon 0 \leq i < n\colon \underline{\text{non}}\ (P\ i)) \tag{4}$$

Since (3) is of the same structure as (2), why not prove it by mathematical induction? I.e. why not prove (3) by proving

$$(\underline{A}\ n\colon n \geq 0\colon P''\ n)$$

with --as in (4)-- P" defined by

$$(P''\ n) = (P'\ n) \vee (\underline{E}\ j\colon 0 \leq j < n\colon \underline{\text{non}}\ (P'\ j)) \qquad ?$$

By substituting (4) into the right-hand side of the definition of P" , however, we deduce P" = P' . In other words, we can apply that trick only once, and the decision to conduct a proof by means of mathematical induction is idempotent. Is this observation important? I don't know, but from a methodological point of view I think it is illuminating; it is in any case very fundamental. Admittedly my sample was small: I showed this observation to about a dozen mature mathematicians, and learned that it was new --even a surprise!-- for <u>all</u> of them. In my more optimistic moments such an experience makes me believe that a greater methodological awareness could be one of the currencies in which computing science could repay its debts.

More in particular, I believe that mathematics could benefit from more explicit applications of mathematical induction. By way of example I refer the reader to my article in Acta Informatica (Jan. 1980), which derives by means of mathematical induction over n the multiplicity of a prime factor p in n! ; this inductive proof is much simpler than the traditional argument --see, for instance, D.E.Knuth's "The Art of Computer Programming", Vol. I-- which relies on the shrewd rearrangement of factors. (Eventually, it even needs the summation formula for the geometric series!)

<u>Note</u>. That traditional argument worries me very much since I met an otherwise gifted mathematician who particularly liked it precisely for its shrewdness. (End of Note.)

* * *

Assuming, for the sake of the argument, that the usual style of mathematical reasoning leaves considerable room for improvement, we may raise the question "Does it matter?". I think it does.

I remember vividly how I showed to one of my colleagues in the Department how we had learned to deal formally and intellectually with all sorts of little algorithms. My colleague wasn't impressed at all and only asked "But does this enable you to find new algorithms?" I was shocked by that reaction.

Firstly, because I would have answered in full confidence "Of course it will.", hadn't it been for the fact that I could answer truthfully "Yes, it has already done so.".

Secondly, because he allowed the "quod" to prevail so strongly over the "quo modo". This utilitarian view of science is very much en vogue today. I am afraid that it is the prevailing view among today's mathematicians, but I have a lurking suspicion that exactly that attitude is mainly responsible for the fact that mathematics is no longer regarded as an integral part of our culture and --to quote Morris Kline-- "Indeed, ignorance of mathematics has attained the status of a social grace.".

I am daily faced with the havoc created by the wide-spread profound ignorance of what mathematics is about. By the cultural isolation of mathematics we all lose, and mathematicians had better remain aware that Elegance is the very core of their business! (And, please, remember that this remark is made by a pragmatic industrial mathematician.)

Part II
DENOTATIONAL SEMANTICS

The ability to solve recursive equations of domains and functions is a main prerequisite for the definition of semantics of programming languages. The first of the following two articles contains a new presentation of the lattice-theoretic tools needed in order to avoid paradoxes in connection with self-application and with the solution of domain equations. Domains thus obtained replace the naive notion of a set for the purpose of specification of a programming language. The language considered in the second article, includes as typical concepts stores and assignment, scopes and environments, continuations and sequencing.

LECTURES ON A MATHEMATICAL THEORY OF COMPUTATION

Dana S. Scott

Carnegie-Mellon University
Department of Computer Science

Pittsburgh

M. Broy and G. Schmidt (eds.), Theoretical Foundations of Programming Methodology, 145–292.

INTRODUCTION

These notes were originally written for lectures on the semantics of programming languages delivered at Oxford during Michaelmas Term 1980. The purpose of the course was to provide the foundations needed for the method of denotational semantics; in particular I wanted to make the connections with recursive function theory more definite and to show how to obtain explicit, effectively given solutions to domain equations. Roughly, these chapters cover the first half of the book by Stoy, and he was able to continue the lectures the next term discussing semantical concepts following his text.

When I started writing Lecture I in October, I did not know exactly what the other lectures would contain; in fact, in the beginning I could see no further ahead than the first part of Lecture III. Of course I had given somewhat similar lectures several times in previous years, but the present course is based on a new approach to the exposition. The lectures had to be typed in advance of the class meetings, and so at the time of composition there was no opportunity for second thoughts. During the spring, after receiving many helpful comments, I was able to introduce a few changes in the text and to make some necessary corrections; however, a complete retyping was impossible, and the text remains rough in many places. I hope to expand the presentation soon into a text book, which can then be less condensed and can contain more examples and easier exercises. The eight-week Oxford term also did not allow me to cover enough topics, and I also hope to add much new material. Even in the present form, though, this preliminary exposition seems sufficient to exhibit the scope of the approach and to function as the basis for several applications.

The idea of using neighbourhood systems to give set-theoretical representations of domains had been in the back of my mind for some time in connection with specific examples. But the thought that a systematic development using the idea might be easier to follow than the more abstract lattice-theoretic and topological approach used in many other publications only came to me during the meeting of the IFIP Working Group 2.2 in Copenhagen in mid-June of 1980. I also gave a brief public presentation at ICALP '80 in Holland in mid-July. After the lectures in Oxford, the next public airing was at a series of lectures at Boston University in mid-December of 1980, at the invitation of Professor Rohit Parikh and the Computer Science Department there. The following spring I attended the *Sixth IBM Symposium on Mathematical Foundations of Computer Science* at Hakone, Japan, May 25–27, 1981, and I also lectured at Kyoto University. IBM have kindly reproduced the present text in the proceedings volume from that conference.

It was especially sad for me that I was unfortunately kept from taking part in the *Marktoberdorf Summer School* by my having to move my house and family from Oxford to Pittsburgh, but the notes were distributed at the conference, and everyone was particularly grateful to Joe Stoy for stepping in and doing my part of the lectures so well and at such short notice.

The title used at Hakone was "Domain Equations" and the title at Marktoberdorf was "Domains and Infinite Objects". Currently at Carnegie- Mellon I am lecturing on the same material under the title "A Theory of Domains and Computability". I think I prefer this last title over the others and over the original "Lectures on a Mathematical Theory of Computation" which I used at Oxford. During part of the period August 28 to September 12, 1981, I was lucky to be able to participate in the *Symposium on Ordered Sets*, which was organized by Professor Ivan Rival of the University of Calgary and held at the Banff Centre. I gave two lectures there under the title "Some Ordered Sets in Computer Science". The text of these talks, which contains many second thoughts about how to explain the programme, has been published by D. Reidel in the proceedings of that conference.

In the present version of the notes, I made the mistake of de- emphasizing the partial orderings too much, since at the right point the concepts and the language are in fact helpful. I discuss the order- theoretic aspects more fully in the Banff lectures, and in particular I show the easy connections between the representations of domains with the aid of neighbourhood systems and as (algebraic) closure systems. In the final text, this will need a full, elementary presentation. Nevertheless, even from these notes it should be be clear that the basic plan works in this way: instead of axiomatizing the theory using partial orderings, the definitions using neighbourhood systems (or closure systems) make the necessary facts come out as *theorems*. The set of elements of a system is naturally partially ordered, because elements are a certain kind of set often called a filter. Approximable mappings preserve this order and can easily be proved to be continuous. As you need facts, they can be proved one by one, and you do not have to say everything you know about these structures all at once. The pedagogic value is clear, and I also think the more explicit approach makes it easier to construct examples of domains and to see what they are like.

I would like to record here a warm word of thanks to the many people and institutions who invited me to lecture, and to all the people that commented on the lectures and the text. I am especially indebted to Steve Brookes and Steve Comer, who spent many hours in Oxford proof reading the typescript and in making corrections. The biggest word of thanks, however, is reserved for Elsie Hinkes, who, under considerable pressure over several weeks, did a wonderful job of typing.

Dana S. Scott
Department of Computer Science
Carnegie-Mellon University
Pittsburgh, Pennsylvania

February 1982

LECTURE I

DOMAINS GIVEN BY NEIGHBOURHOODS

Often an object (or element) can be determined by a selection of its properties. Often it is also the case that it is easier (more convenient, more elementary) to think of these properties than it is to think of the elements themselves. Let us term the properties under consideration *neighbourhoods*, the family of those allowed a *neighbourhood system*. Generally, the collection of these neighbourhoods is, for one reason or another, somewhat restricted; that is, a completely arbitrary property may not be allowable as a neighbourhood. Therefore, the elements determined by selections of neighbourhoods may not be as separable into the discrete objects common to the classical view of set theory. This is particularly true in working with infinite objects: it is hard to specify an infinite element completely. The theory of elements to be studied here, then, is going to permit *partial* elements as well as *total* elements, and each neighbourhood system will define a *domain* of such elements.

Since we may wish to use a neighbourhood system to introduce elements not previously investigated, the neighbourhoods do not have to be regarded as sets of the as-yet-to-be-defined objects. We can take a non-empty set Δ of *tokens* (or "traces") that function as "parts" of elements - or even as parts of "descriptions" of elements. Then a neighbourhood is a subset $X \subseteq \Delta$ containing all those tokens that provide sufficient information when taken together to "approximate" a possible element up to a certain "degree". All these words in inverted commas are vague, and in any case we shall have at the start only a *qualitative* theory of "degree of approximation". A token should be considered as a very "rough" representative of an element, and a neighbourhood should be regarded as "smoothing out" irrelevant details by grouping together *all* those representatives sharing some common feature. One neighbourhood, then, may be only a very incomplete specification of an (ideal) element; fuller specifications can be secured by taking "convergent" sequences of neighbourhoods. Even then convergence need not be to a total element.

Let us call the family of allowed neighbourhoods $\mathcal{D}$; it is a family of subsets of the set Δ. An obvious first question is: when are two neighbourhoods $X, Y \in \mathcal{D}$ neighbourhoods of the "same" element? This question of course generalizes to a (finite) sequence of neighbourhoods. This property we will call the *consistency* of the sequence of neighbourhoods. By definition this will mean that the given neighbourhoods all contain a common neighbourhood in $\mathcal{D}$. That is, for X, Y to be consistent, there must be a $Z \in \mathcal{D}$ with $Z \subseteq X$ and $Z \subseteq Y$. This is not a very informative definition, but it has something of the flavour of a notion of consistency insofar as it can be expressed within $\mathcal{D}$. When consistency holds it seems reasonable enough at first glance to say that the *intersection* $X \cap Y$ is also an approximation to this common element. If this is reasonable, then $X \cap Y$ should also be regarded as a neighbourhood. This assumption has many consequences, but as a preliminary theory of approximation we will find it quite workable with many natural instances. Taking intersections just means taking more and more properties of the element and putting them together "conjunctively." It is something we do all the time. We therefore accept the idea for the present for giving our first principal definition.

DEFINITION 1.1. A family $\mathcal{D}$ of subsets of a given set Δ is called a *neighbourhood system* (over Δ) iff it is a non-empty family closed under the intersection of finite consistent sequences of neighbourhoods. That is to say, $\mathcal{D}$ must fulfill these two conditions:

(i) $\Delta \in \mathcal{D}$;

(ii) whenever $X, Y, Z \in \mathcal{D}$ and $Z \subseteq X \cap Y$, then $X \cap Y \in \mathcal{D}$. □

We remark that by convention Δ corresponds to the intersection of an *empty* sequence of neighbourhoods; in particular,

$$\bigcap_{i<n} X_i = \Delta, \text{ if } n = 0;$$
$$= \Big(\bigcap_{i<n-1} X_i\Big) \cap X_{n-1}, \text{ if } n > 0.$$

Of course, from (ii), we can extend the intersection property to any finite sequence. Consequently, we can say $X_0, \ldots, X_{n-1}$ is consistent in $\mathcal{D}$ iff

$$\bigcap_{i<n} X_i \in \mathcal{D}.$$

Some examples will help us understand the notions better.

EXAMPLE 1.2. Let $\Delta = \{0,1\}$ and let

$$\mathcal{D} = \{\{0,1\}, \{0\}, \{1\}\}.$$

In pictures we have:

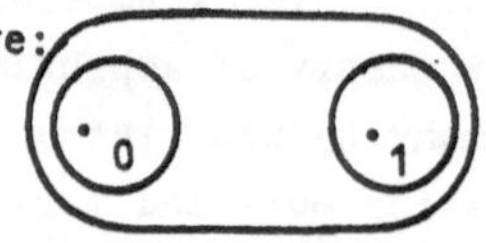

The intention is that 0 and 1 can be completely specified and that they can be identified with the total elements. As we shall see, there is only one partial element: either we give no information (the neighbourhood {0,1}), or we decide between 0 and 1 (by giving {0} or {1}). □

EXAMPLE 1.3. Let $\Delta = \{0,1,2\}$ and let

$$\mathcal{D} = \{\{0,1,2\}, \{1,2\}, \{2\}\}.$$

In pictures we have:

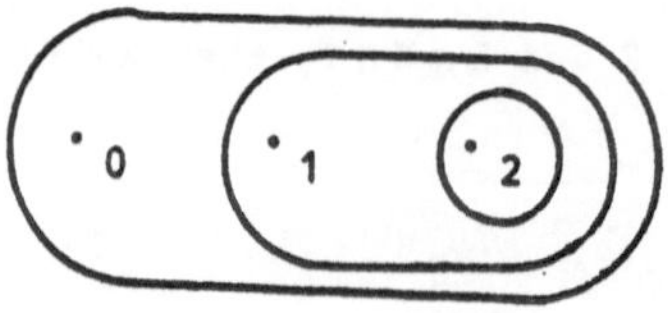

Instead of stepping to the total element (here represented by 2) in one big step, the passage is divided into two steps. (Note 0 and 1 cannot be taken as representing total elements.) This example is not very interesting because the direction of approximation in unique. We need an example with some choice. □

EXAMPLE 1.4 Let

$$\Delta = \{\Lambda,0,1,00,01,10,11\}$$

$$\mathcal{D} = \{\Delta,\{0,00,01\}, \{1,10,11\}, \{00\}, \{01\}, \{10\}, \{11\}\}.$$

Or more understandably in pictures:

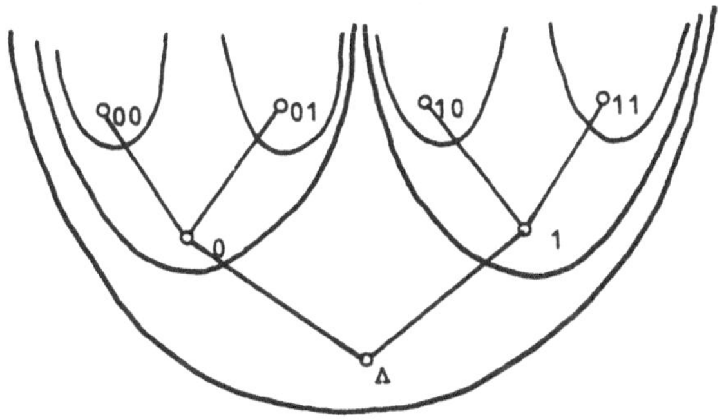

The tokens are finite sequences of 0's and 1's (up to length 2) with Λ the empty sequence; they form - in the picture - the binary tree with the sequences as the nodes. The neighbourhoods are the subtrees of all nodes above a given node. Obviously this can be generalized to sequences of any length (and to trees less regular than the binary tree). The total elements of the example correspond to the top nodes 00, 01, 10, 11 and the lower nodes to the partial elements. When we are not at a top node we have only partially determined a sequence, and the branching indicates that we have some choice as to how the sequence can be extended. □

It should be noted that, in these three examples, the reason that we have a neighbourhood system is a simple consequence of a very special circumstance: in these systems two neighbourhoods are either disjoint or one is included in the other. This arrangement of neighbourhoods is by no means necessary.

EXAMPLE 1.5. Let $\Delta = \{0,1,2,3\}$ and let $\mathcal{D}$ be the family of all non-empty subsets of Δ.

This system is a direct generalization of Example 1.2., which was special owing to the small number of tokens. (The other examples were special by virtue of the choice of neighbourhoods.) The verification that the present $\mathcal{D}$ is a neighbourhood system rests on nothing more than the remark that sets are consistent in $\mathcal{D}$ iff they have a non-empty intersection. Clearly the arrangement of neighbourhoods in $\mathcal{D}$ can be as varied as a four-element set will allow; if Δ were made larger, the possible combinations of neighbourhoods could be made as complex as you wish. □

Having some idea now of the variety of neighbourhood systems, we have to discuss what it is they do. As stressed before, the tokens do not have to correspond directly to the elements; but where, we ask, do the elements come from? One obvious suggestion for determining an element is to produce a sequence of "better and better" neighbourhoods:

$$X_0 \supseteq X_1 \supseteq \cdots \supseteq X_n \supseteq \cdots$$

Trivially, any finite initial segment of this sequence is consistent, and so each X_n is a partial approximation to the "limit". If $\mathcal{D}$ were always to be taken as *finite*, of course, there would be no point in discussing limits since any such sequence would eventually be constant. The elements in the finite case would therefore be completely represented by neighbourhoods with the *minimal* neighbourhoods corresponding to the total elements. But there are many reasons to go beyond the finite (though perhaps not too far beyond).

Suppose $\langle Y_n \rangle_{n=0}^{\infty}$ is another "convergent" sequence with

$Y_{n+1} \subseteq Y_n$ for all indices: when do the two sequences of neighbourhoods determine the *same* limit? The two sequences can surely be different; for example, $\langle Y_n \rangle_{n=0}^{\infty}$ could be a subsequence of $\langle X_n \rangle_{n=0}^{\infty}$, say, $Y_n = X_{2n}$. Still we would want to say that the same limit is obtained. Without being given any further structure on the neighbourhoods, a simple answer is just to say that each sequence goes "equally deep" as the other:

for each m there is an n with $X_n \subseteq Y_m$, and

for each n there is an m with $Y_m \subseteq X_n$.

This definition obviously puts sequences into equivalence classes, and so elements could be identified with these. But such a definition is clumsy for two reasons: it is always tiresome to work with equivalence classes, and there is no reason to think that simple infinite sequences are adequate for determining elements without some rather drastic assumptions on $\mathcal{D}$. Nevertheless, the idea is suggestive; we just have to find some construct to represent elements in a unique way and to phrase it in a general enough manner.

Start with $\langle X_n \rangle_{n=0}^{\infty}$ again, which "converges" as before. Think of all the other sequences equivalent to this one in the sense just defined. We can define the class of all terms of all such sequences very easily as being the family:

$$x = \{Z \in \mathcal{D} \mid X_n \subseteq Z \text{ for some } n\}.$$

It is easy to prove that if we form the analogous class for $\langle Y_n \rangle_{n=0}^{\infty}$, then the two families are *equal* if and only if the sequences are *equivalent*. Thus, we seem justified in letting x represent the limit of $\langle X_n \rangle_{n=0}^{\infty}$. All we have to do now is to remark on what sort of class x is as a subfamily of $\mathcal{D}$; what we abstract from the construction, however, will be just a bit more general than taking those x that result from sequences.

DEFINITION 1.6. The (ideal) *elements* of a neighbourhood system $\mathcal{D}$ are those subfamilies $x \subseteq \mathcal{D}$ where:

(i) $\Delta \in x$;

(ii) $X, Y \in x$ always implies $X \cap Y \in x$; and

(iii) whenever $X \in x$ and $X \subseteq Y \in \mathcal{D}$, then $Y \in x$.

The *domain* of all such elements is written as $|\mathcal{D}|$. □

The idea of 1.6 is a well-known mathematical device: the families x satisfying (i) - (iii) are usually called *filters*. Most frequently the emphasis is put on the *maximal* filters, and these would be our *total* elements; however, in general, the proof that maximal filters exist is non-constructive, so for our purposes it is better not to neglect the partial filters. When maximal filters can be found, well and good, but we do not have to insist on them. Note that the generality of 1.6 is achieved by not requiring that there is a sequence of neighbourhoods that "generates" the filter x. (See Exercise 1.22.)

We have often said that neighbourhoods determine partial elements by themselves; we now make this remark precise.

DEFINITION 1.7. For $X \in \mathcal{D}$, the *principal filter* determined by X is defined by:

$$\uparrow X = \{Y \in \mathcal{D} \mid X \subseteq Y\}.$$

The principal filters form what we shall call the *finite elements* of the domain $|\mathcal{D}|$. □

It is obvious that the correspondence between X and $\uparrow X$ is one-one and inclusion *reversing*, in the sense that

$$X \subseteq Y \quad \text{iff} \quad \uparrow Y \subseteq \uparrow X$$

for all $X, Y \in \mathcal{D}$. But, except in very special cases, there is much more to $|\mathcal{D}|$ than just the finite elements. Much of our investigation will be concerned with finding out how much more. The finite elements are, in a certain sense, "dense" in $|\mathcal{D}|$, however, because it is also obvious from the definitions that for each $x \in |\mathcal{D}|$

$$x = \bigcup \{\uparrow X \mid X \in x\}.$$

That is, every element is a certain type of "limit" of finite elements. (This statement is made more precise in Exercise 1.21)

We note that we have now had several occasions to use inclusion relationships between elements; this is an important relationship, and we give it a special name.

DEFINITION 1.8. For $x, y \in |\mathcal{D}|$, we say that x *approximates* y iff $x \subseteq y$. The element that approximates all others, $\{\Delta\}$, is called $\perp$ (read: *bottom*) ; it is the "least defined" element, or the "most partial" element. Elements maximal with respect to the approximation relation are called *total* elements. □

EXAMPLES 1.2 - 1.5 (Revisited). The examples as given were all finite, so any explicitly given filter x is principal, the element is finite, the minimal $X \in x$ tells us all we need to know. In such simple situations there is essentially no difference between elements and neighbourhoods -- except for the reversal of the order as noted. This (necessary) reversal should not, however, become a matter of confusion: the smaller the neighbourhood has become, the more it has "converged", and so the better defined the element has become. In the approximation relation the "poorer" elements are placed below the "better" with the total up at the top. This will become clearer in discussing "infinite" elements.

Example 1.3 will be generalized in Exercise 1.1. Let us here generalize first 1.4. We let

$$\Delta = \Sigma^* ,$$

where $\Sigma = \{0,1\}$ and Σ^* means the set of all finite sequences of 0's and 1's, with Λ being the empty sequence. We write $\sigma\tau$ for the *concatenation* or *juxtaposition* of two sequences $\sigma, \tau \in \Sigma^*$. Define

$$\mathcal{B} = \{\sigma\Sigma^* \mid \sigma \in \Sigma^*\}, \text{ where}$$

$$\sigma X = \{\sigma\tau \mid \tau \in X\} ,$$

for an arbitrary set $X \subseteq \Sigma^*$. In other words, a neighbourhood in $\mathcal{B}$ consists of all *extensions* of a given sequence σ. (Refer back to the finite version of 1.4.) We use the letter "$\mathcal{B}$" to remind us of "binary", and this is an example we shall refer to many times. The proof (if it is not obvious) that $\mathcal{B}$ is a neighbourhood system should be done as an exercise.

What do we find in $|\mathcal{B}|$? Of course $\perp = \{\Delta\} \in |\mathcal{B}|$. For any $x \in |\mathcal{B}|$ and $\sigma \in \Sigma^*$ define

$$\sigma x = \{Y \mid \sigma X \subseteq Y \text{ some } X \in x\}.$$

Again there is an exercise here to show $\sigma x \in |\mathcal{B}|$. In particular $\sigma\bot \in |\mathcal{B}|$ for all $\sigma \in \Sigma^*$, and these are just the finite elements. The minimal element of $\sigma\bot$ is $\sigma\Delta$. Note that $\sigma_0\bot \subseteq \sigma_1\bot$ if and only if σ_0 is an *initial segment* of the sequence σ_1.

If now $x \in |\mathcal{B}|$ is any explicitly given element (that is, if we know for any $X \in \mathcal{B}$ whether or not $X \in x$), we have but to work out from these definitions that

$$x = \bigcup_{n=0}^{\infty} \sigma_n \bot ,$$

where the $\sigma_n \in \Sigma^*$ and each σ_n is an initial segment of the next σ_{n+1}. In general, in any domain, an element is *uniquely determined by its finite approximations*, and we are just making this explicit in $|\mathcal{B}|$. When we have complete knowledge of x, then there are two cases: either the approximations $\sigma_n\bot$ become constant from some point on (where $n \geq n_0$), or not. In the first case x is finite and equal to $\sigma_{n_0}\bot$; in the second case x is *infinite* and the σ_n fill out an infinite (one-way) sequence.

The generalization of 1.5 to the infinite case where

$$\Delta = \mathbb{N} = \{0,1,2,3, \ldots, n, \ldots\}$$

can be made in more than one way: for instance either we use as neighbourhoods *all* non-empty subsets of Δ or just those omitting but a finite number of integers. And, as will become apparent, there are other choices giving domains of quite different characters. □

Many constructions (choices of $\mathcal{D}$) lead to the "same" domain; "sameness" is an important notion and it is to be defined in terms of "isomorphism", which in turn is to be defined in terms of approximation preserving correspondences.

DEFINITION 1.9. Two neighbourhood systems $\mathcal{D}_0$ and $\mathcal{D}_1$ determine *isomorphic domains* iff there is a one-one correspondence between $|\mathcal{D}_0|$ and $|\mathcal{D}_1|$ which preserves inclusion between the elements of the domains. In symbols we write $\mathcal{D}_0 \cong \mathcal{D}_1$. □

It is certain that the property of 1.9 is necessary, but it may not be so clear that it is sufficient. We shall in fact prove in the next lecture that an isomorphism between domains always maps finite elements to finite elements, so it always results from a one-one inclusion-preserving correspondence between neighbourhoods. This is surely as strong as could be hoped. This general result is not needed to see that particular domains *are* isomorphic.

In some of the examples tokens corresponded to total elements and in some to partial elements; it is not difficult to see (ex post facto) that every domain can be presented with tokens exactly corresponding to partial elements.

THEOREM 1.10. Given any neighbourhood system $\mathcal{D}$, define for $X \in \mathcal{D}$

$$[X] = \{x \in |\mathcal{D}| \mid X \in x\}.$$

The subsets $[X] \subseteq |\mathcal{D}|$ for $X \in \mathcal{D}$ form a neighbourhood system over $|\mathcal{D}|$ which determines a domain isomorphic to $|\mathcal{D}|$.

Proof: We note first that

(1) $[\Delta] = |\mathcal{D}|$.

Next note that

(2) X, Y are consistent in $\mathcal{D}$ iff $[X] \cap [Y] \neq \emptyset$;

and that for $X, Y \in \mathcal{D}$

(3) $[X] \cap [Y] = [X \cap Y]$ if $X \cap Y \in \mathcal{D}$.

Inasmuch as

(4) $\uparrow X \in [X]$ for all $X \in \mathcal{D}$,

it easily follows that $\mathcal{D}$ and the family

$$\{[X] \mid X \in \mathcal{D}\}$$

are in a one-one, inclusion-preserving correspondence. Thus, we can induce the desired one-one correspondence between the elements of the two systems. □

The import of 1.10 is that the original tokens in Δ can be replaced by the elements of $|\mathcal{D}|$. This process replaces the neighbourhood $X \subseteq \Delta$ by the subset $[X] \subseteq |\mathcal{D}|$. As the passage is inclusion preserving, the domain has not really changed, only its presentation. Though of some theoretical charm, the theorem is not of much use since we still have to get $\mathcal{D}$ from somewhere. It does emphasize, though, that the rôle of the tokens is simply to keep the inclusions (and intersections) of neighbourhoods sorted out. It is *not* always true that the tokens can be identified with the total elements.

The last theorem in this lecture is a result on *closure properties* of a domain with respect to set-theoretical operations which have interesting meanings with respect to approximation.

THEOREM 1.11. If $\mathcal{D}$ is a neighbourhood system and $x_n \in |\mathcal{D}|$ for $n = 0,1,2,\ldots$, then

(i) $\bigcap_{n=0}^{\infty} x_n \in |\mathcal{D}|$; and

(ii) $\bigcup_{n=0}^{\infty} x_n \in |\mathcal{D}|$, provided

$$x_0 \subseteq x_1 \subseteq x_2 \subseteq \cdots \subseteq x_n \subseteq x_{n+1} \subseteq \cdots .$$

Proof: The conditions of 1.6 have to be checked. For the case of intersection, all of 1.6(i) - (iii) are quite obvious. For the case of union, only 1.6(ii) gives pause and it requires the proviso. If X and Y belong to the union, then $X \in x_n$, say, and $Y \in x_m$. But, either $n < m$ or $m < n$, and if $k = \max(n,m)$, then $X, Y \in x_k$. Since $x_k \in |\mathcal{D}|$, we have $X \cap Y \in x_k$; thus, $X \cap Y$ belongs to the union. This proves (ii). □

In words, the intersection is the best element that is at the same time an approximation to all of the elements x_n; the intersection is exactly what is common to all the given elements. The union on the other hand is just what the (increasing sequence of the) x_n approximates; the union combines contributions from all the x_n into a "better" element -- but no more than that.

In thinking about domains a rough diagram of the partial-ordering relation $\subseteq$ between elements is often helpful. The picture of 1.4 is an example where the nodes represent the elements. Any finite tree growing up from a root node would also be an example. Indeed, any finite partially ordered set with least element would be an example. (Here no distinction between tokens and elements is necessary.) A lattice diagram is also illustrated.

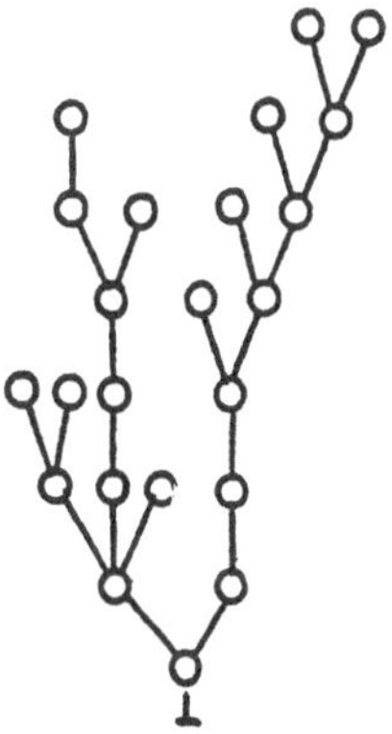

A TREE

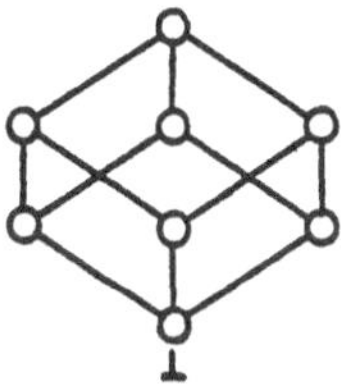

A LATTICE

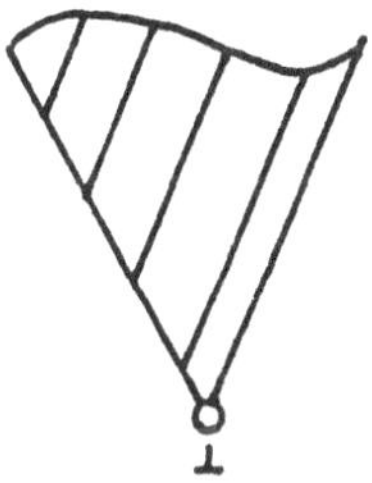

A ROUGH PICTURE

The root node is the element $\perp$ of $|\mathcal{D}|$; there need be no top node $\top$. *Approximation* is represented by a passage from a lower node to a higher node along the rising lines. The system $\mathcal{D}$ of neighbourhoods is the collection of sets each of which is all the nodes above a given node. For *infinite* examples, however, care must be given to introduce *limit* nodes. The first few exercises should be provided with pictures to illustrate the structure.

EXERCISES

EXERCISE 1.12. Let $\Delta = \mathbb{N} = \{0,1,2,\ldots,n,\ldots\}$ be the set of non-negative integers. Use as neighbourhoods final segments:

$$\{m \in \mathbb{N} \mid m > n\}$$

for $n \in \mathbb{N}$. Verify that this is a neighbourhood system. What are the total elements? What are the finite elements? Draw a picture of the approximation relation in this domain. (Hint: there is only one limit element.)

EXERCISE 1.13. Verify all the assertions made about the system $\mathcal{B}$ defined as the infinite generalization of Example 1.4. Draw a picture similar to that given in the text which includes nodes for all $\sigma \in \Sigma^*$. Show the neighbourhoods, how the approximation relation behaves, and where the total elements lie. (The picture is closely related to the "binary tree", but has to have limit nodes all along the top.)

EXERCISE 1.14. Let $\Delta = \mathbb{N}$ and let $\mathcal{D}$ be the family of finite non-empty subsets of Δ plus the set Δ. Show that this is a neighbourhood system. What are the total elements? What are the finite elements? Draw a picture.

EXERCISE 1.15. Construct non-isomorphic infinite domains where all elements are finite but where there are no infinite chains $\langle x_n \rangle_{n=0}^{\infty}$ of elements with $x_n \sqsubseteq x_{n+1}$ but $x_n \neq x_{n+1}$ for all n.

EXERCISE 1.16. Let $\Delta = \mathbb{N}$ and let $\mathcal{D}$ be the family of *cofinite* subsets of $\mathbb{N}$. Show that $|\mathcal{D}|$ is isomorphic to the partially ordered set of *all* subsets of $\mathbb{N}$ under inclusion. Construct some other neighbourhood systems where $\mathcal{D}$ is closed under finite intersection. What happens to the total elements in such systems?

EXERCISE 1.17. Let $\Delta = \mathbb{R}$ be the real line. Let $\mathcal{D}$ be the set of non-empty open intervals with rational end points plus the set Δ. Show that this is a neighbourhood system. For any real $t \in \mathbb{R}$, show that

$$\{X \in \mathcal{D} \mid t \in X\}$$

is a filter. Is it always total? What are the total elements of $|\mathcal{D}|$? (Hint: When t is rational consider all intervals with t as a right-hand end point.)

EXERCISE 1.18. Let $\mathcal{D}$ be a neighbourhood system. Call a subset $C \subseteq \mathcal{D}$ *consistent* iff every finite subset of C is consistent in $\mathcal{D}$. Give an example where C is a subset with more than two elements, every pair of neighbourhoods in C is consistent, but C is *not* consistent. Show that if C is consistent, then there is a *least* filter $x \in |\mathcal{D}|$ with $C \subseteq x$. Show generally that the *intersection* of any non-empty collection of filters is again a filter.

EXERCISE 1.19. Define a *positive neighbourhood system* to be a family $\mathcal{D}$ where (ii) of 1.1 is replaced by

(ii′) whenever $X, Y \in \mathcal{D}$, then $X \cap Y \neq \emptyset$ iff $X \cap Y \in \mathcal{D}$.

Prove that a positive neighbourhood system is indeed a neighbourhood system in the sense of the earlier definition. Give an example of a neighbourhood system that is *not* positive. (Hint: (suggested by C.A.R. Hoare). Let $\Delta = \mathbb{N} \times \mathbb{N}$, in the plane. Let $\mathcal{D}$ be the family of subsets $X \subseteq \mathbb{N} \times \mathbb{N}$ where all but a finite number of places the *vertical* sections of X are the whole of $\mathbb{N}$ but at the other places the sections are finite and nonempty. Smaller examples are of course possible.)

EXERCISE 1.20. Let $\mathcal{D}$ be any neighbourhood system over a set Δ. Let $\Delta' = \mathcal{D}$ and define

$$\mathcal{D}' = \{\uparrow X \mid X \in \mathcal{D}\}$$

where

$$\uparrow X = \{Y \in \mathcal{D} \mid Y \subseteq X\}.$$

Show that $\mathcal{D}'$ is a positive neighbourhood system and that $|\mathcal{D}|$ and $|\mathcal{D}'|$ are isomorphic. Note that for $\mathcal{D}'$ finite elements and tokens are in a one-one correspondence.

EXERCISE 1.21. Work out in greater detail the proof of 1.10. Remark that the neighbourhood system over $|\mathcal{D}|$ so constructed is positive, thereby obtaining in a different way the same kind of conclusion as in 1.20. Show also that the system over $|\mathcal{D}|$ is *complete* in the sense that every filter is fixed by a *unique* member of the underlying set. (A filter is *fixed by a point* iff it is the filter of *all* neighbourhoods containing that point.) Remark that a complete system is one where tokens and (partial) elements can always be identified (under a suitable one-one correspondence). Show also that consistency of a set $\{X_i \mid i<n\}$ of neighbourhoods in $\mathcal{D}$ is equivalent to saying

$$\bigcap_{i<n} [X_i] \neq \emptyset$$

EXERCISE 1.22. (For topologists). Show that the neighbourhoods $[X]$ for $X \in \mathcal{D}$ make $|\mathcal{D}|$ into a topological space where the open subsets $\mathcal{U} \subseteq |\mathcal{D}|$ can be characterized by the following two conditions:

(i) whenever $x \in \mathcal{U}$ and $x \subseteq y \in |\mathcal{D}|$, then $y \in \mathcal{U}$; and

(ii) whenever $x \in \mathcal{U}$, then $\uparrow X \in \mathcal{U}$ for some $X \in x$.

Prove also that the inclusion relation on $|\mathcal{D}|$ can be defined topologically as:

(iii) $x \subseteq y$ iff for all open $\mathcal{U} \subseteq |\mathcal{D}|$, if $x \in \mathcal{U}$ then $y \in \mathcal{U}$.

Is $|\mathcal{D}|$ ever a Hausdorff space? Show that if $\langle x_n \rangle_{n=0}^{\infty}$ is a sequence of elements of $|\mathcal{D}|$ with $x_n \subseteq x_{n+1}$ for all n, then

$$\bigcup_{n=0}^{\infty} x_n$$

is not only in $|\mathcal{D}|$ but is a topological limit point of the sequence. Show that any element x is a limit point of the set $\{\uparrow X \mid X \in x\}$. Are there other limit points?

EXERCISE 1.23. Suppose that the neighbourhood system $\mathcal{D}$ is countable, say,

$$\mathcal{D} = \{X_0, X_1, X_2, \ldots, X_n, \ldots\}.$$

Suppose further that the property of consistency of finite sequences of neighbourhoods is decidable (or "effectively known"). Then the following sequence is well defined:

$$Y_0 = X_0$$

$$Y_{n+1} = X_{n+1}, \text{ if this set is consistent with } Y_0, Y_1, \ldots, Y_n;$$

$$= Y_n, \text{ if not.}$$

Show that $\{Y_0, Y_1, \ldots, Y_n, \ldots\}$ is a total element of $|\mathcal{D}|$. (Hint: Show first that $Y_0, Y_1, \ldots, Y_{n-1}$ is consistent for all n.) In such a system show that all filters can be determined by sequences.

EXERCISE 1.24. (For set theorists). Prove, using the Axiom of Choice, or some equivalent principle, that in every domain a partial element can always be extended to a total element. Is this assertion equivalent to the Axiom of Choice? (Hint: Remember to prove that the union of every (transfinite) chain of filters is again a filter.)

EXERCISE 1.25. (For set theorists). Let Δ be any well-ordered set (ordinal). (Even small ordinals like $\omega.3$ or ω^5 are interesting.) Let $\mathcal{D}$ be the family of non-empty *final* segments of Δ. What is $|\mathcal{D}|$? Are all elements finite? Is every approximation to a finite element finite?

EXERCISE 1.26. (For algebraists). Let A be a commutative ring with unit. Let Δ be the set of finite subsets $F \subseteq A$. Define

$$I(F) = \{G \in \Delta \mid F \subseteq \text{ the ideal generated by } G\}.$$

Prove that the sets of the form I(F) form a neighbourhood system, and that the corresponding domain is isomorphic to the set of ring-theoretic ideals of A partially ordered by inclusion. What would happen if we excluded from Δ all F with $I(F) = I(\{1\})$, where 1 is the unit of A?

EXERCISE 1.27. Further closure properties of domains can be proved for bounded sets. We say $X \subseteq |\mathcal{D}|$ is *bounded* iff for some $y \in |\mathcal{D}|$ we have $x \subseteq y$ for all $x \in X$. This y is called an *upper bound*. We let

$$\bigsqcup X = \bigcap \{y \in |\mathcal{D}| \mid x \subseteq y \text{ all } x \in X\}.$$

Prove that if X is bounded, then $\bigsqcup X$ is the *least upper bound* for X in $|\mathcal{D}|$. Prove also: if $U, V \in \mathcal{D}$ are neighbourhoods, then $\{U, V\}$ is consistent in $\mathcal{D}$ iff $\{\uparrow U, \uparrow V\}$ is bounded in $|\mathcal{D}|$. (That is, boundedness is for elements what consistency is for neighbourhoods.) Prove finally with the aid of 1.18 that $X \subseteq |\mathcal{D}|$ is bounded iff every finite subset of X is bounded.

LECTURE II

MAPPINGS BETWEEN DOMAINS

The elements of a domain are regarded as being specified by approximations: the neighbourhoods. With the idea of approximation as the dominant notion, therefore, it is natural to look for a concept of mapping (transformation of domains) that in some suitable sense preserves the spirit of the approximations. In a theory of computability, where the (finite) approximations to the elements are all we can ever know at one time, the only mappings that can be computed are those that proceed by approximation, somehow passing from the neighbourhoods of one domain over to the neighbourhoods of the other.

Suppose $X \in \mathcal{D}_0$ is given - it is an approximation to certain elements of $|\mathcal{D}_0|$. (More precisely $\uparrow X$ is the approximation *in* the domain, but it is easier to speak of the neighbourhood X.) What can be said about the approximations of the images of these elements under the mapping we will call f? If X is not a very sharp approximation, then not very much can be said about the image in the other domain $|\mathcal{D}_1|$. Trivially, of course, we can say that Δ_1 is an approximation - because it approximates everything in its domain. Suppose, however, that we could say more. Suppose we could say that both Y and Y' approximate the image of X. If the mapping f is coherent, then it is reasonable to suppose that such a statement would imply that Y and Y' are *consistent* in $\mathcal{D}_1$. But if this is so, then since the two neighbourhoods are meant to cluster around the same images, we can feel some confidence in saying that $Y \cap Y'$ approximates these images., In other words to specify f we do not supply a *unique* image of X, but we say which of the $Y \in \mathcal{D}_0$ approximate the (ideal) image. To make this idea work a *monotonicity condition* is also needed since we are trying to express the idea that "*if* we give at least X as an approximate input to f, *then* we can expect at least Y as output." Thus, a mapping is taken as a kind of relation between neighbourhoods.

DEFINITION 2.1. An *approximable mapping* $f: \mathcal{D}_0 \to \mathcal{D}_1$ between domains is a binary relation $f \subseteq \mathcal{D}_0 \times \mathcal{D}_1$ between neighbourhoods such that

(i) $\Delta_0 f \Delta_1$;

(ii) $X f Y$ and $X f Y'$ always imply $X f (Y \cap Y')$;

(iii) $X f Y$, $X' \subseteq X$, and $Y \subseteq Y'$ always imply $X' f Y'$. □

Condition (i) we have already discussed; in a sense it means "ask me no questions and I shall tell you no lies." In other words "zero input can expect at least zero output." The other conditions are compatible with having

$$f = \{ \langle X, \Delta_1 \rangle \mid X \in \mathcal{D}_0 \};$$

that is, f might be the least informative relation and nothing more. But if it is more, then (ii) is, as we explained, a consistency condition. To explain monotonicity in (iii), suppose a mapping relationship is already known, $X f Y$, say. If we *improve* the accuracy of X to $X' \subseteq X$ and if we *degrade* the accuracy of Y to $Y' \supseteq Y$, then we can still assert $X' f Y'$ since this relationship is *less informative* than the former relationship, which was already known. Thus, we see that conditions (i) - (iii) are all reasonably argued as necessary.

One indication that the conditions of 2.1 are sufficient for a definition is that they are exactly what we need to show that f as a neighbourhood relation determines an equivalent elementwise mapping from $|\mathcal{D}_0|$ into $|\mathcal{D}_1|$. (Owing to the equivalence, we use the same symbol f for both.)

PROPOSITION 2.2. Given neighbourhood systems $\mathcal{D}_0$ and $\mathcal{D}_1$, an approximable mapping $f: \mathcal{D}_0 \to \mathcal{D}_1$ always determines a function $f: |\mathcal{D}_0| \to |\mathcal{D}_1|$ between domains by virtue of the formula:

(i) $f(x) = \{Y \in \mathcal{D}_1 \mid \exists X \in x.\ X f Y\}$

for all $x \in |\mathcal{D}_0|$. Conversely, this function uniquely determines the original relation by the equivalence:

(ii) $X f Y$ iff $Y \in f(\uparrow X)$

for all $X \in \mathcal{D}_0$ and $Y \in \mathcal{D}_1$. Approximable functions are always *monotone* in the following sense:

(iii) $x \subseteq y$ always implies $f(x) \subseteq f(y)$,

for $x, y \in |\mathcal{D}_0|$; moreover two approximable functions $f : \mathcal{D}_0 \to \mathcal{D}_1$ and $g : \mathcal{D}_0 \to \mathcal{D}_1$ are identical as relations iff

(iv) $f(x) = g(x)$, for all $x \in |\mathcal{D}_0|$.

Proof: The argument that formula (i) always gives us $f(x) \in |\mathcal{D}_1|$ when $x \in |\mathcal{D}_0|$ can be safely left to the reader. Note, however, that all the conditions of 2.1 are required to show this. As for (ii), the implication from left to right follows directly from (i) because $X \in \uparrow X$. In the other direction $Y \in f(\uparrow X)$ means that $Z f Y$ holds for some $Z \in \uparrow X$. But from $X \subseteq Z$ it follows that $X f Y$, as we wished.

To prove monotonicity, assume $x \subseteq y$. Now $X \in x$ and $X f Y$ always imply $X \in y$ and $X f Y$. This means $Y \in f(x)$ always implies $Y \in f(y)$; that is, $f(x) \subseteq f(y)$.

Finally, to check that (iv) means $f = g$ as relations, all that has to be remarked that this follows from formulae (i) and (ii). □

Note that the right-hand side of (ii) can be written:

$$\uparrow Y \subseteq f(\uparrow X),$$

which can be read as saying that the partial element determined by the neighbourhood Y approximates the function value at the element determined by X. This precise relationship of course fits the informal discussion of mapping given earlier. Indeed whenever $x \in [X]$ and $X f Y$ hold, then $f(x) \in [Y]$ always follows, which is another way to construe the mapping character of f. Some examples of mappings are now called for.

EXAMPLE 2.3. Let T be the neighbourhood system of the two-token domain of Example 1.2. To avoid confusion with some other domains, we will call the two total elements of $|T|$ respectively true and false. There is only one other finite element here, namely $\bot$ = undefined. We often use these elements as indicators of results: true indicates a positive outcome; false, a negative outcome; and $\bot$ indicates that there is not enough information to decide the outcome totally.

Let B be the system for the binary tree as in the last chapter. What we wish to define is an approximable mapping $f : B \to T$. The intuitive idea of the mapping we have in mind is that the binary sequence is being read from left to right, and we are counting the number of 0's seen before the first 1 is encountered. We then test the parity of this count; if it is even, the output is true; if not, false. Using a suggestive informal notation with three dots, some results of the function that does the counting and testing can be written as:

$$f\ (0000101\cdots) = \text{true}$$
$$f\ (1101110\cdots) = \text{true}$$
$$f\ (0111011\cdots) = \text{false}$$
$$f\ (0000000\cdots) = \bot$$

The last equation is necessary, because 0000000 as a partial element *cannot* be counted as either even or odd since it can have *inconsistent* extensions:

$$0000000\bot \subseteq 00000001\bot$$
$$0000000\bot \subseteq 00000000001\bot .$$

So, as far as f is concerned, a plain string of 0's is *indefinite*. The same answer holds if the 0's go on infinitely.

To be more precise we want

$$f\ (0^n 1\bot) = \text{true} \quad \text{if n is even;}$$
$$= \text{false} \quad \text{if n is odd.}$$

As a binary relation $f \subseteq B \times T$ we will have

$X f Y$ iff $Y \in \bot$ or $X \subseteq 0^n 1 \Delta$ for some $n \in \mathbb{N}$ and either n is even and $Y \in$ true or n is odd and $Y \in$ false.

It should be checked that 2.1(i)-(ii) are satisfied. □

EXAMPLE 2.4. Let us briefly describe an approximable mapping $g: \mathcal{B} \to \mathcal{B}$. Informally, g can be said to "read a sequence from left to right and eliminate the first consecutive run of 1's while copying all the other digits as read." We will have

$$g(0^n 1^k 0x) = 0^{n+1}x$$

provided $k > 0$. (Here 1^k means a string of 1's of length k.) However, if 1^∞ is the infinite sequences of 1's, then

$$g(1^\infty) = \bot, \text{ and}$$
$$g(0^n 1^\infty) = 0^n$$

This example is instructive, since it shows that a non-trivial mapping can transform a total element into a partial element. □

Aside from our being able to define particular functions outright, we can combine functions in many different ways; the idea of composition is probably the most basic scheme of combination, and there is a technical name for a family of structures with mappings that can be so combined.

THEOREM 2.5. The class of neighbourhood systems and approximable mappings form a *category*, where the *identity mapping* $I_\mathcal{D} : \mathcal{D} \to \mathcal{D}$ relates $X, Y \in \mathcal{D}$ as follows:

(i) $X \, I_\mathcal{D} \, Y$ iff $X \subseteq Y$.

If $f : \mathcal{D}_0 \to \mathcal{D}_1$ and $g : \mathcal{D}_1 \to \mathcal{D}_2$ are given, then the *composition* $g \circ f : \mathcal{D}_0 \to \mathcal{D}_2$ relates $X \in \mathcal{D}_0$ and $Z \in \mathcal{D}_2$ as follows:

(ii) $X \, g \circ f \, Z$ iff $\exists Y \in \mathcal{D}_1 .\ X f Y$ and $Y g Z$.

Proof : (We may use MacLane [1971] as the standard reference on category theory, but we require hardly more than the basic definitions at this stage.) To check that we have a category, we need to know that the identity and composition maps really are maps in the category and that certain identity and associative laws hold. Now it is obvious that $I_\mathcal{D}$ satisfies 2.1 (i)-(iii). Moreover if $f : \mathcal{D}_0 \to \mathcal{D}_1$, all we have to prove is:

$$f \circ I_{\mathcal{D}_0} = I_{\mathcal{D}_1} \circ f = f$$

Checking one of these equations is enough. Thus, for $X \in \mathcal{D}_0$ and $Z \in \mathcal{D}_1$ we find

$$X\, f \circ I_{\mathcal{D}_0}\, Z \text{ iff } \exists Y \in \mathcal{D}_0.\ X \subseteq Y \text{ and } Y\, f\, Z$$
$$\text{iff } X\, f\, Z.$$

So, f and $f \circ I_{\mathcal{D}_0}$ are the same mapping.

Suppose now that $f : \mathcal{D}_0 \to \mathcal{D}_1$ and $g : \mathcal{D}_1 \to \mathcal{D}_2$. We have to verify that $g \circ f$ is an approximable mapping. First off, there is no trouble in seeing that $\Delta_0 g \circ f \Delta_2$ holds. Next, suppose that $X\, g \circ f\, Z$ and $X\, g \circ f\, Z'$ hold. Then we have $X\, f\, Y$ and $Y\, g\, Z$ for some choice of $Y \in \mathcal{D}_1$. Also $X\, f\, Y'$ and $Y'\, g\, Z'$ hold for some choice of $Y' \in \mathcal{D}_1$. By 2.1 (ii) it follows that $X\, f\, (Y \cap Y')$. Since $Y \cap Y' \subseteq Y$, we conclude $(Y \cap Y')\, g\, Z$ by 2.1 (iii); similarly $(Y \cap Y')\, g\, Z'$. Invoking 2.1 (ii) again, we obtain $(Y \cap Y')\, g\, (Z \cap Z')$, and $X\, g \circ f\, (Z \cap Z')$ is proved.

Suppose finally that $X' \subseteq X\, g \circ f\, Z \subseteq Z'$. Now $X\, f\, Y$ and $Y\, g\, Z$ for some $Y \in \mathcal{D}_1$. But then $X'\, f\, Y$ holds; for a similar reason $Y\, g\, Z'$ holds also. Therefore, $X'\, g \circ f\, Z'$ is established, which means that we have checked 2.1 (iii) for $g \circ f$ and have completed the proof that $g \circ f : \mathcal{D}_0 \to \mathcal{D}_1$.

The verification of associativity is a purely logical deduction. Thus suppose that in addition to f and g we have $h : \mathcal{D}_2 \to \mathcal{D}_3$. If $X \in \mathcal{D}_0$ and $W \in \mathcal{D}_3$ we find

$$\begin{aligned} X\, h \circ (g \circ f)\, W \quad & \text{iff } \exists Z \in \mathcal{D}_2.\ X\, g \circ f\, Z \text{ and } Z\, h\, W \\ & \text{iff } \exists Z \in \mathcal{D}_2\ \exists Y \in \mathcal{D}_1.\ X\, f\, Y \text{ and } Y\, g\, Z \text{ and } Z\, h\, W \\ & \text{iff } \exists Y \in \mathcal{D}_1\ \exists Z \in \mathcal{D}_2.\ X\, f\, Y \text{ and } Y\, g\, Z \text{ and } Z\, h\, W \\ & \text{iff } \exists Y \in \mathcal{D}_1.\ X\, f\, Y \text{ and } Y (h \circ g)\, W \\ & \text{iff } X\, (h \circ g) \circ f\, W. \end{aligned}$$

So, as relations, $h \circ (g \circ f) = (h \circ g) \circ f$. □

It may seem as though we have, in the definition of composition, written things backwards. But the reason is that when mappings are taken as elementwise functions, then the order is preserved in expressions involving the usual function value notation. We have, for example:

PROPOSITION 2.6. Given $f : \mathcal{D}_0 \to \mathcal{D}_1$ and $g : \mathcal{D}_1 \to \mathcal{D}_2$, the following equations hold:

(i) $I_{\mathcal{D}_0}(x) = x$, and

(ii) $(g \circ f)(x) = g(f(x))$,

for all $x \in |\mathcal{D}_0|$. □

The proof is not troublesome and is left as an exercise. In technical language the result shows that the category defined in Theorem 2.5 is equivalent to a "concrete category" of sets and functions, namely the domains and elementwise transformations of 2.2.

Toward the end of the last lecture (see 1.9) we promised to show that isomorphisms of domains always come from approximable mappings, and this we now do. It means that the category contains all the isomorphisms it should have.

THEOREM 2.7. Every isomorphism between domains results from an approximable mapping between the neighbourhood systems. Moreover, finite elements are always transformed into finite elements.

Proof: Suppose that $f : |\mathcal{D}_0| \to |\mathcal{D}_1|$ is a one-one, inclusion-preserving function defined on elements, where the range of the function is the whole of $|\mathcal{D}_1|$, of course. Taking the hint from 2.2, there is only one way we could define a neighbourhood mapping; namely, we consider the relation $Y \in f(\uparrow X)$ for $X \in \mathcal{D}_0$ and $Y \in \mathcal{D}_1$. What has to be shown is that this is an approximable mapping which determines the original function via the formula 2.2 (i).

The first part is easy; indeed, there is a general result that monotone functions on finite elements of one domain to arbitrary elements of another domain always determine approximable mappings (cf. Exercise 2.8). What remains, then, is to show that the relation re-defines the function. This comes down to showing that for $x \in |\mathcal{D}_0|$

$$f(x) = \{Y \in \mathcal{D}_1 \mid \exists X \in x .\ Y \in f(\uparrow X)\}.$$

Consider the right-hand side of this equation: it is a filter. (This either can be proved directly or Exercise 2.11 can be used.) Because f is an onto-function, we can call the right-hand side $f(x')$ for some $x' \in |\mathcal{D}_0|$. But since $X \in x$ implies $\uparrow X \subseteq x$ and $f(\uparrow X) \subseteq f(x)$, the right-hand side is included in the left-hand side. In other words $f(x') \subseteq f(x)$. But, since f is an isomorphism $x' \subseteq x$ follows.

In the other direction, if $X \in x$, then $f(\uparrow X) \subseteq f(x')$ holds by definition, so $\uparrow X \subseteq x'$. This implies $X \in x'$; and, as X is arbitrary, $x \subseteq x'$ follows. So $x = x'$, and $f(x) = f(x')$ as desired.

Finally, consider any finite element $\uparrow X \in |\mathcal{D}_0|$ where $X \in \mathcal{D}_0$. What we have to show is that $f(\uparrow X)$ is finite in $|\mathcal{D}_1|$. Because f is an isomorphism, we can associate uniquely to every $Y \in f(\uparrow X)$ an element $y_Y \subseteq \uparrow X$ in $|\mathcal{D}_0|$ where $f(y_Y) = \uparrow Y$. (Just apply the inverse of the function f.) Define

$$z = \bigcup \{y_Y \mid Y \in f(\uparrow X)\}.$$

Because $Y' \subseteq Y$ always implies $y_{Y'} \subseteq y_Y$ and each $y_Y \in |\mathcal{D}_0|$, it is easy to show z is a filter and hence is in $|\mathcal{D}_0|$ also (cf. Exercise 2.11). Because each $y_Y \subseteq \uparrow X$, then $z \subseteq \uparrow X$, too. But each $y_Y \subseteq z$, so $\uparrow Y = f(y_Y) \subseteq f(z)$ and hence $Y \in f(z)$. As this holds for all $Y \in f(\uparrow X)$, the inclusion $f(\uparrow X) \subseteq f(z)$ follows, as well as $\uparrow X \subseteq z$. Therefore, $z = \uparrow X$ and so $X \in z$. But then $X \in y_Y$ for some $Y \in f(\uparrow X)$, by definition of Z. Since $\uparrow X \subseteq y_Y$, we obtain $f(\uparrow X) \subseteq \uparrow Y$. But of course the opposite inclusion is also true from the choice of Y. This means that $f(\uparrow X) = \uparrow Y$ is finite in $|\mathcal{D}_1|$ as claimed. We can apply the same argument to the inverse function; and, thus, the finite elements of $|\mathcal{D}_0|$ and $|\mathcal{D}_1|$ are in a one-one inclusion-preserving correspondence under the isomorphism. □

EXERCISES

EXERCISE 2.8. With reference to the proof of 2.2 show that an approximable mapping is uniquely determined by its elementwise effect on finite elements. Moreover any arbitrary monotone function on finite elements of $|\mathcal{D}_0|$ with values in $|\mathcal{D}_1|$ comes from an approximable $f : \mathcal{D}_0 \to \mathcal{D}_1$.

EXERCISE 2.9. Prove that if $f: \mathcal{D}_0 \to \mathcal{D}_1$ is an approximable mapping, then the elementwise mapping $f: |\mathcal{D}_0| \to |\mathcal{D}_1|$ satisfies the equation

$$f(x) = \bigcup\{f(\uparrow X) \mid X \in x\}$$

for all $x \in |\mathcal{D}_0|$. Conversely, show that every elementwise function satisfying this equation comes from an approximable mapping as defined in 2.2.

EXERCISE 2.10. Carry out the proof of Proposition 2.6; and in addition show that, if $f, g : \mathcal{D}_0 \to \mathcal{D}_1$ are two approximable mappings, there exists $h : \mathcal{D}_0 \to \mathcal{D}_1$ such that

$$h(x) = f(x) \cap g(x)$$

for all $x \in |\mathcal{D}_0|$.

EXERCISE 2.11. Let $\langle I, \leq \rangle$ be a non-empty abstract partially ordered set; suppose it is *directed* in the sense that whenever $i, j \in I$, then $i \leq k$ and $j \leq k$ for some $k \in I$. Suppose that $a: I \to |\mathcal{D}|$ is such that

$$i \leq j \text{ implies } a_i \subseteq a_j$$

for all $i, j \in I$. Prove that

$$\bigcup\{a_i \mid i \in I\}$$

is always a filter in $|\mathcal{D}|$. (Note the ways this lemma could be used in the proof of 2.7; but be careful in defining the partially ordered set and do not confuse $\subseteq$ and $\supseteq$.) In words we could say that the domain of filters is *closed under directed unions*. Prove also that if $f : \mathcal{D} \to \mathcal{D}'$ is an approximable mapping, then for any directed union

$$f(\bigcup\{a_i \mid i \in I\}) = \bigcup\{f(a_i) \mid i \in I\};$$

that is, *approximable mappings always preserve directed unions*. If an elementwise function preserves directed unions, must it come from an approximable mapping? (Hint: Invoke 2.9.)

EXERCISE 2.12. Suppose $\langle I, \leq \rangle$ is a directed, partially ordered set and $f_i : \mathcal{D}_0 \to \mathcal{D}_1$ is a family of approximable mappings indexed by $i \in I$, where we assume

$$i \leq j \text{ implies } f_i(x) \subseteq f_j(x)$$

for all $i, j \in I$ and all $x \in |\mathcal{D}_0|$. Prove that there is an approximable mapping $g : \mathcal{D}_0 \to \mathcal{D}_1$ where

$$g(x) = \bigcup \{f_i(x) \mid i \in I\}$$

for all $x \in |\mathcal{D}_0|$.

EXERCISE 2.13. (For topologists.) Recall Exercise 1.22 where it was shown that any domain $|\mathcal{D}|$ is a topological space. Prove from Exercise 2.9 that the functions $f : |\mathcal{D}_0| \to |\mathcal{D}_1|$ determined by approximable mappings are exactly *the continuous functions between these spaces*. (Hint: To prove continuity, remark that by 2.9

$$f^{-1}[Y] = \bigcup \{[X] \mid Y \in f(\uparrow X)\};$$

hence, the inverse image of any open set is open. In the other direction, suppose that $f : |\mathcal{D}_0| \to |\mathcal{D}_1|$ is topologically continuous. Argue that for all $x \in |\mathcal{D}_0|$ and all open subsets $\mathcal{U} \subseteq |\mathcal{D}_1|$ we have

$$f(x) \in \mathcal{U} \text{ iff } \exists X \in x \,.\, f(\uparrow X) \in \mathcal{U}.$$

This holds because an open subset of $|\mathcal{D}_0|$ is always a union of basic open subsets of the form $[X']$ for $X \in \mathcal{D}_0$ and because

$$x = \bigcup \{\uparrow X \mid X \in x\}$$

for all $x \in |\mathcal{D}_0|$.)

EXERCISE 2.14. Let $f : |\mathcal{D}_0| \to |\mathcal{D}_1|$ be an isomorphism between domains. Let $\varphi : \mathcal{D}_0 \to \mathcal{D}_1$ be the one-one correspondence between neighbourhoods provided by Theorem 2.7 where

$$f(\uparrow X) = \uparrow \varphi(X)$$

for all $X \in \mathcal{D}_0$. Show that the approximable mapping determined by f is just the relationship $\varphi(X) \subseteq Y$. In addition prove that if $X, X' \in \mathcal{D}_0$ are consistent, then

$$\varphi(X \cap X') = \varphi(X) \cap \varphi(X').$$

Remark that the isomorphisms between domains correspond exactly to the isomorphisms between neighbourhood systems (in the sense of one-one inclusion preserving correspondences).

EXERCISE 2.15. (For topologists). Consider the one-token system with

$$\mathcal{O} = \{ \{0\}, \emptyset\}$$

We can regard $|\mathcal{O}|$ as having just two finite elements $\perp$ (bottom) and $\top$ (top), where $\perp \subseteq \top$. For any system $\mathcal{D}$, show that the open subsets $\mathcal{U}$ of $|\mathcal{D}|$ are in a one-one correspondence with the approximable mappings $f : \mathcal{D} \to \mathcal{O}$, where the correspondence is given by the equation

$$\mathcal{U} = \{x \in |\mathcal{D}| \mid f(x) = \top\}.$$

What are the open subsets of $|\mathcal{O}|$? of $|\mathcal{T}|$? of $|\mathcal{B}|$?

EXERCISE 2.16. In the discussion of $\mathcal{B}$ in Chapter 1 we defined a mapping $x \mapsto \sigma x$ for any given $\sigma \in \Sigma^*$. Is this (elementwise) mapping approximable? Show in addition that the mapping $f : \mathcal{B} \to \mathcal{T}$ of 2.3 is uniquely determined among approximable mappings by the equations:

$$f(1x) = \text{true},$$
$$f(01x) = \text{false, and}$$
$$f(00x) = f(x).$$

EXERCISE 2.17. Establish in detail that the mapping $g : \mathcal{B} \to \mathcal{B}$ of Exercise 2.4 is approximable. Is it uniquely determined by these equations:

$$g(0x) = 0g(x),$$
$$g(11x) = g(1x),$$
$$g(10x) = 0x,$$
$$g(1) = \perp,$$

or are some missing?

30

EXERCISE 2.18. What is the meaning in words of the approximable mapping $h : \mathcal{B} \to \mathcal{B}$, where

$$h(0x) = 00h(x), \text{ and}$$
$$h(1x) = 10h(x),$$

for all elements $x \in |\mathcal{B}|$? Is h an isomorphism? Does there exist a map $k : \mathcal{B} \to \mathcal{B}$ where

$$k \circ h = I_{\mathcal{B}},$$

and is k one-one?

EXERCISE 2.19. Generalize Definition 2.1 in an appropriate way in order to define the concept of *an approximable mapping*

$$f : \mathcal{D}_0 \times \mathcal{D}_1 \to \mathcal{D}_2$$

of two variables. (Hint: f can be taken to be a certain kind of ternary relation

$$f \subseteq \mathcal{D}_0 \times \mathcal{D}_1 \times \mathcal{D}_2,$$

where we can write

$$X, Y f Z$$

for the relationship among neighbourhoods.) What is the corresponding version of Proposition 2.2 for functions of two variables?

EXERCISE 2.20. Discuss again the example of Exercise 1.15 where the domain turns out to be the powerset (set of all subsets) of $\mathbb{N}$. Show how the finite elements can be taken to be the finite subsets of $\mathbb{N}$ and can be identified with the tokens of a suitable neighbourhood system $\mathcal{P}$. (Hint: Define $\uparrow F$ for finite sets $F \subseteq \mathbb{N}$.) Show that both union and intersection ($x \cup y$ and $x \cap y$) are functions on $|\mathcal{P}|$ that are approximable in the sense of Exercise 2.19. (The elements of $|\mathcal{P}|$ are being identified with arbitrary sets $x \subseteq \mathbb{N}$.) Show also the following transformations approximable:

$$x + 1 = \{n + 1 \mid n \in x\}, \text{ and}$$
$$x - 1 = \{n \mid n + 1 \in x\}.$$

EXERCISE 2.21. The system $\mathcal{B}$ of 2.3 has as its total elements only the infinite sequences. Modify the construction of $\mathcal{B}$ to another neighbourhood system C which has *both* the finite and infinite sequences as total elements. (Hint: $\mathcal{B} \subseteq C$.) Show that there is an approximable map $x\,y$ on elements naturally extending ordinary juxtaposition of sequences. (Hint: Write 01001 for a *total* finite sequence and 01001⊥ for the corresponding finite partial element. Remember to distinguish between Λ (the total empty sequence) and ⊥ (the undefined sequence). The definition should work out so that if x is an infinite sequence (hence, total), then $x\,y = x$ for all y. What will $x\,y$ equal if x is not total? In other words, the construction possesses a rather strong left-to-right bias.)

EXERCISE 2.22. (For set theorists). We have remarked in Exercise 1.18 and in Exercise 2.11 that any domain $|\mathcal{D}|$, as a family of sets (in fact, a family of subsets of the set $\mathcal{D}$ itself), is closed under the intersection of an arbitrary non-empty sub family and under the union of any directed sub family. For those familiar with the subject matter, the example of the (proper) ideals of a commutative ring (with unit) is also seen to be such a family. What is the abstract situation? Let $\mathbf{C}$ be *any* family of sets with these closure properties. It is to be shown that $\mathbf{C}$ is inclusion-isomorphic to a domain. (Hint: Let Δ be the set of finite sets included in sets in $\mathbf{C}$. For $F \in \Delta$, define its "closure" by the equation:

$$\overline{F} = \bigcap\{X \in \mathbf{C} \mid F \subseteq X\}.$$

Every $\overline{F} \in \mathbf{C}$, and these will prove to be the "finite" elements of C. The neighbourhood system $\mathcal{D}$ over Δ can be taken to be the sets of the form

$$C(F) = \{G \in \Delta \mid F \subseteq \overline{G}\}$$

for $F \in \Delta$. Notice that for all $X \in \mathbf{C}$

$$X = \bigcup\{\overline{F} \mid F \subseteq X \text{ and } F \in \Delta\}.)$$

Check that approximable functions on these families are just those preserving directed unions.

LECTURE III

DOMAIN CONSTRUCTS

Having now seen a number of domains presented through their neighbourhood systems, we need next to introduce general constructs for forming new domains from old. There are an unlimited number of such constructs (technically called *functors*), but we have time only to single out a few of the more important ones. Outstanding among all of them is the notion of product of systems, which in our chosen category has all the expected properties. For the time being in order to simplify notation we assume of the underlying sets Δ_0 and Δ_1 of systems $\mathcal{D}_0$ and $\mathcal{D}_1$ that they are disjoint. There is no loss of generality as $\mathcal{D}_1$ can always be replaced by an isomorphic system disjoint from $\mathcal{D}_0$ in the required sense.

DEFINITION 3.1. Let neighbourhood systems $\mathcal{D}_0$ and $\mathcal{D}_1$ be given over disjoint sets Δ_0 and Δ_1. The *product system* over $\Delta_0 \cup \Delta_1$ is defined by:

$$\mathcal{D}_0 \times \mathcal{D}_1 = \{X \cup Y \mid X \in \mathcal{D}_0 \text{ and } Y \in \mathcal{D}_1\}.$$

For elements $x \in |\mathcal{D}_0|$ and $y \in |\mathcal{D}_1|$ we also define:

$$\langle x,y\rangle = \{X \cup Y \mid X \in x \text{ and } Y \in y\}. \quad \square$$

PROPOSITION 3.2. The construct $\mathcal{D}_0 \times \mathcal{D}_1$ always gives a neighbourhood system where for elements $x,x' \in |\mathcal{D}_0|$ and $y,y' \in |\mathcal{D}_1|$ we have

(i) $\langle x,y\rangle \subseteq \langle x',y'\rangle$ iff $x \subseteq x'$ and $y \subseteq y'$.

Moreover, there is a one-one correspondence between the elements of $|\mathcal{D}_0 \times \mathcal{D}_1|$ and pairs of elements of $|\mathcal{D}_0|$ and $|\mathcal{D}_1|$ since all elements of $|\mathcal{D}_0 \times \mathcal{D}_1|$ are of the form $\langle x,y\rangle$.

Proof: Owing to the disjointness of Δ_0 and Δ_1, we note that for $X, X' \in \mathcal{D}_0$ and $Y, Y' \in \mathcal{D}_1$ we have

(1) $X \cup Y \subseteq X' \cup Y'$ iff $X \subseteq X'$ and $Y \subseteq Y'$.

Thus, $\{X \cup Y, X' \cup Y'\}$ is consistent in $\mathcal{D}_0 \times \mathcal{D}_1$ iff $\{X, X'\}$ is consistent in $\mathcal{D}_0$ and $\{Y, Y'\}$ is consistent in $\mathcal{D}_1$. In the consistent case we find

(2) $$(X \cup Y) \cap (X' \cup Y') = (X \cap X') \cup (Y \cap Y'),$$

and so $\mathcal{D}_0 \times \mathcal{D}_1$ is closed under consistent intersection. As $\Delta_0 \cup \Delta_1 \in \mathcal{D}_0 \times \mathcal{D}_1$, it is certainly a neighbourhood system.

It is easy to check by the previous calculations that $\langle x, y\rangle \in |\mathcal{D}_0 \times \mathcal{D}_1|$ if $x \in |\mathcal{D}_0|$ and $y \in |\mathcal{D}_1|$. The proof of 3.2(i) follows directly from the definition and (1).

Suppose $z \in |\mathcal{D}_0 \times \mathcal{D}_1|$. Define as a temporary notation:

$$z_0 = \{X \in \mathcal{D}_0 \mid X \cup \Delta_1 \in z\}, \text{ and}$$

$$z_1 = \{Y \in \mathcal{D}_1 \mid \Delta_0 \cup Y \in z\}.$$

Clearly, both $z_0 \in |\mathcal{D}_0|$ and $z_1 \in |\mathcal{D}_1|$. In view of the formula

(3) $$(X \cup \Delta_1) \cap (\Delta_0 \cup Y) = X \cup Y,$$

we can calculate that

$$z = \langle z_0, z_1\rangle.$$

Moreover, if $z = \langle x, y\rangle$, then

$$\langle x, y\rangle_0 = x \text{ and } \langle x, y\rangle_1 = y.$$

The one-one correspondence required is thus established. □

There is more going on in the proof of 3.2 than just a one-one correspondence between elements and pairs. The extra information is best formalized by introducing a notation for mappings.

DEFINITION 3.3. *Projection mappings*

$$p_0 : \mathcal{D}_0 \times \mathcal{D}_1 \to \mathcal{D}_0 \text{ and } p_1 : \mathcal{D}_0 \times \mathcal{D}_1 \to \mathcal{D}_1$$

are defined as relations where

$$(X \cup Y)\ p_0\ X' \text{ iff } X \subseteq X', \text{ and } (X \cup Y)\ p_1\ Y' \text{ iff } Y \subseteq Y'$$

hold for all $X, X' \in \mathcal{D}_0$ and $Y, Y' \in \mathcal{D}_1$. Given $f : \mathcal{D}_2 \to \mathcal{D}_0$ and $g : \mathcal{D}_2 \to \mathcal{D}_1$, the *paired mapping*

$$\langle f, g\rangle : \mathcal{D}_2 \to \mathcal{D}_0 \times \mathcal{D}_1$$

is defined as a relation where

$$Z \langle f, g\rangle (X \cup Y) \text{ iff } Z\ f\ X \text{ and } Z\ g\ Y$$

holds for all $X \in \mathcal{D}_0$, $Y \in \mathcal{D}_1$, and $Z \in \mathcal{D}_2$. □

PROPOSITION 3.4. The mappings p_0, p_1 and $\langle f,g \rangle$ are approximable mappings, provided f and g are, and we have:

(i) $p_0 \circ \langle f, g \rangle = f$ and $p_1 \circ \langle f, g \rangle = g$.

Moreover, for $z \in |\mathcal{D}_0 \times \mathcal{D}_1|$, we have:

(ii) $p_0(z) = z_0$ and $p_1(z) = z_1$,

in the notation of the proof of 3.2. Further if $h : \mathcal{D}_2 \to \mathcal{D}_0 \times \mathcal{D}_1$ is any approximable mapping, then

(iii) $h = \langle p_0 \circ h, p_1 \circ h \rangle$.

Moreover, for all $w \in |\mathcal{D}_2|$, we have:

(iv) $\langle f,g \rangle (w) = \langle f(w), g(w) \rangle$,

where again on the right-hand side the notation of the proof of 3.2 is used. □

The proof of this result is left as an exercise. Note the consequence that there is a one-one correspondence between pairs of approximable mappings $f : \mathcal{D}_2 \to \mathcal{D}_0$ and $g : \mathcal{D}_2 \to \mathcal{D}_1$ and mappings $h : \mathcal{D}_2 \to \mathcal{D}_0 \times \mathcal{D}_1$. It is clear that we generalize all this to products

$$\mathcal{D}_0 \times \mathcal{D}_1 \times \cdots \times \mathcal{D}_{n-1}$$

of several systems.

The product construct also neatly explains functions of several variables. In Exercise 2.19 we used the informal notation

$$f : \mathcal{D}_0 \times \mathcal{D}_1 \to \mathcal{D}_2$$

and suggested regarding f as a ternary relation

$$X, Y \, f \, Z \, .$$

But now with $\mathcal{D}_0 \times \mathcal{D}_1$ given an independent meaning, all we have to do is to regard f as a binary relation with

$$(X \cup Y) \, f \, Z$$

equivalent to the old relationship. We can also employ an elementwise notation as in $f(\langle x, y \rangle)$, which can more easily be written $f(x, y)$. Similar remarks apply to functions of more than two arguments.

We have discussed several times what it means for a function $f(x)$ to come from an approximable mapping. It is interesting to ask the analogous question for functions of several arguments.

THEOREM 3.5. An elementwise function

$$f : |\mathcal{D}_0 \times \mathcal{D}_1| \to |\mathcal{D}_2|$$

of two arguments comes from an approximable mapping iff for each fixed $a \in |\mathcal{D}_0|$ and each fixed $b \in |\mathcal{D}_1|$ the transformations

$$x \mapsto f(x, b) \text{ and } y \mapsto f(a, y)$$

come from approximable mappings of one argument.

Proof: As this is the first time we have had to deal with constants in functions, a lemma is useful.

LEMMA 3.6. Given $b \in |\mathcal{D}_1|$, the constant function

$$b : |\mathcal{D}_0| \to |\mathcal{D}_1|$$

where $b(x) = b$ for all $x \in |\mathcal{D}_0|$, comes from the approximable mapping such that

$$X\, b\, Y, \text{ iff } Y \in b,$$

for all $X \in \mathcal{D}_0$ and $Y \in \mathcal{D}_1$. □

There is no real confusion here in using "b" both for function and value. Returning, then, to the proof of 3.5, we see that the reason that $x \mapsto f(x, b)$ comes from an approximable mapping is that the mapping in question is the composition of two approximable mappings, namely $f \circ \langle I_{\mathcal{D}_0}, b\rangle$. Clearly we can interchange the rôles of $\mathcal{D}_0$ and $\mathcal{D}_1$ to get at $y \mapsto f(a, y)$.

Conversely, assume that both these functions come from approximable mappings no matter the choice of a and b. Clearly the mapping to determine f is the relation from $X \cup Y$ to Z where

$$Z \in f(\uparrow X, \uparrow Y) = f(\uparrow(X \cup Y)).$$

To prove that this determines f we calculate by the formula of Exercise 2.9:

$$\begin{aligned} f(x,y) &= \bigcup\{f(\uparrow X, y) \mid X \in x\} \\ &= \bigcup\{\bigcup\{f(\uparrow X, \uparrow Y) \mid Y \in y\} \mid X \in x\} \\ &= \bigcup\{f(\uparrow X, \uparrow Y) \mid X \in x \text{ and } Y \in y\} \\ &= \bigcup\{f(\uparrow(X \cup Y)) \mid (X \cup Y) \in \langle x, y\rangle\}. \end{aligned}$$

And, again by 2.9, this is what was needed. □

Said more informally, a function of several arguments is approximable in all the variables *jointly* if it is approximable in each of the variables *separately*.

The type of argument used in 3.5 in the first half of the proof also provides a generalization of 2.6 to functions of several arguments. When we form a function like

$$f(g(x, z, \ldots), h(y, x, \ldots), k(z, w, \ldots), \ldots)$$

from given functions $f, g, h, k, \ldots$; we call the process *substitution*.

PROPOSITION 3.7. The functions of several arguments between domains coming from approximable mappings are closed under substitution.

Proof: An example will establish the method. Suppose there are four variables involved taking values in domains provided by systems $\mathcal{D}_0, \mathcal{D}_1, \mathcal{D}_2, \mathcal{D}_3$. We might have a substitution like:

$$f(g(x_0, x_1), h(x_1, x_2), k(x_3, x_0, x_2)).$$

Here it might be that the values of the functions inside come from quite other systems; for instance,

$$k : \mathcal{D}_3 \times \mathcal{D}_0 \times \mathcal{D}_2 \rightarrow \mathcal{D}_4$$

might be possible. By using projections

$$p_i : \mathcal{D}_0 \times \mathcal{D}_1 \times \mathcal{D}_2 \times \mathcal{D}_3 \rightarrow \mathcal{D}_i,$$

where $i < 4$, we can assure that we have several functions all on the same product; thus,

$$k \circ \langle p_3, p_0, p_2\rangle : \mathcal{D}_0 \times \mathcal{D}_1 \times \mathcal{D}_2 \times \mathcal{D}_3 \rightarrow \mathcal{D}_4.$$

Now no matter on what domains f is defined, the following composition makes sense:

$f \circ \langle g \circ \langle p_0, p_1 \rangle, h \circ \langle p_1, p_2 \rangle, k \circ \langle p_3, p_0, p_2 \rangle\rangle$;

and in fact this is the desired function. Writing it this way makes it clear that the function comes from an approximable mapping: we apply 3.3 (generalized, of course, to products with several terms) to construe the parts between brackets < and > as approximable mappings, and then by this trick the composition $\circ$ is the ordinary composition of 2.6. □

It has to be admitted that there is a slight point overlooked in forming products like $\mathcal{D} \times \mathcal{D}$ with two identical domains. This is discussed in Exercise 3.14, invoking explicit isomorphisms.

The construct that makes the whole theory of domains work so smoothly is the function - space construct: it is possible to regard functions as *objects* which form a domain. Look back at Definition 2.1 and compare it with the original definition of element in 1.6. There are obvious formal similarities, except that filters are sets of neighbourhoods and mappings are sets of pairs of neighbourhoods (relations). But as we saw in 1.10 it is possible to turn the filters into tokens *via* a simple definition of neighbourhood. We apply the same kind of definition to the mappings.

DEFINITION 3.8. Given neighbourhood systems $\mathcal{D}_0$ and $\mathcal{D}_1$, the *function space* $(\mathcal{D}_0 \to \mathcal{D}_1)$ is the system whose set of tokens is the set of approximable mappings of Definition 2.1 and whose neighbourhoods are finite non-empty intersections of sets of the form

$$[X, Y] = \{f : \mathcal{D}_0 \to \mathcal{D}_1 \mid X f Y\},$$

where $X \in \mathcal{D}_0$ and $Y \in \mathcal{D}_1$. □

We have been calling our mappings "approximable" for a long time now without saying exactly how they can be approximated! Definition 3.8 supplies the missing key, because once a domain has been defined, then the general theory gives an explicit meaning to the word approximation. We still have to verify, however, that the mappings do correspond to the elements of the domain.

PROPOSITION 3.9. Let neighbourhoods $X_i \in \mathcal{D}_0$ and $Y_i \in \mathcal{D}_1$ be given for $i < n$. Then the set of $[X_i, Y_i]$ for $i < n$ is consistent in $(\mathcal{D}_0 \to \mathcal{D}_1)$ iff the following condition holds:

(i) whenever $I \subseteq \{0, 1, \ldots, n-1\}$ and $\{X_i \mid i \in I\}$ is consistent in $\mathcal{D}_0$, then $\{Y_i \mid i \in I\}$ must be consistent in $\mathcal{D}_1$.

Moreover, when consistency holds, the least approximable mapping f_0 belonging to the intersection of the $[X_i, Y_i]$ is defined by:

(ii) $X f_0 Y$ iff $\bigcap\{Y_i \mid X \subseteq X_i\} \subseteq Y$

for $X \in \mathcal{D}_0$ and $Y \in \mathcal{D}_1$.

Proof: Suppose the $[X_i, Y_i]$ are consistent in $(\mathcal{D}_0 \to \mathcal{D}_1)$. Since the function space is being defined outright as a positive system, consistency means

$$f \in \bigcap\{[X_i, Y_i] \mid i < n\}$$

for some $f : \mathcal{D}_0 \to \mathcal{D}_1$. Now, with f in hand, let us check condition (i). Suppose $\{X_i \mid i \in I\}$ is consistent. This means

$$x \in \bigcap\{[X_i] \mid i \in I\}$$

for some $x \in |\mathcal{D}_0|$. Suppose $i \in I$, so $x \in [X_i]$. Since $X_i f Y_i$ holds, $f(x) \in [Y_i]$. This means, therefore, that

$$f(x) \in \bigcap\{[Y_i] \mid i \in I\},$$

and so $\{Y_i \mid i \in I\}$ is consistent.

For the converse, suppose (i) is the case. We take (ii) as the definition of a mapping and remark that for an arbitrary $X \in \mathcal{D}_0$, the set $\{X_i \mid X \subseteq X_i\}$ is automatically consistent in $\mathcal{D}_0$. By our assumption, the set $\{Y_i \mid X \subseteq X_i\}$ is therefore consistent. This means that

$$\bigcap\{Y_i \mid X \subseteq Y_i\} \in \mathcal{D}_1 .$$

(Keep in mind that i is restricted to those $i < n$, and there are only finitely many neighbourhoods being considered here.) It is thus almost immediate that the relation f_0 defined by (ii) satisfies conditions of 2.1 and so is an approximable mapping $f_0 : \mathcal{D}_0 \to \mathcal{D}_1$. By construction

$$X_i \; f_0 \; Y_i$$

holds trivially for all $i < n$; therefore,

$$f_0 \in \bigcap \{[X_i, Y_i] \mid i < n\}$$

and the desired consistency is established.

Finally suppose that f is any mapping in the neighbourhood under discussion; this means $X_i \, f \, Y_i$ holds for all $i < n$. Suppose $X \, f_0 \, Y$ holds. We have for $X \subseteq X_i$, $X \, f \, Y_i$; so

$$X \, f \, \bigcap \{Y_i \mid X \subseteq X_i\} \subseteq Y.$$

Thus, $X \, f \, Y$ follows; hence, as relations, $f_0 \subseteq f$. In other words f_0 is the minimal element of the neighbourhood. □

We note that, as a consequence of what we have just proved, when the neighbourhood is consistent, then

$$\bigcap \{[X_i, Y_i] \mid i < n\} \subseteq [X, Y]$$

is exactly equivalent to

$$\bigcap \{Y_i \mid X \subseteq X_i\} \subseteq Y.$$

Note also that a single neighbourhood $[X_0, Y_0]$ is always consistent since it contains the *constant mapping* k where

$$X \, k \, Y \quad \text{iff} \quad Y_0 \subseteq Y,$$

for all $X \in \mathcal{D}_0$ and $Y \in \mathcal{D}_1$. Some other simple observations about these neighbourhoods are just translations of the conditions of Definition 2.1:

$$[\Delta_0, \Delta_1] = |\mathcal{D}_0 \to \mathcal{D}_1|;$$

$$[X, Y] \cap [X, Y'] = [X, Y \cap Y']; \text{ and}$$

$$X' \subseteq X \text{ and } Y \subseteq Y' \text{ imply } [X, Y] \subseteq [X', Y'],$$

for all $X, X' \in \mathcal{D}_0$ and $Y, Y' \in \mathcal{D}_1$. We are now ready to prove the main result about the construct.

THEOREM 3.10. Given neighbourhood systems $\mathcal{D}_0$ and $\mathcal{D}_1$, the function space system $(\mathcal{D}_0 \to \mathcal{D}_1)$ is complete in the sense that every filter in $|\mathcal{D}_0 \to \mathcal{D}_1|$ is fixed by a unique approximable mapping.

Proof: Let $f : \mathcal{D}_0 \to \mathcal{D}_1$ be an approximable mapping. By the very definition of $(\mathcal{D}_0 \to \mathcal{D}_1)$ it determines a filter by the definition:

$$\hat{f} = \{F \in (\mathcal{D}_0 \to \mathcal{D}_1) \mid f \in F\}.$$

Trivially $[X,Y] \in \hat{f}$ iff $f \in [X,Y]$ iff $X f Y$; so this filter uniquely determines the relation f. What we have to show is that every filter in $|\mathcal{D}_0 \to \mathcal{D}_1|$ is of this form.

Suppose $\varphi \in |\mathcal{D}_0 \to \mathcal{D}_1|$ is any filter. A relation can be defined at once by

$$X \hat{\varphi} Y \text{ iff } [X,Y] \in \varphi.$$

In view of the remarks we made just before stating this theorem, there is no problem in showing that $\hat{\varphi}$ is an approximable mapping. Since the neighbourhoods of the function space are in any case finite intersections of sets like $[X,Y]$, those $[X,Y] \in \varphi$ generate φ. This means that $\hat{\hat{\varphi}} = \varphi$. By definition $\hat{\hat{f}} = f$, so there is a one-one correspondence between mappings and filters. (This correspondence is obviously inclusion preserving, too.) □

We now know just about everything about $|\mathcal{D}_0 \to \mathcal{D}_1|$ as a domain: the elements correspond isomorphically to the approximable mappings; the finite elements are explained completely by 3.9; and we have seen how to calculate with neighbourhoods. The final step is to relate the function space to other domains by appropriate mappings. In doing this we shall freely construe elements of $|\mathcal{D}_0 \to \mathcal{D}_1|$ as approximable mappings in view of 3.10.

THEOREM 3.11. Given neighbourhood systems $\mathcal{D}_1$ and $\mathcal{D}_2$, there is a uniquely determined approximable mapping

$$\mathrm{eval} : (\mathcal{D}_1 \to \mathcal{D}_2) \times \mathcal{D}_1 \to \mathcal{D}_2 ,$$

where for all $f : \mathcal{D}_1 \to \mathcal{D}_2$ and all $x \in |\mathcal{D}_1|$ we have

(i) $\mathrm{eval}(f, x) = f(x)$.

Proof: For $F \in (\mathcal{D}_1 \to \mathcal{D}_2)$ and $X \in \mathcal{D}_1$ and $Y \in \mathcal{D}_2$ define eval as a relation by:

$$F \cup X \ \mathrm{eval}\ Y \text{ iff } X f Y \text{ for all } f \in F.$$

Remember that neighbourhoods in the function space are sets of approximable mappings. It is easily checked that this definition makes eval approximable. We now calculate the function values by the formula of 2.2 (i):

$$\text{eval}\ (f,\ x) = \{Y \in \mathcal{D}_2 \mid \exists F \in (\mathcal{D}_1 \to \mathcal{D}_2)\ \exists X \in x.\ f \in F \text{ and } F \cup X \text{ eval } Y\}.$$

Because, again by 2.2 (i), we have

$$f\ (x) = \{Y \in \mathcal{D}_2 \mid \exists X \in x\ .\ X f Y\},$$

we can see from the definition of eval that eval $(f, x) \subseteq f\ (x)$. Suppose that $Y \in f(x)$. Then $X f Y$ holds for some $X \in x$. We can write $f \in [X, Y] \in (\mathcal{D}_1 \to \mathcal{D}_2)$ and it is clear that

$$[X, Y] \cup X \text{ eval } Y$$

holds by definition. Therefore, $Y \in$ eval (f, x), and so $f(x) \subseteq$ eval (f, x). □

This theorem is essential for our programme: it shows that in taking functions as objects the very basic operation of forming the function value is an approximable mapping. In other words we can treat the expression $f(x)$ not just as a function of x, as we have done from the start, but also as a function of f as well. The result also indicates that there are useful maps defined on domains that themselves are function spaces; we shall meet many more of these. The next theorem provides further examples.

THEOREM 3.12. Given neighbourhood systems $\mathcal{D}_0$, $\mathcal{D}_1$, $\mathcal{D}_2$ there is associated with every approximable mapping $g : \mathcal{D}_0 \times \mathcal{D}_1 \to \mathcal{D}_2$ a uniquely determined approximable mapping

$$\text{curry}\ (g) : \mathcal{D}_0 \to (\mathcal{D}_1 \to \mathcal{D}_2)$$

such that for $x \in |\mathcal{D}_0|$ and $y \in |\mathcal{D}_1|$

(i) curry $(g)(x)(y) = g(x,y)$.

Moreover we have these functional equations:

(ii) eval $\circ \langle$ curry $(g) \circ p_0,\ p_1\rangle = g$, and

(iii) curry (eval $\circ \langle h \circ p_0,\ p_1\rangle) = h$, where the $p_i : \mathcal{D}_0 \times \mathcal{D}_1 \to \mathcal{D}_i$ are the projection mappings and $h : \mathcal{D}_0 \to (\mathcal{D}_1 \to \mathcal{D}_2)$ is any approximable mapping. This provides an isomorphism between the domains $|\mathcal{D}_0 \times \mathcal{D}_1 \to \mathcal{D}_2|$ and $|\mathcal{D}_0 \to (\mathcal{D}_1 \to \mathcal{D}_2)|$ and so we can regard

$$\text{curry} : (\mathcal{D}_0 \times \mathcal{D}_1 \to \mathcal{D}_2) \to (\mathcal{D}_0 \to (\mathcal{D}_1 \to \mathcal{D}_2))$$

as itself being an approximable mapping.

Proof : Given g as indicated, we can define curry (g) as a relation and as an approximable mapping by:

$$X \text{ curry } (g) \ [Y,Z] \ \text{ iff } X \cup Y \, g \, Z \qquad \text{(but see Ex. 3.21)}$$

for all $X \in \mathcal{D}_0$, $Y \in \mathcal{D}_1$, $Z \in \mathcal{D}_2$. This is sufficient because an approximable mapping is intersective in the right-hand neighbourhood, so we know from the above exactly what $X \text{ curry}(g) \bigcap \{[Y_i, Z_i] \mid i < n\}$ means for all finite intersections. The remark after 3.9 is then helpful in checking that by this definition curry (g) satisfies the monotonicity condition and so is indeed approximable. We now calculate :

$$\begin{aligned} \text{curry } (g) \ (x) \ (y) &= \{Z \in \mathcal{D}_2 \mid \exists Y \in y \,.\, Y \text{ curry } (g) \ (x) \ Z\} \\ &= \{Z \in \mathcal{D}_2 \mid \exists Y \in y \ \exists X \in x.\ X \text{ curry}(g)[Y,Z]\} \\ &= \{Z \in \mathcal{D}_2 \mid \exists Y \in y \ \exists X \in x.\ X \cup Y \, g \, Z\} \\ &= \{Z \in \mathcal{D}_2 \mid \exists W \in \langle x,y\rangle.\ W \, g \, Z\} \\ &= g(\langle x,y\rangle) = g(x,y). \end{aligned}$$

This proves (i). We also see, that if we take the left-hand side of (ii) and apply it to a pair $\langle x,y\rangle$, it reduces to $g(x,y)$ by virtue of (i). Thus, the two functions in (ii) are the same.

Turning to (iii), call the left-hand side k. Using (i) again, we find

$$\begin{aligned} k(x)(y) &= \text{eval} \circ \langle h \circ p_0, p_1\rangle \ (\langle x,y\rangle) \\ &= \text{eval} \ (\langle h \circ p_0 \ (\langle x,y\rangle), p_1(\langle x,y\rangle)\rangle) \\ &= \text{eval} \ (\langle h(x), y\rangle) \\ &= h(x)(y). \end{aligned}$$

As this is true for all $y \in |\mathcal{D}_1|$, then $k(x) = h(x)$ follows. As this is true for all $x \in |\mathcal{D}_0|$, then $k = h$ follows, and (iii) is proved.

Taking (ii) and (iii) together, it is clear that the domains $|\mathcal{D}_0 \times \mathcal{D}_1 \to \mathcal{D}_2|$ and $|\mathcal{D}_0 \to (\mathcal{D}_1 \to \mathcal{D}_2)|$ are in a one-one correspondence. But from the very definition of curry it is clear that

$$\text{curry}\ (g) \subseteq \text{curry}\ (g') \text{ iff } g \subseteq g'.$$

Hence, curry is an isomorphism, and we can invoke 2.7 to conclude that it comes from an approximable mapping. □

We close this lecture with some order-theoretic properties of function spaces that characterize inclusion and upper bounds of functions in a "pointwise" manner.

THEOREM 3.13. For approximable functions $f, g : \mathcal{D}_0 \to \mathcal{D}_1$ we have

(i) $f \subseteq g$ iff $f(x) \subseteq g(x)$ for all $x \in |\mathcal{D}_0|$.

For subsets $F \subseteq |\mathcal{D}_0 \to \mathcal{D}_1|$ we have

(ii) F is bounded in $|\mathcal{D}_0 \to \mathcal{D}_1|$ iff $\{f(x) \mid f \in F\}$ is bounded in $|\mathcal{D}_1|$ for each $x \in |\mathcal{D}_0|$;

and in that case for all $x \in |\mathcal{D}_0|$:

(iii) $(\bigsqcup F)(x) = \bigsqcup \{f(x) \mid f \in F\}$.

Proof. The implication in (i) from left to right follows because evaluation is monotone in the function as well as the argument. The converse implication is a consequence of 2.2(ii).

For the proof of (ii) and (iii) we see that by (i) if F is bounded, so is every set $\{f(x) \mid f \in F\}$. For the converse direction, it is clear that (iii) defines *some* pointwise mapping; we have only to prove that it is *approximable*. The calculation that $\bigsqcup F$ preserves directed unions (see 2.9 and 2.11) is probably the simplest way to reach the conclusion. □

EXERCISES

EXERCISE 3.14. For the most part we can assume that there is at most a *countable number* of tokens; thus, without loss of generality the underlying sets Δ_i of given systems $\mathcal{D}_i$ could be assumed to be subsets of Σ^* where $\Sigma = \{0,1\}$. (Any denumerable set would do.) Show that the product $\mathcal{D}_0 \times \mathcal{D}_1$ could be defined as the system over the set $0\Delta_0 \cup 1\Delta_1$ where

$$\mathcal{D}_0 \times \mathcal{D}_1 = \{0X \cup 1Y \mid X \in \mathcal{D}_0 \text{ and } Y \in \mathcal{D}_1\}.$$

In other words, the assumption of the disjointness of Δ_0 and Δ_1 is unnecessary. Give, therefore, the revised definition of $\langle x,y \rangle$ for elements, and prove that for a single system $\mathcal{D}$, there exists an approximable mapping

$$\mathrm{diag} : \mathcal{D} \to \mathcal{D} \times \mathcal{D}$$

where $\mathrm{diag}(x) = \langle x,x \rangle$ for all $x \in |\mathcal{D}|$. Also extend the definition to a product of n-factors

$$\mathcal{D}_0 \times \mathcal{D}_1 \times \cdots \times \mathcal{D}_{n-1}$$

which will be a system over the set

$$\bigcup_{i<n} 1^i 0\Delta_i$$

Note that for a 2-termed product we simplify $10\Delta_1$ to $1\Delta_1$.

EXERCISE 3.15. Establish the usual isomorphisms:

(i) $\mathcal{D}_0 \times \mathcal{D}_1 \cong \mathcal{D}_1 \times \mathcal{D}_0$;

(ii) $\mathcal{D}_0 \times (\mathcal{D}_1 \times \mathcal{D}_2) \cong (\mathcal{D}_0 \times \mathcal{D}_1) \times \mathcal{D}_2 \cong \mathcal{D}_0 \times \mathcal{D}_1 \times \mathcal{D}_2$.

How does the product of no factors fit in? Prove also:

(iii) $\mathcal{D}_0 \cong \mathcal{D}'_0$ and $\mathcal{D}_1 \cong \mathcal{D}'_1$ imply $\mathcal{D}_0 \times \mathcal{D}_1 \cong \mathcal{D}'_0 \times \mathcal{D}'_1$.

EXERCISE 3.16. Let $\mathcal{D}$ be a given neighbourhood system over $\Delta \subseteq \Sigma^*$. Define

$$\Delta^\infty = \bigcup_{n=0}^{\infty} 1^n 0\, \Delta$$

so that Δ^∞ is split into infinitely many disjoint copies of Δ. Let $\mathcal{D}^\infty$ be the least family of subsets of Σ^* where

(1) $\Delta^\infty \in \mathcal{D}^\infty$, and

(2) whenever $X \in \mathcal{D}$ and $Y \in \mathcal{D}^\infty$, then $0X \cup 1Y \in \mathcal{D}^\infty$.

Show that $\mathcal{D}^\infty$ is a neighbourhood system over Δ^∞. Prove the isomorphism

$$\mathcal{D}^\infty \cong \mathcal{D} \times \mathcal{D}^\infty .$$

Show, moreover, that the elements of $|\mathcal{D}^\infty|$ are in a one-one correspondence with arbitrary *infinite sequences* $\langle x_n \rangle_{n=0}^{\infty}$ of elements $x_n \in |\mathcal{D}|$ by using combinations of neighbourhoods

$$0X_0 \cup 10X_1 \cup \cdots \cup 1^n 0X_n \cup \cdots$$

where from some point on all the X_m are equal to Δ.

EXERCISE 3.17. Using the $\mathcal{B}$ and $\mathcal{T}$ of Example 2.3 show there is a one-one approximable mapping

$$f : \mathcal{B} \to \mathcal{T}^\infty$$

and another approximable mapping

$$g : \mathcal{T}^\infty \to \mathcal{B}$$

such that

$$g \circ f = I_{\mathcal{B}} \text{ and } f \circ g \subseteq I_y .$$

Are $\mathcal{B}$ and $\mathcal{T}^\infty$ isomorphic? Are $\mathcal{B}$ and $\mathcal{T} \times \mathcal{B}$ isomorphic?

EXERCISE 3.18. Let $\mathcal{D}_0$ and $\mathcal{D}_1$ be neighbourhood systems over Δ_0 and Δ_1, where we again assume that these are subsets of Σ^*. We assume that in addition *no neighbourhood is empty*. Why is this possible without loss of generality? Define the *sum system* by:

$$\mathcal{D}_0 + \mathcal{D}_1 = \{\{\Lambda\} \cup 0\Delta_0 \cup 1\Delta_1\} \cup \{0X \mid X \in \mathcal{D}_0\} \cup \{1Y \mid Y \in \mathcal{D}_1\}.$$

Prove that this is a neighbourhood system over $\{\Lambda\} \cup 0\Delta_0 \cup 1\Delta_1$. (Throwing in $\{\Lambda\}$ was not all that necessary, but note that

$$\mathcal{B} = \mathcal{B} + \mathcal{B} ,$$

and this is an equality of sets not just an isomorphism of systems.) Prove that in general there are mappings

$$\mathrm{in}_i : \mathcal{D}_i \to \mathcal{D}_0 + \mathcal{D}_1 \quad \text{and} \quad \mathrm{out}_i : \mathcal{D}_0 + \mathcal{D}_1 \to \mathcal{D}_i$$

where $\mathrm{out}_i \circ \mathrm{in}_i = I_{\mathcal{D}_i}$. Where does the assumption $\emptyset \notin \mathcal{D}_i$ come in here? How can these sums be generalized to n-terms? (Hint: As for products use sets $1^i0\Delta_i$.) Draw some pictures.

EXERCISE 3.19. Suppose we are given systems and approximable mappings

$$f : \mathcal{D}_0 \to \mathcal{D}_0' \quad \text{and} \quad g : \mathcal{D}_1 \to \mathcal{D}_1' .$$

Prove there are approximable mappings

$$f \times g : \mathcal{D}_0 \times \mathcal{D}_1 \to \mathcal{D}_0' \times \mathcal{D}_1' \quad \text{and} \quad f + g : \mathcal{D}_0 + \mathcal{D}_1 \to \mathcal{D}_0' + \mathcal{D}_1'$$

such that

(i) $(f \times g)(x,y) = \langle f(x), g(y) \rangle$

for all $x \in |\mathcal{D}_0|$ and $y \in |\mathcal{D}_1|$, and rewrite this as:

(ii) $f \times g = \langle f \circ p_0, g \circ p_1 \rangle$.

In addition prove that

(iii) $\mathrm{out}_0 \circ (f + g) \circ \mathrm{in}_0 = f$, and

(iv) $\mathrm{out}_1 \circ (f + g) \circ \mathrm{in}_1 = g$.

Do equations (iii) and (iv) uniquely determine $f + g$?

EXERCISE 3.20. (For category theorists). Show that the result of 3.19 can be used to prove that + and x on the category of domains and approximable maps are indeed functors. Show further that x *is* the categorical product for this category.

EXERCISE 3.21. In the proofs of 3.12 in the definition of curry (g) it is rather cavalierly assumed that the neighbourhood [Y,Z] uniquely determines Y and Z. Show that this is true *if* $Z \neq \Delta_2$. (Hint: Find explicitly the least of $f \in [Y,Z]$.) Show that if $Z = \Delta_2$ the biconditional stated at the start of the proof is still valid even though Y is not uniquely determined. (Hint: Remember that $\Delta_1\, g\, \Delta_2$ must hold.) For arbitrary pairs of neighbourhoods of $(\mathcal{D}_1 \to \mathcal{D}_2)$ is there a simple criterion for identity?

EXERCISE 3.22. Prove that there is an approximable mapping

$$\text{comp}: (\mathcal{D}_1 \to \mathcal{D}_2) \times (\mathcal{D}_0 \to \mathcal{D}_1) \to (\mathcal{D}_0 \to \mathcal{D}_2)$$

where for all $g : \mathcal{D}_1 \to \mathcal{D}_2$ and $f : \mathcal{D}_0 \to \mathcal{D}_1$ we have

$$\text{comp}\,(g, f) = g \circ f.$$

Show this directly by writing down the neighbourhood relation and by building the mapping up from eval and curry (on suitable domains) using $\circ$ and $\langle\, ,\rangle$. (Hint: Fill in maps in the following sequence of domains:

$$(\mathcal{D}_0 \to \mathcal{D}_1) \times \mathcal{D}_0 \to \mathcal{D}_1$$

$$(\mathcal{D}_1 \to \mathcal{D}_2) \times ((\mathcal{D}_0 \to \mathcal{D}_1) \times \mathcal{D}_0) \to (\mathcal{D}_1 \to \mathcal{D}_2) \times \mathcal{D}_1$$

$$((\mathcal{D}_1 \to \mathcal{D}_2) \times (\mathcal{D}_0 \to \mathcal{D}_1)) \times \mathcal{D}_0 \to (\mathcal{D}_1 \to \mathcal{D}_2) \times \mathcal{D}_1$$

$$((\mathcal{D}_1 \to \mathcal{D}_2) \times (\mathcal{D}_0 \to \mathcal{D}_1)) \times \mathcal{D}_0 \to \mathcal{D}_2$$

$$(\mathcal{D}_1 \to \mathcal{D}_2) \times (\mathcal{D}_0 \to \mathcal{D}_1) \to (\mathcal{D}_0 \to \mathcal{D}_2).$$

The maps are of course not uniquely determined, but the shifting of brackets ought to suggest the right choice.)

EXERCISE 3.23. (For category theorists.) Show that the results of 3.11 and 3.12 prove that the category of domains and approximable mappings is a *cartesian closed category*. (Mac Lane [1971] pp. 95-96 may be consulted for a very brief introduction.) What is the *terminal domain* in this category? What sort of functor is $(\mathcal{D}_0 \to \mathcal{D}_1)$?

EXERCISE 3.24. Establish some more isomorphisms :

(i) $(\mathcal{D}_0 \to (\mathcal{D}_1 \times \mathcal{D}_2)) \cong (\mathcal{D}_0 \to \mathcal{D}_1) \times (\mathcal{D}_0 \to \mathcal{D}_2)$

(ii) $(\mathcal{D}_0 \to \mathcal{D}_1^{\infty}) \cong (\mathcal{D}_0 \to \mathcal{D}_1)^{\infty}$

(iii) $\mathcal{D}_0 \times (\mathcal{D}_1 + \mathcal{D}_2) \cong (\mathcal{D}_0 \times \mathcal{D}_1) + (\mathcal{D}_0 \times \mathcal{D}_2)$

(iv) $(\mathcal{D}_0 + \mathcal{D}_1) \to \mathcal{D}_2 \cong (\mathcal{D}_0 \to \mathcal{D}_2) \times (\mathcal{D}_1 \to \mathcal{D}_2)$..

If some of the above are *not* true, perhaps at least some mapping relationships can be established.

EXERCISE 3.25. (For topologists.) Recall from Exercises 1.21 and 2.13 on how to regard a domain $|\mathcal{D}|$ as a topological space. Using 3.10 show that the family of open subsets of $|\mathcal{D}|$ is isomorphic to a domain.

EXERCISE 3.26. Show that for every domain $\mathcal{D}$ there is an approximable mapping

$$\text{cond} : T \times \mathcal{D} \times \mathcal{D} \to \mathcal{D},$$

called the *conditional operator*, satisfying

(i) cond (true, x, y) = x

(ii) cond (false, x, y) = y

(iii) cond $(\bot, x, y) = \bot$.

(Hint: Recalling that $T = \{\{0\}, \{1\}, \{0,1\}\}$, define cond as a relation by

$$0C \cup 10X \cup 110Y \text{ cond } Z \text{ iff } 0 \in C \text{ and } X \subseteq Z \text{ or}$$
$$1 \in C \text{ and } Y \subseteq Z \text{ or}$$
$$0,1 \in C \text{ and } \Delta \subseteq Z ,$$

where $C \in T$ and $X \in \mathcal{D}$ and $Y \in \mathcal{D}$ and where we are using the construction of Exercise 3.14.) Find a similar operator in the domain

$$T \times \mathcal{D}_0 \times \mathcal{D}_1 \to \mathcal{D}_0 + \mathcal{D}_1 \ .$$

Show also there is an approximable mapping

$$\text{which} : \mathcal{D}_0 + \mathcal{D}_1 \to T$$

such that for all $x \in |\mathcal{D}_0 + \mathcal{D}_1|$

$$\text{cond}\ (\text{which}(x),\ \text{in}_0(\text{out}_0(x)),\ \text{in}_1(\text{out}_1(x))) = x.$$

EXERCISE 3.27. (For set theorists.) Give another proof that the family of approximable mappings $f : \mathcal{D}_0 \to \mathcal{D}_1$ is isomorphic to a domain by employing the general argument of Exercise 2.22. How does this compare with the proof method of 3.9 and 3.10? Can the general remarks also be employed to show that

$$\text{eval} : (\mathcal{D}_1 \to \mathcal{D}_2) \times \mathcal{D}_1 \to \mathcal{D}_2$$

is approximable without bringing in the neighbourhoods in such an explicit way? (Hint: Use 3.5 and the idea of Exercise 2.12.)

EXERCISE 3.28. In the function space $(\mathcal{D}_0 \to \mathcal{D}_1)$ let

$$\bigcap \{[X_i,\ Y_i] \mid i < n\}$$

be a (non-empty) neighbourhood. In 3.9 the minimal element of this neighbourhood is characterized as a relation f_0. Show that as an elementwise mapping it can be defined by the formula

$$f_0(x) = \bigsqcup \{\uparrow Y_i \mid x \in [X_i]\},$$

for $x \in |\mathcal{D}_0|$. Try to draw a picture of $|\mathcal{D}_0|$ with neighbourhoods $[X_i]$ and the corresponding values of the function f_0.

LECTURE IV

FIXED POINTS AND RECURSION

Having at this point a large supply of examples of domains (and further constructs of new domains), we now have to consider some other ways of defining functions - other than by explicit compositions of the very basic functions already mentioned. One of the most fruitful techniques is an infinitely *iterated* composition that is at the back of the idea of *recursion*. We will use the process over and over again in these lectures, not only to define new functions but also to define new domains. The heart of the matter lies in the so-called "Fixed-point Theorem":

THEOREM 4.1. For any approximable mapping $f : \mathcal{D} \to \mathcal{D}$ on any domain, there exists a *least* element $x \in |\mathcal{D}|$ where

$$f(x) = x.$$

Proof: Let f^n for $n \in \mathbb{N}$ stand for the n-fold composition of f with itself. That is,

$$f^0 = I_{\mathcal{D}}, \text{ and}$$
$$f^{n+1} = f \circ f^n.$$

Define

$$x = \{X \in \mathcal{D} \mid \Delta f^n X, \text{ for some } n \in \mathbb{N}\}.$$

We see $X \in x$ iff there is a finite sequence $\Delta = X_0, X_1, \ldots, X_n = X$ where $X_i f X_{i+1}$ holds for all $i < n$. Now since $\Delta f \Delta$ automatically holds, a sequence for an $X \in x$ can always be extended to a longer sequence just by adding more Δ's on the front.

We want to prove $x \in |\mathcal{D}|$. Clearly $\Delta \in x$; and if $X \subseteq Y$ and $X \in x$, then $Y \in x$. All that remains to be shown is the closure of x under intersection. Note that if

$$U f V \text{ and } U' f V'$$

hold and U, U' are consistent in $\mathcal{D}$, then V and V' are consistent and

$$(U \cap U') \ f \ (V \cap V')$$

must hold. Generalizing this to sequences, if

$$\Delta = X_0 \ f \ X_1 \ f \ \cdots \ f \ X_n = X, \text{ and}$$

$$\Delta = Y_0 \ f \ Y_1 \ f \ \cdots \ f \ Y_n = Y$$

both hold (and note we have arranged the lengths of the two sequences to be equal), then each pair X_i, Y_i is consistent and we have

$$\Delta = (X_0 \cap Y_0) \ f \ (X_1 \cap Y_1) \ f \ \cdots \ f (X_n \cap Y_n) = X \cap Y.$$

This establishes the desired closure.

We also note that if $X \in x$ and XfY then $Y \in x$. Therefore, $f(x) \subseteq x$ and indeed by its very construction x is the *least* element of $|\mathcal{D}|$ with this property. (Why?) But f is monotone, so $f(f(x)) \subseteq f(x)$; hence, $x = f(x)$. By what we have already said it must be the least such element. □

Because the element we have shown to exist in 4.1 is a least element, it is *unique*. That is, we have associated with each $f : \mathcal{D} \to \mathcal{D}$ a special element $x_f \in |\mathcal{D}|$ determined by the choice of f. A function has therefore been defined mapping the set $|\mathcal{D} \to \mathcal{D}|$ into $|\mathcal{D}|$. The next result shows that this function, or operator on functions, is in fact approximable.

THEOREM 4.2. For any domain $\mathcal{D}$, there is an approximable mapping

$$\text{fix} : (\mathcal{D} \to \mathcal{D}) \to \mathcal{D}$$

such that if $f : \mathcal{D} \to \mathcal{D}$ is any approximable mapping, then

(i) $\text{fix}(f) = f(\text{fix}(f))$.

Furthermore, if $x \in |\mathcal{D}|$, then

(ii) $f(x) \subseteq x$ implies $\text{fix}(f) \subseteq x$.

And this last property implies that fix is unique. Explicitly we can characterize fix by the equation:

(iii) $$\text{fix}(f) = \bigcup_{n=0}^{\infty} f^n(\perp) ,$$

for all $f : \mathcal{D} \to \mathcal{D}$.

Proof: Formula (iii) can be put in a more elementary form:

$$\mathrm{fix}\ (f) = \{X \mid \Delta f^n X, \text{ for some } n \in \mathbb{N}\}.$$

To show an elementwise mapping approximable we can use the formula of Exercise 2.9, applied to the above as the definition of fix:

$$(*) \qquad \mathrm{fix}\ (f) = \bigcup \{\mathrm{fix}\ (\uparrow F) \mid f \in [F]\},$$

where F ranges over the neighbourhoods of $(\mathcal{D} \to \mathcal{D})$, and where $\uparrow F$ can be considered to be the least element of F as calculated in 3.9.

Now from the definition of fix, it is clear that whenever $f \subseteq g$, then $\mathrm{fix}\ (f) \subseteq \mathrm{fix}\ (g)$, because $f^n \subseteq g^n$. (That is, fix is obviously monotone.) Next, if $f \in F$, then $\uparrow F$ is a (finite) approximation to f; so $\uparrow F \subseteq f$ and $\mathrm{fix}\ (\uparrow F) \subseteq \mathrm{fix}\ (f)$. This means that half of equation (*) already holds by monotonicity. All that is left is to prove the other half.

So suppose $X \in \mathrm{fix}\ (f)$. Then, as we have already remarked, there is a finite sequence of neighbourhoods where

$$\Delta = X_0\ f\ X_1\ \cdots\ X_{n-1}\ f\ X_n = X.$$

Let the function-space neighbourhood be defined as

$$F = \bigcap \{[X_i, X_{i+1}] \mid i < n\},$$

and note that since $f \in [F]$ we have at once consistency. But, by 3.9, $\uparrow F \in [F]$, so the *same* sequence of X_i is sufficient to show that

$$X \in \mathrm{fix}\ (\uparrow F).$$

In other words, if X belongs to the left-hand side of (*), it also belongs to the right-hand side. This completes the proof of (*).

Formula (i) is just a restatement of what we proved in 4.1. And (ii) follows easily, because $f(x) \subseteq x$ implies that $\Delta \in x$ and whenever $X \in x$ and $X\,f\,Y$, then $Y \in x$. Thus, by induction, if $\Delta f^n X$, then $X \in x$. So $\mathrm{fix}\ (f) \subseteq x$.

Finally, if $\mathrm{fax} : (\mathcal{D} \to \mathcal{D}) \to \mathcal{D}$ were any other operator satisfying (i) and (ii), we would prove at once that

$$\mathrm{fix}\ (f) \subseteq \mathrm{fax}\ (f) \quad \text{and}$$
$$\mathrm{fax}\ (f) \subseteq \mathrm{fix}\ (f).$$

That is to say, the two operators are identical. □

The reader may have noticed that we used recursion in the proof of 4.1 (we had to define f^n for all $n \in \mathbb{N}$). But 4.1 and 4.2 can be used to justify definitions by recursion on a large number of domains - definitions where the process of iteration is far from being as straightforward. In discussing this point, let us start with some basic examples.

EXAMPLE 4.3. The infinite generalization of our original example 1.2 is the system

$$N = \{\{n\} \mid n \in \mathbb{N}\} \cup \{\mathbb{N}\}.$$

The total elements are clearly in a one-one correspondence with the integers in $\mathbb{N}$. We can apply the construction of Exercise 3.16 to obtain a domain

$$F = N^\infty.$$

So we already know quite a bit about this domain - but it has a much more familiar presentation.

Let Φ be the set of all *finite partial functions* $\varphi \subseteq \mathbb{N} \times \mathbb{N}$ (that is, finite sets of ordered pairs of integers where, if $(n,m) \in \varphi$ and $(n,m') \in \varphi$, then $m = m'$). Define

$$\uparrow\varphi = \{\psi \in \Phi \mid \varphi \subseteq \psi\}.$$

Consider the neighbourhood system

$$F' = \{\uparrow\varphi \mid \varphi \in \Phi\}.$$

It is an easy exercise to show that F and F' are isomorphic and that the elements of these domains correspond exactly to the (possibly infinite) *partial functions* $\pi \subseteq \mathbb{N} \times \mathbb{N}$. Moreover, the *total* elements just correspond to the *total* functions $\tau : \mathbb{N} \to \mathbb{N}$ ("function" in the ordinary, set-theoretical sense of the word).

Another easy exercise is to show that the domains

$$F \text{ and } (N \to N)$$

by our definitions are *NOT* isomorphic; though the two domains are closely related. We can define a mapping

$$\mathrm{val} : F \times N \to N$$

by the relationship

$$\uparrow\varphi \cup \{n\} \ \mathrm{val}\ \{m\} \text{ iff } (n, m) \in \varphi.$$

(Of course val has to relate other neighbourhoods such as:

$$\uparrow\varphi \cup \mathbb{N} \ \mathrm{val}\ \mathbb{N},$$

but these are all.) It is then simple to prove that if $\pi \in |F|$ is regarded as a partial function $\pi : \mathbb{N} \to \mathbb{N}$ and if for $n \in \mathbb{N}$ we define $\hat{n} \in |N|$ by

$$\hat{n} = \{ \{n\}, \mathbb{N} \},$$

then we have

$$\begin{aligned} \mathrm{val}\ (\pi, \hat{n}) &= \widehat{\pi(n)}, \text{ if } \pi \text{ is defined at } n; \\ &= \{\mathbb{N}\}, \text{ otherwise.} \end{aligned}$$

(Remember that $\{\mathbb{N}\} \in |N|$ is the "undefined" element.)
This means that

$$\mathrm{curry}\ (\mathrm{val}) : F \to (N \to N)$$

is a one-one function on elements. (The rather slight trouble with $(N \to N)$ is that it has *more* elements than F.)

So much for the construction of F, we now wish to consider mappings

$$f : F \to F$$

and their uses. Consider the possibility

$$\begin{aligned} f(\pi)(n) &= 0, & &\text{if } n = 0; \\ &= \pi(n-1) + n-1, & &\text{if } n > 0. \end{aligned}$$

If π were a total function, then $f(\pi)$ would be total. But if π is partial, and if it is, say, undefined at k, then $f(\pi)$ becomes undefined at $k+1$. Note that $f(\pi)$ is always defined at 0. Note, too, that f is an approximable mapping because it is completely determined by what it does to finite (partial) functions. Indeed,

$$f(\pi) = \bigcup \{f(\varphi) \mid \varphi \subseteq \pi \},$$

where φ ranges over Φ.

Well, we have proved that every approximable map of a domain into itself has a (least) fixed point. What is the least fixed point of this f? Suppose $\sigma = f(\sigma)$. Then $\sigma(0) = 0$, and

$$\begin{aligned}\sigma(n+1) &= f(\sigma)(n+1)\\ &= \sigma(n) + n.\end{aligned}$$

By induction, then

$$\sigma(n) = \sum_{i<n} i$$

and σ is a total function. (Therefore, f has a *unique* fixed point.)

Actually, we can make the procedure more systematic by defining as fixed points elements of $(N \to N)$ rather than F. In the first place we have $\hat{0} \in |N|$, and from now on we will not distinguish between n and $\hat{n}$. Next we have two mappings:

$$\text{succ, pred} : N \to N$$

where, as approximable mappings we have

$$X \text{ succ } Y \text{ iff } \exists n \in \mathbb{N}\,.\, n \in X \text{ and } n+1 \in Y,$$
$$X \text{ pred } Y \text{ iff } \exists n \in \mathbb{N}\,.\, n+1 \in X \text{ and } n \in Y,$$

for all $X, Y \in N$. This is *correct* , but what we mean in more understandable terms is:

$$\begin{aligned}\text{succ}(n) &= n+1;\\ \text{pred}(n) &= n-1, \text{ if } n>0;\\ &= \bot, \quad \text{ if } n=0.\end{aligned}$$

Here, n has been identified with $\hat{n} \in |N|$ and $\bot = \{\mathbb{N}\} \in |N|$. Moreover, we have a mapping

$$\text{zero} : N \to T$$

which is such that

$$\begin{aligned}\text{zero}(n) &= \text{true, if } n=0;\\ &= \text{false, if } n>0.\end{aligned}$$

The *structured domain*

$$\langle N, 0, \text{succ, pred, zero}\rangle$$

can be called "THE domain of integers" for our present theory. We shall meet many other structured domains in the sequel.

Now the iterated summation function σ can be completely characterized - as a map $\sigma : N \to N$ rather than as an element $\sigma \in |F|$ - by the following equation:

$$\sigma(n) = \text{cond}\,(\text{zero}(n), 0, \sigma(\text{pred}(n)) + \text{pred}(n)).$$

The only problem is that we have not defined $+ : N \times N \to N$. (A direct definition is left to the reader; general remarks are given later.) But + could be any function of two variables in order to make the point about the form of the definition of σ. Remember

$$\text{cond} : T \times N \times N \to N,$$

as defined in Exercise 3.26. We do not put cond in as part of the structure of N because (as should be clear from 3.26) it is part of the structure of T.

The above equation for σ is properly called a *functional equation*; it will be written as a fixed-point equation in Lecture V when we have the notation for the λ-calculus. □

EXAMPLE 4.4. The domain C of finite or infinite binary sequences mentioned in Exercise 2.21 may be regarded as a generalization of N. This can be made plain by saying how we wish to regard C as a structured domain. To do this we should recall what C is as a neighbourhood system. In the first place

$$B = \{\sigma\Sigma^* \mid \sigma \in \Sigma^*\}$$

where $\Sigma = \{0,1\}$. To form the system C we have

$$C = B \cup \{\{\sigma\} \mid \sigma \in \Sigma^*\}.$$

The total elements of B correspond to *infinite* binary sequences; while the total elements of C to *finite or infinite* sequences. To simplify notation let us write for $\sigma \in \Sigma^*$

$$\sigma = \uparrow\{\sigma\} \qquad \text{(a total element);}$$

$$\sigma\bot = \uparrow\sigma\Sigma^* \qquad \text{(a partial element).}$$

In other words we identify σ with the corresponding total element in $|C|$.

We wish now to think of C as a structured domain seen as a kind of generalization of N. The empty sequence Λ will play the rôle of $0 \in |N|$; the map succ has two different analogues for C, however. Just as for B we define for $x \in |C|$ and $\sigma \in \Sigma^*$:

$$\sigma x = \{Y \mid \sigma X \subseteq Y \text{ some } X \in x\},$$

where of course now X and Y range over C. It should be checked that $\sigma\tau$ has the right meaning whether we think of $\tau \in \Sigma^*$ or $\tau \in |C|$. The two "successor" mappings we are looking for are

$$x \mapsto 0\,x \quad \text{and} \quad x \mapsto 1\,x.$$

All the maps $x \mapsto \sigma x$ can be obtained as compositions of these iterated as many times as needed.

Here are two questions which we now should ask:

What plays the rôle of pred? The mapping will be called tail, and it is characterized by:

$$\begin{aligned} &\text{tail } (0x) = x, \\ &\text{tail } (1x) = x, \text{ and} \\ &\text{tail } (\Lambda) = \bot. \end{aligned}$$

It is left to the reader to show that tail exists as an approximable mapping.

What plays the rôle of zero? The answer is not unique, because in C there are several distinctions that have to be made; in fact we will define three maps:

$$\text{empty, zero, one} : C \to T$$

where the three maps take on truth-values to distinguish various kinds of elements in $|C|$ as follows:

$$
\begin{aligned}
\text{empty}(\Lambda) &= \text{true},\\
\text{empty}(0x) &= \text{false},\\
\text{empty}(1x) &= \text{false},\\
\text{zero}(\Lambda) &= \text{false}\\
\text{zero}(0x) &= \text{true}\\
\text{zero}(1x) &= \text{false}\\
\text{one}(\Lambda) &= \text{false}\\
\text{one}(0x) &= \text{false}\\
\text{one}(1x) &= \text{true}.
\end{aligned}
$$

Again, it is an exercise to show these are approximable. The structured domain is therefore

$$\langle C, \Lambda, 0, 1, \text{tail}, \text{empty}, \text{zero}, \text{one}\rangle.$$

Note that we have changed the meaning of some of the symbols in passing from N to C. Note too that there is a confusion between 0 as an element and 0 as the map $x \mapsto 0x$. There are just too few symbols! In any case this is only an example and not a philosophy of life, so the reader can be expected not to suffer too much.

An example of a definition of an *element* of $|C|$ by a fixed-point equation is:

$$a = 0\,1\,a.$$

This equation has one and only one solution in $|C|$, the infinite sequence that alternates 0's and 1's. Note that a is also characterized by:

$$a = 0101a.$$

Another element is

$$b = 010\,b,$$

which is quite different from a.

An example of a *map* in $|C \to C|$ has the characterization

$$
\begin{aligned}
d(\Lambda) &= \Lambda\\
d(0x) &= 00d(x), \text{ and}\\
d(1x) &= 11d(x).
\end{aligned}
$$

We can write:

```
d(x) = cond (empty (x), Λ,
       cond (zero(x), 00d(tail(x)), 11d(tail(x)))).
```

As we shall see in due course, this can be regarded as a fixed-point definition of d.

An example of a map in $|C \times C \to C|$ was suggested in 2.21. We can write:

```
x y = cond (empty (x), y,
      cond (zero (x), 0(tail(x) y), 1 (tail ( x) y ))).
```

It should be checked that this equation exactly characterizes the intended mapping. □

The examples we have given with N and C are examples of definitions of functions by *recursion* . The literal meaning of "recursion" is "running backwards", and a look at the equations for our examples will show that the functions are characterized by giving their values either *outright* (e.g. at 0 or at Λ) or at *earlier* arguments (e.g. at pred(x) or at tail(x)). The reader should keep in mind that a recursive "definition" is not really a definition in the sense of *explicit definition* but rather is a characterization; a theorem has to be proved to show that such functions exist. Now we have a general definition of domain and a general theorem on fixed points and a general construction of function-space domain; *THEREFORE*, we know that there are solutions to our equations *PROVIDED THAT* the variables range over elements of a domain and that the other, given functions that appear in the equations are already known to be approximable (continuous). This proviso is very important, and we shall remark on it time after time.

But, as is well known, recursion also can be done over *sets* like $\mathbb{N}$, and we should examine now the connection between the familiar kind of recursion and what we are doing over domains. Of course, one simple connection is already provided by the way we regard $\mathbb{N}$ as a subset of N. But there are other useful connections that can be employed in a way that may seem more direct.

DEFINITION 4.5. A structured set $\langle \mathbf{N}, 0, {}^{+} \rangle$, where $0 \in \mathbf{N}$ and ${}^{+}: \mathbf{N} \to \mathbf{N}$ is a unary function, is said to be *a model for Peano's Axioms* if the following conditions are satisfied:

(i) $0 \neq n^{+}$, for all $n \in \mathbf{N}$;

(ii) $n^{+} = m^{+}$ implies $n = m$, for all $n, m \in \mathbf{N}$;

(iii) whenever $x \subseteq \mathbf{N}$ and $0 \in x$ and $x^{+} \subseteq x$, then $x = \mathbf{N}$.

Here $x^{+} = \{n^{+} \mid n \in x\}$. □

Clause (iii) is recognized as the principle of mathematical induction stated in terms of sets. We usually think of $\mathbf{N}$ as being "God given", and (i) - (iii) as known without question. Suppose God, however, decides to withdraw His set of integers and substitute another. We can ask: "Oh! Why did You take from us our beloved numbers? Why must we now live with these strange new beasts?" God will probably reply "Trust Me!" Perhaps we should in view of the theorem:

THEOREM 4.6. All models of Peano's Axioms are isomorphic.

Proof: There are several ways to give the proof, but, for the sake of illustration, an application of the fixed-point theorem is appropriate here. Let $\langle \mathbf{N}, 0, {}^{+} \rangle$ be one model, and let $\langle \mathbf{M}, \square, {}^{\#} \rangle$ be another. Let $\mathbf{N} \times \mathbf{M}$ be the ordinary cartesian product of the two sets and let

$$P(\mathbf{N} \times \mathbf{M})$$

be the powerset (set of all subsets) of $\mathbf{N} \times \mathbf{M}$. As in Exercises 1.15 and 2.20, we regard this set of elements as a domain, whose finite elements are just the finite subsets of the given set $\mathbf{N} \times \mathbf{M}$. The following mapping on $u \subseteq \mathbf{N} \times \mathbf{M}$ is easily proved approximable:

$$u \mapsto \{(0, \square)\} \cup \{(n^{+}, m^{\#}) \mid (n, m) \in u\}.$$

(This assertion should be checked as an exercise.) We thus let r be the (least) fixed point:

$$r = \{(0, \square)\} \cup \{(n^{+}, m^{\#}) \mid (n, m) \in r\}.$$

This $r \subseteq \mathbb{N} \times \mathbb{M}$ as a binary relation will turn out to be a one-one correspondence giving the required isomorphism.

First of all we see by construction that

(i) $0 \; r \; \square$;

(ii) $n \; r \; m$ implies $n^+ \; r \; m^\#$.

So, if r proves to be a one-one correspondence, it will then be the desired isomorphism. Now, the two sets shown in the equation

$$\{(0,\square)\} \cap \{(n^+,m^\#) \mid (n,m) \in r\} = \emptyset$$

are disjoint by virtue of axiom 4.5(i). Therefore, 0 in $\mathbb{N}$ corresponds by r to one and only one element of $\mathbb{M}$, namely the element $\square$. Let $x \subseteq \mathbb{N}$ be the set of all elements of $\mathbb{N}$ corresponding by r to a unique element of $\mathbb{M}$. We have just shown $0 \in x$. Suppose $n \in x$, and let $m \in \mathbb{M}$ be the unique element with $n \; r \; m$. Now $n^+ \; r \; m^\#$ holds, so n^+ corresponds to at least one element of $\mathbb{M}$. If $n^+ \; r \; k$ also holds, then since $(n^+, k) \neq (0,\square)$, the fixed-point equation implies

$$n^+ = n_0^+ \quad \text{and} \quad k = m_0^\#$$

for some $(n_0,m_0) \in r$. By axiom 4.5(ii), $n = n_0$, and, by uniqueness (remember $n \in x$), $m = m_0$; thus, $m^\#$ is the unique correspondent for n^+. We have proved $n^+ \in x$. Therefore, $x^+ \subseteq x$; so by 4.5(iii), $x = \mathbb{N}$ holds. Otherwise said, *every* element in $\mathbb{N}$ corresponds to a unique element of $\mathbb{M}$.

Note that the rôles of $\mathbb{N}$ and $\mathbb{M}$ are completely symmetric, *and* they satisfy the same axioms as structured sets. It follows, then, that every element of $\mathbb{M}$ corresponds to a unique element of $\mathbb{N}$. The proof that r is a one-one correspondence is now complete. □

EXERCISES

EXERCISE 4.7. Formula 4.2(iii) shows how to find the *least* fixed point of $f : \mathcal{D} \to \mathcal{D}$. Suppose on the other hand that $a \in |\mathcal{D}|$ is such that $a \subseteq f(a)$. Will there be a fixed point $x = f(x)$ with $a \subseteq x$?

(Hint: How do we know $\bigcup_{n=0}^{\infty} f^n(a) \in |\mathcal{D}|$?)

EXERCISE 4.8. Suppose $f : \mathcal{D} \to \mathcal{D}$ and $S \subseteq |\mathcal{D}|$ are such that

(i) $\bot \in S$;

(ii) $x \in S$ always implies $f(x) \in S$;

(iii) whenever $\{x_n\}_{n=0}^{\infty} \subseteq S$ and $x_n \subseteq x_{n+1}$ for all n, then $\bigcup_{n=0}^{\infty} x_n \in S$.

Conclude that $fix(f) \in S$. (This could be called the principle of *fixed-point induction*.) Apply the method to a set of the form

$$S = \{x \in |\mathcal{D}| \mid a(x) = b(x)\},$$

where $a, b : \mathcal{D} \to \mathcal{D}$ are approximable, and where we know $a(\bot) = b(\bot)$, and $f \circ a = a \circ f$ and $f \circ b = b \circ f$.

EXERCISE 4.9. Show that there is an approximable operator

$$\Psi : ((\mathcal{D} \to \mathcal{D}) \to \mathcal{D}) \to ((\mathcal{D} \to \mathcal{D}) \to \mathcal{D})$$

such that for $\Theta : (\mathcal{D} \to \mathcal{D}) \to \mathcal{D}$ and $f : \mathcal{D} \to \mathcal{D}$ we have

$$\Psi(\Theta)(f) = f(\Theta(f)).$$

Prove further that $fix : (\mathcal{D} \to \mathcal{D}) \to \mathcal{D}$ is the *least* fixed point of Ψ.

EXERCISE 4.10. Given a domain $\mathcal{D}$ and an element $a \in |\mathcal{D}|$, construct a domain $\mathcal{D}_a$ where

$$|\mathcal{D}_a| = \{x \in |\mathcal{D}| \mid x \subseteq a\}.$$

Show that if $f : \mathcal{D} \to \mathcal{D}$ is approximable, then f can be *restricted* to an approximable map $f' : \mathcal{D}_{fix(f)} \to \mathcal{D}_{fix(f)}$ where $f'(x) = f(x)$ for all $x \in |\mathcal{D}_{fix(f)}|$.

How many fixed points does f' have in $|\mathcal{D}_{fix(f)}|$?

EXERCISE 4.11. (Suggested by G. Plotkin). We can regard fix as assigning a fixed-point operator to each domain $\mathcal{D}$. Show that fix is uniquely determined by the following general conditions on an assignment $\mathcal{D} \mapsto F_{\mathcal{D}}$:

(i) $F_{\mathcal{D}} : (\mathcal{D} \to \mathcal{D}) \to \mathcal{D}$;

(ii) $F_{\mathcal{D}}(f) = f(F_{\mathcal{D}}(f))$ for all $f : \mathcal{D} \to \mathcal{D}$;

(iii) whenever $f_0 : \mathcal{D}_0 \to \mathcal{D}_0$ and $f_1 : \mathcal{D}_1 \to \mathcal{D}_1$ are given and $h : \mathcal{D}_0 \to \mathcal{D}_1$ is such that $h(\bot) = \bot$ and $h \circ f_0 = f_1 \circ h$, then

$$h(F_{\mathcal{D}_0}(f_0)) = F_{\mathcal{D}_1}(f_1).$$

(Hint: Apply 4.7 to prove fix satisfies (iii). In the other direction use 4.10.)

EXERCISE 4.12. Need an approximable $f : \mathcal{D} \to \mathcal{D}$ have a *maximum* fixed point? Give an example where there are *many* fixed points.

EXERCISE 4.13. The proof of 4.1 uses the integers, whereas the proof of 4.6 uses 4.1. There is a hint of circularity here! It can be eliminated by the following steps:

(1) If a domain $\mathcal{D}$ has an element a where, for $f : \mathcal{D} \to \mathcal{D}$ the relation $f(a) \subseteq a$ holds, then the least fixed point can be defined by

$$\mathrm{fix}(f) = \bigcap \{x \in |\mathcal{D}| \mid f(x) \subseteq x\}.$$

Note that $\mathrm{fix}(f) \subseteq a$. (Hint: Remark that by 1.17 the formula gives a well-defined element. Call the element b. Prove that $f(b) \subseteq b$ by showing that $f(b) \subseteq x$ whenever $f(x) \subseteq x$. Then note that $f(f(b)) \subseteq f(b)$ so that $b \subseteq f(b)$ also. Conclude $b = \mathrm{fix}(f)$ as least fixed point.)

(2) Remark that this proof uses only the monotonicity property of $f : |\mathcal{D}| \to |\mathcal{D}|$. Remark, too, that (1) can always be applied to powerset domains $\mathcal{P}A$ for any set A.

(3) Review the proof of 4.6 and establish by a fixed-point method that for any structured set $\langle Z, z, ^{\bullet} \rangle$ there is a *unique* function $s : \mathbb{N} \to Z$ such that

(i) $s(0) = z$;

(ii) $s(n^+) = s(n)^{\bullet}$, for $n \in \mathbb{N}$.

(4) Employ (3) for the proof of 4.1 by identifying $\langle Z, z, ^{\circ} \rangle$.

EXERCISE 4.14. Need a *monotone* function $f : \mathcal{P}A \to \mathcal{P}A$ always have a *maximum* fixed point?

EXERCISE 4.15. (For set theorists.) Let $f : |\mathcal{D}| \to |\mathcal{D}|$ be a monotone function on (the elements of) a domain. Show that f has a *maximal* fixed point (i.e. a fixed point that cannot be extended to a larger fixed point). (Hint: By Zorn's Lemma consider a maximal chain

$$C \subseteq \{x \in |\mathcal{D}| \mid x \subseteq f(x)\},$$

and use 2.11 to remark that $\bigcup C \in |\mathcal{D}|$.) Now argue that f has a least fixed point.

EXERCISE 4.16. (For fixed-point nuts). Show that a monotone function as in 4.15 has an "optimal" fixed point in the sense that it is the greatest fixed point below all the maximal fixed points and at the same time it is the largest fixed point consistent with all other fixed points. *Consistency* for sets of *elements* means having a common upper bound. (Hint: Follow these steps:

(1) Show that any non-empty set S of fixed points has a largest fixed point below by using the formula

$$f(\bigcap S) \subseteq \bigcap S$$

and finding the least fixed point over $\bigcap S$.

(2) Letting a be the fixed point of (1) constructed from the set of maximal fixed points, remark that a is consistent with any other fixed point $x = f(x)$, since x can be extended to a maximal one. Suppose b is consistent with all fixed points, then $b \subseteq y$ if y is maximal. (Why?).)

EXERCISE 4.17. (For algebraists). Suppose $\langle S, 1, \cdot \rangle$ is a semi-group with unit (sometimes called a *monoid*). Remark that $\mathcal{P}S$ is a domain. For $a, b \in S$, what is the least $x \in \mathcal{P}S$ such that

$$x = \{1\} \cup \{a, b\} \cup x \cdot x,$$

where in general for $x, y \subseteq S$

$$x \cdot y = \{t \cdot u \mid t \in x \text{ and } u \in y\}?$$

Need the fixed point be unique?

EXERCISE 4.18. In Example 4.3 there are many unproved assertions about N and F. These should be checked. In particular, the isomorphism theorem of 4.6 could be proved by constructing a simple domain M from $\mathbb{M}$ in the way N is constructed from $\mathbb{N}$.

EXERCISE 4.19. There are many unproved assertions in Example 4.4! In particular discuss "Peano's Axioms" for $\{0,1\}^*$. Show, moreover, that one : $C \to T$ can be defined from the rest of the structure by a fixed-point equation.

EXERCISE 4.20. For approximable $f, g : \mathcal{D} \to \mathcal{D}$ prove that

$$\mathrm{fix}\,(f \circ g) = f(\mathrm{fix}\,(g \circ f)).$$

EXERCISE 4.21. Show that the less-than-or-equal-to relation $\ell \subseteq \mathbb{N} \times \mathbb{N}$ is uniquely determined by the fixed point equation

$$\ell = \{(n,n) \mid n \in \mathbb{N}\} \cup \{(n,m^+) \mid (n,m) \in \ell\}.$$

Consider the structured set $\langle P\mathbb{N}, \mathbb{N}, {}^+\rangle$ where, as before,

$$x^+ = \{n^+ \mid n \in x\}.$$

What is the unique function $[\cdot] : \mathbb{N} \to P\,\mathbb{N}$ given by 4.13(3)? Prove that the structures $\langle \mathbb{N}, 0, {}^+\rangle$ and $\langle [m], m, {}^+\rangle$ are uniquely isomorphic for each $m \in \mathbb{N}$, and connect the isomorphism with ordinary addition of integers. Can the same be done for multiplication? (Hint: Consider the fixed-point equation:

$$n \cdot \mathbb{N} = \{0\} \cup \{n + m \mid m \in n \cdot \mathbb{N}\},$$

where $n \in \mathbb{N}$ is fixed.)

EXERCISE 4.22. Suppose $\mathbb{N}^*$ is a structured set satisfying only axioms (i) and (ii) of 4.5. Must there be a subset $\mathbb{N} \subseteq \mathbb{N}^*$ that satisfies (i), (ii), and (iii)? (Hint: Use a least fixed point in $P\,\mathbb{N}^*$.) (For set theorists): How do we know from the axioms of set theory that there exists such a set $\mathbb{N}^*$?

EXERCISE 4.23. (Suggested by S. Eilenberg). Suppose $f : \mathcal{D} \to \mathcal{D}$ is approximable on a given domain $\mathcal{D}$. Suppose $a_n : \mathcal{D} \to \mathcal{D}$ is a sequence of approximable maps where

(i) $a_0(x) = \bot$, for all $x \in |\mathcal{D}|$;

(ii) $a_n \subseteq a_{n+1}$ in $\mathcal{D} \to \mathcal{D}$, for all $n \in \mathbb{N}$;

(iii) $\bigcup_{n=0}^{\infty} a_n = I_{\mathcal{D}}$ in $\mathcal{D} \to \mathcal{D}$;

(iv) $a_{n+1} \circ f = a_{n+1} \circ f \circ a_n$, for all $n \in \mathbb{N}$.

Prove that f has a *unique* fixed point. (Hint: Show that if $x = f(x)$, then $a_n(x) \subseteq a_n(\mathrm{fix}(f))$ for all $n \in \mathbb{N}$ by induction on n.)

EXERCISE 4.24. (For set theorists). Let $f : A \to B$ and $g : B \to A$ be one-one functions (*into*, not necessarily *onto* !) Prove the Schroeder - Bernstein theorem to the effect that there exists a one-one correspondence $h : A \leftrightarrow B$. (Hint: (Suggested by A. Tarski). By the fixed-point theorem find $X \subseteq A$ where

$$X = (A - g(B)) \cup g(f(X))$$

where $f(X)$ = the image of the set f under the function f. Define $h \subseteq A \times B$ as a union of two restrictions:

$$h = f \upharpoonright X \cup g^{-1} \upharpoonright (A - X).$$

A picture helps.)

EXERCISE 4.25. Perhaps the domains N and C are not exactly analogous? C was based on {0,1} as the underlying set of tokens. Construct a system C_1 based on $\{1\}^*$ (= finite strings of 1's) with neighbourhoods:

$$C_1 = \{\{1^m \mid m \geq n\} \mid n \in \mathbb{N}\} \cup \{\{1^n\} \mid n \in \mathbb{N}\}.$$

What structure should be put on C_1 strictly analogous to that on C (= C_2)? What kinds of approximable maps relate N, C_1, and C_2? Draw some pictures.

LECTURE V

TYPED λ - CALCULUS

In Examples 4.3 and 4.4, after suitable domains have been constructed, functions are characterized by recursion equations whose form of expression is - basically - a composition or substitution of known functions together with the function to be defined. This method can be made more precise and more easily usable by expanding our notation for functions - particularly by inventing a "temporary" notation for a function as a thing in itself without having to have special letters for functions. The device is called *λ - abstraction*. It is related to ordinary set abstraction (the $\{x \mid \cdots\}$ - notation already much used in these lectures), but we gear the approach to domains and their elements, and especially to function spaces.

At this stage it would not be so helpful to produce a rigorously formal definition of the syntax of the typed λ - calculus; we shall try to suggest what is needed by example. There are so many examples at hand, the less formal discussion ought to be sufficient.

In the first place we should set aside, in the notational store room as it were, a stock of variables

$$x, y, z, w, \ldots \quad .$$

These variables will be required in different "sizes" or "types". Roughly speaking there should be an infinite number of variables to range over the elements of each domain $\mathcal{D}$. We could perhaps write

$$x_0^{\mathcal{D}}, x_1^{\mathcal{D}}, x_2^{\mathcal{D}}, \ldots,$$

but the subscripts to insure an infinity of variables and the superscripts to record the typing of the variables lead to a notation as tiresome to write as it is to read. We simply agree that we can have as many variables as we need and that they come in all the types.

Strictly speaking we should also introduce type *symbols* and not confuse types with domains. But if the reader will simply keep in mind that *form* in language has always to be kept distinct from *content*, the confusion at the type level will not matter so very much. A point at which the confusion might cause a real confusion concerns *compound types*. Given $\mathcal{D}_0$ and $\mathcal{D}_1$ we can form such compounds as

$$\mathcal{D}_0 + \mathcal{D}_1, \quad \mathcal{D}_0 \times \mathcal{D}_1, \quad \mathcal{D}_0 \to \mathcal{D}_1.$$

What has to be remembered is that a compound domain (neighbourhood system), $\mathcal{D}_0 \times \mathcal{D}_1$ say, does not uniquely determine the "parts" $\mathcal{D}_0$ and $\mathcal{D}_1$. (We could make it do so, but it would cost some effort.) Of course, the *symbol* "$\mathcal{D}_0 \times \mathcal{D}_1$" has well defined parts. The point is that *different* ways of forming a compound domain could lead to the *same* result, meaning that a domain does not let us retrace its exact history of construction. Compound symbols, however, always carry their histories around with them, since otherwise they would not be readable. What we want, of course, are *both* domain symbols *and* domains, the latter being the meanings of the former. Most of the time we can happily pretend that it is only the domains themselves we have to think about.

Besides variables, we will also need certain *constants*. For instance, the symbol 0 (perhaps, better $0^{\mathcal{N}}$) denotes a certain element of $|\mathcal{N}|$. Similarly, in view of Theorem 4.2, for each domain $\mathcal{D}$ there is a well-determined element $\mathrm{fix}^{\mathcal{D}}$ of the compound type $((\mathcal{D} \to \mathcal{D}) \to \mathcal{D})$ denoting the least fixed-point operator. We have considered any number of similar constants of a great variety of types already (cf. 4.3 and 4.4; cond is an especially good one). We can say that the variables and constants are *atomic terms*, where "atomic" here means non-compound.

To form compound terms, there are several means: for example, if $\tau, \ldots, \sigma$ is a list of already obtained terms (including variables or constants), then we can form an ordered *tuple*

$$\langle \tau, \ldots, \sigma \rangle .$$

We have already done so in 3.1. If the types of $\tau, \ldots, \sigma$ are $\mathcal{D}, \ldots, \mathcal{D}'$, respectively, then the type of the tuple is the product domain

$$\mathcal{D} \times \cdots \times \mathcal{D}' ,$$

because we intend that the tuple denote an element of this domain. (The tuple notation for *functions* as in 3.3 is being forgotten for the time being.)

Next suppose that τ has type $(\mathcal{D}_0 \to \mathcal{D}_1)$ and σ has type $\mathcal{D}_0$, then the usual *function-value notation*

$$\tau\ (\sigma)$$

is a compound term of type $\mathcal{D}_1$. We also use

$$\tau\ (\sigma_0, \ldots, \sigma_{n-1})$$

as an abbreviation of

$$\tau\ (\langle\sigma_0, \ldots, \sigma_{n-1}\rangle),$$

where, if the types of $\sigma_0, \ldots, \sigma_{n-1}$ are $\mathcal{D}_0, \ldots, \mathcal{D}_{n-1}$, then the type of τ has to be of the form

$$((\mathcal{D}_0 \times \cdots \times \mathcal{D}_{n-1}) \to \mathcal{D}_n)$$

where $\mathcal{D}_n$ is the type of the compound. In this manner, with functions applied to tuples, we have the full facility of substitution into functions of many variables just by iterating the notation.

Having taken into account function *value*, it remains to provide for function *definition*. Suppose that $x_0, \ldots, x_{n-1}$ is a list of distinct variables of types $\mathcal{D}_0, \ldots, \mathcal{D}_{n-1}$. Suppose further that τ is a term - no matter how complicated - of type $\mathcal{D}_n$. Then we can regard τ as defining a function of n - variables of type

$$((\mathcal{D}_0 \times \cdots \times \mathcal{D}_{n-1}) \to \mathcal{D}_n).$$

What we have not done is to reward our regard by, as yet, providing a quick-to-write "name" for that function. This we now do; it is called

$$\lambda x_0, \ldots, x_{n-1} .\ \tau ,$$

where we stress that the x_i must be *distinct* variables and that this expression denotes the *whole function.* That is why we provide it with a special symbol.

Here is an example of the λ-notation

$$\lambda\, x, y.\, x,$$

which is read "lambda ex wye ... (pause) ... ex". If the types of x and y are $\mathcal{D}_0$ and $\mathcal{D}_1$, then the type of the above is

$$((\mathcal{D}_0 \times \mathcal{D}_1) \rightarrow \mathcal{D}_0).$$

Indeed, we know this function very well: it is the *first projection function* p_0 of 3.3 and the equation

$$p_0 = \lambda\, x, y.\, x$$

is true, as is the equation

$$p_1 = \lambda\, x, y.\, y.$$

In the notation of 3.3, we also find the true equation

$$\langle f, g\rangle = \lambda\, w. \langle f(w),\ g(w)\rangle,$$

where on the right-hand side we are using "official" λ-notation for a function of type

$$(\mathcal{D}_2 \rightarrow (\mathcal{D}_0 \times \mathcal{D}_1)).$$

The notation on the left is just an *abbreviation* and it should not be confused with the pair (2-tuple) of type

$$((\mathcal{D}_2 \rightarrow \mathcal{D}_0) \times (\mathcal{D}_2 \rightarrow \mathcal{D}_1)).$$

(Since the two domains just mentioned are isomorphic, the possible confusion is not all that serious. On the other hand, one confusion we will completely overlook is that between 1-tuples $\langle x\rangle$ and elements x. Strictly speaking they are different, but we shall not bother to make the distinction.)

Here are some other examples of true equations:

$$\mathrm{eval} = \lambda\, f, x.\ f(x) \qquad \text{(cf. 3.11)}$$

$$\mathrm{curry} = \lambda\, g\, \lambda\, x\, \lambda\, y.\ g(x, y) \quad \text{(cf. 3.12)}$$

The first should be immediately clear; while the second is particularly instructive. What is being illustrated is that the λ-notation can be *iterated*. The distinction being drawn is between

$$\lambda x_0, x_1, \ldots, x_{n-1} . \tau \quad \text{and} \quad \lambda x_0 \lambda x_1 \cdots \lambda x_{n-1}. \tau .$$

The first has type

$$((\mathcal{D}_0 \times \mathcal{D}_1 \times \cdots \times \mathcal{D}_{n-1}) \to \mathcal{D}_n) ;$$

while the second has type

$$(\mathcal{D}_0 \to (\mathcal{D}_1 \to (\cdots (\mathcal{D}_{n-1} \to \mathcal{D}_n) \cdots))).$$

This is related also to the true equation

$$\text{curry} (\lambda x,y .\tau) = \lambda x \lambda y.\tau ,$$

which shows that there are operators relating to the two notations. The first is the *multivariate* form; the second is the *curried* form.

Here is another true equation

$$\text{fix} = \text{fix} (\lambda F \lambda f. f(F(f))),$$

where the fix on the left has type $((\mathcal{D} \to \mathcal{D}) \to \mathcal{D})$ and that on the right type

$$((((\mathcal{D} \to \mathcal{D}) \to \mathcal{D}) \to ((\mathcal{D} \to \mathcal{D}) \to \mathcal{D})) \to ((\mathcal{D} \to \mathcal{D}) \to \mathcal{D})).$$

This is the content of Exercise 4.9. (This also shows why type superscripts are tiresome.)

The combination

$$\text{fix} (\lambda x.\tau)$$

occurs so often, that from time to time we abbreviate it as

$$\mathsf{I}\, x . \tau,$$

but remember it only makes sense if x and τ have the *same* type. For example in 4.3 we could have written

$$\sigma = \mathsf{I}\, f \lambda n. \text{ cond } (\text{zero}(n), 0, f(\text{pred}(n)) + \text{pred}(n))$$

and read this as

"σ is the least (recursively defined) function f whose value at n is cond $(\cdots)$."

We note that in the so-called "body" of the expression inside the cond-part the variable f occurs again. That is just the point! This is a *recursive* definition; it is made into an *explicit* definition by invoking the least fixed-point operator.

In a λ-expression, $\lambda x, y, z . \tau$, say, the variables x, y, z are being *bound* in τ; but τ may have other variables that are nowhere bound in τ and these remain *free variables* of the whole expression. Bound variables are dummy variables and may be rewritten by other variables; thus

$$\lambda x . \tau = \lambda y . \tau [y/x]$$

is a true equation *PROVIDED* the variable y does not occur in τ. In the equation the notation $\tau[y/x]$ means the result of *substituting* (rewriting) the variable y for the variable x throughout the term τ. We can also write $\tau[\sigma/x]$ for substituting a whole term σ for a variable in the other term.

We have already spoken of "true equations", but how do we know that these curious equations are meaningful at all? They are, but this is something that has to be proved.

THEOREM 5.1. Every typed λ-term τ defines an approximable function of its free variables.

Proof: We argue by an induction on the complexity of τ; there will only be a few cases to consider since the "syntax" of λ-terms is limited — even though terms can be of any length.

If τ is a variable or a constant there is nothing to prove. We already know that

$$x \mapsto x \quad \text{and} \quad x \mapsto k$$

are approximable functions.

Suppose τ has the form

$$\langle \sigma_0, \ldots, \sigma_{n-1} \rangle .$$

Then the σ_i are less complex terms, and so we can assume — as our induction hypothesis — that they define approximable functions of the free variables. Having said this, we just apply the already proved 3.4 to conclude (after a suitable generalization to the multivariate case) that τ, which takes on tuples as values, also defines an approximable function.

Next, suppose τ has the form

$$\sigma_0 (\sigma_1),$$

where we are sure that the types of all the terms match properly. Again we can assume the σ_i to be well behaved. But the values we seek can also be written as

$$\mathrm{eval} (\sigma_0, \sigma_1).$$

Since eval is approximable by 3.11, we just have to invoke an instance of 3.7 to gain the desired conclusion.

Finally, suppose that τ has the form

$$\lambda x . \sigma.$$

By a judicious choice of the order of the variables in σ (including x), we can assume that σ defines an approximable function

$$g : \mathcal{D}_0 \times \cdots \times \mathcal{D}_{n-1} \times \mathcal{D}_n \to \mathcal{D}'$$

where $\mathcal{D}'$ is the type of σ, $\mathcal{D}_n$ is the type of x, and $\mathcal{D}_0, \ldots, \mathcal{D}_{n-1}$ are the types of the remaining free variables of σ. We apply 3.12 and obtain an approximable function

$$\mathrm{curry}\ (g) : \mathcal{D}_0 \times \cdots \times \mathcal{D}_{n-1} \to (\mathcal{D}_n \to \mathcal{D}').$$

But, this is just exactly the function defined by τ.

We leave as an exercise the more general case of a term τ of the form

$$\lambda x_0, \ldots, x_{k-1} . \sigma ,$$

which has a string of bound variables. □

We can now say more precisely what it means to call $\sigma = \tau$ a "true equation". This means that, if we employ the method of the proof of 5.1, the two terms define the *same function* of the free variables. For example,

$$\lambda x.\tau = \lambda y.\tau\,[y/x]$$

is true, provided y does not occur free in the term τ, since the systematic generation of the function defined by $\lambda x.\tau$ does not depend on what the variable x *looks like* but only on its *position* in the term τ. Some other obviously desirable rules for generating true equations are stated in the exercises. But one rule is so basic that we state it here in full generality.

THEOREM 5.2. For suitably typed λ-terms the following equation is true:

$$(\lambda x_0, \ldots, x_{n-1}.\tau)(\sigma_0, \ldots, \sigma_{n-1}) = \tau\,[\sigma_0/x_0, \ldots, \sigma_{n-1}/x_{n-1}].$$

Proof: It will be sufficient to carry out the proof for $n = 1$. The proof proceeds by induction on the complexity of the term τ. In case τ is a *constant* k, the result reads

$$(\lambda x.k)(\sigma) = k,$$

and this is a true equation.

In case τ is a *variable* (in particular, the variable x), the result reads

$$(\lambda x.x)(\sigma) = \sigma,$$

and again this is a true equation.

In case τ is a *tuple* (say, $\langle \tau_0, \tau_1 \rangle$) the result reads

$$(\lambda x.\langle \tau_0, \tau_1 \rangle)(\sigma) = \langle \tau_0[\sigma/x], \tau_1[\sigma/x] \rangle.$$

This is true, because the left-hand side can be transformed by the true equation

$$(\lambda x.\langle \tau_0, \tau_1 \rangle)(\sigma) = \langle (\lambda x.\tau_0)(\sigma), (\lambda x.\tau_1)(\sigma) \rangle;$$

and then we apply the inductive assumption for τ_0 and for τ_1.

In case τ is an *application*, we want (supposing the term is $\tau_0(\tau_1)$),

$$(\lambda x.\tau_0(\tau_1))(\sigma) = \tau_0[\sigma/x](\tau_1[\sigma/x]).$$

We can proceed as in the last case, noting that the left-hand side equals

$$\mathrm{eval}\ ((\lambda x . \langle \tau_0, \tau_1 \rangle)(\sigma)).$$

In case τ is an *abstract* (say, $\lambda y . \tau_0$), we want

$$(\lambda x . \lambda y . \tau_0)(\sigma) = \lambda y . \tau_0 [\sigma / x]$$

PROVIDED the variable y is not free in σ. For this we require the true equation

$$(\lambda x . \lambda y . \tau)(\sigma) = \lambda y . (\lambda x . \tau)(\sigma).$$

We argue for this by letting g be the function of $n+2$ free variables defined by τ. Then, by 5.1, the λ-term $\lambda x . \lambda y . \tau$ defines the function curry (curry (g)) of n arguments. We can call this function h for the moment. We can write

$$h(v)(\sigma)(y) = g(v, \sigma, y),$$

where v is a *list* of arguments. But, with an appropriate combinator inv, which applied to g inverts the order of the last two arguments, we can write

$$h(v)(\sigma)(y) = \mathrm{curry}\ (\mathrm{inv}\ (g))(v, y)(\sigma).$$

But, curry (inv(g)) is just the function defined by $(\lambda x . \tau)$. So what we have proved as true is

$$(\lambda x . \lambda y . \tau)(\sigma)(y) = (\lambda x . \tau)(\sigma).$$

But if y is not free in α and

$$\alpha(y) = \beta$$

is true, then so is

$$\alpha = \lambda y . \beta .$$

This completes the proof. □

We note that if τ' is the term $\lambda x, y . \tau$, then $\tau'(x,y)$ means the same as τ. This gives a convenient way of indicating free variables: we just write $\sigma(x,y)$ - where x, y are *not* free in σ - and this will have the same values as any term τ which does involve the extra free variables x and y. We use this notational device in the next theorem.

PROPOSITION 5.3. The least fixed point of

$$\lambda x, y. \langle \tau(x,y), \sigma(x,y) \rangle$$

is the pair with coordinates

$$!x.\tau(x, !y.\sigma(x,y)) \quad \text{and}$$

$$!y.\sigma(!x.\tau(x,y), y).$$

Proof: (We are assuming that x and y are not free in τ and σ.) The purpose of the fixed-point search is to find the least solution of the *pair* of equations

$$x = \tau(x,y) \quad \text{and} \quad y = \sigma(x,y).$$

In other words, we are generalizing the fixed-point equation from one to two variables - and, of course, we could go much further to any number of variables. To this end, let

$$y_* = !y.\sigma(!x.\tau(x,y), y), \quad \text{and}$$

$$x_* = !x.\tau(x, y_*).$$

Then

$$x_* = \tau(x_*, y_*),$$

and

$$y_* = \sigma(!x.\tau(x,y_*), y_*)$$

$$= \sigma(x_*, y_*).$$

This proves that $\langle x_*, y_* \rangle$ is one fixed-point pair.

Suppose, then, that $\langle x_0, y_0 \rangle$ is the least solution. (Why does a least solution have to exist? Hint: Consider a suitable mapping of type

$$\mathcal{D}_0 \times \mathcal{D}_1 \to \mathcal{D}_0 \times \mathcal{D}_1,$$

where $\mathcal{D}_0$ is the type of x and $\mathcal{D}_1$ the type of y.) Then we know

$$x_0 = \tau(x_0, y_0) \text{ and } y_0 = \sigma(x_0, y_0),$$

and also $x_0 \subseteq x_*$ and $y_0 \subseteq y_*$. But from

$$\tau(x_0, y_0) \subseteq x_0,$$

it follows that

$$!x.\tau(x, y_0) \subseteq x_0.$$

Consequently

$$\sigma(!x.\tau(x,y_0),y_0) \subseteq \sigma(x_0,y_0) \subseteq y_0.$$

By the fixed-point definition of y_*, we have $y_* \subseteq y_0$, so $y_* = y_0$; whence,

$$x_* = !x.\tau(x,y_*) = !x.\tau(x,y_0) \subseteq x_0.$$

So also $x_* = x_0$. **We have the right formula for y_0, and a similar argument gives** x_0. □

The purpose of giving the above proof was to illustrate the use of the least-fixed-point operator in *proofs*. We have such true principles as:

$$!x.\tau(x) = \tau(!x.\tau(x));$$

and

$$\tau(y) \subseteq y \text{ implies } !x.\tau(x) \subseteq y,$$

provided, of course, that x is not free in τ. These, together with the monotonicity of all the functions, were just the methods used in the above proof. Here is another example.

PROPOSITION 5.4. Let x, y, and $\tau(x,y)$ be of the same type $\mathcal{D}$ and let g be of type $(\mathcal{D} \to \mathcal{D})$, then the equation

$$\lambda x\, !y.\tau(x,y) = !g\,\lambda x.\tau(x,g(x))$$

is true.

Proof: Let f be the function on the left-hand side. We can write

$$f(x) = !y.\tau(x,y) = \tau(x,f(x)).$$

Therefore

$$f = \lambda x.\tau(x,f(x)),$$

and it follows that

$$g_0 = !g.\lambda x.\tau(x,g(x)) \subseteq f.$$

Then we have at once, by definition of g_0,

$$g_0(x) = \tau(x,g_0(x)),$$

for any given x. But by definition of f we find

$$f(x) = !y.\tau(x,y) \subseteq g_0(x).$$

As this holds for all x, then $f \subseteq g_0$ follows. So the equation is true. □

The last proof is instructive as it uses equations and inclusions between *functions* . In particular we have just made use of the principle:

if $\tau \subseteq \sigma$ holds for all values of x,
then $\lambda x.\, \tau \subseteq \lambda x.\, \sigma$ holds.

This is another form of Theorem 3.13(i).

TABLE 5.5. In the displayed table we give a summary of uses of the λ-notation to define various *combinators* . We have mentioned some of these equations before, and there are some combinators here we have not mentioned before - their meanings, however, should be clear.

$$p_0 = \lambda x, y.\, x$$

$$p_1 = \lambda x, y.\, y$$

$$\mathrm{pair} = \lambda x \lambda y.\, \langle x, y\rangle$$

$$n\text{-tuple} = \lambda x_0 \lambda x_1 \ldots \lambda x_{n-1}.\, \langle x_0, x_1, \ldots, x_{n-1}\rangle$$

$$\mathrm{diag} = \lambda x.\, \langle x, x\rangle$$

$$\mathrm{funpair} = \lambda f \lambda g \lambda x.\, \langle f(x), g(x)\rangle$$

$$\mathrm{proj}_i^n = \lambda x_0, x_1, \ldots, x_{n-1}.\, x_i$$

$$\mathrm{inv}_{i,j}^n = \lambda x_0, \ldots, x_i, \ldots, x_j, \ldots, x_{n-1}.\, \langle x_0, \ldots, x_j, \ldots, x_i, \ldots, x_{n-1}\rangle$$

$$\mathrm{eval} = \lambda f, x.\, f(x)$$

$$\mathrm{curry} = \lambda g \lambda x \lambda y.\, g(x,y)$$

$$\mathrm{comp} = \lambda g, f \lambda x.\, g(f(x))$$

$$\mathrm{const} = \lambda k \lambda x.\, k$$

$$\mathrm{fix} = \lambda f\, !\, x.\, f(x)$$

A TABLE OF COMBINATORS

It is important to note that since we have not typed the variables, these equations are ambiguous: they only become precise when the types are specified. It follows, therefore, that what we find in the table are *schemes* for combinators; there are actually infinitely many distinct combinators corresponding to any one equation depending on how the variables have types chosen for them. Clearly it is better to imagine this variety of combinators than it is to try to notate them with type superscripts.

One interest of combinators is that it is often possible to write expressions without variables - if enough combinators are used. This is sometimes useful, but it can become clumsy. On the other hand , if the same combination occurs over and over, it is sometimes useful to give it a name. This is what we do with, say, *composition* where

$$\text{comp}\ (g, f) = g \circ f.$$

On the one side we have the prefix notation, and on the other, the more common infix notation. With either notation the variable seen in $\lambda x.g(f(x))$ has been got rid of. The choice between equivalent notations ought to be based on a desire for readability. □

The reader will have noted that there are some combinators not appearing in Table 5.5. The reason is that combinators like cond, succ, pred, zero, 0 cannot be defined in the pure λ-notation but are specific to domains like T and N; we, thus, have to regard them as primitive. But once they are in hand, a very large number of other functions can be defined from these combined with λ-expressions. The next theorem gives an indication of the possibilities.

THEOREM 5.6. For every partial recursive function $h : \mathbb{N} \to \mathbb{N}$, there is a λ-term τ of type $(N \to N)$ such that the only constants occurring in τ are

$$\text{cond, succ, pred, zero, } 0$$

and where if $h(n) = m$, then

$$\tau(n) = m$$

is true; and if h (n) is undefined, then

$$\tau(n) = \bot$$

is true. The equation $\tau(\bot) = \bot$ is also true.

Proof: We have only formulated the theorem for functions of one variable - but to give the proof, it is convenient to pass through functions of any number of (integer) variables. We shall also have to recall the precise definition of the notion of partial recursive function.

It is also convenient to work with *(very) strict* functions

$$f : N^k \to N.$$

These are functions such that if $n_0, \ldots, n_{k-1} \in |N|$ and $n_i = \bot$ for at least one $i < k$, then

$$f(n_0, \ldots, n_{k-1}) = \bot.$$

It is easy to check that compositions of strict functions are strict. It is also easy to see that any *partial* function

$$g : \mathbb{N}^k \to \mathbb{N}$$

extends to a strict (approximable) function

$$\bar{g} : N^k \to N,$$

which takes the same values as g as long as g is defined; otherwise $\bar{g}$ takes the value $\bot$. What we want to show for *partial recursive* g is that the corresponding $\bar{g}$ is defined by a λ-expression.

In the first place we have to check that *primitive recursive* functions have λ-definitions in this sense. We recall that primitive recursive functions are generated from certain elementary starting functions by multi-variate composition and the scheme of primitive recrusion. The starting functions are the constant function with value zero and the "identity" or "projection" functions. For example, $g(n_0, n_1, n_2) = n_1$ for all $n_0, n_1, n_2 \in \mathbb{N}$ is one of the starting functions. Now we cannot just use the λ-term

$$\lambda x_0, x_1, x_2 . x_1$$

to represent $\bar{g}$, because the function so defined is not strict. But any function in $|N^k \to N|$ can be cut down to a strict function by a simple device. Consider

$$\lambda x.\ \mathrm{cond}\ (\mathrm{zero}(x), x, x)$$

with x of type N. This is the strict version of the identity function of one argument. The strict projection function of two arguments can be defined by

$$\lambda x_0, x_1.\ \mathrm{cond}\ (\mathrm{zero}(x_1), x_0, x_0).$$

The one of three arguments by:

$$\lambda x_0, x_1, x_2.\ \mathrm{cond}\ (\mathrm{zero}(x_0),\ \mathrm{cond}\ (\mathrm{zero}(x_2), x_1, x_1),\ \mathrm{cond}\ (\mathrm{zero}(x_2), x_1, x_1)).$$

This is not done very elegantly, and the reader can find for himself a general solution based on perhaps a better notation for the required compositions of functions.

As we remarked, strict functions are closed under substitution, and any substitution of a batch of functions into another function can be given by a λ-term, if the various functions can themselves be so defined. It only remains to λ-define functions obtained by primitive recursion. Thus, suppose, for the sake of argument, that

$$f : \mathbf{N} \to \mathbf{N} \text{ and } g : \mathbf{N} \to \mathbf{N}$$

are given as total functions with $\bar{f}$ and $\bar{g}$ being λ-definable. From them, we obtain by primitive recursion $h : \mathbf{N} \to \mathbf{N}$ where

$$h(0,m) = f(m),$$

$$h(n+1, m) = g(n,m,h(n,m))$$

for all $n, m \in \mathbf{N}$. The λ-term defining $\bar{h}$ is

$$!\,k\,\lambda x, y.\ \mathrm{cond}\ (\mathrm{zero}(x), \bar{f}(y), \bar{g}\,(\mathrm{pred}(x), y, k(\mathrm{pred}(x), y))).$$

Here we have had to use the fixed-point operator on a variable k of type $(N^2 \to N)$. The variables x, y are of type N and the cond-construction puts the two traditional equations into two clauses of one expression. It is easy to see that the fixed-point function *is* strict and is nothing more than $\bar{h}$.

That completes the representation of primitive recursive functions. To obtain the partial recursive functions, the idea is to use the so-called μ-scheme (least number operator) and, further, to close up under substitution. We need only treat the μ-scheme. Suppose, by way of example, f(n,m) is given as a primitive recursive function. We then define h (generally, a partial function) by

$$h(m) = \text{the least } n \text{ where } f(n,m) = 0.$$

This is often written

$$h(m) = \mu n.\ f(n,m) = 0.$$

Supposing, as we may, $\bar{f}$ is λ-definable, we introduce first

$$\bar{g} = \mathsf{!}\, g\, \lambda x, y.\ \text{cond}\ (\text{zero}\,(\bar{f}(x,y)), x, g\ (\text{succ}(x), y)).$$

Then $\bar{h} = \lambda y.\ \bar{g}(0,y)$. This is easily seen to be strict. Also easy to see is that if h(m) is defined, then $\bar{g}(0,m) = h(m)$. But, if h(m) is not defined, it takes some argument to make sure that the least fixed-point construction forces $\bar{g}(0,m) = \bot$. However, the argument is not very difficult. □

What is *not* said in 5.6 is that every λ-term defines a partial recursive function. This is true (with suitable control over the constants and types in the expression), but the proof requires a full analysis of computability properties of domain constructions. This is the topic of Lecture VII.

It should be remarked that the types of variables needed for the proof of 5.6 never get very high. In fact, types like N, N^k, and $(N^k \to N)$ were the only ones needed (with perhaps T thrown in also).

Recursion on N was the topic of 5.6; further examples of recursion on other domains are included in the exercises.

EXERCISES

EXERCISE 5.7. Find definitions of

$$\lambda x, y. \tau \quad \text{and} \quad \sigma(x, y)$$

which use only λv with one variable and applications only to one argument at a time. Note that use must be made of the combinators p_0, p_1, pair. Generalize the result to functions of many variables.

EXERCISE 5.8. (For combinator nuts.) Table 5.5 was meant to show how combinators could be defined in terms of λ-expressions. Can the tables be turned to show that with enough combinators available, every λ-expression can be defined by combining combinators, using $\sigma(\tau)$ as the only mode of combination?

EXERCISE 5.9. Suppose that $f, g : \mathcal{D} \to \mathcal{D}$ are approximable and $f \circ g = g \circ f$. Show that f and g have a least *common* fixed point $x = f(x) = g(x)$. (Hint: Refer back to Exercise 4.20) If in addition $f(\bot) = g(\bot)$, show that $\mathrm{fix}(f) = \mathrm{fix}(g)$. In particular will $\mathrm{fix}(f) = \mathrm{fix}(f^2)$? What if we only assume $f \circ g = g^2 \circ f$?

EXERCISE 5.10. Suppose $\mathcal{D}_0$ and $\mathcal{D}_1$ are neighbourhood systems over disjoint sets Δ_0 and Δ_1. Define the *smash product* $\mathcal{D}_0 \otimes \mathcal{D}_1$ with neighbourhoods

$$\{\Delta_0 \cup \Delta_1\} \cup \{X \cup Y \mid X \in \mathcal{D}_0 \setminus \{\Delta_0\} \text{ and } Y \in \mathcal{D}_1 \setminus \{\Delta_1\}\}.$$

Show that this *is* a neighbourhood system. Define $(\mathcal{D}_0 \to_\bot \mathcal{D}_1)$ so that $|\mathcal{D}_0 \to_\bot \mathcal{D}_1|$ consists exactly of the *strict functions*. By introducing appropriate combinators, show that

$$(\mathcal{D}_0 \to_\bot (\mathcal{D}_1 \to_\bot \mathcal{D}_2)) \text{ and } ((\mathcal{D}_0 \otimes \mathcal{D}_1) \to_\bot \mathcal{D}_2)$$

are isomorphic.

EXERCISE 5.11. For any domain $\mathcal{D}$ we may regard $\mathcal{D}^\infty$ as consisting of (bottomless) *stacks* of elements of $\mathcal{D}$. With this image in mind, define appropriate combinators with the obvious meanings:

$$\mathrm{head} : \mathcal{D}^\infty \to \mathcal{D};$$
$$\mathrm{tail} : \mathcal{D}^\infty \to \mathcal{D}^\infty;$$
$$\mathrm{push} : \mathcal{D} \times \mathcal{D}^\infty \to \mathcal{D}^\infty.$$

Using the fixed-point theorem argue that there is a combinator

$$\mathrm{diag} : \mathcal{D} \to \mathcal{D}^\infty$$

where for all $x \in |\mathcal{D}|$ we have

$$\mathrm{diag}(x) = \langle x \rangle_{n=0}^{\infty}.$$

(Hint: Try a recursive definition, say

$$\text{diag}(x) = \text{push}\ (x,\ \text{diag}(x)),$$

but be sure to prove *all* terms of diag(x) equal x.) Also introduce by an appropriate recursion a combinator

$$\text{map} : (\mathcal{D} \to \mathcal{D})^{\infty} \times \mathcal{D} \to \mathcal{D}^{\infty}$$

where for elements of the suitable types:

$$\text{map}\ (\langle f_n \rangle_{n=0}^{\infty}, x) = \langle f_n(x) \rangle_{n=0}^{\infty}.$$

EXERCISE 5.12. On any domain $\mathcal{D}$ introduce (as a least fixed point) a combinator

$$\text{while} : (\mathcal{D} \to T) \times (\mathcal{D} \to \mathcal{D}) \to (\mathcal{D} \to \mathcal{D})$$

by the recursion

$$\text{while}(p,f)(x) = \text{cond}\ (p\ (x),\ \text{while}\ (p,f)\ (f(x)),\ x).$$

Prove that

$$\text{while}\ (p,\ \text{while}\ (p,f)) = \text{while}\ (p,f).$$

Show how while could have been used to obtain the least number operator mentioned in the proof of 5.6. Generalize the idea to define a combinator

$$\text{find} : \mathcal{D}^{\infty} \times (\mathcal{D} \to T) \to \mathcal{D}$$

with the meaning "find the first term of the sequence (if any) which satisfies the given precicate."

EXERCISE 5.13. Prove the existence of a one-one function $\text{num} : \mathbb{N} \times \mathbb{N} \leftrightarrow \mathbb{N}$ such that

$$\text{num}\ (0,0) = 0\ ;$$

$$\text{num}\ (n,m+1) = \text{num}(n+1,m) + 1\ ;$$

$$\text{num}\ (n+1,0) = \text{num}(0,n) + 1.$$

Draw a picture (i.e. an infinite matrix) for the function and find a closed form for its values, if possible. Use the function to prove the isomorphism of the domains

$$P\ \mathbb{N},\ P(\mathbb{N} \times \mathbb{N}),\ P\ \mathbb{N} \times P\ \mathbb{N}.$$

EXERCISE 5.14. Show that there are approximable mappings

$$\text{graph} : (P\mathbb{N} \to P\mathbb{N}) \to P\mathbb{N} \text{ and}$$

$$\text{fun} \quad : P\mathbb{N} \to (P\mathbb{N} \to P\mathbb{N}),$$

where we have

$$\text{fun} \circ \text{graph} = \lambda f.\, f, \text{ and}$$

$$\text{graph} \circ \text{fun} \supseteq \lambda x.\, x.$$

(Hint: Using the notation

$$[n_0, n_1, \ldots, n_k] = \text{num}(n_0, [n_1, \ldots, n_k])$$

two such combinators can be given by formulae

$$\text{fun}(u)(x) = \{m \mid \exists\, n_0, \ldots, n_{k-1} \in x.\, [n_0+1, \ldots, n_{k-1}+1, 0, m] \in u\}$$

$$\text{graph}(f) = \{[n_0+1, \ldots, n_{k-1}+1, 0, m] \mid m \in f(\{n_0, \ldots, n_{k-1}\})\},$$

where k is variable - meaning all finite sequences are to be considered.)

EXERCISE 5.15. (For algebraists.) We can regard $\langle\{0,1\}^*, \Lambda, \cdot\rangle$ as the free semigroup on two generators 0 and 1. The powerset $P\{0,1\}^*$ is taken as a domain as in Exercise 4.17. For "words" $e \in \{0,1\}^*$ define

$$e^* = \{\Lambda, e, e^2, e^3, \ldots, e^n, \ldots\}.$$

Show that the least fixed point of

$$z = \{e\} \cdot z \cup \{e'\}$$

in $P\{0,1\}^*$ is $z = e^* \cdot \{e'\}$. Show further (as suggested by David Park) that the least solution of

$$x = a \cdot x \cup b \cdot y \cup c$$

$$y = b \cdot x \cup a \cdot y \cup d$$

has

$$x = (a \cup b \cdot a^* \cdot b)^* \cdot (c \cup b \cdot a^* \cdot d),$$

where the $\{\cdot\}$ has been dropped off $\{a\}$, $\{b\}$ etc., and where the *-notation has been extended to the whole domain, so that

$$z^* = \Lambda \cup z^* \cdot z.$$

(Hint: Apply 5.3.)

EXERCISE 5.16. Return to the discussion of Example 4.4 and the construction of the domain of finite and infinite binary sequences. Give a fixed-point definition of neg: $C \to C$, where

$$\text{neg}(0x) = 1\,\text{neg}(x);$$
$$\text{neg}(1x) = 0\,\text{neg}(x).$$

Prove that neg (neg (x)) = x for all $x \in |C|$. Also define merge : $C \times C \to C$, where for $\varepsilon, \delta \in \{0,1\}$ we have:

$$\text{merge}(\varepsilon x, \delta y) = \varepsilon\,\delta\,\text{merge}(x,y).$$

(Note: There may be a little trouble with merge (x,y) when x is finite and total and y is infinite - you have to decide what you want in e.g. merge (Λ,y).) Prove that

$$\text{merge}(x,x) = d(x),$$

in the notation of 4.4. Consider also the infinite non-periodic sequence

$$t = 0\,\text{merge}(\text{neg}(t), \text{tail}(t)).$$

Prove that the nth digit of t is the sum mod 2 of the digits of the number n written in the binary scale (a suggestion of J. Lambek). Show also that $t \neq u\,a\,a\,a\,v$ where a is any finite sequence $\neq \Lambda$, and where u is finite.

LECTURE VI

INTRODUCTION TO DOMAIN EQUATIONS

The major reason for introducing the theory of domains is to have a notion of *computability* incorporating both finite and infinite elements. In our many examples already explored we have seen how functions (functionals, operators, combinators) can be defined on domains; owing to the property of approximability (continuity) of these functions, we have also seen how they can be "calculated" by finite approximation. In this lecture further examples of domains will be constructed -- especially domains having infinite elements, which can be introduced in a variety of ways giving rise to interesting structural possibilities. The next lecture then treats a precise notion of computability appropriate to these domains; while the last lecture opens up new methods of domain construction.

EXAMPLE 6.1. Let $\mathcal{D}$ be fixed as a given domain. We are now familiar with a useful construct like $\mathcal{D}\times\mathcal{D}$ whose elements are ordered pairs $\langle x,y\rangle$ of elements x, y of $\mathcal{D}$. The question is: can this construct be iterated? The answer is obviously yes, since $\mathcal{D}\times(\mathcal{D}\times\mathcal{D})$ and $(\mathcal{D}\times\mathcal{D})\times(\mathcal{D}\times\mathcal{D})$ and so on can be formed with elements $\langle x,\langle y,z\rangle\rangle$ and $\langle\langle u,v\rangle, \langle x,y\rangle\rangle$ and the like. But the real question is: can the construct be iterated *indefinitely?* AND can the results be collected together into a *single domain*? The answer is yes, but it requires a bit of work to get it right. The method to be introduced will be open to many variations, so more than one answer is possible, giving non-isomorphic domains.

In order to collect all the iterates into one large domain we give ourselves first a very big domain inside of which the desired family of neighbourhoods will be found. There are many ways to make this choice, and we are fixing on one that will keep the notation simple. We have often used binary sequences for examples and constructions, but for this example let us use *ternary* sequences. Let $\Sigma = \{0,1,2\}$ and let Σ^* be all finite sequences from this three-letter alphabet. We will select subsets of Σ^* for our neighbourhoods. As Σ^* is countably infinite, it is without much loss of generality to assume that $\mathcal{D}$ is a neighbourhood system over Δ where we take $\Delta \subseteq \Sigma^*$. Also without loss of generality we can assume $\emptyset \notin \mathcal{D}$. (Why?) We wish to find another set $\Gamma \subseteq \Sigma^*$ to be the set of tokens for the new domain. After we find it, we will still have to say just which $X \subseteq \Gamma$ are appropriate for the structure we want.

The totality $\{X \mid X \subseteq \Sigma^*\}$ is, as a powerset, isomorphic to the set of elements of a domain: a point we have remarked several times. So, by the Fixed-Point Theorem we know there is a set $\Gamma \subseteq \Sigma^*$ where

$$\Gamma = 0\Delta \cup 1\Gamma \cup 2\Gamma .$$

In fact $\Gamma = \{1,2\}^* 0\Delta$, because we can say:

$$\{1,2\}^* = \{\Lambda\} \cup 1\{1,2\}^* \cup 2\{1,2\}^* .$$

The domain we are looking for will be found as a domain $\mathcal{D}^\S$ over Γ. The reason for splitting Γ up, as shown in the equation above, is to ensure that if $X, Y \in \mathcal{D}^\S$ are two neighbourhoods in the system $\mathcal{D}^\S$, then $1X \cup 2Y$ has a chance of being also in $\mathcal{D}^\S$ because

$$1X \cup 2Y \subseteq \Gamma .$$

This will make $\mathcal{D}^\S \times \mathcal{D}^\S$ isomorphic to a part of $\mathcal{D}^\S$. If we make $\mathcal{D}$ also isomorphic to a part of $\mathcal{D}^\S$, then all the iterated products will be contained in $\mathcal{D}^\S$.

What is a neighbourhood system? Just a set of sets. But $\mathcal{P}\mathcal{P}\Sigma^*$ is a domain (as a powerset) and because $\Gamma \subseteq \Sigma^*$, we find

$$\mathcal{D}^\S \in \mathcal{P}\mathcal{P}\Sigma^*$$

as an element. But elements of domains can often be defined by fixed-point equations. Indeed we will introduce $\mathcal{D}^\S$ this way:

$$\mathcal{D}^\S = \{\Gamma\} \cup \{0X \mid X \in \mathcal{D}\} \cup \{1X \cup 2Y \mid X, Y \in \mathcal{D}^\S\}.$$

The reader should stop to think why $\mathcal{D}^\S$ can be immediately seen to exist by writing such an equation. Of course another way to describe $\mathcal{D}^\S$ is to say it is the least family of sets containing (i) the set Γ, (ii) the sets $0X$ for X in the given system $\mathcal{D}$, and (iii) sets $1X \cup 2Y$ whenever it already contains X and Y (closure under a set-forming operation). By saying "least", we mean (iv) nothing else belongs to $\mathcal{D}^{\S}$ except as allowed by (i)-(iii); this makes the truth of the *equation* for $\mathcal{D}^{\S}$ clear. So $\mathcal{D}^{\S}$ exists as a family of sets, but what good is it?

By our construction of Γ, all the sets we put into $\mathcal{D}^{\S}$ are subsets of Γ (why?), so $\mathcal{D}^{\S}$ has a chance of being a system over Γ if we can check the closure under intersection. So suppose $Z \subseteq X \cap Y$ where $Z, X, Y \in \mathcal{D}^{\S}$; we want to show $X \cap Y \in \mathcal{D}^{\S}$. We argue by induction on the number of steps required to put X and Y into $\mathcal{D}^{\S}$ by (i)-(iii). There are several cases.

If $X = \Gamma$ or $Y = \Gamma$, there is nothing to prove, because both sets are subsets of Γ. We note that $\emptyset \notin \mathcal{D}^{\S}$, because (i)-(iii) cannot introduce $\emptyset$ as a member of $\mathcal{D}^{\S}$. So, if $X = 0A$ for $A \in \mathcal{D}$, then Y must have this form also (if it is not Γ), because

$$0A \cap (1B \cup 2C) = \emptyset .$$

(That is, if Y had the form (iii), then $Z = \emptyset$ would be a consequence, which is impossible.) Thus, if $X = 0A$ for $A \in \mathcal{D}$, then $Y = 0B$ for some $B \in \mathcal{D}$. But by the same reasoning $Z = 0C$ for some $C \in \mathcal{D}$ also. But the relationship $0C \subseteq 0A \cap 0B$ is equivalent to $C \subseteq A \cap B$. We see, therefore, that $A \cap B \in \mathcal{D}$, and so

$$X \cap Y = 0A \cap 0B = 0(A \cap B)$$

must belong to $\mathcal{D}^{\S}$.

The final case has X,Y,Z all of the form (iii):

$$X = 1A_1 \cup 2A_2 ,$$

$$Y = 1B_1 \cup 2B_2 , \text{ and}$$

$$Z = 1C_1 \cup 2C_2 .$$

We can think of the A_i and B_i put into $\mathcal{D}^{\S}$ *earlier* and the intersection result as being already established for them. But the relationship $Z \subseteq X \cap Y$ is equivalent to $C_i \subseteq A_i \cap B_i$ for $i = 1,2$. Therefore $A_i \cap B_i \in \mathcal{D}^{\S}$, and so does

$$X \cap Y = (1A_1 \cup 2A_2) \cap (1B_1 \cup 2B_2) = 1(A_1 \cap B_1) \cup 2(A_2 \cap B_2) .$$

We have now seen that $\mathcal{D}^\S$ is a neighbourhood system, but why was it constructed that way? The reason is simply this isomorphism (or *domain equation*):

$$\mathcal{D}^\S \cong \mathcal{D} + (\mathcal{D}^\S \times \mathcal{D}^\S) \ ,$$

as can be seen by reference to the equation for $\mathcal{D}^\S$ and the definitions of + and ×. What are the elements of $\mathcal{D}^\S$? There is always

$$\bot = \{\Gamma\}.$$

Next if $x \in |\mathcal{D}|$ we define

$$x^\S = \{\Gamma\} \cup \{0\,X \mid X \in x\}.$$

That gives an isomorphic injection

$$\lambda x.x^\S : \mathcal{D} \to \mathcal{D}^\S.$$

Then for $x, y \in |\mathcal{D}^\S|$ we can define

$$\langle x,y\rangle = \{\Gamma\} \cup \{1X \cup 2\,Y \mid X \in x \text{ and } Y \in y\}.$$

We have another isomorphic injection

$$\lambda x,y.\langle x,y\rangle : \mathcal{D}^\S \times \mathcal{D}^\S \to \mathcal{D}^\S.$$

Indeed by looking at the neighbourhood definition of $\mathcal{D}^\S$ we conclude that the *finite* elements of $\mathcal{D}^\S$ are exactly those that are either of the form (i) $\bot$, or (ii) $a^\S$, where a is finite in $|\mathcal{D}|$ or (iii) $\langle a,b\rangle$, where a and b are previously obtained finite elements of $|\mathcal{D}^\S|$.

Suppose a,..., f are finite in $|\mathcal{D}|$. We can picture the element

$$u = \langle\langle a^\S, \langle\langle b^\S, c^\S\rangle, d^\S\rangle\rangle, \langle e^\S, f^\S\rangle\rangle$$

in $|\mathcal{D}^\S|$ as a tree:

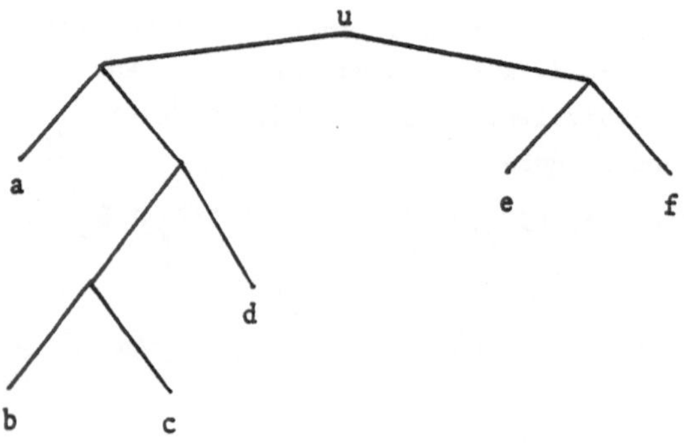

Note that the tree has binary branching with the elements of $|\mathcal{D}|$ at the ends of the branches. Any such tree could be given a notation as an element of $|\mathcal{D}^{\S}|$. The finite elements of $|\mathcal{D}^{\S}|$ correspond exactly to such finite trees.

What of the infinite elements of $|\mathcal{D}^{\S}|$? Are there infinite trees? Let $a, b \in |\mathcal{D}^{\S}|$ be any elements of $|\mathcal{D}^{\S}|$. Since pairing is an approximable mapping, we can solve the fixed-point equation

$$v = \langle a, \langle b, v \rangle \rangle .$$

In pictures we can diagram v roughly as:

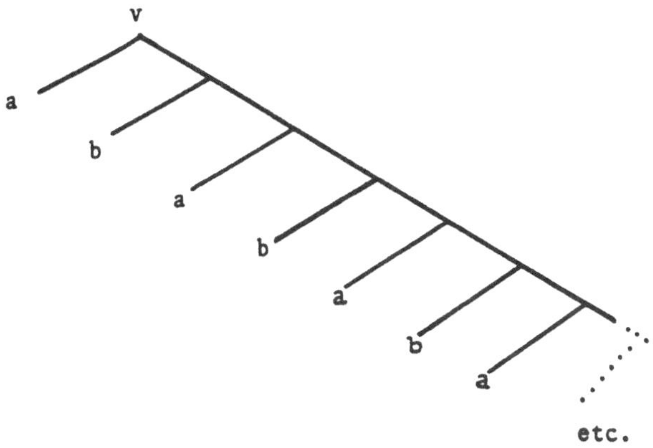

The word is "roughly" here, since if a or b were not in the $|\mathcal{D}|$ part of $|\mathcal{D}^{\S}|$, then in the diagram the letters "a" and "b" should be replaced by the corresponding tree diagrams for a and b.

Suppose that a and b are finite. Then we can easily see that the infinite tree v is the limit of the following sequence of finite trees:

$$v_0 = \bot ,$$

$$v_{n+1} = \langle a, \langle b, v_n \rangle \rangle \text{ , and}$$

$$v = \bigcup_{n=0}^{\infty} v_n .$$

The reader should think how to explain from tree diagrams the approximation relation $v_n \subseteq v$ and more general such relationships.

We could call $\mathcal{D}^§$ a *tree algebra* over $\mathcal{D}$. There may be others. A general one is a structure of the form

$$\langle \mathcal{E}, \text{in}, \text{pair} \rangle ,$$

where

$$\text{in} : \mathcal{D} \to \mathcal{E} , \text{ and}$$

$$\text{pair} : \mathcal{E} \times \mathcal{E} \to \mathcal{E} .$$

The algebra

$$\langle \mathcal{D}^§, \lambda x.x^§, \lambda x,y. \langle x,y \rangle \rangle ,$$

however, is a very special one: it is "minimal" among all tree algebras over $\mathcal{D}$ in a sense we shall have to make precise.

To do this think of how $\mathcal{E}$ and $\mathcal{D}^§$ can differ. In view of the isomorphism that $\mathcal{D}^§$ satisfies the injection of $\mathcal{D}$ and the pairing are one-one, so no "information" is lost by these mappings. The same may not at all be true of $\mathcal{E}$, but it is reasonable to think that at least we can define an approximable mapping $g : \mathcal{D}^§ \to \mathcal{E}$ where

(1) $g(\bot) = \bot_{\mathcal{E}}$,

(2) $g(x^§) = \text{in}(x)$, for $x \in |\mathcal{D}|$, and

(3) $g(\langle x,y \rangle) = \text{pair}(g(x), g(y))$, for $x,y \in |\mathcal{D}^§|$.

By what we said earlier, g will be uniquely determined by (1)-(3), because these equations tell us exactly how to calculate g on all finite elements of $|\mathcal{D}^§|$. An approximable mapping is always determined by its action on the finite elements. But why does g exist?

It would not be too hard to give an inductive construction of g as a neighbourhood relation, but a fixed-point equation is easier to write down for the same purpose. We need, though, to have the inverse ("predecessor") functions:

$$\text{out}: \mathcal{D}^{\S} \to \mathcal{D}$$

$$\text{proj}_i: \mathcal{D}^{\S} \to \mathcal{D}^{\S}, \text{ for } i = 0,1,$$

where

$$\text{out}(x^{\S}) = x\ ,$$

$$\text{proj}_0(\langle x,y\rangle) = x, \quad \text{and}$$

$$\text{proj}_1(\langle x,y\rangle) = y.$$

We also need

$$\text{atom}: \mathcal{D}^{\S} \to \mathcal{T},$$

where

$$\text{atom}(x^{\S}) = \text{true, and}$$

$$\text{atom}(\langle x,y\rangle) = \text{false}\ .$$

We can then write

$$g(x) = \text{cond}\ (\ \text{atom}(x)\ ,\ \text{in}\ (\ \text{out}(x))\ ,\ \text{pair}(g(\text{proj}_0(x)),\ g(\ \text{proj}_1(x)))).$$

This g exists by fixed-point theory, and it satisfies (1)-(3) by what we know about the structure of $|\mathcal{D}^{\S}|$. As we said, g is unique because the values on finite elements are fixed.

In algebraic language g is a *homomorphism* of tree algebras; and $\mathcal{D}^{\S}$ is called an *initial algebra*, because for any tree algebra $\mathcal{E}$ there is a unique homomorphism $g: \mathcal{D}^{\S} \to \mathcal{E}$. We note at once that any two initial algebras are isomorphic. For if $\mathcal{D}^{\star}$ were another, there would exist homomorphisms in both directions between $\mathcal{D}^{\S}$ and $\mathcal{D}^{\star}$. But the compositions of homomorphisms are again homomorphisms, and in the case of $\mathcal{D}^{\S}$ if we go from $\mathcal{D}^{\S}$ $\mathcal{D}^{\star}$ and back to $\mathcal{D}^{\S}$, the result must be the identity. The reason is that the identity can be the only homomorphism of an initial algebra into itself. We thus have a precise meaning of the minimal character of $\mathcal{D}^{\S}$. But note it still took a construction to show that the domain $\mathcal{D}^{\S}$ *exists*. □

EXAMPLE 6.2. Our staple examples $\mathcal{B}$ and $\mathcal{C}$ satisfy "domain equations" in the form of isomorphisms as we have previously seen. Indeed

$$\mathcal{B} \cong \mathcal{B} + \mathcal{B}, \text{ and}$$

$$\mathcal{C} \cong \{\{\Lambda\}\} + \mathcal{C} + \mathcal{C} \ ,$$

where if we liked we could construct both systems over $\{0,1\}^*$ and have :

$$\mathcal{B} = \{\{0,1\}^*\} \cup \{0\,X \mid X \in \mathcal{B}\} \cup \{1\,X \mid X \in \mathcal{B}\}, \quad \text{and}$$

$$\mathcal{C} = \{\{0,1\}^*\} \cup \{\{\Lambda\}\} \cup \{0\,X \mid X \in \mathcal{C}\} \cup \{1\,X \mid X \in \mathcal{C}\}.$$

We leave to the exercises the explanations of what kinds of algebras $\mathcal{B}$ and $\mathcal{C}$ are and why they are initial. Here we want to propose a simple, yet interesting generalization of $\mathcal{B}$.

Consider this domain equation :

$$\mathcal{A} \cong \mathcal{A}^n + \mathcal{A}^n \ ,$$

where $\mathcal{A}^n$ atands for the n-fold cartesian power of $\mathcal{A}$. We can, with the aid of some encoding solve this equation as a neighbourhood system over $\{0,1\}^*$ as follows:

$$\mathcal{A} = \{\{0,1\}^*\} \cup \bigcup_{i=0,1} \{\ i \bigcup_{j<n} 1^j 0 X_j \mid X_j \in \mathcal{A} \text{ all } j<n\} \ .$$

For instance, if $n = 3$, then a typical neighbourhood in $\mathcal{A}$ is something like

$$00X_0 \cup 010X_1 \cup 0110X_2 \ ,$$

where $X_0, X_1, X_2 \in \mathcal{A}$. The first '0' could also be a '1' in front of each of the terms.

In words, an element of $\mathcal{A}$ (other than $\perp$) is an n-tuple of elements of $\mathcal{A}$: but there are two separate copies of these, the left one and the right one. We can write for $a \in |\mathcal{A}|$

$$a = \pm\langle a_0, a_1, \ldots, a_{n-1}\rangle,$$

where + is chosen if a is on the right, and − if on the left. As a tree diagram a might look like this for $n = 3$:

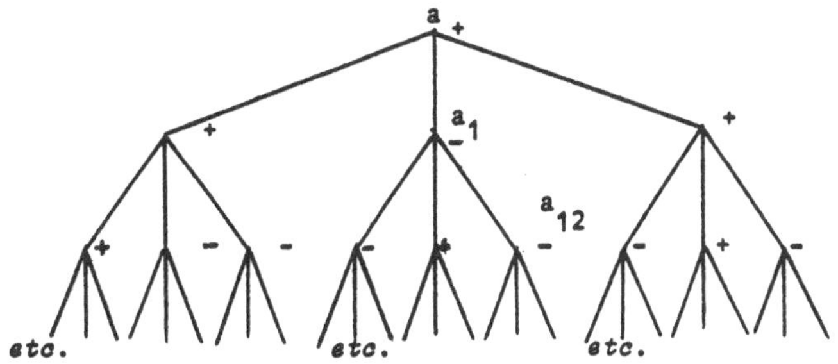

That is, a is an infinite ternary tree with + or − labels at each node. If each node (subtree) is truly infinite, the element is *total*; if ⊥ is ever encountered, it is only *partial*; if every branch ends with ⊥, the tree is a *finite* element of $|A|$.

What can be done with such trees? Let $\sigma \in \{0,1,\ldots,n-1\}^*$ be a finite sequence of "digits" each less than n. We let $\Sigma = \{0,1,\ldots,n-1\}$. We can define for $a \in |A|$ the operation $\sigma \mapsto a\sigma$ by recursion on σ:

$$a\Lambda = a\ , \text{ and}$$

$$a\, i\, \sigma = (a_i)\,\sigma\ .$$

The $a\sigma$ are just the *subtrees* of a with σ as a *selector*. We also have a map

$$\text{pos} : A \to T$$

where

$$\text{pos}(+\langle a_0, a_1, \ldots, a_{n-1}\rangle) = \text{true, and}$$

$$\text{pos}(-\langle a_0, a_1, \ldots, a_{n-1}\rangle) = \text{false.}$$

We say that a (total) tree a is *eventually periodic* iff the set $\{a\sigma \mid \sigma \in \Sigma^*\}$ is finite. The result is that the "language"

$$L_a = \{\sigma \in \Sigma^* \mid \text{pos}(a\sigma) = \text{true}\}$$

corresponding to an eventually periodic tree is always a *regular event* of automata theory, and every such language has this form. In fact, a just represents the initial state of an automaton, and $a\sigma$ represents the state after "reading" a tape σ. □

In order to formulate more generally the idea of a domain equation and initial algebra, we must introduce a small amount of the terminology of category theory. To be as specific as possible, think of systems $\mathcal{D}$ over sets $\Delta \subseteq \Sigma^*$ with $\Sigma = \{0,1\}$, say. They form quite an interesting category with respect to the approximable maps $f : \mathcal{D} \to \mathcal{D}'$. Recall that to be a category of "domains" and "maps" all that is required is an associative composition $g \circ f$ of maps with identity maps $I : \mathcal{D} \to \mathcal{D}$ for each domain of the category. And this we certainly have for the systems indicated. And there are many other categories waiting around: for instance, restrict systems to those where $\emptyset \notin \mathcal{D}$. This is not much of a restriction, as every system is isomorphic to one like this. Or restrict the maps to being the strict maps $f : \mathcal{D} \to \mathcal{D}'$ where $f(\bot_{\mathcal{D}}) = \bot_{\mathcal{D}'}$. This is an essentially different, though related category. We shall find many others.

What examples 6.1 and 6.2 suggest is the notion of a construct which makes new domains out of old. For example, with $\mathcal{D}$ fixed, 6.1 suggests for any domain X over $\Gamma \subseteq \Sigma^*$ a domain

$$T(X) = \mathcal{D} + (X \times X).$$

More specifically (converting from $\Sigma = \{0,1,2\}$ to $\Sigma = \{0,1\}$) we could write

$$T(X) = \{\Gamma'\} \cup \{0X \mid X \in \mathcal{D}\} \cup \{10X \cup 11Y \mid X,Y \in X\},$$

where we have $\Gamma' = 0\Delta \cup 10\Gamma \cup 11\Gamma$. (By the way, here we definitely want to assume $\emptyset \notin \mathcal{D}$ and $\emptyset \notin X$ and to get $\emptyset \notin T(X)$.) This construct is an example of a *functor*, a notion that can be defined abstractly on any category.

DEFINITION 6.3. A *functor* on a category (into itself) associates with every domain X in the category another domain $T(X)$ *and* to every map

$$f : X \to Y$$

another map

$$T(f) : T(X) \to T(Y)$$

in such a way that identity maps and compositions are preserved:

$$T(I_{\mathcal{X}}) = I_{T(\mathcal{X})}, \text{ and}$$
$$T(g \circ f) = T(g) \circ T(f),$$

whenever $f : \mathcal{X} \to \mathcal{Y}$ and $g : \mathcal{Y} \to \mathcal{Z}$. □

In the example from 6.1 we have not checked how the special T is a functor. The hint is that whenever $f : \mathcal{X} \to \mathcal{Y}$, then there is a map

$$f \times f : \mathcal{X} \times \mathcal{X} \to \mathcal{Y} \times \mathcal{Y}.$$

But there is also a map

$$I_{\mathcal{D}} + f \times f : \mathcal{D} + (\mathcal{X} \times \mathcal{X}) \to \mathcal{D} + (\mathcal{Y} \times \mathcal{Y})$$

and this suggests the definition of T(f). The details are left to the exercises. Note that the map T(f) just suggested is always strict, so T is a functor also for the category of strict maps.

One good reason for a little of the category-theoretic language is that the next definition becomes very neat indeed.

DEFINITION 6.4. A T-*algebra* is a domain $\mathcal{E}$ in the category together with a map

$$k : T(\mathcal{E}) \to \mathcal{E}.$$

If $m : T(\mathcal{F}) \to \mathcal{F}$ is another T-algebra, then a *homomorphism* is a map $h : \mathcal{E} \to \mathcal{F}$ in the category such that the diagram

$$\begin{array}{ccc} T(\mathcal{E}) & \xrightarrow{k} & \mathcal{E} \\ {\scriptstyle T(h)}\downarrow & & \downarrow{\scriptstyle h} \\ T(\mathcal{F}) & \xrightarrow{m} & \mathcal{F} \end{array}$$

commutes; that is, the equation

$$h \circ k = m \circ T(h)$$

holds. □

In our example from 6.1 a T-algebra is a *strict* map

$$k : \mathcal{D} + (E \times E) \to E .$$

But such strict maps are in a one-one correspondence with pairs of (not necessarily strict) maps

$$n : \mathcal{D} \to E \quad \text{and} \quad p : E \times E \to E .$$

And the structure $\langle E,n,p\rangle$ is what we called a tree algebra. Definition 6.4 just makes this abstract. The reader should also work out the details showing that 6.4's definition of homomorphism is just what we ought to expect.

Note that the T-algebras and homomorphisms form a category. (Why?) The following definition is so abstract that it could be given for *any* category.

DEFINITION 6.5. A T-algebra is *initial* if and only if there is a unique homomorphism from it into any other T-algebra. □

The word "other" here is not meant to imply "distinct". For an initial algebra there is one and only one homomorphism into itself: the identity map. As we already indicated in 6.1 it is a general fact that the next proposition holds.

PROPOSITION 6.6. Any two initial T-algebras are uniquely isomorphic. □

Slightly more interesting is the behaviour of T on initial algebras.

PROPOSITION 6.7. If $i : T(\mathcal{D}) \to \mathcal{D}$ is an initial T-algebra, then so is $T(i) : T^2(\mathcal{D}) \to T(\mathcal{D})$ and i is the isomorphism from $T(\mathcal{D})$ to $\mathcal{D}$.

Proof: Clearly since T is a functor, the map T(i) has the right mapping character to make $T(\mathcal{D})$ a T-algebra. Since $\mathcal{D}$ is initial, we have a commuting diagram :

$$\begin{array}{ccc} T(\mathcal{D}) & \xrightarrow{\ i\ } & \mathcal{D} \\ {\scriptstyle T(j)}\downarrow & & \downarrow{\scriptstyle j} \\ T^2(\mathcal{D}) & \xrightarrow{T(i)} & T(\mathcal{D}) \end{array}$$

But we also have the trivial diagram:

$$\begin{array}{ccc} T^2(\mathcal{D}) & \xrightarrow{T(i)} & T(\mathcal{D}) \\ {\scriptstyle T(i)}\downarrow & & \downarrow{\scriptstyle i} \\ T(\mathcal{D}) & \xrightarrow{\ i\ } & \mathcal{D} \end{array}$$

It follows that $i \circ j$ is a homomorphism, so

$$i \circ j = I_{\mathcal{D}}.$$

But then because T is a functor we find:

$$T(i) \circ T(j) = I_{T(\mathcal{D})},$$

and, since j is a homomorphism, we have

$$j \circ i = I_{T(\mathcal{D})}.$$

This shows that i is an isomorphism. □

From 6.7 we see that if we are going to have initial algebras at all we have to satisfy the domain equation

$$\mathcal{D} \cong T(\mathcal{D}).$$

But generally that is not enough to assure that $\mathcal{D}$ is initial. There is a condition that our functors satisfy, however, which guarantees the existence of homomorphisms.

DEFINITION 6.8. On the category of domains and strict approximable maps a functor T is *continuous on maps* if for any systems $\mathcal{D}$ and E the induced mapping

$$\lambda f.\ T(f) : (\mathcal{D} \to_{\perp} E) \to (T(\mathcal{D}) \to_{\perp} T(E))$$

is approximable.

THEOREM 6.9. If the functor T is continuous on maps and if $\mathcal{D} \cong T(\mathcal{D})$, so in particular $\mathcal{D}$ is a T-algebra, then for any T-algebra $k : T(\mathcal{E}) \to \mathcal{E}$ there is a homomorphism $h : \mathcal{D} \to \mathcal{E}$.

Proof: Let $i : T(\mathcal{D}) \to \mathcal{D}$ make $\mathcal{D}$ a T-algebra, where $j : \mathcal{D} \to T(\mathcal{D})$ is the inverse so that i is an isomorphism of domains. Suppose that $k : T(\mathcal{E}) \to \mathcal{E}$ is any T-algebra. A homomorphism $h : \mathcal{D} \to \mathcal{E}$ would satisfy

$$h \circ i = k \circ T(h).$$

Rewrite this equation as

$$h = k \circ T(h) \circ j.$$

In the domain of strict maps $(\mathcal{D} \to \mathcal{E})$ this is a fixed-point equation for an approximable map

$$\lambda h.\, k \circ T(h) \circ j$$

by our assumption on T. Thus, the desired homomorphism exists. □

The final question we have to answer is why in our category the minimal $\mathcal{D}$ exist. The reason is that the functors T that we have in mind possess further continuity properties on domains. This is conveniently expressed in terms of a notion of "subdomain".

DEFINITION 6.10. For two neighbourhood systems $\mathcal{D}$ and $\mathcal{E}$ we write

$$\mathcal{D} \lhd \mathcal{E}$$

to mean that these are neighbourhood systems over the same set of tokens Δ and not only is $\mathcal{D} \subseteq \mathcal{E}$ but whenever $X, Y \in \mathcal{D}$ and $X \cap Y \in \mathcal{E}$, then $X \cap Y \in \mathcal{D}$. □

For the subdomain relation $\mathcal{D} \lhd \mathcal{E}$ to hold, $\mathcal{D}$ has to be a smaller family of neighbourhoods, but the notion of consistency in $\mathcal{D}$ also has to be the same as in $\mathcal{E}$. Note that if $\mathcal{D}_0 \lhd \mathcal{E}$ and $\mathcal{D}_1 \lhd \mathcal{E}$ then

$$\mathcal{D}_0 \lhd \mathcal{D}_1 \text{ iff } \mathcal{D}_0 \subseteq \mathcal{D}_1 .$$

It is also easy to prove that the union of a directed family of subdomains of $\mathcal{E}$ is again a subdomain. As a consequence of this remark we have:

PROPOSITION 6.11. For a given neighbourhood system $\mathcal{E}$, the set of subsystems

$$\{\mathcal{D} \mid \mathcal{D} \lhd \mathcal{E}\}$$

forms a domain in its own right. □

The subdomain relationship implies a mapping relationship between the domains.

PROPOSITION 6.12. If $\mathcal{D} \lhd \mathcal{E}$, then there exists a projection pair of approximable mappings:

$$i : \mathcal{D} \to \mathcal{E} \quad \text{and } j : \mathcal{E} \to \mathcal{D}$$

where $j \circ i = I_{\mathcal{D}}$ and $i \circ j \subseteq I_{\mathcal{E}}$, which are determined as element-wise functions by these equations:

$$i(x) = \{Y \in \mathcal{E} \mid \exists X \in x . X \subseteq Y\}, \quad \text{and}$$

$$j(y) = y \cap \mathcal{D},$$

for all $x \in |\mathcal{D}|$ and $y \in |\mathcal{E}|$. □

The proof is left for the exercises.

DEFINITION 6.13. A functor T is *monotone on domains* iff whenever $\mathcal{D} \lhd \mathcal{E}$, then not only do we have $T(\mathcal{D}) \lhd T(\mathcal{E})$ but the projection pair i, j of 6.12 is mapped to the same kind of projection pair $T(i)$, $T(j)$. A monotone functor is *continuous on domains* iff whenever $\mathcal{E}$ is a domain, then the mapping

$$\lambda \mathcal{D}. T(\mathcal{D}) : \{\mathcal{D} \mid \mathcal{D} \lhd \mathcal{E}\} \to \{\mathcal{D}' \mid \mathcal{D}' \lhd T(\mathcal{E})\}$$

is approximable. □

We can now state an existence theorem that covers in fairly wide generality the examples of this lecture.

THEOREM 6.14. If the functor T is continuous on maps and monotone and continuous on domains, and if there is a set Γ such that

$$\{\Gamma\} \lhd T(\{\Gamma\}),$$

then there exists an initial T-algebra.

Proof: We proceed as in the proof of the fixed-point theorem by iterating the functor. The assumption about Γ means that, as a neighbourhood system, $T(\{\Gamma\})$ is a system over the *same* set Γ. Thus, if we iterate T to form $T^n(\{\Gamma\})$, all these systems are over Γ and indeed

$$T^n(\{\Gamma\}) \lhd T^{n+1}(\{\Gamma\})$$

for all n. We can thus introduce

$$\mathcal{D} = \bigcup_{n=0}^{\infty} T^n(\{\Gamma\}),$$

and it is easy to check that $\mathcal{D}$ is a system over Γ and

$$T^n(\{\Gamma\}) \lhd \mathcal{D}$$

holds for all n. But then we have for all n:

$$T^n(\{\Gamma\}) \lhd T^{n+1}(\{\Gamma\}) \lhd T(\mathcal{D}),$$

which implies $\mathcal{D} \lhd T(\mathcal{D})$. But T is continuous on domains, so

$$\begin{aligned} T(\mathcal{D}) &= T\Big(\bigcup_{n=0}^{\infty} T^n(\{\Gamma\})\Big) \\ &= \bigcup_{n=0}^{\infty} T^{n+1}(\{\Gamma\}) \\ &= \mathcal{D}\,. \end{aligned}$$

Thus, not only is $\mathcal{D}$ a T-algebra, but the isomorphism we get for $\mathcal{D}$ and $T(\mathcal{D})$ is just the identity mapping. We know by 6.9 that homomorphisms exist; what remains to show is that homomorphism from $\mathcal{D}$ are unique. As in the examples, we will show in effect they are determined uniquely on the finite elements of $\mathcal{D}$.

Since each $T^n(\{\Gamma\}) \lhd \mathcal{D}$, there are projection mappings

$$i_n : T^n(\{\Gamma\}) \to \mathcal{D} \text{ and } j_n : \mathcal{D} \to T^n(\{\Gamma\}).$$

Define $\rho_n : \mathcal{D} \to \mathcal{D}$ by $\rho_n = i_n \circ j_n$. Projection pairs are always pairs of strict mappings (Why?), and so are in the category. By assumption and 6.13, the functor T preserves these maps, so we have

$$T(\rho_n) = T(i_n) \circ T(j_n) = i_{n+1} \circ j_{n+1} = \rho_{n+1} \,.$$

As a neighbourhood relation ρ_n can be characterized by :

$$X \rho_n Y \text{ iff } \exists Z \in T^n(\{\Gamma\}).\ X \subseteq Z \subseteq Y.$$

We thus see that $\rho_n \subseteq \rho_{n+1}$ and

$$\bigcup_{n=0}^{\infty} \rho_n = I_{\mathcal{D}}.$$

Now suppose $k : T(E) \to E$ is any T-algebra and $h : \mathcal{D} \to E$ is a homomorphism. The mapping will satisfy the fixed-point equation

$$h = k \circ T(h),$$

where no other mappings need be written in because $\mathcal{D} = T(\mathcal{D})$ and so

$$T(h) : \mathcal{D} \to T(E) \,.$$

We wish to show that h really is the least fixed point of this equation.

Define $h_n = h \circ \rho_n : \mathcal{D} \to E$. For $n = 0$, the map ρ_0 is the trivial map where $\rho_0(x) = \bot_{\mathcal{D}}$ for all $x \in |\mathcal{D}|$. But h must be strict, so $h_0(x) = \bot_E$ for all $x \in |\mathcal{D}|$; that is, h_0 is the least element of $|\mathcal{D} \to_{\bot} E|$. Now calculate :

$$k \circ T(h_n) = k \circ T(h) \circ T(\rho_n)$$
$$= h \circ \rho_{n+1}$$
$$= h_{n+1} .$$

This shows that the union of the h_n is the least fixed point of $\lambda h.k \circ T(h)$. But

$$\bigcup_{n=0}^{\infty} h_n = \bigcup_{n=0}^{\infty} h \circ \rho_n$$
$$= h \circ \bigcup_{n=0}^{\infty} \rho_n$$
$$= h \circ I_{\mathcal{D}} = h,$$

so the given h is in fact the least fixed point. The homomorphism is uniquely determined, and $\mathcal{D}$ is the initial T-algebra. □

Having the existence of initial T-algebras, we can prove one more result that shows just how minimal they are. We need a lemma about projection pairs, first, that shows where subdomains fit it. We write $\mathcal{D} \trianglelefteq\!\!\!\sim \mathcal{E}$ as short for $\mathcal{D} \cong \mathcal{D}'$ for some $\mathcal{D}' \triangleleft \mathcal{E}$ in the following. The lemma gives a converse to 6.12.

LEMMA 6.15. For two neighbourhood systems $\mathcal{D}$ and $\mathcal{E}$, if there exist a projection pair

$$i : \mathcal{D} \to \mathcal{E} \text{ and } j : \mathcal{E} \to \mathcal{D}$$

with $j \circ i = I_{\mathcal{D}}$ and $i \circ j \subseteq I_{\mathcal{E}}$, then $\mathcal{D} \trianglelefteq\!\!\!\sim \mathcal{E}$.

Proof. What we want to show is that i maps finite elements to finite elements, and that the desired $\mathcal{D}'$ is the image of $\mathcal{D}$ in $\mathcal{E}$.

Suppose $X \in \mathcal{D}$. We can write:

$$i(\uparrow X) = \bigcup\{\uparrow Y \mid Y \in i\ (\uparrow X)\}.$$

Applying j to both sides we have:

$$\uparrow X = j \circ i\,(\uparrow X) = \bigcup \{ j(\uparrow Y) \mid Y \in i(\uparrow X) \}.$$

But then, since $X \in \uparrow X$, we find $X \in j\,(\uparrow Y)$ for some $Y \in i\,(\uparrow X)$. This implies

$$\uparrow X \subseteq j\,(\uparrow Y)\ ;\ \text{and so } i(\uparrow X) \subseteq i \circ j\,(\uparrow Y) \subseteq \uparrow Y.$$

Since $\uparrow Y \subseteq i\,(\uparrow X)$ in any case, we conclude $i\,(\uparrow X) = \uparrow Y$. This proves finite elements *are* mapped to finite elements.

What of Δ; that is, what is $i\,(\uparrow \Delta)$? We find, supposing $\mathcal{E}$ to be a neighbourhood system over a set Δ', that since $\uparrow \Delta \subseteq j\,(\uparrow \Delta')$, then $i\,(\uparrow \Delta) \subseteq \uparrow \Delta'$ and so $i\,(\uparrow \Delta) = \uparrow \Delta'$. This means that Δ corresponds to Δ'. So we have established that $\mathcal{D}$ is in an inclusion preserving one-one correspondence with a subset $\mathcal{D}'$ of $\mathcal{E}$ where $\Delta' \in \mathcal{D}'$. But it remains to show that $\mathcal{D}'$ is a neighbourhood system and that $\mathcal{D}' \lhd \mathcal{E}$ holds. All we really have to show is that $\mathcal{D}'$ is closed under intersection whenever the intersection belongs to $\mathcal{E}$.

Suppose $Y', Z' \in \mathcal{D}'$ and $Y' \cap Z' \in \mathcal{E}$. Let $X' = Y' \cap Z'$. We have, for suitable $Y, Z \in \mathcal{D}$,

$$i(\uparrow Y) = \uparrow Y',\ \text{and so } \uparrow Y = j(\uparrow Y');\ \text{ and}$$

$$i(\uparrow Z) = \uparrow Z',\ \text{and so } \uparrow Z = j(\uparrow Z').$$

But $\uparrow Y' \subseteq \uparrow X'$ and $j(\uparrow Y') \subseteq j(\uparrow X')$; thus $Y \in j(\uparrow X')$. For similar reasons $Z \in j\,(\uparrow X')$. But then $X = Y \cap Z \in j(\uparrow X')$, and therefore $Y \cap Z \in \mathcal{D}$. (The element $j(\uparrow X')$ must be a filter.) Notice, however, that

$$\uparrow Y \subseteq \uparrow X,\ \text{and so } \uparrow Y' \subseteq i\,(\uparrow X)\ ;\ \text{and}$$
$$\uparrow Z \subseteq \uparrow X,\ \text{and so } \uparrow Z' \subseteq i\,(\uparrow X)\ .$$

It follows that $Y' \cap Z' = X' \in i(\uparrow X)$. On the other hand we already knew $X \in j\,(\uparrow X')$, which implies $i(\uparrow X) \subseteq \uparrow X'$. We may thus conclude that $i(\uparrow X) = \uparrow X'$. In other words $X' \in \mathcal{D}'$. □

THEOREM 6.16. If on the category of domains and strict approximable maps the functor T is continuous on maps, and if $\mathcal{D}$ is an initial T-algebra, then for any system $\mathcal{E} \cong T(\mathcal{E})$ we have $\mathcal{D} \trianglelefteq \mathcal{E}$.

Proof: There is a homomorphism $h : \mathcal{D} \to \mathcal{E}$. By 6.9 there is a homomorphism $g : \mathcal{E} \to \mathcal{D}$. Now $g \circ h : \mathcal{D} \to \mathcal{D}$ is also a homomorphism, so $g \circ h = I_{\mathcal{D}}$ because $\mathcal{D}$ is initial. In view of 6.15, all we have to prove now is that $h \circ g \subseteq I_{\mathcal{E}}$.

Let the maps $i : T(\mathcal{D}) \to \mathcal{D}$ and $j : \mathcal{D} \to T(\mathcal{D})$ give the isomorphism for $\mathcal{D}$, and let $u : T(\mathcal{E}) \to \mathcal{E}$ and $v : \mathcal{E} \to T(\mathcal{E})$ do the same for $\mathcal{E}$. By the proof of 6.9 we know

$$g = i \circ T(g) \circ v \text{ and } h = u \circ T(h) \circ j$$

and each of these maps is the least fixed point of its respective equation. Let

$$g_0 = \bot_{\mathcal{E} \to \mathcal{D}} \quad \text{and} \quad h_0 = \bot_{\mathcal{D} \to \mathcal{E}}$$

and define by recursion

$$g_{n+1} = i \circ T(g_n) \circ v \text{ and } h_{n+1} = u \circ T(h_n) \circ j.$$

By the fixed-point calculation

$$g = \bigcup_{n=0}^{\infty} g_n \quad \text{and} \quad h = \bigcup_{n=0}^{\infty} h_n.$$

Now we see that

$$h_0 \circ g_0 = \bot_{\mathcal{E} \to \mathcal{E}},$$

and for each n that

$$\begin{aligned} h_{n+1} \circ g_{n+1} &= u \circ T(h_n) \circ j \circ i \circ T(g_n) \circ v \\ &= u \circ T(h_n) \circ T(g_n) \circ v \\ &= u \circ T(h_n \circ g_n) \circ v. \end{aligned}$$

But this means that

$$h \circ g = \bigcup_{n=0}^{\infty} (h_n \circ g_n)$$

is the least fixed point for the equation

$$k = u \circ T(k) \circ v.$$

But $I_{\mathcal{E}}$ is *one* of the fixed points; whence $h \circ g \subseteq I_{\mathcal{E}}$ must follow. □

EXERCISES

EXERCISE 6.17. What are the algebras for which C is initial? If A of 6.2 is a generalization of B, what is the corresponding generalization of C? Prove that it exists and explain what are the algebras involved.

EXERCISE 6.18. With reference back to Exercise 3.16 discuss the construction of $\mathcal{D}^\infty$ as an initial algebra and as a solution to the domain equation

$$\mathcal{D}^\infty \cong \mathcal{D} \times \mathcal{D}^\infty .$$

(I do not know whether all solutions must be of the form $\mathcal{D}^\infty \times E$.)

EXERCISE 6.19. For the sake of uniformity restrict attention to systems $\mathcal{D}$ on sets $\Delta \subseteq \{0,1\}^*$, where $\Lambda \in \Delta$ and $\emptyset \notin \mathcal{D}$, and to the category of strict maps. Define sum and product by:

$$\mathcal{D}_0 + \mathcal{D}_1 = \{\{\Lambda\} \cup 0\Delta_0 \cup 0\Delta_1\} \cup \{0X \mid X \in \mathcal{D}_0\} \cup \{1Y \mid Y \in \mathcal{D}_1\},$$

$$\mathcal{D}_0 \times \mathcal{D}_1 = \{\{\Lambda\} \cup 0X \cup 1Y \mid X \in \mathcal{D}_0 \text{ and } Y \in \mathcal{D}_1\}.$$

Are these correct up to isomorphism? Now generate all constructs T(X) formed by the constants (that is, $T(X) = \mathcal{D}$ for a fixed $\mathcal{D}$), by the identity ($T(X) = X$), and by sums and products ($T_0(X) + T_1(X)$, etc.) Show that these are all functors, continuous on maps, and monotone and continuous on domains.

EXERCISE 6.20. For any system $\mathcal{D}$ let tok($\mathcal{D}$) be the underlying set of tokens, so that $\mathcal{D}$ is a system over tok($\mathcal{D}$). For the category of Exercise 6.19 show that the function

$$\lambda\Gamma.\,\mathrm{tok}(T(\{\Gamma\}))$$

is continuous on the domain $\{\Gamma \subseteq \{0,1\}^* \mid \Lambda \in \Gamma\}$, where T is any of the functors generated in 6.19. Conclude that there must exist a set

$$\Gamma = \mathrm{tok}(T(\{\Gamma\})) ,$$

so that $\{\Gamma\} \lhd T(\{\Gamma\})$, and so 6.14 applies.

EXERCISE 6.21. Do the same as 6.19 and 6.20 when the functors are also allowed to be generated by the operations:

$$\mathcal{D}_0 \oplus \mathcal{D}_1 = \{\{\Lambda\} \cup 0\Delta_0 \cup 1\Delta_1\} \cup \{0X \mid X \in \mathcal{D}_0 \setminus \{\Delta_0\}\} \cup \{1Y \mid Y \in \mathcal{D}_1 \setminus \{\Delta_1\}\},$$

$$\mathcal{D}_0 \otimes \mathcal{D}_1 = \{\{\Lambda\} \cup 0\Delta_0 \cup 1\Delta_1\} \cup \{\{\Lambda\} \cup 0X \cup 1Y \mid X \in \mathcal{D}_0 \setminus \{\Delta_0\} \text{ and } Y \in \mathcal{D}_1 \setminus \{\Delta_1\}\}.$$

Generalize all of $+$, $\times$, $\oplus$, $\otimes$ to combinations of several terms, not just the binary sums and products.

EXERCISE 6.22. Comment on these domain equations:

$$N \cong \{\{0\},\{0,\Lambda\}\} \oplus N,$$

$$M \cong \{\{\Lambda\}\} + M,$$

$$N^* \cong N \oplus (N \otimes N^*).$$

EXERCISE 6.23. Construe the initial solution to

$$Exp \cong N \oplus ((Exp \times Exp) + (Exp \times Exp))$$

as a "syntactical domain" of *expressions* generated from infinitely many "variables" by means of two binary "operation symbols". Given an algebra $\mathcal{D}$ with two operations

$$u : \mathcal{D} \times \mathcal{D} \to \mathcal{D} \text{ and } v : \mathcal{D} \times \mathcal{D} \to \mathcal{D},$$

show how any strict map $s : N \to \mathcal{D}$ determines a unique map

$$\mathrm{val}(s) : Exp \to \mathcal{D}$$

that can be regarded as the "evaluation of an expression".

EXERCISE 6.24. Show that there must exist domains satisfying:

$$\mathcal{D} \cong \mathcal{D} + (\mathcal{D} \times E), \text{ and}$$

$$E \cong \mathcal{D} + E,$$

by using a double fixed-point method. First decide what the underlying set of tokens should be, and then define $\mathcal{D}$ and E by simultaneous fixed points. (Syntactical domains as in 6.23 may very well require several simultaneous equations.)

EXERCISE 6.25. For a projection pair $g : \mathcal{D} \to E$ and $h : E \to \mathcal{D}$ show that for $x \in |\mathcal{D}|$ and $y \in |E|$ we have:

$$g(x) \subseteq y \text{ iff } x \subseteq h(y).$$

Thus, conclude that:

$$h(y) = \bigcup\{x \in |\mathcal{D}| \mid g(x) \subseteq y\}, \text{ and}$$

$$g(x) = \bigcap\{y \in |E| \mid x \subseteq h(y)\},$$

for all $x \in |\mathcal{D}|$ and $y \in |E|$. So each of the functions determines the other. In the first equation check that the set on the right is directed, and in the second equation that the set on the right is non empty. Prove also that g maps consistent sets to consistent sets and preserves $\bigsqcup$ (not just directed unions).

EXERCISE 6.26. For systems $\mathcal{D}$ as in 6.19 define

$$\mathcal{D}_\perp = \{\{\Lambda\} \cup 0\,\Delta\} \cup \{0X \mid X \in \mathcal{D}\}.$$

Describe the construct in terms of elements. Is this a suitable functor? Prove that

$$\mathcal{D}_\perp \oplus E_\perp \cong \mathcal{D} + E .$$

What is

$$\mathcal{D}_\perp \otimes E_\perp \cong \,??$$

EXERCISE 6.27. Which of the following relationships are true:

$$(\mathcal{D} \otimes E) \trianglelefteq (\mathcal{D} \times E) \ ; \ \mathcal{D} \trianglelefteq \mathcal{D} \times E \ ;$$

$$(\mathcal{D} \oplus E) \trianglelefteq (\mathcal{D} + E) \ ; \ \mathcal{D} \trianglelefteq \mathcal{D} \oplus E \ ;$$

$$(\mathcal{D} \to_\perp E) \trianglelefteq (\mathcal{D} \to E); \ \mathcal{D} \trianglelefteq \mathcal{D} \otimes E \ ?$$

EXERCISE 6.28. (Suggested by G. Plotkin). Show that if $\mathcal{D}$ and E are *finite* systems and

$$\mathcal{D} \trianglelefteq E \trianglelefteq \mathcal{D} ,$$

then $\mathcal{D} \cong E$. Need the same be true of infinite systems?

EXERCISE 6.29. Generalize + and × to infinitary operations on domains:

$$\sum_{n=0}^{\infty} \mathcal{D}_n \quad \text{and} \quad \prod_{n=0}^{\infty} \mathcal{D}_n \ .$$

Would a similar generalization be possible for ⊕ and ⊗ ?

LECTURE VII

COMPUTABILITY IN EFFECTIVELY GIVEN DOMAINS

For the domain N the strict functions from N into N, the strict maps $f : N \to N$, correspond exactly to the partial functions $g : \mathbf{N} \to \mathbf{N}$ (as we wrote in 5.6 we had $f = \bar{g}$). For such functions there is a standard theory of computability: g is called computable if it can be defined as a partial recursive function with its "program" written down in a certain standard form. The non-strict maps $h : N \to N$ are all constant, and so are intuitively computable; so we know all about computable maps in $|N \to N|$ in general. The question is: what are the computable maps on (elements of) other domains?

The answer will of course depend on how the domain is presented to us. Even with N, there are continuum many isomorphisms $\pi : N \to N$ of N onto itself, not all of which can be computable. That is, if we permute N and, so to speak, present the integers in a different order, then a well-behaved computable function $f : N \to N$ may well be transformed into a non-computable function,

$$\pi \circ f \circ \pi^{-1} : N \to N.$$

(Hint: Consider the characteristic function e of the even numbers. Take $f = \bar{e}$ and let π be very horrid.) The reason we imagined we knew which were the computable $f : N \to N$ is that N is always thought of in a standard presentation. We must thus define "in general" a concept of an *effectively given domain*, that is to say, one with a sufficiently computable presentation to represent the additional knowledge about the domain.

The main idea will be that the *finite* elements of $|D|$ should be regarded as the ones initially known. Abstractly, to know a finite element is to know how it is *related to* other finite elements.

Of course, this will mean that we will allow at most a countable infinity of finite elements - but this restriction well accords with intuition. To make precise the terminology "related to" it proves most convenient to go back to the neighbourhoods (in any case they are in a one-one correspondence with the finite elements).

DEFINITION 7.1. A neighbourhood system $\mathcal{D}$ has a *computable presentation* provided we can write

$$\mathcal{D} = \{ X_n \mid n \in \mathbb{N} \},$$

where the following two relations

(i) $X_n \cap X_m = X_k$; and

(ii) $\exists k \in \mathbb{N} .\ X_k \subseteq X_n$ and $X_k \subseteq X_m$

are recursively decidable (in integer indices n, m, k and in n, m, respectively). □

More strictly the sequence,

$$\langle X_n \rangle_{n=0}^{\infty},$$

is the presentation. Even more strictly, when it is required to cope with infinitely many domains at a time, it would be necessary to give the actual Gödel numbers of the recursive relations (i) and (ii) (rather than just saying there exists some way of showing them to be recursively decidable).

The intuitive idea of 7.1 is that the system is effectively given if you know how to do elementary "calculations" with neighbourhoods. The basic calculations are the forming of intersections. The neighbourhoods have to be laid out in a systematic way; and, if we are asked for an intersection of two given neighbourhoods, we have to be able to locate it in the standard sequence. Relation (ii) is the *consistency condition*, which is the necessary and sufficient condition for the intersection to exist in $\mathcal{D}$. When (ii) is true, therefore, we have only to try $k = 0,1,2,$ ... until we discover that we have found the intersection. We are assuming that these basic decisions can be carried out in "finite time". Note that the obvious biconditional,

$$X_n \subseteq X_m \text{ iff } X_n \cap X_m = X_n,$$

assures us that the inclusion relation between neighbourhoods is itself decidable in terms of the indices. So in (ii) *if* k exists, *then* it (or the first one) can indeed be found in finite time. The rub is that if it *does not exist*, no finite number of inclusion checks will determine that fact. That is why we have to *assume* that (ii) is always decidable. The information contained in (ii) is a fundamental part of the neighbourhood structure. (An axiomatic characterization of neighbourhood structures is given in Exercise 7.13, which may make clearer what we are assuming and what a presentation is.)

DEFINITION 7.2. Given two recursively presented domains,

$$\mathcal{D} = \{X_n \mid n \in \mathbb{N}\} \text{ and } \mathcal{E} = \{Y_m \mid m \in \mathbb{N}\},$$

an approximable mapping $f: \mathcal{D} \to \mathcal{E}$ is said to be *computable* iff the relation

$$X_n \, f \, Y_m$$

is recursively enumerable in n and m. □

The question to ask first is why "recursively enumerable" rather than "recursive" (= "recursively decidable")? The answer will become clear when we let $\mathcal{D}$ degenerate to the one-element domain, $\mathcal{D} = \{\Delta\}$. Then what we are considering is merely a single element

$$y = f(\{\Delta\}) \in |\mathcal{E}|.$$

Therefore, 7.2 incorporates the notion of a *computable element* of a domain. And the condition reduces to the statement that the filter $y \in |\mathcal{E}|$ is such that the set

$$\{m \in \mathbb{N} \mid Y_m \in y\}$$

is a recursively enumerable set of integers. The point is that the elements of $|\mathcal{E}|$ are finite or infinite. If y were finite, the set of indices above would indeed be recursive in view of our assumptions on E. But an infinite element can in general only be approximated "a little at a time". We cannot expect to know the whole story of its approximations in a flash. What it means to be recursively enumerable is that there is a primitive recursive function (hence, a *total* function), $r : \mathbf{N} \to \mathbf{N}$, such that

$$y = \{Y_{r(i)} \mid i \in \mathbf{N}\}.$$

That is to say, *all* the approximations to y can *eventually* be listed. In the case of the mapping f we could write

$$f = \{(X_{s(i)}, Y_{r(i)}) \mid i \in \mathbf{N}\},$$

for a suitable pair of primitive recursive functions s and r.

Definitions 7.1 and 7.2 may very well irritate the person hearing them for the first time: instead of explaining computability in direct terms, the whole question is thrown into the lap of recursion theory! There are several answers. "You have to start somewhere" is one thing I always say. Recursion on the integers is a well-understood theory, and we shall not need the refined parts of the development, fortunately. In any case, our definitions apply to *many* domains of quite different structure, not just to the domain N. And the next step we shall take is to show how to build up computable functions (and also effectively given domains) from simpler ones. Thus, often it will not be necessary to go back to the seemingly over-precise definitions involving the indices but to appeal to some broad general principles.

PROPOSITION 7.3. The identity map on an effectively given domain is computable; the composition of computable mappings on effectively given domains is again computable. □

The proofs for 7.3 are so trivial they are hardly worth an exercise. Note the immediate and useful consequence: if $f : \mathcal{D} \to E$ is computable and $x \in |\mathcal{D}|$ is computable, then $f(x) \in |E|$ is also computable. The next result is, however, worth working out even though it is quite easy.

THEOREM 7.4. If $\mathcal{D}_0$ and $\mathcal{D}_1$ are effectively given, then so are

$$(\mathcal{D}_0 + \mathcal{D}_1) \text{ and } (\mathcal{D}_0 \times \mathcal{D}_1).$$

Moreover the combinators in_i and out_i and proj_i^2 are all computable; further if f and g are computable maps, then so are $f + g$ and $f \times g$.

Proof: Let the computable presentations be given as:

$$\mathcal{D}_i = \{X_n^i \mid n \in \mathbb{N}\}.$$

We can assume that the sets of tokens Δ_0 and Δ_1 are disjoint and $\emptyset \notin \mathcal{D}_i$. Then the construction of the sum is just

$$\mathcal{D}_0 + \mathcal{D}_1 = \{\Delta_0 \cup \Delta_1\} \cup \mathcal{D}_0 \cup \mathcal{D}_1.$$

As an enumeration we define for $n \in \mathbb{N}$:

$$Z_0 = \Delta_0 \cup \Delta_1 \ ; \ Z_{2n+1} = X_n^0 \ ; \ Z_{2n+2} = X_n^1 \ .$$

We leave as an exercise the check of 7.1(i)-(ii).

For the product we want:

$$\mathcal{D}_0 \times \mathcal{D}_1 = \{X_n^0 \cup X_m^1 \mid n, m \in \mathbb{N}\}$$

What we then need are recursive functions $p : \mathbb{N} \to \mathbb{N}$, $q : \mathbb{N} \to \mathbb{N}$, and $r : \mathbb{N} \times \mathbb{N} \to \mathbb{N}$ where for $m, n, k \in \mathbb{N}$ we have:

$$p(r(n, m)) = n \text{ and } q(r(n, m)) = m, \text{ and } r(p(k), q(k)) = k.$$

Thus r is a "one-one pairing function"; there are many ways to find such functions (see Exercise 5.13). We can then define for $k \in \mathbb{N}$:

$$W_k = X_{p(k)}^0 \cup X_{q(k)}^1 \ .$$

Again we leave as an exercise the check that this provides a computable presentation of $\mathcal{D}_0 \times \mathcal{D}_1$.

As for the combinators, the neighbourhood relations have to be worked out in terms of the indices. For example

$$X_n^0 \ \mathrm{in}_0 \ Z_m \text{ iff either } m = 0 \text{ or for some } k$$
$$m = 2k + 1 \text{ and } X_n^0 \subseteq X_k^0 \ .$$

$$W_k \ \mathrm{proj}_1^2 \ X_m^1 \text{ iff } X_{q(k)}^1 \subseteq X_m^1 \ .$$

The reader needs to check that these are recursively enumerable relations in the indices. For this purpose it may be convenient to recall some closure properties of these relations: taking conjunctions, disjunctions, substituting recursive functions, applying an existential quantifier to the front. □

Products give us a way of providing an immediate meaning to the notion of a computable function of several variables. Note that the proof of 3.7 is "effective" and shows that substitution of computable functions of several variables into each other always gives computable functions. We turn next to the function spaces.

THEOREM 7.5. If $\mathcal{D}_0$ and $\mathcal{D}_1$ are effectively given, then so is $(\mathcal{D}_0 \to \mathcal{D}_1)$. The combinators eval and curry are computable, provided all the domains involved are effectively given. The computable elements $f \in |\mathcal{D}_0 \to \mathcal{D}_1|$ are exactly the computable maps $f : \mathcal{D}_0 \to \mathcal{D}_1$.

Proof : The proofs of 3.9, 3.11, and 3.12 were set up with this theorem in mind. If

$$\mathcal{D}_0 = \{X_n | n \in \mathbb{N}\} \text{ and } \mathcal{D}_1 = \{Y_m | m \in \mathbb{N}\}$$

are two effectively given neighbourhood systems, then the neighbourhoods of $(\mathcal{D}_0 \to \mathcal{D}_1)$, by Definition 3.8, are non-empty intersections like

$$\bigcap_{i<q} [X_{n_i}, Y_{m_i}],$$

where $<n_0, n_1, \ldots, n_{q-1}>$ and $<m_0, m_1, \ldots, m_{q-1}>$ are two finite sequences of integers determining the choice of the function-space neighbourhood. In 3.9(i) the test for nonemptiness is given. Assuming the decidability of relations in $\mathcal{D}_0$ and $\mathcal{D}_1$, one remarks that the consistency of *finite sequences* of neighbourhoods is also decidable. (Hint: Test the first two, then form their intersection. Next test the third given neighbourhood against this one set; if consistent, form the intersection, and carry on.) By 3.9(i) at most 2.2^q such sequential checks must be carried out to determine whether the function-space neighbourhood is non empty.

It may not be fun, but the checks can be carried out in finite time. Owing to this decidability, we can therefore enumerate in a systematic way *all* the pairs of finite sequences $\langle n_0, \ldots \rangle$ and $\langle m_0, \ldots \rangle$ that determine neighbourhoods: that is the way that $(\mathcal{D}_0 \to \mathcal{D}_1)$ obtains its enumeration.

Concerning the decidability of the required relations on $(\mathcal{D}_0 \to \mathcal{D}_1)$, we remark first off that consistency is more of the same: to test two finite intersections against each other, just form one big intersection and test it for non-emptiness as before. Secondly, the testing for intersection comes down in the end to testing one typical intersection of $[X, Y]$ - neighbourhoods for equality with another. But equality amounts to two inclusions; inclusion in an intersection amounts to inclusion in each term. Therefore, what we need to do is to check a finite number of statements of the form:

$$\bigcap_{i<q} [X_{n_i}, Y_{m_i}] \subseteq [X_k, Y_\ell].$$

As we pointed out after the proof of 3.9, this inclusion is equivalent to

$$\bigcap \{Y_{m_i} \mid X_k \subseteq X_{n_i}\} \subseteq Y_\ell.$$

By decidability in $\mathcal{D}_0$, we can effectively find the n_i that are needed. Then in $\mathcal{D}_1$ we form the intersection of the corresponding Y_{m_i}. Finally, we check the inclusion. Again, one check in $(\mathcal{D}_0 \to \mathcal{D}_1)$ requires a whole sequence of checks in $\mathcal{D}_0$ and in $\mathcal{D}_1$, but the process is finite. So we have argued that $(\mathcal{D}_0 \to \mathcal{D}_1)$ is effectively given.

In showing that the combinators are computable, we refer first to the proof of 3.11. The typical pair of neighbourhoods possibly belonging to eval is

$$\bigcap_{i<q} [X_{n_i}, Y_{m_i}] \cup X_k \text{ eval } Y_\ell.$$

As we needed not to be so specific, we expressed the holding of this relationship in terms of *all* the functions in the function-space neighbourhood. But we know that the neighbourhood, by 3.9(ii), has a minimal element; it is then sufficient to test for the holding of $X_k f_0 Y_\ell$ at this minimal function f_0. But this test, we have already seen, is decidable. So the pairs in eval actually form a recursive set, not just a recursively enumerable set; thus, eval is a computable function.

The case of curry involves three domains and is a bit more messy. But again, if the required neighbourhoods are written out in full, it will be seen that curry, too, is computable. We leave this minor struggle to the exercises.

The final statement is an easy consequence of the fundamental connection between approximable $f : \mathcal{D}_0 \to \mathcal{D}_1$ as relations and as elements. Recall, as in the proof of 3.10, that we have

$$f \in [X, Y] \text{ iff } X f Y,$$

for all $X \in \mathcal{D}_0$ and $Y \in \mathcal{D}_1$. Therefore,

$$f \in \bigcap_{i<q} [X_{n_i}, Y_{m_i}] \text{ iff } \forall i < q.\, X_{n_i} f Y_{m_i}.$$

It follows that if f is recursively enumerable as a set of pairs, then, by forming all the non-empty intersections (as shown), we get an enumeration of all the neighbourhoods to which f belongs; and this is the same as the filter corresponding to f as an element of the function space. The converse direction is clear. □

We have nearly all our favourite combinators computable, but perhaps the most important one - since it is the key to recursive definitions - is the fixed-point combinator. It is not left out.

THEOREM 7.6. For any effectively given domain D, the combinator $\mathrm{fix} : (\mathcal{D} \to \mathcal{D}) \to \mathcal{D}$ is computable.

Proof : Referring back to the proof of Theorem 4.2 and thinking of

$$\mathcal{D} = \{ X_n \mid n \in \mathbb{N} \}$$

as effectively given, fix as a relation comes down to

$\bigcap_{i<q}[X_{n_i}, X_{m_i}]$ fix X_ℓ iff for some finite sequence
$\Delta = X_{k_0}, \ldots, X_{k_p} = X_\ell$
we have, for each $j<p$,

$$\bigcap \{X_{m_i} \mid X_{k_j} \subseteq X_{n_i}\} \subseteq X_{k_{j+1}}.$$

Inside the "for some finite sequence" all the checks are decidable by assumption on $\mathcal{D}$. But the existential quantification of a decidable predicate always gives a recursively enumerable predicate. (And, as there is no implied bound on the size of the finite sequence we are looking for, this really *is* an enumerable set and not generally a recursive set.) □

The major consequence of what we have done up to this point concerns typed λ-calculus. *Any* expression involving only *effectively given types* and, perhaps, some basic *computable constants* using only the λ, ! -notation defines a computable function of its free variables. And such functions applied to computable arguments give computable values. And such functions have computable least fixed points. Etc., etc. In a definite sense then we have in the "metalanguage", as people say, a quite precise and fully *mathematical programming language* for defining computable operators. It is not a machine implemented language, but it is a mathematically well-defined and easy-to-use language. And when we combine the usual type-definition facility together with *domain equations*, we have an especially powerful language.

PROPOSITION 7.7. For any effectively given domain $\mathcal{D}$, the domain $\mathcal{D}^{\S}$ is also effectively given, and all the combinators of Example 6.1 prove to be computable.

Proof: This proof is essentially an exercise, but it is useful to have an easy-to-grasp example. Indeed, to make things easy to reason about, we can assume that $\mathcal{D}$ is a system over $\Delta = \mathbb{N}$, and that in the presentation where

$$\mathcal{D} = \{X_n \mid n \in \mathbb{N}\},$$

the relation $k \in X_n$ is *recursive* in k and n. (It is worth thinking why this is so.) Of course, a lot of other things are recursive also.

Now what kind of a system is $\mathcal{D}^{\S}$? The construction of 6.1 made it a system over a certain set of strings Γ. For the sake of checking various assertions about computability, we are transposing everything back to $\mathbb{N}$. (These are all denumerable sets in any case.) The set Γ is divided into three equally big parts, and we can do the same for $\mathbb{N}$. Let us write for any $m, k \in \mathbb{N}$ and subset $X \subseteq \mathbb{N}$: $mX + k = \{m.n + k \mid n \in X\}$.
Then by splitting the integers modulo 3 we have:

$$\mathbb{N} = 3\mathbb{N} \cup (3\mathbb{N} + 1) \cup (3\mathbb{N} + 2),$$

and this equation is quite analogous to that for Γ. We then propose this definition for $\mathcal{D}^{\S}$

$$\mathcal{D}^{\S} = \{\mathbb{N}\} \cup \{3X \mid X \in \mathcal{D}\} \cup \{(3X + 1) \cup (3Y + 2) \mid X, Y \in \mathcal{D}^{\S}\},$$

but this does not make the enumeration of $\mathcal{D}^{\S}$ all that obvious. This is one way to do it:

$$V_0 = \mathbb{N} \; ; \; V_{2n+1} = 3X_n \; ; \; V_{2n+2} = (3V_{p(n)} + 1) \cup (3V_{q(n)} + 2).$$

Here p and q are the inverse of the pairing functions mentioned in 7.4 They must be chosen so that $p(n) < n$ and $q(n) < n$ for all $n \in \mathbb{N}$. Thus, in calculating V_k where $k = 2n+2$ we will be using $V_{p(n)}$ and $V_{q(n)}$ where *both* subscripts are strictly less than k. This observation is required so that $m \in V_k$ is going to be a recursive relation. What we claim is that

$$\mathcal{D}^{\S} = \{V_k \mid k \in \mathbb{N}\}.$$

It should be clear that everything on the right belongs to $\mathcal{D}^{\S}$. What needs an inductive argument is that everything in $\mathcal{D}^{\S}$ is eventually of the form V_k. But this should be fairly obvious owing to the properties of $r : \mathbb{N} \times \mathbb{N} \leftrightarrow \mathbb{N}$.

The reader also has to check that 7.1(i)-(ii) hold for the V_k. The idea is that any such check is either (1) trivial, or (2) something already assumed about $\mathcal{D}$ and the X_n, or (3) can be thrown back to some sets V_m with strictly smaller subscripts. Therefore, the checks will give an answer in finite time according to an effective reduction.

Next for the combinators, we have to translate neighbourhood relations into relations among integer indices. A selection of examples must suffice.

$$X_n (\lambda x.x^{\S}) \; V_k \text{ iff } V_{2n+1} \subseteq V_k$$

$$V_m \ \mathrm{proj}_0 \ V_k \text{ iff } k = 0 \text{ or } \exists n \in \mathbb{N}.\ m = 2n+2 \text{ and } V_{p(n)} \subseteq V_k.$$

The reader should write out other cases. □

EXAMPLE 7.8. We have often made reference to the powerset $\mathcal{P}\mathbb{N}$ as a domain and we should check here that it is effectively given. One easy way to see this is to note

$$\mathcal{P}\,\mathbb{N} \cong |T^{\infty}|.$$

The (slight) trouble with $\mathcal{P}\,\mathbb{N}$ is that we usually think of it in terms of *elements* rather than *neighbourhoods*. Going back to Exercise 1.16, we can argue that the neighbourhoods of $\mathcal{P}\mathbb{N}$ are ordered not like the finite sets of integers but in the partial ordering *converse* to that. But this is of no trouble, since all will be decidable. What we need first is an enumeration of all finite sets of integers. We can do this by:

$$E_n = \{k \mid \exists i, j\ .\ i < 2^k \text{ and } n = i + 2^k + j.2^{k+1}\}.$$

The idea is that $k \in E_n$ means that the exponent k does occur in the binary expansion of n as a sum of powers of 2. All finite subsets of $\mathbb{N}$ are of the form E_n. We then find that as a neighbourhood system

$$(\mathcal{P}\mathbb{N}) = \{\mathbb{N} \setminus E_n \mid n \in \mathbb{N}\}.$$

As the relationship $E_n \cup E_m = E_k$ is recursive, there is no trouble in proving that this is a computable presentation. In this system, of course, any two neighbourhoods are consistent. Various combinators on $\mathcal{P}\mathbb{N}$ are suggested in Exercise 7.23. □

We end this chapter with an example of another kind of domain construct. This construct is known as the *Smyth Power Domain*. It is defined for any neighbourhood system $\mathcal{D}$ and results in a new system we shall call here $\mathbb{P}\,\mathcal{D}$. The elements of $\mathbb{P}\,\mathcal{D}$ behave rather like *sets of elements* of $\mathcal{D}$, but since our elements can be either partial or total, there are certain dangers to pushing the analogy too far. For some purposes a rival construct called the *Plotkin Power Domain* is better, but it leads outside the category of neighbourhood systems as defined in these lectures. Do not confuse $\mathcal{P}\mathbb{N}$ with $\mathbb{P}\,\mathcal{D}$.

DEFINITION 7.9. Let $\mathcal{D}$ be any neighbourhood system and define

$$\mathbb{P}\,\mathcal{D} = \{\bigcup_{i<n}(\downarrow X_i) \mid \forall\, i<n.\ X_i \in \mathcal{D}\}.$$

We recall that for any $X \in \mathcal{D}$

$$\downarrow X = \{Y \in \mathcal{D} \mid Y \subseteq X\}.$$

The finite unions in $\mathbb{P}\,\mathcal{D}$ can be empty (i.e. if $n = 0$). □

Formally, the system $\mathbb{P}\,\mathcal{D}$ is just more or less the closure of $\mathcal{D}$ under finite unions; however, this would not be an isomorphism-invariant construct unless $\mathcal{D}$ is "prepared". The preparation consists of replacing $\mathcal{D}$ by the isomorphic domain

$$\mathcal{D}^{\downarrow} = \{\downarrow X \mid X \in \mathcal{D}\}.$$

(In this connection refer back to Exercise 1.20.) We remark that

$$\downarrow X \cap \downarrow Y \neq \emptyset \text{ iff } \{X,Y\} \text{ is consistent in } \mathcal{D},$$

and in that case

$$\downarrow X \cap \downarrow Y = \downarrow (X \cap Y).$$

PROPOSITION 7.10. The power domain $\mathbb{P}\,\mathcal{D}$ is a neighbourhood system if $\mathcal{D}$ is, and it is effectively given if $\mathcal{D}$ is.

Proof: The system $\mathcal{D}^{\downarrow}$ is a neighbourhood system as we just remarked; indeed it is a positive neighbourhood system. It is easy to prove that the closure of any positive system under finite unions is a neighbourhood system, because the resulting family of sets is closed under *all* finite intersections. (If we left out the empty union, the result would be a positive system.) The proof is obvious since intersection of sets distributes over finite union. So $\mathbb{P}\,\mathcal{D}$ is a neighbourhood system.

For the second half of the proposition, we just have to constructivize the previous argument. Thus, if

$$\mathcal{D} = \{X_n \mid n \in N\},$$

then the elements of $\mathbb{P}\,\mathcal{D}$ can be written as:

$$\bigcup_{i<q} (\downarrow X_{n_i}),$$

and hence are indexed by the finite sequences $\langle n_0, \ldots, n_{q-1}\rangle$ of integers. Now one of the standard devices of recursion theory is to put the finite sequences of integers into a recursive one-one correspondence with the integers themselves. This is the start of the recursive presentation of $\mathbb{P}\mathcal{D}$, since it means we can list effectively all the required neighbourhoods.

Next consider an intersection

$$\bigcup_{i<q}(\downarrow X_{n_i}) \cap \bigcup_{j<r}(\downarrow X_{m_j}) = \bigcup_{\substack{i<q\\ j<r}} \downarrow(X_{n_i} \cap X_{m_j}) \; .$$

Some of the terms which are $\emptyset$ have to be thrown out - but this requires only a finite number of decisions all computable by assumption. Now we have to rewrite

$$X_{n_i} \cap X_{m_j} = X_{k_{ij}} \; ,$$

but the finding of k_{ij} is also computable. *Finally*, we have to re-order the doubly indexed sequence into a singly indexed sequence of length q.r, but this is easily seen to be computable also. Therefore, intersections can be "calculated".

It remains to be shown that equality between neighbourhoods in $\mathbb{P}\mathcal{D}$ is decidable. The question really comes down to deciding something like:

$$\downarrow X_k \subseteq \bigcup_{i<q} \downarrow X_{n_i} \; .$$

Now since $X_k \in \downarrow X_k$, we find that the above is just equivalent to:

$$\exists\, i < q.\ X_k \subseteq X_{n_i} \, .$$

By our assumptions on $\mathcal{D}$, this is decidable. (It is this part of the argument that required the passage to $\mathcal{D}^{\downarrow}$. It does not seem to be generally true that the closure under finite unions of an effectively given system is again effectively given.) □

One of the main reasons that $\mathbb{P}\mathcal{D}$ is like a power domain is the possibility of forming "finite sets".

DEFINITION 7.11. For elements $x_0, \dots, x_{n-1} \in |\mathcal{D}|$ we define

$$\{x_0, \dots, x_{n-1}\} = \{Z \in \mathbb{P}\mathcal{D} \mid \exists X_0 \in x_0 \dots \exists X_{n-1} \in x_{n-1} \bigcup_{i<n} (\downarrow X_i) \subseteq Z\}.$$

(Note, we could also write $\forall i < n . X_i \in Z$). □

PROPOSITION 7.12. The mapping

$$\lambda x_0, \dots, x_{n-1} . \{x_0, \dots, x_{n-1}\} : \mathcal{D}^n \to \mathbb{P}\mathcal{D}$$

is approximable and is computable if $\mathcal{D}$ is effectively given. Moreover, the map $\lambda x . \{x\}$ shows that $\mathcal{D} \trianglelefteq \mathbb{P}\mathcal{D}$, and we also have the law:

$$\{x_0, \dots, x_{n-1}\} = \{x_0\} \cap \dots \cap \{x_{n-1}\}$$

as an intersection of filters.

Proof : The second part shows that everything reduces to $\lambda x . \{x\}$. We see that

$$X_k \, (\lambda x . \{x\}) \bigcup_{i<q} (\downarrow X_{n_i}) \text{ iff } \exists i < q . \ X_k \subseteq X_{n_i} .$$

Thus, $\lambda x . \{x\}$ is an approximable mapping and is computable in the effectively given case.

The proof of the law can be reduced to the special case

$$\{x\} \cap \{y\} = \{x, y\}$$

for the sake of illustration. In terms of finite elements of the two domains $\mathcal{D}$ and $\mathbb{P}\mathcal{D}$ we find

$$\{\uparrow X\} = \uparrow\downarrow X ,$$

and so,

$$\begin{aligned} \{\uparrow X\} \cap \{\uparrow Y\} &= \uparrow\downarrow X \cap \uparrow\downarrow Y \\ &= \uparrow(\downarrow X \cup \downarrow Y) \\ &= \{\uparrow X, \uparrow Y\} . \end{aligned}$$

An equation between approximable functions that checks for finite elements also holds for all elements.

Finally, we note that

$$\mathcal{D} \cong \mathcal{D}^{+} \triangleleft \mathbb{P}\mathcal{D}$$

and that the isomorphism involved is just $\lambda x.\{x\}$ by what we saw on the finite elements. □

Further combinators on the power domain are given in the exercises.

EXERCISES

EXERCISE 7.13. Show that an effectively given domain can always be identified with a relation

$$\mathrm{INCL}(n, m)$$

on integers, where the two derived relations

$\mathrm{CONS}(n, m)$ iff $\exists k.\ \mathrm{INCL}(k, n)$ and $\mathrm{INCL}(k, m)$;

$\mathrm{MEET}(n,m,k)$ iff $\forall j\,[\mathrm{INCL}(j,k)$ iff $\mathrm{INCL}(j,n)$ and $\mathrm{INCL}(j,m)]$

are both recursively decidable, and where the following axioms hold:

(i) $\forall n.\ \mathrm{INCL}(n,n)$;

(ii) $\forall n,m,k.\ \mathrm{INCL}(n,m)$ and $\mathrm{INCL}(m,k)$ imply $\mathrm{INCL}(n,k)$;

(iii) $\exists m\ \forall n.\ \mathrm{INCL}(n,m)$

(iv) $\forall n,m.\ \mathrm{CONS}(n,m)$ implies $\exists k.\ \mathrm{MEET}(n,m,k)$.

(Hint: Consider the neighbourhood system

$$\mathcal{D} = \{\{m \in \mathbb{N} \mid \mathrm{INCL}(m,n)\} \mid n \in \mathbb{N}\}.$$

Is this essentially any effectively given system?)

EXERCISE 7.14. (For recursive-function theorists.) Prove the statements after definition 7.2 about the existence of primitive recursive functions for showing things recursively enumerable. (Recall that a non-empty set is r.e. iff it is the range of a primitive recursive function.) Show also that every computable element $y \in |E|$ can be written

$$y = \bigcup\{\uparrow Y_{t(i)} \mid i \in \mathbb{N}\},$$

where $t : \mathbb{N} \to \mathbb{N}$ is primitive recursive and where we may assume

$$Y_{t(i+1)} \subseteq Y_{t(i)}$$

for all $i \in \mathbb{N}$.

EXERCISE 7.15. Finish the proof of 7.4 and establish similar results for the constructs $(\mathcal{D}_0 \otimes \mathcal{D}_1)$, $(\mathcal{D}_0 \oplus \mathcal{D}_1)$ and $\mathcal{D}^\infty$. Take into account the various appropriate combinators.

EXERCISE 7.16. Let $\mathcal{D}_0 = \{X_n | n \in \mathbb{N}\}$, $\mathcal{D}_1 = \{Y_m | m \in \mathbb{N}\}$ and $\mathcal{D}_2 = \{Z_k | k \in \mathbb{N}\}$ be three effectively given domains. Complete the proof of 7.5 by writing out curry as a relation between neighbourhoods. Is it a recursive set or only a recursively enumerable set?

EXERCISE 7.17. Complete the proof of 7.7 for showing that $\mathcal{D}^\S$ is effectively given if $\mathcal{D}$ is. Include all the combinators of 6.2. Prove also that if E is effectively given and

$$u : \mathcal{D} \to E \text{ and } v : E \times E \to E$$

are computable, then the unique strict mapping

$$g : \mathcal{D}^\S \to E ,$$

where, for $x \in |\mathcal{D}|$ and $y, z \in |E|$,

$$g(\mathrm{in}(x)) = u(g(x)) \text{, and}$$

$$g(\mathrm{pair}(y,z)) = v(g(y), g(z)),$$

is a computable mapping.

EXERCISE 7.18. Two effectively given systems $\mathcal{D}$ and E are *effectively isomorphic* iff ... (complete the sentence!). Show that if $\mathcal{D}$ is effectively given then the isomorphism

$$\mathcal{D}^\infty \cong (\mathcal{D}^\infty)^\infty$$

is effective.

EXERCISE 7.19. Prove that $\mathcal{D} \mapsto \mathbb{P}\,\mathcal{D}$ is a functor by defining for each $f : \mathcal{D} \to \mathcal{E}$ a mapping

$$\mathbb{P}\, f : \mathbb{P}\,\mathcal{D} \to \mathbb{P}\,\mathcal{E}$$

by the formula

$$\bigcup_{i<n} \uparrow X_i \quad \mathbb{P}\, f \quad \bigcup_{j<m} \uparrow Y_j \quad \text{iff } \forall i<n \exists j<m.\ X_i f Y_j\ .$$

Be sure to check that $\mathbb{P}\, f$ is approximable and that $\mathbb{P}$ preserves identity maps and composition. If f is computable is $\mathbb{P}\, f$? Is there a combinator $\lambda f.\mathbb{P}\, f$? What is

$$\mathbb{P}\, f(\{x,y\}) = ??$$

EXERCISE 7.20. Show that there is a combinator

$$\text{union} : \mathbb{P}\,(\mathbb{P}\,\mathcal{D}) \to \mathbb{P}\,\mathcal{D}$$

where for suitable neighbourhoods

$$\bigcup_{i<n} \uparrow(\bigcup_{j<m_i} \uparrow X_{ij}) \text{ union } \bigcup_{k<q} \uparrow Y_k \text{ iff } \forall i<n\, \forall j<m_i\, \exists\, k<q.\ X_{ij} \subseteq Y_k\ .$$

Is union computable if $\mathcal{D}$ is effectively given? What is

$$\text{union}\ (\{\{x\},\ \{y,z\}\}) = ??$$

Are $\mathbb{P}\,(\mathbb{P}\,\mathcal{D})$ and $\mathbb{P}\,\mathcal{D}$ generally isomorphic??

EXERCISE 7.21. Is there a non-trivial combinator of type

$$\mathbb{P}\,(\mathcal{D} \to \mathcal{E}) \to (\mathbb{P}\,\mathcal{D} \to \mathbb{P}\,\mathcal{E})\ ?$$

Are there in general any isomorphisms between the systems

$$(\mathcal{D} \to \mathbb{P}\,\mathcal{E}),\ \mathbb{P}\,(\mathcal{D} \times \mathcal{E}),\ \ \mathbb{P}\,\mathcal{D} \times \mathbb{P}\,\mathcal{E}\ ??$$

Is there a non-trivial combinator of type

$$\mathbb{P}\,(\mathcal{D} \times \mathcal{E}) \times \mathbb{P}\,(\mathcal{E} \times \mathcal{F}) \to \mathbb{P}\,(\mathcal{D} \times \mathcal{F})\ ???$$

Is there any connection between

$$\mathbb{P}\,\mathcal{N} \text{ and } \mathcal{P}\,\mathbb{N}\ ????$$

EXERCISE 7.22. (For algebraists.) Let $\Sigma = \{0,1\}^*$ be the free semigroup. A new domain is constructed by defining a family of sets by the least fixed point theorem as follows

$$S = \{\Sigma\} \cup \{\{\sigma\} \mid \sigma \in \Sigma\} \cup \{XY \mid X, Y \in S\} \cup$$
$$\{X \cap Y \mid X, Y \in S \text{ and } X \cap Y \neq \emptyset\}.$$

Here we write:

$$XY = \{\sigma\tau \mid \sigma \in X \text{ and } \tau \in Y\}.$$

Prove that S is an effectively given, positive neighbourhood system. (Hint: The sets in S are each "regular events" in the terminology of automata theory, and we have a decision method for the set algebra of regular events.) Define multiplication on $|S|$ by

$$xy = \{Z \in S \mid \exists X \in x\ \exists Y \in y.\ XY \subseteq Z\},$$

and show $|S|$ becomes a semigroup with Σ embedded into $|S|$ by the homomorphism $\sigma \mapsto \{X \in S \mid \sigma \in X\}$. Investigate some *infinite words* in S, say those defined by least fixed points such as:

$$\vec{\sigma} = \sigma\vec{\sigma} \text{ and } \overleftarrow{\sigma} = \overleftarrow{\sigma}\sigma.$$

Are these equations true:

$$\vec{0}\vec{0} = \vec{0},\quad \vec{0}\overleftarrow{0}\vec{0} = \vec{0},\quad \vec{0}\vec{1}\vec{0}\vec{1} = \vec{0}\vec{1},$$
$$\text{and } \overleftarrow{01}\,\overrightarrow{01}\,\overleftarrow{01}\,\overrightarrow{01} = \overleftarrow{01}\,\overrightarrow{01}\ ?$$

EXERCISE 7.23. Complete the discussion of $P\mathbb{N}$ of Example 7.8. Show that the combinators fun and graph of Exercise 5.14 are computable. Also do the same for

$$\lambda x, y.\ x \cap y,\quad \lambda x, y.\ x \cup y,\quad \text{and } \lambda x, y.\ x + y,$$

where for $x, y \in P\mathbb{N}$ we define

$$x + y = \{n+m \mid n \in x \text{ and } m \in y\}.$$

What are the computable elements of $P\mathbb{N}$?

EXERCISE 7.24. (Suggested by the LUCID language of Ashcroft and Wadge: SIAM Jour. Comp. vol. 5 (1976).) Define a set Γ by

$$\Gamma = \bigcup_{i=0}^{\infty} (\{i\} \times \Gamma) \cup \{\star\}.$$

Define a system

$$L = \{\Gamma\} \cup \{\{i\} \times X \mid i \in \mathbb{N} \text{ and } X \in L\}.$$

Show that L is effectively given. Show that the elements of $|L|$ can be identified with the finite and infinite sequences of natural numbers. What is the connection between B and L? Show that the combinators of LUCID can be construed as computable mappings of type

$$(L \to T) \to (L \to T)$$

or of type

$$(L \to T) \times (L \to T) \to (L \to T)$$

Conclude that programs in LUCID define computable maps.

LECTURE VIII

RETRACTS OF THE UNIVERSAL DOMAIN

In order to be able to have a fully flexible method of solving domain equations *and* to be able to see why the domains obtained are effectively given, we shall embed all the desired domains in one "largest" domain. This universal domain will be easily shown to be effectively given, and the mappings needed to extract the other domains will be found to be computable. In order to be able to carry out this programme, we investigate first how certain subdomains correspond to mappings - the so-called *retracts*. An advantage of this analysis is that all the necessary definitions can be written out in λ - calculus notation, thus demonstrating the power of our mathematical programming language.

DEFINITION 8.1. A *retraction* of a given domain $\mathcal{E}$ is an approximable mapping $a : \mathcal{E} \to \mathcal{E}$ such that $a \circ a = a$. □

PROPOSITION 8.2. If $\mathcal{D} \lhd \mathcal{E}$ and if $a : \mathcal{E} \to \mathcal{E}$ is defined by

$$X \, a \, Z \quad \text{iff} \quad \exists\, Y \in \mathcal{D}.\ X \subseteq Y \subseteq Z$$

for all $X, Z \in \mathcal{E}$, then a is a retraction and $|\mathcal{D}|$ is isomorphic to the fixed-point set of a, the set $\{y \in |\mathcal{E}| \mid a(y) = y\}$, under inclusion.

Proof: That a is an approximable mapping is a direct consequence of Definition 6.10. Indeed, in the notation of Proposition 6.12, we have

$$a = i \circ j,$$

and this is another proof that a is approximable. This remark is also convenient, since we know from 6.10

$$j \circ i = I_{\mathcal{D}}.$$

Therefore, we find:

$$a \circ a = i \circ j \circ i \circ j = i \circ j = a;$$

and so a is a retraction.

We can also employ i and j to give the isomorphism on $|\mathcal{D}|$. If $x \in |\mathcal{D}|$, then $i(x) \in |\mathcal{E}|$ and we calculate:

$$a(i(x)) = i \circ j \circ i(x) = i(x).$$

Thus, i(x) belongs to the fixed-point set of a. In the other direction, if a(y) = y, then i(j(y)) = y. But $j(y) \in |\mathcal{D}|$, so i maps $|\mathcal{D}|$ one-one and onto the fixed-point set of a. As i and j are monotone, the map is an isomorphism with respect to $\subseteq$. □

Not every retraction comes from a relationship like $\mathcal{D} \lhd E$; in fact, we can see from the definition of a above that $a \subseteq I_E$. But, as is indicated in Exercise 8.11, even this condition is not sufficient to characterize the kind of retractions provided by 8.2. The characterization is as follows.

DEFINITION 8.3. A retraction $a : E \to E$ is called a *projection* provided

$$a \subseteq I_E ;$$

it is *finitary* iff its fixed-point set is isomorphic to a domain. □

EXAMPLES 8.4. If a system $\mathcal{D}$ over Δ is not trivial, then the two element system $\mathcal{O} = \{\{0\}, \{0,1\}\}$ comes from a retraction on $\mathcal{D}$. Specifically, define a combinator

$$\text{check} : \mathcal{D} \to \mathcal{O}$$

by the relation

$$X \text{ check } Y \text{ iff either } Y = \{0,1\} \text{ or } X \neq \Delta.$$

We see $\text{check}(x) = \bot_{\mathcal{O}}$ iff $x = \bot_{\mathcal{D}}$. We leave to the reader the definition of a combinator:

$$\text{fade} : \mathcal{O} \times \mathcal{D} \to \mathcal{D} ,$$

where we have for $t \in |\mathcal{O}|$ and $x \in |\mathcal{D}|$:

$$\text{fade}(t,x) = \bot_{\mathcal{D}}, \text{ if } t = \bot_{\mathcal{O}} ;$$
$$= x, \text{ if not.}$$

Now, take any $u \in |\mathcal{D}|$ with $u \neq \bot$, and define

$$a(x) = \text{fade}(\text{check}(x), u).$$

Then a is a retraction (not a projection in general) and the range of a is isomorphic to $\mathcal{O}$.

Another way of using these combinators is to find $(\mathcal{D} \to_{\perp} E)$ as a retraction of $(\mathcal{D} \to E)$. Specifically, define a combinator

$$\text{strict} : (\mathcal{D} \to E) \to (\mathcal{D} \to E)$$

by the equation

$$\text{strict}(f) = \lambda x.\ \text{fade}\ (\text{check}(x),\ f(x)),$$

where this time

$$\text{fade} : \mathcal{O} \times E \to E\ .$$

The range of strict consists exactly of the strict functions and this time strict is a projection whose range is indeed a domain.

Similarly, we can find a projection on $\mathcal{D} \times E$ with a range isomorphic to $\mathcal{D} \otimes E$ by the combinator such that:

$$\text{smash}(x,y) = \text{fade}\ (\text{check}(x),\text{fade}\ (\text{check}(y),\langle x,y\rangle)),$$

for $x \in |\mathcal{D}|$ and $y \in |E|$. □

THEOREM 8.5. For an approximable mapping $a : E \to E$ the following are equivalent:

(i) a is a finitary projection;

(ii) $a(x) = \{Y \in E \mid \exists X \in x.\ \ X\, a\, X \subseteq Y\}$, for all $x \in |E|$.

Proof: Suppose a satisfies (ii) first. Inasmuch as

$X \in x$ and $X \subseteq Y$ always imply $Y \in x$,

for all $x \in |E|$, we see $a(x) \subseteq x$ must always hold. Moreover, it is obvious that

$X \in x$ and $X\, a\, X$ always imply $X \in a(x)$;

therefore, $a(x) \subseteq a\,(a\,(x)\,)$ for all $x \in |E|$. This shows that a is indeed a projection.

Let $\mathcal{D} = \{X \in E \mid X\, a\, X\}$, then it is easy to check that $\mathcal{D} \lhd E$ and that a is determined from $\mathcal{D}$ exactly as in 8.2; thus, the fixed-point set of a is isomorphic to a domain, by what we have already proved. So we have shown (ii) implies (i).

In the converse direction, assume that a is a finitary projection. And let the system $\mathcal{D}$ be isomorphic to the fixed point set of a. We have the situation of Theorem 6.15 There is a projection pair,

$$i : \mathcal{D} \to E \text{ and } j : E \to \mathcal{D},$$

where the connection with a gives:

$$j \circ i = I_{\mathcal{D}} \text{ and } i \circ j = a \subseteq I_E.$$

By 6.15 $\mathcal{D} \cong \mathcal{D}' \lhd E$ and we want to identify $\mathcal{D}'$ in terms of a as follows:

$$\mathcal{D}' = \{X \in E \mid X\, a\, X\}.$$

Now from a reading of the proof of 6.15 the neighbourhoods of $\mathcal{D}'$ are just those corresponding to the finite elements of $\mathcal{D}$. But any such element is a fixed point of a. We have

$$X \in \mathcal{D}' \text{ implies } a(\uparrow X) = \uparrow X \text{ implies } X\, a\, X.$$

Conversely, if $X\, a\, X$ holds, then $\uparrow X \subseteq a(\uparrow X)$. But a is a projection, so $\uparrow X$ is a fixed point. But $i(j(\uparrow X)) = \uparrow X$ means $j(\uparrow X)$ is a finite element of $|\mathcal{D}|$. So $X \in \mathcal{D}'$, and we have $\mathcal{D}'$ identified as desired.

Finally, if we calculate $a = i \circ j$ by the formulae of 6.12 (with $\mathcal{D}'$ for $\mathcal{D}$, of course), we obtain our formula (ii). □

The criterion for being a finitary projection just obtained provides us with a very interesting new combinator.

THEOREM 8.6. For any domain E define

$$\text{sub} : (E \to E) \to (E \to E)$$

by the formula

$$X \text{ sub}(f)\, Z \text{ iff } \exists Y \in E.\ X \subseteq Y\, f\, Y \subseteq Z,$$

for all $X, Z \in E$ and all $f : E \to E$. Then the range of sub consists exactly of the finitary projections on E, and moreover sub itself is a finitary projection on $(E \to E)$. If E is effectively given, then sub is computable.

Proof : It is trivial to check that sub(f) is always approximable. Also, it is obvious from the definition that the correspondence

$$f \mapsto \mathrm{sub}(f)$$

preserves directed unions of f's. Thus, sub is itself approximable. We note that

$$X \subseteq Y \; f \; Y \subseteq Z \text{ always implies } X \; f \; Z;$$

hence, $\mathrm{sub}(f) \subseteq f$ holds. Also

$$Y \; f \; Y \text{ always implies } Y \; \mathrm{sub}(f) \; Y,$$

hence, $\mathrm{sub}(f) \subseteq \mathrm{sub}(\mathrm{sub}(f))$ holds. This shows sub to be a projection on $(E \to E)$. The effectiveness of the definition makes it also clear that sub is computable when E has a computable presentation.

Since, sub is a projection, its range is the same as its fixed-point set. If

$$\mathrm{sub}(a) = a,$$

then there is no problem in checking that a satisfies 8.5(ii) *and conversely*. So the range of sub picks out exactly the finitary projections in view of 8.5.

Finally, to prove that sub is a finitary projection of $(E \to E)$, we invoke 6.11 and remark that, in view of 8.2, the fixed point set (range) of sub is in a one-one inclusion-preserving correspondence with the domain $\{D \mid D \lhd E\}$. □

These results have almost completely translated the theory of $\lhd$ - subdomains into λ - calculus *via* the sub-combinator. One last step will complete the passage, and then we shall be able to return to solving domain equations.

DEFINITION 8.7. Let $\mathbb{Q}$ be the set of rational numbers, and let

$$[0, 1) = \{q \in \mathbb{Q} \mid 0 \leq q < 1\},$$

and similarly for $[r, s)$ for any $r < s$ in $\mathbb{Q}$. The neighbourhood system $\mathcal{U}$ over $[0,1)$ is the set of all non-empty finite *unions* of intervals of rational intervals $[r, s)$ with $0 \leq r < s \leq 1$. □

A picture of a typical element of $\mathcal{U}$ could be drawn like this:

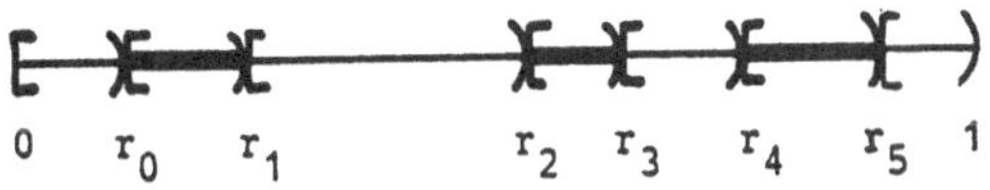

Note that any union can be taken as a *disjoint* union of the form

$$\bigcup_{i \leq n} [r_{2i}, r_{2i+1})$$

where $0 \leq r_0 < r_1 < r_2 < \cdots < r_{2n} < r_{2n+1} \leq 1$. (Hint: Any overlapping intervals or abutting intervals can always be combined into one long interval.) It is a most elementary exercise to show that, by virtue of this representation, the system $\mathcal{U}$ has a computable presentation. (Some isomorphic versions of $\mathcal{U}$ - equally effective - are recorded in the exercises.) Note that $\mathcal{U}$ has no minimal neighbourhoods: every set in $\mathcal{U}$ can be written as the union of two disjoint sets in $\mathcal{U}$. (Hint: Use the density of the ordering of $\mathbb{Q}$.) The significance of $\mathcal{U}$ can now be explained.

THEOREM 8.8. The system $\mathcal{U}$ is universal in the sense that, for every *countable* neighbourhood system $\mathcal{D}$, we have

$$\mathcal{D} \trianglelefteq \mathcal{U}.$$

Moreover, if $\mathcal{D}$ is effectively given, then the projection pair making the embedding can be taken as computable. Indeed there is a correspondence between effectively presented domains and the computable, finitary projections of $\mathcal{U}$.

Proof: As $\mathcal{D}$ is countable, we can assume that

$$\mathcal{D} = \{X_n \mid n \in \mathbb{N}\},$$

where $\mathcal{D}$ is a system over a set Δ (say, $X_0 = \Delta$). We shall do the effective and general cases together, where for the latter all remarks on recursiveness are just left out. So, if we want $\mathcal{D}$ effectively given, the above enumeration should be taken as the computable presentation.

Without loss of generality we can assume $\mathcal{D} = \mathcal{D}^+$, since otherwise we would just replace $\mathcal{D}$ by $\mathcal{D}^+$. The advantage of this preparation is that unions in $\mathcal{D}^+$ keep things rather *separate* (as we noticed in constructing $\mathbb{P}\mathcal{D}$). In particular, we can be sure of this equivalence:

$$(\blacklozenge) \qquad X_m \subseteq \bigcup_{i<k} X_{n_i} \quad \text{iff} \quad \exists i<k.\ X_m \subseteq X_{n_i}.$$

This property, for example, fails for the system $\mathcal{U}$ as presented in Definition 8.7. However, that observation is of no moment, because we are employing the assumption with respect to $\mathcal{D}$ not $\mathcal{U}$.

The reason for the assumption is this: for $\delta \in \{+,-\}$ define for $X \in \mathcal{D}$:

$$\begin{aligned} \delta X &= X \quad \text{if } \delta = + \,; \\ &= \Delta \setminus X \quad \text{if } \delta = - \,. \end{aligned}$$

(A similar notation will be used for $Y \in \mathcal{U}$.) Then for $\delta \in \{+,-\}^n$ the sets of the form

$$\bigcap_{i<n} \delta_i X_i \quad (= X_\delta, \text{ for short})$$

form a partition of Δ into (at most) 2^n parts. The reason for assumption $(\blacklozenge)$ is that we can effectively decide for each $\delta \in \{+,-\}^n$ whether one of these intersections is empty or not. (Why? - assuming that $\mathcal{D}$ is effectively given, of course). If for some reason we had not wanted to pass to $\mathcal{D}^+$, we could have made this stronger assumption of decidability on the (positive) system $\mathcal{D}$. ($\mathcal{U}$, for example, satisfies it.)

Suppose, corresponding to $X_0, X_1, \ldots, X_{n-1}$, we have selected $Y_0, Y_1, \ldots, Y_{n-1}, \in \mathcal{U}$ so that, for all $\delta \in \{+,-\}^n$,

$$(\blacksquare) \qquad \bigcap_{i<n} \delta_i X_i = \emptyset \quad \text{iff} \quad \bigcap_{i<n} \delta_i Y_i = \emptyset .$$

We wish to show - effectively - how to choose Y_n corresponding to X_n, so that (■) holds with n+1 replacing n. Proceeding inductively, we obtain a recursive enumeration of sets $Y_n \in \mathcal{U}$ so that

$$\mathcal{D} \cong \{Y_n \mid n \in \mathbb{N}\} \triangleleft \mathcal{U}.$$

Clearly the isomorphism (matching X_i to Y_i) will be computable and the projection is computable. (It will then remain only to consider the arbitrary finitary computable projection to complete the proof of the theorem.)

So, consider X_n; for each $\delta \in \{+,-\}^n$ there are four cases:

$$X_\delta \cap X_n = \emptyset\,, \quad X_\delta \cap -X_n = \emptyset\,,$$
$$X_\delta \cap X_n \neq \emptyset\,, \quad X_\delta \cap -X_n \neq \emptyset\,.$$

Corresponding to X_δ is a similar intersection Y_δ. If X_δ were $\emptyset$, then Y_δ would be also. If not, $Y_\delta \subseteq [0,1)$ is a union of rational intervals that can be written down explicitly. (Why?) In our four cases on X_n, the first implies the fourth. (Why?) Thus, we need only make some choices in these circumstances:

$$\begin{aligned} X_\delta \cap X_n = \emptyset &: \text{ choose } I_{\delta,n} = \emptyset\,; \\ X_\delta \cap -X_n = \emptyset &: \text{ choose } I_{\delta,n} = Y_\delta; \\ \text{otherwise} &: \text{ choose } I_{\delta,n} \subseteq Y_\delta, \text{ with } \emptyset \neq I_{\delta,n} \neq Y_\delta. \end{aligned}$$

All these cases are decidable by assumption on $\mathcal{D}$, and the effective choice of (unions of) intervals is effective by construction of $\mathcal{U}$. Now set

$$Y_n = \bigcup_{\delta \in \{+,-\}^n} I_{\delta,n} \neq \emptyset\,.$$

The set $Y_n \in \mathcal{U}$, it can be found effectively, and (■) is obviously satisfied for n+1 .

Finally, suppose that a is a computable, finitary projection of $\mathcal{U}$. As we have seen in the proof of 8.5, the domain corresponding to the range of a is isomorphic to the neighbourhood system

$$\{Y \in \mathcal{U} \mid Y\, a\, Y\} \triangleleft \mathcal{U}\,.$$

Clearly, if a as a set of ordered pairs of neighbourhoods is recursively enumerable, then the above set is also recursively enumerable (because equality between neighbourhoods is decidable). It follows easily that the subsystem is effectively given as a neighbourhood system in its own right. □

We have now proved that $\mathcal{U}$ is a nice and big domain that is nicely behaved with respect to computable mappings. It has some very interesting subdomains; to name a few:

$$\mathcal{U}+\mathcal{U},\ \ \mathcal{U}\oplus\mathcal{U},\ \ \mathcal{U}\times\mathcal{U},\ \ \mathcal{U}\otimes\mathcal{U}$$

$$\mathcal{U}_{\perp},\ \mathcal{U}^{\infty},\ \mathcal{U}^{\S},\ \mathbb{P}\,\mathcal{U},\ \mathcal{U}\rightarrow\mathcal{U}.$$

That all of these are $\trianglelefteq \mathcal{U}$ follows from knowing that they are all effectively presented. What we wish to check next is that they all combine well with respect to projections. To this end the explicit definitions are given for the constructs +, ×, and →, and the details of the others are left for the exercises.

DEFINITION 8.9. Let the computable projection pairs

$$i_{+} : \mathcal{U}+\mathcal{U}\rightarrow\mathcal{U} \text{ and } j_{+} : \mathcal{U}\rightarrow\mathcal{U}+\mathcal{U}$$

be fixed. Similarly choose $i_{\times}, j_{\times}$ and $i_{\rightarrow}, j_{\rightarrow}$ for $\mathcal{U}\times\mathcal{U}$ and $\mathcal{U}\rightarrow\mathcal{U}$. Define:

$$a{+}b = \mathrm{cond}\circ\langle \mathrm{which},\ i_{+}\circ \mathrm{in}_0\circ a\circ \mathrm{out}_0,\ i_{+}\circ \mathrm{in}_1\circ b\circ \mathrm{out}_1\rangle\circ j_{+}\ ;$$

$$a{\times}b = i_{\times}\circ\langle a\circ \mathrm{proj}_0,\ b\circ \mathrm{prog}_1\rangle\circ j_{\times}\ ;$$

$$a{\rightarrow}b = i_{\rightarrow}\circ(\lambda f.\, b\circ f\circ a)\circ j_{\rightarrow}\ ,$$

for all $a, b : \mathcal{U}\rightarrow\mathcal{U}$. □

These interesting (computable!) combinators on elements of $\mathcal{U}\rightarrow\mathcal{U}$ have many, many properties. We shall, however, only see what they do to projections.

PROPOSITION 8.10. If $a, b : \mathcal{U}\rightarrow\mathcal{U}$ are projections, then so are a+b, a×b, and a→b. If a and b are finitary, then so are the others; for the fixed-point set of each of them is isomorphic to the corresponding construct applied to the domains determined by a and b.

Proof: Suppose that $a, b \subseteq I_{\mathcal{U}}$ (= I for short). Then

$$a+b \subseteq I + I = i_+ \circ j_+ \subseteq I.$$

The other cases are similar.

Suppose $a = a \circ a$ and $b = b \circ b$, then, for example,

$$(a\times b) \circ (a\times b) = i_\times \circ \langle a \circ proj_0, b \circ proj_1\rangle \circ \langle a \circ proj_0, b \circ proj_1\rangle \circ j_\times$$

$$= i_\times \circ \langle a \circ a \circ proj_0, b \circ b \circ proj_1\rangle \circ j_\times$$

$$= a \times b.$$

The other cases are similar.

Now in case the fixed-point sets of a and b are domains, they are respectively isomorphic to

$$\mathcal{D}_a = \{X \in \mathcal{U} \mid X\, a\, X\} \text{ and}$$

$$\mathcal{D}_b = \{Y \in \mathcal{U} \mid Y\, b\, Y\}.$$

We have to show, for example, that

$$\mathcal{D}_a \to \mathcal{D}_b \cong \mathcal{D}_{a \to b}.$$

Now to simplify matters, remark that the fixed-point set of $a \to b$ on $\mathcal{U}$ is isomorphic to the fixed-point set of $\lambda f.\, b \circ f \circ a$ on $(\mathcal{U} \to \mathcal{U})$. (Hint: use $i_\to$ and $j_\to$ to set up the isomorphism.) So we have to think what it is for an $f : \mathcal{U} \to \mathcal{U}$ to satisfy

$$f = b \circ f \circ a.$$

Notice that we might as well say that $a : \mathcal{U} \to \mathcal{D}_a$ and that this map is the other half of an obvious projection pair where

$$i_a : \mathcal{D}_a \to \mathcal{U},$$

and $i_a \circ a = a$ and $a \circ i_a = i_a$. So if $g : \mathcal{D}_a \to \mathcal{D}_b$, let

$$f = i_b \circ g \circ a,$$

then $b \circ f \circ a = f$. Conversely, if f is like this, then let

$$g = b \circ f \circ i_a.$$

Thus, $i_b \circ g \circ a = b \circ f \circ a = f$; so there is an order-preserving isomorphism between the $g : \mathcal{D}_a \to \mathcal{D}_b$ and the $f = b \circ f \circ a$.

The isomorphism proofs for + and × are similar. □

Well, this was a lot of work, but the pay-off is rather handsome. What we have done is transpose all the

$$\mathcal{D}_a \lhd \mathcal{U}$$

over to finitary projections $a : \mathcal{U} \to \mathcal{U}$. This transposition is an isomorphism, because

$$\mathcal{D}_a \lhd \mathcal{D}_b \quad \text{iff} \quad a \subseteq b.$$

Moreover, by the method of 8.9 and 8.10, *all* our favourite constructs have been made into *combinators*, that is, approximable - even computable - maps on the domain of finitary projections. *ALL APPROXIMABLE (COMPUTABLE) MAPS HAVE (COMPUTABLE) FIXED POINTS.* And there you are! The standard fixed-point method is available to obtain computable (i.e. effectively given) solutions to *all* domain equations (even sets of equations) where the constructs can be reworked in this way to be defined on projections. Examples are suggested in the exercises.

Another pay-off concerns the λ-calculus itself. Inasmuch as

$$\mathcal{U} + \mathcal{U}, \quad \mathcal{U} \times \mathcal{U}, \quad \mathcal{U} \to \mathcal{U} \precsim \mathcal{U},$$

we might just as well forget the outside world and regard all these useful domains as being part of $\mathcal{U}$. For example, on the left we have the new notation and on the right the old notation:

$$\text{which}(z) = \text{which}(j_+(z)) \; ;$$

$$\text{in}_i(x) = i_+(\text{in}_i(x)), \quad i = 0, 1 \; ;$$

$$\text{out}_i(x) = \text{out}_i(j_+(x)), \quad i = 0, 1 \; ;$$

$$\langle x, y \rangle = i_\times(\langle x, y \rangle) \; ;$$

$$\text{proj}_i(z) = \text{proj}_i(j_\times(z)), i = 0, 1 \; ;$$

$$u(x) = j_\to(u)(x) \; ;$$

$$\lambda x.\tau = i_\to(\lambda x.\tau).$$

And, there is no reason to stop here. The system

$$T \cong \{[0,1/2), [1/2,1), [0,1)\} \lhd \mathcal{U} \; ,$$

so we might as well think of

$$\text{true, false} \in |U|$$

and think of cond: $U \times U \times U \to U$. No! that is wrong: under the new regime *EVERYTHING IS AN ELEMENT OF* U. With the new meaning of λ, all functions, all pairs, all combinators, all constructs become *elements* of U.

It takes a little time to get used to "universal conscription" with all elements doing (at least) double duty in the same domain, but there are many advantages, both notational and conceptual.

EXERCISES

EXERCISE 8.11. Let $\mathbb{Q}$ be the set of *rational* numbers and define a neighbourhood system by the equation

$$R = \{[0, r) \mid r \in \mathbb{Q} \text{ and } 0 < r \leq 1\}.$$

Show that the following defines an approximable map $a : R \to R$:

$$[0,r)\, a\, [0,s) \text{ iff } r < s \text{ or } r = s = 1.$$

Show in addition that a is a projection where the fixed-point set of a is in a one-one correspondence with the *real* numbers between 0 and 1 inclusive. (Hint: Recall Dedekind cuts and show $\subseteq$ matches $\leq$.) Conclude that a is *NOT* finitary. (Hint: Aside from $\bot$ there areno finite elements for $\{x \mid x = a(x)\}$.)

EXERCISE 8.12. Generalize the notation $2X+1$ for subsets $X \subseteq \mathbb{N}$ to sets of the form

$$2^k X + \ell, \text{ where } \ell < 2^k.$$

Let V be the non-empty finite unions of sets $2^k\, \mathbb{N} + \ell$. Show that $U \cong V$ and that the isomorphism is effective, thus obtaining another presentation of U.

EXERCISE 8.13. (For logicians.) Prove that the universal domain U is isomorphic to the domain of all proper filters of the free Boolean algebra on $\aleph_0$-generators (= the Lindenbaum algebra of propositional calculus). (For topologists.) Connect this representation of $\mathcal{U}$ with the collection of non-empty open subsets of the product space $2^{\mathbb{N}}$ (= Cantor space).

EXERCISE 8.14. A retraction $a : \mathcal{D} \to \mathcal{D}$ is called a *closure operator* iff $I_{\mathcal{D}} \subseteq a$. On a domain like $P\mathbb{N}$, give some examples of closure operators. (Hint: Close up a set of integers under addition. Is this continuous on $P\mathbb{N}$?) Prove in general for any closure $a : \mathcal{D} \to \mathcal{D}$ that the fixed-point set of a is always a finitary domain. (Hint: Show that the fixed-point set is closed under intersections and directed unions.) What are the finite elements of the fixed-point set?

EXERCISE 8.15. Give a direct proof that the domain $\{X \mid X \lhd \mathcal{D}\}$ is effectively presented if $\mathcal{D}$ is. (Hint: The finite elements of the domain correspond exactly to the finite systems $X \lhd \mathcal{D}$.) In the case of $\mathcal{D} = \mathcal{U}$, show that the computable elements of the domain correspond exactly to the effectively presented domains (up to effective isomorphism).

EXERCISE 8.16. For finitary projections $a : E \to E$, write

$$\mathcal{D}_a = \{X \in E \mid X\, a\, X\}$$

(cf. 8.5.). Show that for any two such projections $a, b : E \to E$ we have

$$a \subseteq b \quad \text{iff} \quad \mathcal{D}_a \lhd \mathcal{D}_b.$$

(This fills in the gap at the end of the proof of 8.6.) Also finish off the proof of 8.8 by showing that if E is effectively given and $a : E \to E$ is computable, then $\mathcal{D}_a$ is effectively given.

EXERCISE 8.17. Find explicitly (if possible) the projection pairs for $\mathcal{U} + \mathcal{U}$, $\mathcal{U} \times \mathcal{U}$, and $\mathcal{U} \to \mathcal{U}$ needed for 8.9. Are any of these domains isomorphic with $\mathcal{U}$? (The author does not know a really good construction for $\mathcal{U} \to \mathcal{U}$.) Find a universal domain $\mathcal{V} \ncong \mathcal{U}$.

EXERCISE 8.18. Many of the cases of 8.10 were left unproved. Please establish these assertions explicitly.

EXERCISE 8.19. Suppose we know both

$$T \text{ and } E \to E \trianglelefteq E .$$

Does it follow that $E + E$ and $E \times E \trianglelefteq E$?

EXERCISE 8.20. For any system we know $\mathcal{D} \trianglelefteq \mathcal{D} + \mathcal{D}$, but what about

$$\mathcal{D} \trianglelefteq \mathcal{D} \times \mathcal{D} \text{ and } \mathcal{D} \trianglelefteq \mathcal{D} \to \mathcal{D} \;?$$

Would these projections be computable if $\mathcal{D}$ is effectively given? Are there more than one projection pair in each case?

EXERCISE 8.21. Using the fixed-point construction, show that there is a continuous and computable operator $\lambda a. a^{\S}$, such that if a is a finitary projection of $\mathcal{U}$, then

$$\mathcal{D}_{a^{\S}} \cong (\mathcal{D}_a)^{\S} .$$

EXERCISE 8.22. Which of the two relations hold:

$$B \trianglelefteq C \text{ or } C \trianglelefteq B \;?$$

Or do they both hold? In general if we use domain equations

$$\mathcal{D} = T(\mathcal{D}) + S(\mathcal{D}), \text{ and}$$
$$E = T(E) ,$$

will $E \trianglelefteq \mathcal{D}$ hold? What projections do you see in the examples in 6.2?

EXERCISE 8.23. Suppose a construct T on domains can be made into a computable operator $t : (\mathcal{U} \to \mathcal{U}) \to (\mathcal{U} \to \mathcal{U})$ so that whenever $a : \mathcal{U} \to \mathcal{U}$ is a finitary projection, then so is t(a) and

$$\mathcal{D}_{t(a)} \cong T(\mathcal{D}_a).$$

Does it follow that $\| t \| = \text{fix}(t)$ is such that

$$\mathcal{D}_{\|t\|} \cong T(\mathcal{D}_{\|t\|})$$

really is the initial solution of the domain equation with respect to projections? Since t is computable, will this solution be effectively given?

EXERCISE 8.24. Suppose S and T are two (binary-argument) constructs on domains that can be made into computable operators on projections of the universal domain. Show that we can therefore find a pair of effectively presented domains such that

$$\mathcal{D} \cong S(\mathcal{D},\mathcal{E}) \text{ and } \mathcal{E} \cong T(\mathcal{D},\mathcal{E}).$$

EXERCISE 8.25. The problem is to find non-trivial solutions to the domain equation

$$(\spadesuit) \qquad \mathcal{D} \cong \mathcal{D} \to \mathcal{D}.$$

Show that the "obvious" solution by retracts is of no use because

$$\bot \to \bot = \bot$$

for projections. Change the method as follows. Show first

$$\mathcal{U}^{\infty} \times \mathcal{U}^{\infty} \cong \mathcal{U}^{\infty}.$$

Next solve

$$\mathcal{D} \cong \mathcal{D} + \mathcal{U}^{\infty}$$

and remark that $\mathcal{U} \lhd \mathcal{D}$; so $\mathcal{D}$ is universal and non-trivial. Finally prove ($\spadesuit$) for this $\mathcal{D}$. (Hint: First show

$$\mathcal{D} \times \mathcal{D} \cong \mathcal{D},$$

and then show $\mathcal{D}$ satisfies ($\spadesuit$).) Is this $\mathcal{D}$ effectively given?

EXERCISE 8.26. Discuss in more detail the "pay-off" for $\mathcal{U}$, namely the translation of "untyped" λ-calculus into $\mathcal{U}$ as shown by the equations at the end of the lecture after the proof of 8.9. In particular show how the whole of the *typed* λ-calculus can be retranslated back into $\mathcal{U}$ with the aid of projections. (Hint: Whenever you want to write

$$f : \mathcal{D}_a \to \mathcal{D}_b,$$

write instead

$$f = b \circ f \circ a,$$

where a, b are finitary projections. Whenever you want to form a λ - abstraction

$$\lambda x^{\mathcal{D}_a} . \sigma ,$$

where σ is of type $\mathcal{D}_b$, instead form

$$\lambda x.\, b(\sigma'[a(x)/x]),$$

where σ' is the further translation of σ into untyped λ - calculus. Be sure to show that this result "has the right type" in the sense defined above.)

EXERCISE 8.27. (Suggested by James Donahue.) Finite cartesian products of domains are formed by the $\mathcal{D}_0 \times \mathcal{D}_1$- construct we have used so often. The problem is to define - computably - some *infinite* cartesian products. In particular, as applied to the universal domain $\mathcal{U}$, the combinator sub is to be regarded as a finitary projection of $\mathcal{U}$ whose fixed points are exactly *all* the finitary projections. A map

$$d = \text{sub} \circ d \circ \text{sub}$$

can be regarded as a *polymorphic type* (because, whenever t is a finitary projection (= type), then so is d(t)). The *continuous product* of all these types would be the domain of all approximable functions x such that

$$x(t) = d(t)(x(t))$$

for all types t. (Why does this equation mean that x is in the product?) Define Π as a combinator by

$$\Pi = \lambda d\, \lambda x\, \lambda t.\, \text{sub}\,(d\,(\text{sub}\,(t)))\,(x\,(\text{sub}(t))).$$

Show that for d a polymorphic type, Π(d) is a type. (Hint: It is easy to check that Π(d) is a projection; the problem is to show it is *finitary*.)

REFERENCES

Note:*It has not been possible to alter the main text to insert all the necessary references. The following list is hardly complete, but the references cited contain pointers to most of the relevant literature.*

H. P. Barendregt. *The Lambda Calculus: Its Syntax and Semantics*, North-Holland Publishing Co., 1981, *xiv*+615 pp.

G. Gierz, K. H. Hofmann, K. Keimel, J. D.Lawson, M. Mislove, and D. S. Scott. *A Compendium of Continuous Lattices*, Springer-Verlag, 1980, *xx*+371 pp.

M. J. Gordon. *The Denotational Description of Programming Languages*, Springer-Verlag, 1979, 160 pp.

M. J. Gordon, A. J. R. Milner, and C. P. Wadsworth. *Edinburgh LCF*, Springer-Verlag Lecture Notes in Computer Science, vol. 78 (1979), 159 pp.

D. S. Scott. "Data types as lattices," *SIAM Journal on Computing*, vol. 5 (1976), pp. 522–587.

D. S. Scott. "Logic and programming languages," *Communications of the ACM*, vol. 20 (1977), pp. 634–641.

D. S. Scott. "Lambda calculus: some models, some philosophy." In: *The Kleene Symposium* (K. J. Barwise, *et al.* editors), North-Holland Publishing Co., 1980, pp 223–265.

D. S. Scott. "Relating theories of the λ-calculus." In: *To H. B. Curry: Essays on Combinatory Logic, Lambda Calculus and Formalism* (J. P. Seldin and J. R. Hindley, editors), Academic Press, 1980, pp. 403–450.

D. S. Scott. "Lectures on a Mathematical Theory of Computation," *Oxford University PRG Technical Monograph*, No. 19 (1981), *iv* +148 pp. (This is the same text as the present publication. It is also reprinted under the title "Domain Equations" in: *Proceedings of the Sixth IBM Symposium on Mathematical Foundations of Computer Science: Logic Aspects of Programs*, Corporate & Scientific Programs, IBM Japan, 1981, pp. 103–256.)

D. S. Scott. "Some ordered sets in computer science." In: *Ordered Sets*. Proceedings of the NATO Advanced Study Institute held at Banff, Canada, August 28 to September 12, 1981 (I. Rival, editor), D. Reidel Publishing Co., pp. 677–717.

J. E. Stoy. *Denotational Semantics: The Scott-Strachey Approach to Programming Language Theory*, MIT Press, 1977, *xxx*+414 pp.

R. D. Tennent. *Principles of Programming Languages*, Prentice/Hall International, 1981, *xiv*+271 pp.

SEMANTIC MODELS

Joseph E. Stoy

Oxford University Computing Laboratory
Programming Research Group

The "denotational semantics" approach to the modelling of programming language concepts: the store and assignment; scopes and environments; continuations and sequencing. Connections with methodologies for correctness of programs and of implementations.

CHAPTER I - DENOTATIONAL SEMANTICS

In his monograph [6], Dana Scott introduces us to a mathematical theory of computation. All values participating in computation, he says, are to be regarded as elements of certain structures called *domains*, and the existence of these domains in appropriate circumstances is guaranteed by the theory. All this mathematics provides firm foundations on which we may rely when we face the job of specifying the computations we wish to have performed: the theory means that we can be sure we know what we are talking about. In this Chapter we shall consider how these new insights may be exploited in the particular context of the specification of programming languages, by the technique known as *Denotational Semantics*. We indicate how three important features of conventional languages are handled - assignment, declarations and scopes, and sequencing and jumps - and at each stage we take the opportunity to exhibit a complete definition of "the language so far", even though each definition will require considerably reworking to accommodate the next stage.

Abstract Syntax

To specify the semantics of a language it is necessary to state what each syntactic construct in the language means. To avoid becoming

M. Broy and G. Schmidt (eds.), Theoretical Foundations of Programming Methodology, 293–325.

immersed in irrelevant details about the parsing of the language, we define the syntax of the language by means of an *abstract syntax*, as shown in the following example.

Syntactic Domains

N ∈ **Nml**	numerals
I ∈ **Ide**	identifiers
E ∈ **Exp**	expressions
Γ ∈ **Cmd**	commands

Syntax

E ::= **true** | . . . | N | I | −E | . . .
Γ ::= **skip** | **abort** | I:=E | Γ;Γ |
if E **then** Γ | **while** E **do** Γ

This definition has some resemblance to a BNF definition of a concrete syntax, but it is interpreted somewhat differently. There are four different types of syntactic object in this language, and the appropriate Greek upper case letter is conventionally used to stand for an object of a given type. The structure of the composite types is shown by the BNF-like productions. The syntactic variables in the right hand sides stand for the subcomponents of the object. The other symbols are present merely as tags, to indicate which of the possible variants of a composite object is being considered: this does not in any way constrain the concrete syntax of the language. To indicate that all this is happening, syntactic constructs are enclosed in special square quotation marks (⌜ and ⌝) or, when they occur as arguments of functions, in double square brackets (⟦ and ⟧).

Note that all this implies that it does not matter that our example definition, considered as a definition of concrete syntax, would be ambiguous. That problem would be sorted out in the actual definition of the concrete syntax, which is separate from the present exercise.

Semantics

We must now state what each construct in the language means. This is in two parts: we first say to which domains of values the meanings belong, and then we give a function, called a *valuation function*, which associates each construct with its value. The present language contains the assignment command which, of course, is used to change the value associated with a particular identifier. To model this, we define the state of the computer to be a function from identifiers to their values. Then command meanings will simply be state transformation functions. Moreover, since the value associated with an expression may depend on the state (as the expression may contain identifiers), expression meanings will be functions from states to values. All this is defined formally as follows.

Semantic Domains

$\mathbf{T}$	truth values
$\mathbf{N}$	integers
$\epsilon \in \mathbf{E} = \mathbf{N} \oplus \mathbf{T} \oplus \{err\}_\perp$	expressible values
$\sigma \in \mathbf{S} = [\mathbf{Ide} \rightarrow \mathbf{E}] \oplus \{err\}_\perp$	states
$\gamma \in \mathbf{C} = \mathbf{S} \rightarrow \mathbf{S}$	command values

Note that if **A** and **B** are domains **A+B** denotes the disjoint union of the two domains, as defined in Exercise 3.18 of Scott's monograph [6]; and **A⊕B** denotes the "coalesced" version (in which the ⊥ elements of the two component domains are *identified* with the ⊥ element of the sum domain), which is defined in Scott's Exercise 6.21.

There are three semantic valuations for this language. We omit the definition of **N**, the function from numerals to the numbers they stand for. The other two are defined as follows.

Semantic Functions

$\mathbf{E} : \mathbf{Exp} \rightarrow \mathbf{S} \rightarrow \mathbf{E}$

$$\mathbf{E}[\![\mathrm{N}]\!]\sigma = \mathbf{N}[\![\mathrm{N}]\!] \tag{1}$$

$$\mathbf{E}[\![\mathbf{true}]\!]\sigma = true \tag{2}$$

$$\mathbf{E}[\![-\mathrm{E}]\!]\sigma = (\epsilon \mathrm{E} \mathbf{N} \rightarrow -\epsilon,\ err) \quad \text{where } \epsilon = \mathbf{E}[\![\mathrm{E}]\!]\sigma \tag{3}$$

$$\mathbf{E}[\![\mathrm{I}]\!]\sigma = \sigma{=}err_{\mathbf{S}} \rightarrow err_{\mathbf{E}},\ \sigma[\![\mathrm{I}]\!] \tag{4}$$

$\mathbf{C} : \mathbf{Cmd} \rightarrow \mathbf{S} \rightarrow \mathbf{S}$

$$\mathbf{C}[\![\mathbf{skip}]\!]\sigma = \sigma \tag{5}$$

$$\mathbf{C}[\![\mathbf{abort}]\!]\sigma = \sigma{\equiv}\perp \rightarrow \perp,\ err \tag{6}$$

$$\mathbf{C}[\![\mathrm{I}{:=}\mathrm{E}]\!]\sigma = (\sigma \neq err \wedge \epsilon \neq err \rightarrow \sigma + [\mathrm{I} \mapsto \epsilon],\ err) \quad \text{where } \epsilon = \mathbf{E}[\![\mathrm{E}]\!]\sigma \tag{7}$$

$$\mathbf{C}[\![\Gamma_1;\Gamma_2]\!]\sigma = \mathbf{C}[\![\Gamma_2]\!](\mathbf{C}[\![\Gamma_1]\!]\sigma) \tag{8}$$

$$\mathbf{C}[\![\mathbf{if}\ \mathrm{E}\ \mathbf{then}\ \Gamma]\!]\sigma = (\sigma \neq err \wedge \epsilon \mathrm{E} \mathbf{T} \rightarrow (\epsilon \rightarrow \mathbf{C}[\![\Gamma]\!]\sigma, \sigma),\ err) \quad \text{where } \epsilon = \mathbf{E}[\![\mathrm{E}]\!]\sigma \tag{9}$$

$$\mathbf{C}[\![\mathbf{while}\ \mathrm{E}\ \mathbf{do}\ \Gamma]\!]\sigma = \theta\sigma \text{ where } \theta \text{ is given by} \tag{10}$$

$$\theta = \lambda\sigma'.((\sigma' \neq err \wedge \epsilon \mathrm{E} \mathbf{T} \rightarrow (\epsilon \rightarrow \theta(\mathbf{C}[\![\Gamma]\!]\sigma'),\ \sigma'),\ err) \quad \text{where } \epsilon = \mathbf{E}[\![\mathrm{E}]\!]\sigma')$$

These may be viewed as fairly conventional recursion equation definitions of **E** and **C**. Notice that if $\epsilon \in \mathbf{E}$ then $\epsilon \mathrm{E} \mathbf{N}$ is *true* if ϵ is a member of the **N** component of **E**, and *false* if it is a member of some other component. $b \rightarrow x,y$ is the conditional expression; its value is x if b is *true*, y if b is *false* and *err* or ⊥ if b is *err* or ⊥. → and E are both monotonic and continuous in all their arguments. In equation (6), although the test $x \equiv \perp$ is, of course, not monotonic in x, notice that it is used only to define a function which *is* both monotonic and continuous. In (7), $\sigma + [\mathrm{I} \mapsto \epsilon]$ denotes the state which maps I to ϵ and is otherwise the same as σ; that is to say,

$$\sigma + [\mathrm{I} \mapsto \epsilon] = \lambda \mathrm{I}'.\ \mathrm{I}' = \mathrm{I} \rightarrow \epsilon,\ \sigma[\![\mathrm{I}']\!] \tag{11}$$

Notice that equation (10) itself contains a circular equation for θ. As we know from Scott's theory, this equation may be rewritten in a non-circular form as

$$\begin{aligned} &\mathbf{C}[\![\textbf{while}\ \mathrm{E}\ \textbf{do}\ \Gamma]\!] = \mathit{fix}(\lambda\theta\lambda\sigma'. \\ &\quad((\sigma' \neq \mathit{err} \wedge \epsilon \in \mathrm{T} \rightarrow (\epsilon \rightarrow \theta(\mathbf{C}[\![\Gamma]\!]\sigma'), \sigma'), \mathit{err}) \\ &\quad \text{where } \epsilon = \mathbf{E}[\![\mathrm{E}]\!]\sigma')) \end{aligned} \tag{12}$$

Definitional Interpreters

We have seen that the semantic definition of our language looks very like a program written in a functional programming language. We could, indeed, regard our definition as being an *interpreter* for the defined language, intended for evaluation on a computer. Our approach would closely resemble that of Peter Landin several years ago [2], who defined a subset of Algol 60 in terms of Lambda Calculus, implemented with his SECD machine [3].

One problem with this way of looking at things is that assumptions made about the defining language (the functional programming language) are automatically carried over to the defined language, without explicit mention. We are left with the problem of defining the semantics of our functional language, and our present approach would be reduced to just another method of operational semantics, defining the semantics of a language in terms of an abstract machine.

This was not our intention. Our defining language was not to be regarded as intended for mechanical evaluation, but rather as describing various relationships which hold in some appropriate domain, according to the theory described in the previous lecture. The fact that the expressions of these relationships are susceptible to manipulation according to rules is an uncovenanted extra. When, in a particular case, some such method fails to deliver an answer, we do not say that the value of that particular expression is *ipso facto* undefined, but rather that the method we tried does not work in this case and we must look around for some other way to do it. It is sometimes stated that the metalanguage for denotational semantics is a "normal order" or "call by name" language. While it is true that normal order or call by name evaluation more reliably produces the value specified by the theory than other methods, nevertheless the statement is misleading: in the theory syntactic matters, like the order of applying conversion rules, are irrelevant.

Environments

We now slightly extend our programming language to include programs corresponding to the following piece of Algol 60:

```
x := 2;                                                        (13)
y := x;
begin integer x;
      x := 3;
end;
y := x;
x := 4
```

We agree informally that in this program the declaration introduces a new variable called *x*, to which the assignment *x*:=3 refers. The old *x* is unaffected by this and, accordingly, at the end of the program *y* still has the value 2.

To model this we replace the state function of the previous example, which mapped identifiers to their values directly, with a two stage mapping. A function called the *environment* maps identifiers to elements of a new set **L** called *locations*; then the *store* is a function mapping locations to their contents.

The syntax of our extended language is as follows, where we include only the new forms of construct.

Syntactic Domains

N ∈ **Nml**	numerals
I ∈ **Ide**	identifiers
E ∈ **Exp**	expressions
Γ ∈ **Cmd**	commands
Δ ∈ **Dec**	declarations

Syntax

E ::= . . .

Γ ::= . . . | **begin** Δ; Γ **end**

Δ ::= **con** I = E | **loc** I = E

The first form of declaration declares an identifier to have a constant value, while the second introduces an initialised variable. Our environments are therefore required to map identifiers both to expressible values (for the constants) and to locations (for the variables). The semantic domains are therefore as follows. Note that if **D** is a domain then we define $\mathbf{D}_+ = \mathbf{D} \oplus \{err\}$.

Semantic Domains

$\epsilon \in \mathbf{E} = \mathbf{N} + \mathbf{T}$	expressible values
$\delta \in \mathbf{D} = \mathbf{E} + \mathbf{L}$	denotable values
$\rho \in \mathbf{U} = \mathbf{Ide} \rightarrow \mathbf{D}_+$	environments
$\alpha \in \mathbf{L}$	locations
$\beta \in \mathbf{V} = \mathbf{E}$	storable values
$\sigma \in \mathbf{S} = \mathbf{L} \rightarrow [\mathbf{V} \oplus \{free\}_\perp]$	stores
$\gamma \in \mathbf{C} = \mathbf{S}_+ \rightarrow \mathbf{S}_+$	command values

Before giving the semantic valuation functions themselves, by way of modularity we first define the primitive functions on stores, as follows.

Store primitives

$$Contents : \mathbf{L} \rightarrow \mathbf{S}_+ \rightarrow \mathbf{V}_+$$

$$Contents\alpha\sigma = \sigma{=}err \rightarrow err,\ \sigma\alpha{=}free \rightarrow err,\ \sigma\alpha \tag{14}$$

$$Update : \mathbf{L} \rightarrow \mathbf{V} \rightarrow \mathbf{S}_+ \rightarrow \mathbf{S}_+$$

$$Update\alpha\beta\sigma = \sigma{=}err \rightarrow err,\ \sigma{+}[\alpha{\mapsto}\beta] \tag{15}$$

$$New : \mathbf{S}_+ \rightarrow \mathbf{L}_+$$

$$New\sigma = \begin{cases} \alpha \text{ if } \sigma \neq err \text{ and } \alpha \text{ exists such that } \sigma\alpha{=}free, \\ err \text{ otherwise} \end{cases} \tag{16}$$

Notice that *New* is given an implicit definition. We have no wish to go into the messy details of storage allocation (and, indeed, since **L** is an unstructured set we could not do so).

The semantic evaluation functions are as follows.

Semantic Functions

$$\mathbf{E} : \mathbf{Exp} \rightarrow \mathbf{U} \rightarrow \mathbf{S}_+ \rightarrow \mathbf{E}_+$$

$$\mathbf{E}[\![\mathrm{N}]\!]\rho\sigma = \mathbf{N}[\![\mathrm{N}]\!] \tag{17}$$

$$\mathbf{E}[\![\mathbf{true}]\!]\rho\sigma = true \tag{18}$$

$$\mathbf{E}[\![-\mathrm{E}]\!]\rho\sigma = (\epsilon\mathbf{EN} \rightarrow -\epsilon,\ err) \quad \text{where } \epsilon = \mathbf{E}[\![\mathrm{E}]\!]\rho\sigma \tag{19}$$

$$\mathbf{E}[\![\mathrm{I}]\!]\rho\sigma = (\delta\mathbf{EE} \rightarrow \delta,\ \delta\mathbf{EL} \rightarrow Contents\delta\sigma,\ err) \quad \text{where } \delta = \rho[\![\mathrm{I}]\!] \tag{20}$$

$$\mathbf{D} : \mathbf{Dec} \rightarrow \mathbf{U} \rightarrow \mathbf{S}_+ \rightarrow [\mathbf{U} \times \mathbf{S}]_+$$

$$\mathbf{D}[\![\mathbf{con}\ \mathrm{I} = \mathrm{E}]\!]\rho\sigma = (\epsilon\mathbf{EE} \rightarrow \langle[\mathrm{I}{\mapsto}\epsilon],\ \sigma\rangle,\ err) \quad \text{where } \epsilon = \mathbf{E}[\![\mathrm{E}]\!]\rho\sigma \tag{21}$$

$$\mathbf{D}[\![\mathbf{loc}\mathrm{I}{=}\mathrm{E}]\!]\rho\sigma = (\epsilon\mathbf{EE}\wedge\alpha\mathbf{EL} \rightarrow \langle[\mathrm{I}{\mapsto}\alpha], Update\alpha\epsilon\sigma\rangle, err) \quad \text{where } \epsilon = \mathbf{E}[\![\mathrm{E}]\!]\rho\sigma \text{ and } \alpha = New\sigma \tag{22}$$

$$\mathbf{C} : \mathbf{Cmd} \rightarrow \mathbf{U} \rightarrow \mathbf{S}_+ \rightarrow \mathbf{S}_+$$

$$\mathbf{C}[\![\mathbf{skip}]\!]\rho\sigma = \sigma \tag{23}$$

$$\mathbf{C}[\![\mathbf{abort}]\!]\rho\sigma = \sigma{\equiv}\bot \rightarrow \bot,\ err \tag{24}$$

$$\mathbf{C}[\![\mathrm{I}{:=}\mathrm{E}]\!]\rho\sigma = (\sigma{\neq}err\wedge\epsilon\mathbf{EE}\wedge\delta\mathbf{EL} \rightarrow Update\delta\epsilon\sigma,\ err) \quad \text{where } \epsilon = \mathbf{E}[\![\mathrm{E}]\!]\rho\sigma \text{ and } \delta = \rho[\![\mathrm{I}]\!] \tag{25}$$

$$\mathbf{C}[\![\Gamma_1;\Gamma_2]\!]\rho\sigma = \mathbf{C}[\![\Gamma_2]\!]\rho(\mathbf{C}[\![\Gamma_1]\!]\rho\sigma) \tag{26}$$

$$\mathbf{C}[\![\mathbf{if}\ \mathrm{E}\ \mathbf{then}\ \Gamma]\!]\rho\sigma = (\sigma{\neq}err \wedge \epsilon\mathbf{ET} \rightarrow (\epsilon \rightarrow \mathbf{C}[\![\Gamma]\!]\rho\sigma,\ \sigma),\ err) \quad \text{where } \epsilon = \mathbf{E}[\![\mathrm{E}]\!]\rho\sigma \tag{27}$$

$$\mathbf{C}[\![\mathbf{while}\ \mathrm{E}\ \mathbf{do}\ \Gamma]\!]\rho\sigma = \theta\sigma \text{ where } \theta \text{ is given by} \tag{28}$$

$$\theta\sigma' = (\sigma'{\neq}err\wedge\epsilon\mathbf{ET} \rightarrow (\epsilon \rightarrow \theta(\mathbf{C}[\![\Gamma]\!]\rho\sigma'),\ \sigma'),\ err \quad \text{where } \epsilon = \mathbf{E}[\![\mathrm{E}]\!]\rho\sigma'$$

$$\mathbf{C}[\![\mathbf{begin}\ \Delta;\ \Gamma\ \mathbf{end}]\!]\rho\sigma = (\psi\mathbf{EU}{\times}\mathbf{S} \rightarrow (\mathrm{let}\ \langle\rho',\sigma'\rangle = \psi \text{ in } \mathbf{C}[\![\Gamma]\!](\rho{+}\rho')\sigma'),\ err) \quad \text{where } \psi = \mathbf{D}[\![\Delta]\!]\rho\sigma \tag{29}$$

Note that the semantic values of constructs now depend on the environment as well as the store. Commands are transformation functions on the store, and do not affect the environment. A declaration, on the other hand, not only results in a new binding of an identifier (that is, a new little environment) but also possibly transforms the store, since it may allocate and initialise a new location. Thus the meaning of a declaration in a given surrounding environment is a function from stores to pairs which consist of a new environment and a possibly altered store. In equation (29) such a new environment is combined with the surrounding one using an obvious extension of the '+' operator, the definition of which we leave to the reader.

Continuations

In the previous two examples we have dealt with one of the characteristic features of imperative programming languages, the assignment command, and we have described it formally in a purely functional notation. We now do the same for the other characteristic feature, namely commands which alter the sequence of control.

Up to now, we have modelled sequencing by functional composition, by equations such as the following:

$$\mathbf{C}[\![\Gamma_1;\Gamma_2]\!]\rho\sigma = \mathbf{C}[\![\Gamma_2]\!]\rho(\mathbf{C}[\![\Gamma_1]\!]\rho\sigma) \qquad (30)$$

This specifies that the meaning of Γ_1 is applied to the given state and the meaning of Γ_2 applied to the state which results from the first application. How do we prevent Γ_2 from being applied to the state if Γ_1 has terminated abnormally, for example if it caused a jump?

The answer is that we pass the meaning of Γ_2 as an extra parameter to that of Γ_1, merely in order to give Γ_1 the option of ignoring it. So now equation (30) becomes

$$\mathbf{C}[\![\Gamma_1;\Gamma_2]\!]\rho\theta\sigma = \mathbf{C}[\![\Gamma_1]\!]\rho\{\mathbf{C}[\![\Gamma_2]\!]\rho\theta\}\sigma \qquad (31)$$

"command continuations"

where $\mathbf{C} : \mathbf{Cmd}\rightarrow\mathbf{U}\rightarrow[\mathbf{S}\rightarrow\mathbf{S}]\rightarrow[\mathbf{S}\rightarrow\mathbf{S}]$ (we often, as above, enclose continuations in curly brackets, to aid the eye).

The intended meaning of these expressions is indicated by the following:

$$\mathbf{C}[\![\Gamma]\!]\rho\theta\sigma$$

The effect of the whole computation starting with Γ.

The effect of the computation required if Γ "falls off the end".

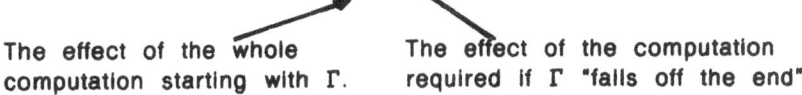

Of course, this "effect" might not be a state transformation (transforming σ to the final state): we might prefer the final answer to be something other than the state. So we make an "effect" an element of a domain **C**:

$\theta \in \mathbf{C} = \mathbf{S} \rightarrow \mathbf{A}$ command continuations

where **A** is a domain of "answers", which may or may not be the same as states. So now:

$$\mathcal{C} : \mathbf{Cmd} \rightarrow \mathbf{U} \rightarrow [\mathbf{S}\rightarrow\mathbf{A}] \rightarrow [\mathbf{S}\rightarrow\mathbf{A}]$$

We could express all this with functions of several arguments (e.g. $\mathcal{C}(\Gamma,\rho,\theta,\sigma)$) instead of using currying; but the various stages each have some kind of meaning:

$\mathcal{C}[\![\Gamma]\!]$	a command *in vacuo*	(the "schema")
$\mathcal{C}[\![\Gamma]\!]\rho$	its identifiers bound, but no continuation	(e.g. a routine)
$\mathcal{C}[\![\Gamma]\!]\rho\theta$	a computation all "ready to go"	(e.g. a label value)
$\mathcal{C}[\![\Gamma]\!]\rho\theta\sigma$	a particular execution	(an answer).

We illustrate all these ideas in a simple example. We add labelled blocks to our example language, together with an ⌜**exit**⌝ command which causes control to leave the specified block.

Syntactic Domains

N $\in$ **Nml**	numerals
I $\in$ **Ide**	identifiers
E $\in$ **Exp**	expressions
Γ $\in$ **Cmd**	commands
Ω $\in$ **Opr**	operators

Syntax

E ::= **true** | . . . | N | I | EΩE | . . .
Γ ::= **skip** | **abort** | I:=E | Γ;Γ | **if** E **then** Γ |
while E **do** Γ | **block** I: Γ **end** | **exit** E

Semantic Domains

$\rho \in \mathbf{U} = \mathbf{Ide} \rightarrow \mathbf{D}$	environments
$\theta \in \mathbf{C} = \mathbf{S} \rightarrow \mathbf{A}$	command continuations
Wrong $\in$ **C** is a particular continuation	
A	answers
$\kappa \in \mathbf{K} = \mathbf{E} \rightarrow \mathbf{C}$	expression continuations
$\epsilon \in \mathbf{E} = \mathbf{N} + \mathbf{T} + \mathbf{C}$	expressible values: the **C** component is for labelled blocks

$\delta \in \mathbf{D} = \mathbf{E} \oplus \mathbf{L} \oplus \{Unset\}_\perp$ denotable values
$\mathbf{N}$ integers
$\mathbf{T}$ truth values
$\alpha \in \mathbf{L}$ locations
$\beta \in \mathbf{B} = \mathbf{E}$ storable values
$\sigma \in \mathbf{S}$ states:
e.g. $\mathbf{S} = [\mathbf{L} \to \mathbf{T}] \times [\mathbf{L} \to [\mathbf{V} \oplus \{err\}_\perp]]$
$Area(\sigma) = \sigma[1]$, $Map(\sigma) = \sigma[2]$
$Contents\alpha\kappa\sigma = Area\sigma\alpha \wedge Map\sigma\alpha \neq err \rightarrow \kappa(Map\sigma\alpha)\sigma,\ Wrong\sigma$
$Assign\alpha\beta\theta\sigma = Area\sigma\alpha \rightarrow \theta\langle Area\sigma,\ Map\sigma + [\alpha \to \beta]\rangle,\ Wrong\sigma$
Allocate . . . etc.

Note that an expression, if it terminates normally, passes a value as well as a state to its continuation. So, whereas a command continuation is in $\mathbf{S} \rightarrow \mathbf{A}$, an expression continuation, κ, is in $\mathbf{E} \rightarrow \mathbf{S} \rightarrow \mathbf{A}$.

Semantic Functions

$\mathbf{E} : \mathbf{Exp} \rightarrow \mathbf{U} \rightarrow \mathbf{K} \rightarrow \mathbf{C}$

$\mathbf{E}[\![\mathrm{N}]\!]\rho\kappa = \kappa(\mathbf{N}[\![\mathrm{N}]\!])$ (32)

$\mathbf{E}[\![\mathbf{true}]\!]\rho\kappa = \kappa(true)$ (33)

$\mathbf{E}[\![\mathrm{E}_1\Omega\mathrm{E}_2]\!]\rho\kappa = \mathbf{E}[\![\mathrm{E}_1]\!]\rho\{\lambda\epsilon_1.\mathbf{E}[\![\mathrm{E}_2]\!]\rho\{\lambda\epsilon_2.\mathbf{O}[\![\Omega]\!]\langle\epsilon_1,\epsilon_2\rangle\kappa\}\}$ (34)

$\mathbf{E}[\![\mathrm{I}]\!]\rho\kappa = (\delta\mathrm{E}\mathbf{L} \rightarrow Contents\delta\kappa,\ \delta{=}Unset \rightarrow Wrong,\ \kappa\delta$ (35)
where $\delta = \rho[\![\mathrm{I}]\!])$

$\mathbf{O} : \mathbf{Opr} \rightarrow [\mathbf{E}\times\mathbf{E}] \rightarrow \mathbf{K} \rightarrow \mathbf{C}$

$\mathbf{O}[\![+]\!]\langle\epsilon_1,\epsilon_2\rangle\kappa = \epsilon_1\mathrm{E}\mathbf{N} \wedge \epsilon_2\mathrm{E}\mathbf{N} \rightarrow \kappa(\epsilon_1+\epsilon_2),\ Wrong$ (36)

$\mathbf{C} : \mathbf{Cmd} \rightarrow \mathbf{U} \rightarrow \mathbf{C} \rightarrow \mathbf{C}$

$\mathbf{C}[\![\mathbf{skip}]\!]\rho\theta = \theta$ (37)

$\mathbf{C}[\![\mathbf{abort}]\!]\rho\theta = Wrong$ (38)

$\mathbf{C}[\![\mathrm{I}{:=}\mathrm{E}]\!]\sigma = (\delta\mathrm{E}\mathbf{L} \rightarrow \mathbf{E}[\![\mathrm{E}]\!]\rho\{\lambda\epsilon.Assign\delta\epsilon\theta\},\ Wrong$ (39)
where $\delta = \rho[\![\mathrm{I}]\!])$

$\mathbf{C}[\![\Gamma_1;\Gamma_2]\!]\rho\theta = \mathbf{C}[\![\Gamma_1]\!]\rho\{\mathbf{C}[\![\Gamma_2]\!]\rho\theta)$ (40)

$\mathbf{C}[\![\mathbf{if}\ \mathrm{E}\ \mathbf{then}\ \Gamma]\!]\rho\theta =$
$\mathbf{E}[\![\mathrm{E}]\!]\rho\{\lambda\epsilon.\epsilon\mathrm{E}\mathbf{T} \rightarrow (\epsilon \rightarrow \mathbf{C}[\![\Gamma]\!]\rho\theta,\theta),Wrong)$ (41)

or

$\mathbf{C}[\![\mathbf{if}\ \mathrm{E}\ \mathbf{then}\ \Gamma]\!]\rho\theta = \mathbf{E}[\![\mathrm{E}]\!]\rho\{Cond\langle\mathbf{C}[\![\Gamma]\!]\rho\theta,\ \theta\rangle\}$ (42)
where $Cond : [\mathbf{C}\times\mathbf{C}] \rightarrow \mathbf{K}$ is defined by
$Cond\langle\theta_1,\theta_2\rangle = \lambda\epsilon.\epsilon\mathrm{E}\mathbf{T} \rightarrow (\epsilon \rightarrow \theta_1,\theta_2),\ Wrong$ (43)

$\mathbf{C}[\![\mathbf{while}\ \mathrm{E}\ \mathbf{do}\ \Gamma]\!]\rho\theta =$
$fix(\lambda\theta'.\mathbf{E}[\![\mathrm{E}]\!]\rho\{Cond\langle\mathbf{C}[\![\Gamma]\!]\rho\theta',\theta\rangle\})$ (44)

$\mathbf{C}[\![\mathbf{block}\ \mathrm{I}{:}\ \Gamma\ \mathbf{end}]\!]\rho\theta = \mathbf{C}[\![\Gamma]\!](\rho+[\mathrm{I}\to\theta])\theta$ (45)

$\mathbf{C}[\![\mathbf{exit}\ \mathrm{E}]\!]\rho\theta = \mathbf{E}[\![\mathrm{E}]\!]\rho\{Exit\}$ (46)
where $Exit\epsilon = \epsilon\mathrm{E}\mathbf{C} \rightarrow \epsilon,\ Wrong$ (47)

An Example Evaluation

Let us evaluate the little program

```
⌜block L:
      x:=y;
      exit L;
      x:= x+1
end⌝
```

in an environment $\rho_1=\rho_0+[\ulcorner x\urcorner\to\alpha_x,\ \ulcorner y\urcorner\to\alpha_y]$, and a state σ such that

$$Area\sigma\alpha_x=Area\sigma\alpha_y=true,\ Map\sigma\alpha_x=\beta_x,\ Map\sigma\alpha_y=\beta_y.$$

We have

$$\mathbf{C}[\![\mathbf{block}\ L\ x{:=}y;\ \mathbf{exit}\ L;\ x{:=}x+1\ \mathbf{end}]\!]\rho_1\theta\sigma$$

$$= \mathbf{C}[\![x{:=}y;\ \mathbf{exit}\ L;\ x{:=}x+1]\!]\underbrace{(\rho_1+[\ulcorner L\urcorner\to\theta])}_{\rho}\theta\sigma$$

$$= \mathbf{C}[\![x{:=}y]\!]\rho\underbrace{\{\mathbf{C}[\![\mathbf{exit}\ L;\ x{:=}x+1]\!]\rho\theta\}}_{\theta_1}\sigma$$

$$= (\underset{true}{\delta\text{E}L} \to \mathbf{E}[\![y]\!]\rho\underbrace{\{\lambda\epsilon.Assign\delta\epsilon\theta_1\}}_{\kappa_\delta},Wrong\ \text{ where }\ \delta=\underset{\alpha_x}{\rho[\![x]\!]})\sigma$$

$$= \mathbf{E}[\![y]\!]\rho\kappa_{\alpha_x}\sigma \qquad (\text{note: } \kappa_{\alpha_x} \text{ is } \kappa_\delta \text{ with } \delta=\alpha_x)$$

$$= (\underbrace{\delta\text{E}L}_{true} \to Contents\delta\kappa_{\alpha_x},\ \ldots\ \text{ where }\ \delta=\underbrace{\rho[\![y]\!]}_{\alpha_y})\sigma$$

$$= Contents\alpha_y\kappa_{\alpha_x}\sigma$$

$$= \underbrace{Area\sigma\alpha_y}_{true} \wedge \underbrace{\underbrace{Map\sigma\alpha_y}_{\beta_y}\neq err}_{true} \to \kappa_{\alpha_x}(\underbrace{Map\sigma\alpha_y}_{\beta_y})\sigma,\ Wrong\sigma$$

$$= (\lambda\epsilon.Assign(\alpha_x)\epsilon\theta_1)(\beta_y)\sigma$$

$$= Assign\alpha_x\beta_y\theta_1\sigma$$

$$= \underbrace{Area\sigma\alpha_x}_{true} \to \theta_1(\underbrace{\langle Area\sigma,\ Map\sigma+[\alpha_x\to\beta_y]\rangle}_{\sigma_1}),\ Wrong\sigma$$

$$= \theta_1\sigma_1$$

$$= \mathbf{C}[\![\mathbf{exit}\ L]\!]\rho\{\mathbf{C}[\![x{:=}x{+}1]\!]\rho\theta\}\sigma_1$$

$$= \mathbf{E}[\![L]\!]\rho\{Exit\}\sigma_1$$

$$= (\underbrace{\delta\mathsf{EL}}_{false} \rightarrow \ldots,\ \underbrace{\delta{=}err}_{false} \rightarrow Wrong,\ Exit\delta \text{ where } \delta{=}\underbrace{\rho[\![L]\!]}_{\theta})\sigma_1$$

$$= (\underbrace{\theta\mathsf{EC}}_{true} \rightarrow \theta,\ Wrong)\sigma_1$$

$$= \theta\sigma_1.$$

Evaluations like this are tedious, but they are, after all, usually left to a computer.

Declaration Continuations

Besides commands and expressions, declarations can also terminate abnormally, so we arrange that they have continuations too. The value supplied to an expression continuation is an expressible value; similarly the value a declaration passes to its continuation is a little environment - the "binding" it has declared. So we may once more add declarations to the syntax:

$\Delta \in$ **Dec** declarations

$\Delta ::=$ **con** $I = E \mid$ **loc** $I = E \mid \ldots$

$\Gamma ::=$ **begin** $\Delta;\ \Gamma$ **end**

and extend the semantics as follows.

$\mathbf{D} : \mathbf{Dec} \rightarrow \mathbf{U} \rightarrow [\mathbf{U} \rightarrow \mathbf{C}] \rightarrow \mathbf{C}$

(cf. $\mathbf{E} : \mathbf{Exp} \rightarrow \mathbf{U} \rightarrow [\mathbf{E} \rightarrow \mathbf{C}] \rightarrow \mathbf{C}$)

$\mathbf{D}[\![\mathbf{con}\ I = E]\!]\rho\chi = \mathbf{E}[\![E]\!]\rho\{\lambda\epsilon.\chi([I \rightarrow \epsilon])\}$ (48)

$\mathbf{D}[\![\mathbf{loc}\ I = E]\!]\rho\chi =$ (49)

$Allocate\{\lambda\alpha.\mathbf{E}[\![E]\!]\rho\{\lambda\epsilon.Assign\alpha\epsilon\{\chi([I \rightarrow \alpha])\}\}\}$

where $Allocate : [\mathbf{L} \rightarrow \mathbf{C}] \rightarrow \mathbf{C}$ is given by

$Allocate\psi\sigma = \psi\alpha\langle Area\sigma + [\alpha \rightarrow true], Map\sigma + [\alpha \rightarrow err]\rangle$ (50)

where α is a location such that $Area\sigma\alpha = false$,

if such a location exists for σ, or

$Wrong\sigma$ otherwise. (Note: $\psi \in \mathbf{L} \rightarrow \mathbf{C}$)

$\mathbf{C}[\![\mathbf{begin}\ \Delta;\ \Gamma\ \mathbf{end}]\!]\rho\theta = \mathbf{D}[\![D]\!]\rho\{\lambda\rho'.\mathbf{C}[\![\Gamma]\!](\rho+\rho')\theta\}$ (51)

Answers

As we have already remarked, we are not always interested in the final state of a program, for example if the program does not have one because it never terminates (e.g. traffic light controllers, or computer operating systems). So the domain **A** of answers may not always be **S** (states). We conclude with an example of this greater generality.

Outputs as Answers. The output of a program is often regarded as a component of the state. However, if the final state of a program is $\perp$ (as it is for all non-terminating programs) we can obtain no information about its output component. So, following a suggestion of Brian Mayoh, we instead define the answer domain **A** by:

$$\mathbf{A} = \mathbf{O}^{\circledast} \qquad \text{(possibly infinite lists of outputs)}$$

The equations for semantic functions are as before; but we add

$$\mathcal{C}[\![\mathbf{write}\ \mathrm{E}]\!]\rho\theta = \mathcal{E}[\![\mathrm{E}]\!]\rho\{\lambda\epsilon.\lambda\sigma.\langle Rep(\epsilon)\rangle \mathbin{\S} (\theta\sigma)\} \tag{52}$$

where *Rep* is, for example, a function mapping numbers to character strings, and $\S$ is the concatenation operator for tuples. The whole program's effect is

$$\mathcal{C}[\![\Gamma_0]\!]\rho_0\{\lambda\sigma.\langle\rangle\}\sigma_0 \tag{53}$$

where ρ_0 is the original environment (perhaps including predefined procedures) and σ_0 the original store (in our model for **S**, perhaps $\langle\lambda\alpha.false,\lambda\alpha.err\rangle$). And we might have

$$Wrong\sigma = \langle\text{'Your program has aborted.'}, \text{'Store dump in hexadecimal:'}\rangle \mathbin{\S} HexDump(\sigma)$$

and so on.

The reader is invited to check that now a program such as

⌜**while true do write** 3⌝

produces an infinite sequence of output as expected. Similarly one could show that a traffic light controller, mentioned above, produced the correct sequence of configurations of lights; but notice that one could say nothing whatever about the *duration* of each configuration: time has been abstracted away in this and almost all denotational definitions.

Gilles Kahn has suggested a nice extension of this idea. He remarks that we are often interested in the input-output behaviour of our programs, and proposes that **A** be a domain of functions from inputs to outputs, both of which are possibly infinite streams of values (that is, elements of $\mathbf{O}^{\circledast}$). All the semantic equations for our language remain exactly as before, except the equations for the ⌜**read**⌝ expression and the ⌜**write** E⌝ command, which are as follows. Note: $\iota \dagger i$, where $\iota = \langle \nu_1, \nu_2, \ldots, \nu_n \rangle$, is $\langle \nu_{i+1}, \nu_{i+2}, \ldots, \nu_n \rangle$ $(0 \leq i \leq n)$.

$$\mathbf{E}[\![\mathbf{read}]\!]\rho\kappa\sigma = \lambda\iota.\kappa(\iota[1])\ \sigma(\iota\dagger 1) \qquad (54)$$

$$\mathbf{C}[\![\mathbf{write}\ \mathrm{E}]\!]\rho\theta = \mathbf{E}[\![\mathrm{E}]\!]\rho\{\lambda\epsilon.\lambda\sigma.\lambda\iota.(Rep\,\epsilon)\ast(\theta\sigma\iota)\} \qquad (55)$$

It is suggested that the reader makes sure that the functionalities of all the sub-expressions in these two equations are clearly understood.

CHAPTER II – CONNECTION WITH PREDICATE TRANSFORMERS

In this Chapter we consider how semantic descriptions like those we have described relate to the weakest precondition semantics of the kind used by Professor Dijkstra. At first, however, we shall confine ourselves to languages which, unlike Dijkstra's, do not exhibit nondeterminism.

We start from the point we reached in the previous chapter, where we were considering various possible domains of answers. Now we wish to judge the effect of a computation by whether or not the computation terminates correctly. So the answer domain, **A**, is simply $\{true, untrue\}$, where $untrue \sqsubseteq true$. The effect of this ordering is that *untrue* corresponds to $\bot_{\mathbf{A}}$, so that the answer for any nonterminating computation will automatically be *untrue*. Thus if we define $Wrong = \lambda\sigma.untrue$ and $Finish = \lambda\sigma.true$, we see that

$$\mathbf{C}[\![\Gamma]\!]\rho\{Finish\}\sigma$$

will have the value *true* if and only if the command Γ terminates normally.

We next introduce pre- and postconditions. We define a new syntactic domain **Cla** ("claims") with typical member Ψ and syntax as follows.

$$\Psi ::= \mathrm{E} \mid {\sim}\mathrm{E} \mid \Psi \vee \Psi \mid \Psi \wedge \Psi \mid \exists n.\Psi_n$$

where E is a programming language expression without side effects; that is to say, an expression such that for any proper ρ and σ there is an ϵ such that

$$\mathbf{E}[\![\mathrm{E}]\!]\rho\kappa\sigma = \kappa\epsilon\sigma. \qquad (56)$$

The semantics of these claims is given by the function

$$\mathbf{V} : \mathbf{Cla} \rightarrow \mathbf{U} \rightarrow \mathbf{S} \rightarrow \mathbf{A}$$

where the domain **A** is as defined above. The definition of **V** is as follows.

$$\mathbf{V}[\![\mathrm{E}]\!]\rho\sigma = (\epsilon \sqsupseteq true \text{ where } \mathbf{E}[\![\mathrm{E}]\!]\rho\kappa\sigma = \kappa\epsilon\sigma) \qquad (57)$$

$$\mathbf{V}[\![\sim\mathrm{E}]\!]\rho\sigma = (\epsilon \sqsupseteq \mathit{false} \text{ where } \mathbf{E}[\![\mathrm{E}]\!]\rho\kappa\sigma = \kappa\epsilon\sigma) \tag{58}$$
$$\mathbf{V}[\![\Psi_1 \vee \Psi_2]\!]\rho\sigma = \mathbf{V}[\![\Psi_1]\!]\rho\sigma \cup \mathbf{V}[\![\Psi_2]\!]\rho\sigma \tag{59}$$
$$\mathbf{V}[\![\Psi_1 \wedge \Psi_2]\!]\rho\sigma = \mathbf{V}[\![\Psi_1]\!]\rho\sigma \cap \mathbf{V}[\![\Psi_2]\!]\rho\sigma \tag{60}$$
$$\mathbf{V}[\![\exists n.\Psi_n]\!]\rho\sigma = \bigcup\{\mathbf{V}[\![\Psi_n]\!]\rho\sigma \mid n{\geqslant}0\} \tag{61}$$

It may now be seen that the claim Ψ_1 would denote a precondition for the command Γ to achieve Ψ_2 if

$$\mathbf{V}[\![\Psi_1]\!]\rho\sigma \sqsubseteq \mathbf{C}[\![\Gamma]\!]\rho\{\mathbf{V}[\![\Psi_2]\!]\rho\}\sigma. \tag{62}$$

If the $\sqsubseteq$ were replaced by a = in this equation then Ψ_1 would be a *weakest* precondition.

Actually for a satisfactory weakest precondition we do not require the previous equation to hold for *all* ρ and σ. It is preferable to rule out nasty states and environments (for example, those containing improper values, or environments in which different names denote the same location). We therefore assume that the criteria for satisfactory environments and stores are expressed by predicates u and s, defined on **U** and **S** respectively. Then we define an equivalence relation $\approx$ such that

$$\theta_1 \approx \theta_2 \text{ if } \theta_1\sigma = \theta_2\sigma \text{ whenever } s\sigma. \tag{63}$$

We may then say that Ψ_1 is a weakest precondition (with respect to u and s) for Γ to achieve Ψ_2 if, whenever $u\rho$,

$$\mathbf{V}[\![\Psi_1]\!]\rho \approx \mathbf{C}[\![\Gamma]\!]\rho\{\mathbf{V}[\![\Psi_2]\!]\rho\}. \tag{64}$$

Now, in Dijkstra's methodology claims are related one to another by means of predicate transformers. In particular, weakest preconditions are generated by the function

$$wp : [\mathbf{Cmd} \times \mathbf{Cla}] \rightarrow \mathbf{Cla}.$$

We say that $wp(\Gamma,\Psi)$, where Γ is a command and Ψ is a claim, is the weakest precondition for the command Γ to achieve the postcondition Ψ. Notice that *wp* is a function mapping syntactic objects to syntactic objects; so we cannot hope that it produces the *unique* weakest precondition for a given Γ and Ψ. We only insist that it produces a suitable claim, in the sense that

$$\mathbf{V}[\![wp(\Gamma,\Psi)]\!]\rho \approx \mathbf{C}[\![\Gamma]\!]\rho\{\mathbf{V}[\![\Psi]\!]\rho\} \tag{65}$$

whenever $u\rho$.

Let us remind ourselves of some of the equations defining *wp* for some standard commands. We have:

$$wp(\mathbf{skip},\ \Psi) = \Psi \tag{66}$$

$$wp(\mathbf{abort}, \Psi) = false \tag{67}$$
$$wp(I:=E, \Psi) = \Psi_{I \to E} \tag{68}$$
$$wp(\Gamma_1;\Gamma_2, \Psi) = wp(\Gamma_1, wp(\Gamma_2, \Psi)) \tag{69}$$
$$wp(\mathbf{if}\ E\ \mathbf{then}\ \Gamma_1\ \mathbf{else}\ \Gamma_2, \Psi) = (E \wedge wp(\Gamma_1, \Psi)) \vee (\sim E \wedge wp(\Gamma_2, \Psi)) \tag{70}$$
$$wp(\mathbf{while}\ E\ \mathbf{do}\ \Gamma, \Psi) = \exists n.C_n \tag{71}$$
$$\text{where}\quad C_0 = false \tag{72}$$
$$C_{n+1} = (\sim E \wedge \Psi) \vee (E \wedge wp(\Gamma, C_n)) \tag{73}$$

In order to prove that these equations define a satisfactory predicate transformer for weakest preconditions, we would have to show that equation (65) holds whenever $u\rho$. The proof is obviously by structural induction on Γ. For equations (66) and (67), with their counterparts (37) and (38), the proof is immediate. For equations (69) and (70) with their counterparts (40) and (41), the results follow straightforwardly after two applications of the inductive hypothesis, namely that

$$\mathbf{C}[\![\Gamma_1]\!]\rho\theta_1 \approx \mathbf{C}[\![\Gamma_1]\!]\rho\theta_2 \quad \text{whenever } \theta_1 \approx \theta_2 \tag{74}$$

and similarly for Γ_2. For the **while**-loop equation (71), we define

$$\psi = \lambda\theta.\mathbf{E}[\![E]\!]\rho\{cond\langle \mathbf{C}[\![\Gamma]\!]\rho\theta, \mathbf{V}[\![\Psi]\!]\rho\rangle\} \tag{75}$$

so that

$$fix\psi = \mathbf{C}[\![\mathbf{while}\ E\ \mathbf{do}\ \Gamma]\!]\rho\{\mathbf{V}[\![\Psi]\!]\rho\}. \tag{76}$$

Then, by a simple induction on n we show that

$$\mathbf{V}[\![C_n]\!]\rho \approx \psi^n\{\lambda\sigma.untrue\}. \tag{77}$$

So, whenever $s\sigma$,

$$\mathbf{U}\{\mathbf{V}[\![C_n]\!]\rho\sigma \mid n \geqslant 0\} = \mathbf{U}\{\psi^n(Wrong)\sigma\} \tag{78}$$

$$= (\mathbf{U}\{\psi^n Wrong\})\sigma \tag{79}$$

by the continuity of application, since $\{\psi^n Wrong\}$ is directed. So, since $Wrong = \bot_{\mathbf{S}\to\mathbf{A}}$,

$$\mathbf{V}[\![\exists n.C_n]\!]\rho = (\mathbf{U}\psi^n(\bot))\sigma \tag{80}$$

$$= (fix\psi)\sigma \tag{81}$$
$$= \mathbf{C}[\![\mathbf{while}\ E\ \mathbf{do}\ \Gamma]\!]\rho\{\mathbf{V}[\![\Psi]\!]\rho\}\sigma \tag{82}$$

as required.

The reader will have noticed that we have left the assignment command until last. For this we must show that

$$\mathbf{C}[\![\mathrm{I}:=\mathrm{E}]\!]\rho\{\mathbf{V}[\![\Psi]\!]\rho\} \approx \mathbf{V}[\![\Psi_{\mathrm{I}\to\mathrm{E}}]\!]\rho \tag{83}$$

whenever $u\rho$. We must assume that u and s constrain ρ and σ respectively in such a way that E is without side effects (as defined above) and I denotes a location (that is, $\rho[\![\mathrm{I}]\!]\in\mathbf{L}$). Then, assuming σ_0 is such that $s\sigma_0$ and $\mathbf{V}[\![\Psi_{\mathrm{I}\to\mathrm{E}}]\!]\rho\sigma_0$ both hold, and defining

$$\sigma_1 = \sigma_0+[\rho[\![\mathrm{I}]\!]\to\epsilon] \tag{84}$$

we must show that $s\sigma_1$ and $\mathbf{V}[\![\Psi]\!]\rho\sigma_1$ are also both satisfied. For this we would try structural induction on Ψ, and indeed for simple languages the proof would go through successfully. For more complicated languages, however, the proof breaks down. For example, Ψ might contain a call of a procedure which uses the identifier I as a non-local variable, an occurence of I which would not be reached by the structural induction analysis of Ψ. To prevent this we might try to strengthen u (and possibly s) to forbid this kind of procedure, but unfortunately it is not possible for us to express this kind of constraint with our present apparatus. The trouble is that although the value of a procedure (an element of $\mathbf{D}\to\mathbf{K}\to\mathbf{C}$) indeed depends on the environment in which it is declared, the identifiers which occur free in the procedure's text are not obtainable from its value. To sort all this out we would need to use a more complicated form of semantics (the "store semantics" of the next Chapter), in which the value of a procedure includes a free-variable list as one of its components.

This complication illustrates the conflict between today's complex modern languages, with their many features, and a methodology for which we wish to assume the validity of simple and powerful proof rules. We need not dwell on this conflict here: the virtues of simplicity are eloquently expounded elsewhere. Instead, let us turn to the much simpler language described in Chapter 4 of Dijkstra's book *A Discipline of Programming* [1].

Dijkstra's Language

For much of this discussion we shall be following Plotkin [5]. The language we shall analyse is a slight variant of that described by Dijkstra, and its syntax is as follows.

Syntactic Domains

A ∈ **ACom**	atomic commands
B ∈ **BExp**	boolean expressions
Σ ∈ **Stmt**	statements
Γ ∈ **GCom**	guarded commands

Syntax

Σ ::= A | **skip** | **abort** | Σ;Σ | **if** Γ **fi** | **do** Γ **od**

Γ ::= **empty** | B→Σ | Γ▯Γ

The domain **BCom** and **BExp** are not further defined: we would expect them to contain commands like the assignment statement, and the usual boolean expressions. Notice that for simplicity's sake we are considering Dijkstra's language at a stage in its evolution before it acquired block structure. Notice, too, that we have to build up lists of guarded commands by pairing.

This language is nondeterministic: in general each program will have several possible results, and the semantics, of course, must not constrain which of the possibilities will eventually be produced by a particular execution. However, the nondeterminism is introduced by the lists of guarded commands, which are finite in length; so, as Dijkstra explains, provided a program is certain to terminate, the number of its nondeterministic possibilities will be finite.

Let us consider how we might capture all this by setting up one of Scott's neighbourhood systems for the domain of results. We recall that each neighbourhood of such a system is a collection of possible attributes (tokens) which might apply to an object being computed: it contains those attributes which have not yet been ruled out. Thus, for example, the neighbourhood giving least information is the one which rules nothing out: it is therefore Δ, containing all possible attributes. In the present case, we are computing one of a finite set of possible results. Since we are not allowed to specify which one we cannot rule the others out, and so each neighbourhood must contain tokens for all of the possibilities. The construct we require is therefore that given in Definition 7.9 of Scott's monograph [6], and known as the Smyth powerdomain. For this example we shall be using ***PD***, a powerdomain on states.

In order to get a little more insight into how the powerdomain behaves, let us assume, as in a previous example, that our states **S** are the domain

$$\mathbf{Ide} \longrightarrow [\mathbf{V} \oplus \{free\}].$$

Let us assume, by way of simplicity, that stored values are incomparable: that is to say, we do not allow $\bot_{\mathbf{V}}$ to be stored (any attempt to do so resulting in the bottom state, $\bot_{\mathbf{S}}$), and that for all other values in **V**, $x \sqsubseteq y$ only if $x = y$. In these circumstances **S** (like **V**) is a *flat* domain: that is to say, for all $\sigma_1, \sigma_2 \in \mathbf{S}$,

$$\sigma_1 \sqsubseteq \sigma_2 \text{ if and only if } \sigma_1 = \bot_{\mathbf{S}} \text{ or } \sigma_1 = \sigma_2. \qquad (85)$$

Thus the neighbourhoods in the system for states are simply the bottom neighbourhood Δ and one neighbourhood for each defined state. The powerdomain ***PS***, therefore, will contain finite unions of such neighbourhoods; and, as usual, smaller neighbourhoods convey more information, since they rule out more possibilities. Notice that if the bottom neighbourhood Δ is one of the components of the finite union, then the whole neighbourhood will also be Δ: since one of the possibilities is undefined there is nothing we can say for certain about the eventual outcome. The domain elements belonging to *PS* are finite sets of states ω where

$$\omega_1 \subseteq_{PS} \omega_2 \text{ if and only if } \perp_S \in \omega_1 \text{ or } \omega_1 \supseteq \omega_2. \tag{86}$$

This apparently contradictory equation needs some interpretation. The $\subseteq$ sign refers to the elements regarded as sets of neighbourhoods, while the $\in$ and the $\supseteq$ refers to them regarded as sets of states. Here too a set of states is better defined if there are fewer possibilities for what might actually happen, although again all situations were the actual outcome might be undefined are equally bad.

We can now give the semantics of our language.

Semantic Domains

$\sigma \in \mathbf{S}$	states
$\mu \in \mathbf{ST} = \mathbf{S} \rightarrow \mathbf{PS}$	nondeterministic state transformations

Semantic Functions

$\mathbf{A} : \mathbf{ACom} \rightarrow \mathbf{S} \rightarrow \mathbf{S}$ $\quad (\alpha \in \mathbf{S} \rightarrow \mathbf{S})$
$\mathbf{B} : \mathbf{BExp} \rightarrow \mathbf{S} \rightarrow \mathbf{T}$
$\mathbf{C} : \mathbf{Stmt} \rightarrow \mathbf{ST}$

Deterministic commands previously mapped states to states. To convert them for the present nondeterministic framework, the output state must be replaced by the singleton set containing just this state, although if this state were $\perp_S$, we would replace this by $\perp_{PS}$ instead. So we define

$$Conv : [\mathbf{S} \rightarrow \mathbf{S}] \rightarrow \mathbf{ST}$$
$$Conv(\alpha)\sigma = \{\!|\, \alpha\sigma \,|\!\} \tag{87}$$

where, for $\sigma \in \mathbf{S}$, $\{\!|\, \sigma \,|\!\} \in PS$ is defined by

$$\{\!|\, \sigma \,|\!\} = \begin{cases} \{\sigma\} & \text{if } \sigma \neq \perp_S \\ \perp_{PS} & \text{otherwise.} \end{cases} \tag{88}$$

Then we may simply write

$$\mathbf{C}[\![A]\!] = Conv(\mathbf{A}[\![A]\!]) \tag{89}$$
$$\mathbf{C}[\![\mathbf{skip}]\!] = \lambda\sigma.\{\!|\, \sigma \,|\!\} \tag{90}$$
$$\mathbf{C}[\![\mathbf{abort}]\!] = \perp_{ST} \tag{91}$$

To form the composition of two nondeterministic state transformations, we must arrange to apply the second transformation to each possible outcome of the first. So we define

$$Comp : \mathbf{ST}^2 \rightarrow \mathbf{ST}$$
$$Comp(\mu,\mu')\sigma = App(\mu',\mu\sigma) \quad (92)$$

where $App : [\mathbf{ST} \times \mathbf{PS}] \rightarrow \mathbf{PS}$ is defined by:

$$App(\mu,\omega) = \begin{cases} \bigcup\{\mu\sigma \mid \sigma\epsilon\omega\} & \text{if } \omega\neq\bot \text{ and for all } \sigma\epsilon\omega \ \mu\sigma\neq\bot \\ \bot_{PS} & \text{otherwise} \end{cases} \quad (93)$$

Then we write

$$\mathbf{C}[\![\Sigma_1;\Sigma_2]\!] = Comp(\mathbf{C}[\![\Sigma_1]\!],\mathbf{C}[\![\Sigma_2]\!]). \quad (94)$$

The remaining two forms of statement involve guarded commands. We shall see that the meaning of a guarded command is a pair; the first component (a member of $\mathbf{S}\rightarrow\mathbf{T}$) tells when the guard is satisfied, and the second (a member of $\mathbf{ST}$) gives the effect of the command. The **if**-command has the effect of the guarded command inside it only if the guard is satisfied, otherwise it fails. So we may write:

$$\mathbf{C}[\![\mathbf{if}\ \Gamma\ \mathbf{fi}]\!] = Cond(\mathbf{G}[\![\Gamma]\!]) \quad (95)$$

where $Cond : [[\mathbf{S}\rightarrow\mathbf{T}] \times \mathbf{ST}] \rightarrow \mathbf{ST}$ is defined by:

$$Cond(\pi,\mu)\sigma = \pi\sigma \rightarrow \mu\sigma,\ \bot_{PS} \quad (96)$$

The **do**-command iterates until the guard is not satisfied. Its semantics is therefore defined by:

$$\mathbf{C}[\![\mathbf{do}\ \Gamma\ \mathbf{od}]\!] = Do(\mathbf{G}[\![\Gamma]\!]) \quad (97)$$

where $Do : [[\mathbf{S}\rightarrow\mathbf{T}] \times \mathbf{ST}] \rightarrow \mathbf{ST}$ is given by:

$$Do(\pi,\mu) = fix(H) \quad (98)$$

where $H(\mu')\sigma = \pi\sigma \rightarrow comp(\mu,\mu')\sigma,\ \{\!|\sigma|\!\}$.

Next we give the semantics of the guarded commands themselves.

$$\mathbf{G} : \mathbf{GCom} \rightarrow [[\mathbf{S} \rightarrow \mathbf{T}] \times \mathbf{ST}]$$

The guard of the empty guarded command is never satisfied; since, therefore, the effect never happens, we can write anything: $\bot_{ST}$ is quite convenient.

$$\mathbf{G}[\![\mathbf{empty}]\!] = \langle \lambda\sigma.false,\ \bot_{ST}\rangle \quad (99)$$
$$\mathbf{G}[\![B\rightarrow\Sigma]\!] = \langle \mathbf{B}[\![B]\!],\ \mathbf{C}[\![\Sigma]\!]\rangle \quad (100)$$

When we join two guarded commands together into one, using the $\square$ symbol, the implicit guard for the composite command is $\pi\vee\pi'$, where

π and π' correspond to the guards of the individual commands and

$$(\pi \vee \pi')\sigma = \pi\sigma \vee \pi'\sigma. \tag{101}$$

The effect of the composite command is the effect of the appropriate arm, if only one guard is satisfied; if neither is satisfied the overall effect is irrelevant and we choose $\perp_{\mathbf{ST}}$; if both guards are satisfied we take the union of the effects. So we write

$$\mathbf{G}[\![\Gamma_1 \,\square\, \Gamma_2]\!] = Bar(\mathbf{G}[\![\Gamma_1]\!], \mathbf{G}[\![\Gamma_2]\!]) \tag{102}$$

where $Bar : [[\mathbf{S}\rightarrow\mathbf{T}] \times \mathbf{ST}]^2 \rightarrow [[\mathbf{S}\rightarrow\mathbf{T}] \times \mathbf{ST}]$ is defined by:

$$\begin{aligned} Bar(\langle\pi,\mu\rangle,\langle\pi',\mu'\rangle) = \langle\pi\vee\pi',\ \lambda\sigma.\\ \pi(\sigma) \rightarrow (\pi'(\sigma) \rightarrow \mu\sigma\cup\mu'\sigma,\ \mu\sigma),\\ (\pi'(\sigma) \rightarrow \mu'\sigma,\ \perp_{\mathbf{ST}})\rangle. \end{aligned} \tag{103}$$

Definition by Predicate Transformers

Having given the denotational semantics for this little language, we now aim to compare it with a definition using predicate transformers. To do this, we take the liberty of recasting the definitions in Dijkstra's book in a form more suitable for the comparison. Taking our cue from Chapter 2 of *A Discipline of Programming*, ("we call two predicates P and Q equal when . . . they characterise the same set of states"), we shall regard a predicate as the set of those states (but never including the undefined state) for which the associated condition is satisfied. So if $P\in\mathbf{Pred}$, then

$$P \subseteq (\mathbf{S}-\{\perp_{\mathbf{S}}\}). \tag{104}$$

Notice that the set of predicates is a complete lattice under the subset ordering ($\subseteq$). Notice, too, that this corresponds with our previous discussion of weakest precondition semantics, in which the minimal predicate was $\lambda\sigma.false$; here, similarly, the weakest predicate is $\emptyset$, the empty set.

Predicate transformers are functions from **Pred** to **Pred**. However, we shall allow only those functions which satisfy Dijkstra's "healthiness conditions". Functions must be *strict*:

$$f\emptyset=\emptyset \tag{105}$$

multiplicative:

$$f(P)\cap f(Q) = f(P\cap Q) \tag{106}$$

and *continuous*:

$$f(\bigcup\{Q_i\}) = \bigcup\{f(Q_i)\} \tag{107}$$

where $Q_0 \subseteq Q_1 \subseteq \ldots$ is an increasing sequence of predicates. (These are, respectively, Dijkstra's properties 1, 3 and 5; properties 2 and 4 are, as he states, consequences of 3.) So the domain of predicate transformers is given by

$$\mathbf{PT} = \{f : \mathbf{Pred} \rightarrow \mathbf{Pred} \mid f \text{ is strict, continuous and multiplicative}\}.$$

It is easy to see that **PT** becomes a domain in our usual sense with the ordering

$$f \subseteq g \text{ if } f(R) \subseteq g(R) \text{ for all } R \in \mathbf{Pred}. \tag{108}$$

Now, following Dijkstra, for each command construct in our language we give the corresponding weakest precondition predicate transformer. Where Dijkstra writes $wp(\Sigma, R)$ we shall write $\mathbf{C}[\![\Sigma]\!](R)$ where, this time,

$$\mathbf{C} : \mathbf{Stmt} \rightarrow \mathbf{PT}.$$

Similarly, this time,

$$\mathbf{G} : \mathbf{GCom} \rightarrow [[\mathbf{S} \rightarrow \mathbf{T}] \times \mathbf{PT}].$$

Firstly, for atomic commands, $\mathbf{C}[\![\mathrm{A}]\!](R)$ must give us just that set of states which the command will map to states in R. So, once again, we write:

$$\mathbf{C}[\![\mathrm{A}]\!] = Conv(\mathbf{A}[\![\mathrm{A}]\!]) \tag{109}$$

where, this time, $Conv$: $[\mathbf{S} \rightarrow \mathbf{S}] \rightarrow \mathbf{PT}$ is defined by:

$$Conv(\alpha)(R) = \alpha^{-1}(R). \tag{110}$$

We must check, of course, that $Conv(\alpha)$ is always an element of **PT**.

Notice that here is the place where we ought to discuss Dijkstra's rule for the assignment statement. We would give such a statement some semantics such as

$$\mathbf{A}[\![\mathrm{I} := \mathrm{E}]\!]\sigma = \sigma + [\mathrm{I} \rightarrow \mathbf{E}[\![\mathrm{E}]\!]\sigma] \tag{111}$$

and then, to justify Dijkstra's rule, we would have to show that

$$\mathbf{B}[\![\mathrm{B}]\!](\sigma + [\mathrm{I} \rightarrow \mathbf{E}[\![\mathrm{E}]\!]\sigma]) = \mathbf{B}[\![\mathrm{B}_{\mathrm{I} \rightarrow \mathrm{E}}]\!]\sigma. \tag{112}$$

This proof would depend, of course, on the syntax of expressions in our language, but would probably be a straightforward structural induction.

Continuing with our language, we write

$$\mathbf{C}[\![\mathbf{skip}]\!] = \lambda R.R \tag{113}$$

$$\mathbf{C}[\![\mathbf{abort}]\!] = \lambda R.\emptyset \tag{114}$$

As Dijkstra remarks, the semicolon operator corresponds simply to the functional composition of predicate transformers. So:

$$\mathbf{C}[\![\Sigma_1;\Sigma_2]\!] = Comp(\mathbf{C}[\![\Sigma_1]\!],\mathbf{C}[\![\Sigma_2]\!]) \tag{115}$$

where $Comp : \mathbf{PT}^2 \rightarrow \mathbf{PT}$ is defined by

$$Comp(f,g) = f \circ g. \tag{116}$$

For $\pi \in [\mathbf{S} \rightarrow \mathbf{T}]$ let us write π^+ for the set of states on which π returns the value *true*, and π^- for the set on which it returns *false*. That is to say,

$$\pi^+ = \pi^{-1}(true); \qquad \pi^- = \pi^{-1}(false). \tag{117}$$

As before, the meaning of a guarded command will be a pair, and again the first component will be such a π, although this time the second component will be a predicate transformer. Since a guarded command has the extra prerequisite that its guard is satisfied, we may write

$$Cond(\pi,f)(R) = \pi^+ \cap f(R) \tag{118}$$

and then, as before,

$$\mathbf{C}[\![\mathbf{if}\ \Gamma\ \mathbf{fi}]\!] = Cond(\mathbf{G}[\![\Gamma]\!]). \tag{119}$$

For the **do**-command, Dijkstra gives a definition having the form

$$\exists k \geqslant 0 : H_k(R). \tag{120}$$

This is equivalent to the infinite disjunction

$$\mathbf{V}\{H_k(R) \mid k \geqslant 0\}. \tag{121}$$

In our framework, his $H_k(R)$ becomes $h_R{}^k(\emptyset)$, where, if $\langle \pi,f \rangle$ is the meaning of the guarded command under consideration,

$$h_R(Q) = (\pi^- \cap R) \cup (\pi^+ \cap f(Q)). \tag{122}$$

Moreover, the disjunction of predicates is, on our view, the least upper bound of the characterised sets. So we may define

$$Do(\pi,f)(R) = \bigsqcup\{h_R{}^k(\emptyset)\} = fix(h_R) \tag{123}$$

where h_R is as defined above. Then, again as before,

$$\mathbf{C}[\![\mathbf{do}\ \Gamma\ \mathbf{od}]\!] = Do(\mathbf{G}[\![\Gamma]\!]). \tag{124}$$

The semantics of the guarded commands is given by

$$\mathbf{G} : \mathbf{GCom} \longrightarrow [[\mathbf{S}\rightarrow\mathbf{T}] \times \mathbf{PT}]$$

defined as follows.

$$\mathbf{G}[\![\mathbf{empty}]\!] = \langle \lambda\sigma.\mathit{false},\ \bot_{\mathbf{PT}}\rangle \tag{125}$$

$$\mathbf{G}[\![B\rightarrow\Sigma]\!] = \langle \mathbf{B}[\![B]\!],\ \mathbf{C}[\![\Sigma]\!]\rangle \tag{126}$$

$$\mathbf{G}[\![\Gamma_1 \ \square\ \Gamma_2]\!] = \mathit{Bar}(\mathbf{G}[\![\Gamma_1]\!],\mathbf{G}[\![\Gamma_2]\!]) \tag{127}$$

where, this time, $\mathit{Bar} : [[\mathbf{S}\rightarrow\mathbf{T}]\times\mathbf{PT}]^2 \rightarrow [[\mathbf{S}\rightarrow\mathbf{T}]\times\mathbf{PT}]$ is defined by:

$$\mathit{Bar}(\langle\pi_1,f\rangle,\langle\pi_2,g\rangle) = \langle\pi_1\vee\pi_2,\ \lambda R. \tag{128}$$
$$(\pi_1^+ \cap [(\pi_2^+\cap f(R)\cap g(R)) \cup (\pi_2^-\cap f(R))]) \cup (\pi_1^- \cap \pi_2^+\cap g(R))\rangle.$$

The reader is invited to check all these definitions against those in Dijkstra's book, to make sure they correspond.

We now check that this predicate transformer definition of our language is compatible with the earlier, denotational one. We connect the domains **ST** (state transformations) and **PT** (predicate transformers) by the function $w : \mathbf{ST} \rightarrow \mathbf{PT}$ defined as follows:

$$w(\mu)(R) = \{\sigma\in\mathbf{S} \mid \bot\notin\mu\sigma \wedge \mu\sigma\subseteq R\}. \tag{129}$$

It turns out that w is monotonic. We recall that the ordering on our state transformations **ST** is such that a weaker μ produces a larger set of output states from a given argument σ. So, arguing informally, if we move to a weaker μ, some value of σ for which previously $\mu(\sigma)\subseteq R$ will no longer be acceptable, if the new set of output states contains some elements outside R. Thus the set of acceptable input states (which is an element of **Pred**) will be less: that is to say, a weaker μ corresponds to a weaker predicate transformer. In fact, it can be shown that w actually determines an *isomorphism* between the two domains **ST** and **PT**.

The reader will have noticed, of course, that the actual semantic equations in our two definitions are very similar. This means that the proof that the predicate transformer semantics is compatible with the denotational semantics is a straightforward structural induction, using a series of little results relating the two definitions of the various auxiliary functions employed. One typical such lemma, for example, is:

$$w(\mathit{Comp}(\mu,\mu')) = \mathit{Comp}(w(\mu),w(\mu')) \quad \text{for all } \mu,\mu' \in \mathbf{ST} \tag{130}$$

where the appropriate *Comp* is used at each occurrence.

In fact, just as with the domains themselves, we can show that our two semantic definitions are isomorphic: each is completely compatible with the other. What this means is that there is no question of one definition building in irrelevant detail which the other leaves free, or one leaving annoying loose ends which the other ties up: each definition is, in a strong sense, saying exactly the same as the other. Which approach we choose will depend on which we find more helpful for the task we are doing at the time, whether that task is writing a program, or designing or implementing a programming language.

CHAPTER III - CORRECTNESS OF IMPLEMENTATIONS

In this chapter we consider the problem of demonstrating the correctness of a programming language implementation; that is to say, of showing that the implementation conforms to the denotational semantic definition of the language. There is no time for more than a general introduction to this area, using the simplest of example languages. The reader must be referred to [7] for a more extensive treatment.

The Example Language

The purpose of this very simple example is to illustrate the general strategy of our method. Its syntax and *standard* (definitive) semantics is as follows.

Syntactic Domains

$N \in \mathbf{Nml}$ numerals

$E \in \mathbf{Exp}$ expressions

Syntax

$E ::= N \mid E{+}E$

Standard Semantics: Semantic Domains

$\nu \in \mathbf{N}$ integers

Semantic Functions

$\mathbf{N} : \mathbf{Nml} \to \mathbf{N}$ (defined as usual, but omitted here)

$\mathbf{E}_0 : \mathbf{Exp} \to \mathbf{N}$

$$\mathbf{E}_0[\![N]\!] = \mathbf{N}[\![N]\!] \tag{131}$$

$$\mathbf{E}_0[\![E_1{+}E_2]\!] = \mathbf{E}_0[\![E_1]\!] + \mathbf{E}_0[\![E_2]\!] \tag{132}$$

This semantics clearly defines the meaning of all expressions, but gives no clue as to a strategy for implementing the language.

Stack Semantics

One obvious implementation strategy is to use a stack. The following is an alternative semantics embodying this decision. We regard a stack as a list (tuple) of integers.

$\zeta \in Z = N^*$ stacks

$\mathbf{E}_1 : \text{Exp} \to Z \to Z$

$\mathbf{E}_1[\![N]\!]\zeta = \langle \mathbf{N}[\![N]\!]\rangle \S \zeta$ (133)

$\mathbf{E}_1[\![E_1+E_2]\!]\zeta = add(\mathbf{E}_1[\![E_2]\!](\mathbf{E}_1[\![E_1]\!]\zeta))$ (134)

where $add\zeta =$ $\langle \zeta[1]+\zeta[2]\rangle \S (\zeta\dagger 2)$ (note: $add : Z \to Z$.)

We must show we are defining the same language as before, so we must always devise an appropriate "congruence condition". In this case it is:

Congruence Condition

$\mathbf{E}_1[\![E]\!]\zeta = \langle \mathbf{E}_0[\![E]\!]\rangle \S \zeta$ (for all E, ζ) (135)

Proof: Structural induction. If E = ⌜N⌝, then

$\mathbf{E}_1[\![N]\!]\zeta = \langle \mathbf{N}[\![N]\!]\rangle \S \zeta = \langle \mathbf{E}_0[\![E]\!]\rangle \S \zeta.$

If E = ⌜E_1+E_2⌝, then

$\mathbf{E}_1[\![E_1+E_2]\!]\zeta = add(\langle \mathbf{E}_0[\![E_2]\!]\rangle \S \langle \mathbf{E}_0[\![E_1]\!]\rangle \S \zeta)$

(using the inductive hypothesis twice)

$= \langle \mathbf{E}_0[\![E_2]\!]+\mathbf{E}_0[\![E_1]\!]\rangle \S \zeta = \langle \mathbf{E}_0[\![E_1+E_2]\!]\rangle \S \zeta.$

So the result holds in both cases as required.

Alternative Stack Semantics

The above definition contains a strong bias towards left-to-right evaluation. We now give an alternative formulation, more suitable for a parallel processing implementation.

$\mathbf{E}_2 : \text{Exp} \to Z \to Z$

$\mathbf{E}_2[\![N]\!]\zeta = \langle \mathbf{N}[\![N]\!]\rangle \S \zeta$ (as before, for $\mathbf{E}_1$) (136)

$\mathbf{E}_2[\![E_1+E_2]\!]\zeta = \langle (\mathbf{E}_2[\![E_1]\!]\langle\rangle)[1] + (\mathbf{E}_2[\![E_2]\!]\langle\rangle)[1]\rangle \S \zeta$ (137)

In fact $\mathbf{E}_2$ is the *same* function as $\mathbf{E}_1$, merely differently formulated. But it is easier to show this by proving each congruent to $\mathbf{E}_0$ as in (135) above. This illustrates the normative role of standard semantics: its lack of bias towards any particular implementation strategy makes it a useful canonical definition.

Note that $\mathbf{E}_1$ and $\mathbf{E}_2$ are still *denotational*: they differ from $\mathbf{E}_0$ only in that the values they involve are more complex.

A Machine Code

$$\mathrm{I} \in \mathbf{Ins} \qquad \text{instructions}$$
$$\Pi \in \mathbf{Prg} = \mathbf{Ins}^* \qquad \text{programs}$$
$$\mathrm{I} ::= \mathbf{load}\ \mathrm{N} \mid \mathbf{add}$$

A Compiler for the Example Language

$$\mathbf{C} : \mathbf{Exp} \to \mathbf{Prg}$$
$$\mathbf{C}[\![\mathrm{N}]\!] = \langle \ulcorner \mathbf{load}\ \mathrm{N} \urcorner \rangle \tag{138}$$
$$\mathbf{C}[\![\mathrm{E}_1 + \mathrm{E}_2]\!] = \mathbf{C}[\![\mathrm{E}_1]\!] \mathbin{\S} \mathbf{C}[\![\mathrm{E}_2]\!] \mathbin{\S} \langle \ulcorner \mathbf{add} \urcorner \rangle \tag{139}$$

Before we can say whether this compiler is correct, we must give some semantics to our machine code. We give it "operational semantics" by defining our machine.

A Machine

Any (non-parallel) machine can be thought of as executing

$$\mathbf{until}\ term(\sigma)\ \mathbf{do}\ \sigma := step(\sigma)$$

where *term* : $\mathbf{S} \to \mathbf{T}$ and *step* : $\mathbf{S} \to \mathbf{S}$ are given. More formally:

$$machine(step, term) = fix(\lambda\phi\lambda\sigma. term(\sigma) \to \sigma,\ \phi(step(\sigma))). \tag{140}$$

In this case:

$$\mathbf{S} = \mathbf{Prg} \times \mathbf{N} \times \mathbf{Z}.$$

So $\sigma = \langle \Pi, \nu, \zeta \rangle \in \mathbf{S}$

stack (ζ)
program counter (ν)
program (Π)

$$term_1 \langle \Pi, \nu, \zeta \rangle = \nu > len(\Pi) \tag{141}$$
$$step_1 \langle \Pi, \nu, \zeta \rangle = \langle \Pi, \nu+1, \mathbf{I}(\Pi[\nu])\zeta \rangle \tag{142}$$

where $\mathbf{I} : \mathbf{Ins} \to \mathbf{Z} \to \mathbf{Z}$ is defined by:

$$\mathbf{I}[\![\mathbf{load}\ \mathrm{N}]\!]\zeta = \langle \mathbf{N}[\![\mathrm{N}]\!] \rangle \mathbin{\S} \zeta \tag{143}$$
$$\mathbf{I}[\![\mathbf{add}]\!]\zeta = add\,\zeta \qquad (add \text{ was defined above}). \tag{144}$$

Then we define

$$\mathbf{M} : \mathbf{Prg} \to \mathbf{Z} \to \mathbf{Z}$$

by

$$\mathbf{M}[\![\Pi]\!]\zeta \;=\; (machine(step_1, term_1)\langle\Pi, 1, \zeta\rangle)[3]. \tag{145}$$

Congruence Condition for the Compiler and Machine

$$\mathbf{M}(\mathbf{C}[\![E]\!])\zeta \;=\; \mathbf{E}_1[\![E]\!]\zeta \qquad \text{(for all } E, \zeta) \tag{146}$$

Proof: Structural induction on E. If E = ⌜N⌝ we merely substitute in the various definitions and work it out (using the fixed point property on the definition of *machine*). If E = $\ulcorner E_1+E_2 \urcorner$ we have to show

$$\mathbf{M}(\mathbf{C}[\![E_1]\!] \dagger \mathbf{C}[\![E_2]\!] \dagger \langle \ulcorner \mathbf{add} \urcorner \rangle \;=\; add \circ (\mathbf{E}_1[\![E_2]\!]) \circ (\mathbf{E}_1[\![E_1]\!]).$$

We can use the inductive hypothesis (twice) for this, provided we can show

$$LHS \;=\; \mathbf{M}(\langle \ulcorner \mathbf{add} \urcorner) \circ \mathbf{M}(\mathbf{C}[\![E_2]\!]) \circ \mathbf{M}(\mathbf{C}[\![E_1]\!]).$$

So we need:

Lemma. For all Π_1, Π_2, ζ, $\mathbf{M}(\Pi_1 \dagger \Pi_2) \;=\; \mathbf{M}(\Pi_2) \circ \mathbf{M}(\Pi_1)$. (147)

Proof: By induction on $len(\Pi_1)$.
Basis: if $len(\Pi_1) = 0$ both sides are $\mathbf{M}(\Pi_2)$.
Induction: suppose $len(\Pi_1) = n+1$, and assume the result for n ($n \geqslant 0$). Let $\mu\sigma = (machine\langle step_1, term_1\rangle\sigma)[3]$; so $\mathbf{M}[\![\Pi]\!]\zeta = \mu\langle\Pi, 1, \zeta\rangle$ and μ is of the form $\lambda\sigma.(fix\ H\ \sigma)[3]$ where $H\phi = \lambda\sigma.term_1\sigma \rightarrow \sigma, \phi(step_1\sigma)$. By use of the fixed point property on this, our lemma becomes

$$\mu\langle\Pi_1 \dagger \Pi_2, 2, \mathbf{I}(\Pi_1[1])\zeta\rangle \;=\; \mathbf{M}[\![\Pi_2]\!] \circ \mu\langle\Pi_1, 2, \mathbf{I}(\Pi_1[1])\zeta\rangle. \tag{148}$$

We can use our inductive hypothesis to show this, provided we have the following result.

Lemma. If $\nu > 1$ then $\mu\langle\Pi, \nu, \zeta\rangle \;=\; \mu\langle\Pi\dagger 1, \nu-1, \zeta\rangle$. (149)

Proof: by fixed point induction.
(1) We prove the result with $\perp$ in place of *fix H*, which is trivial.
(2) We assume the result with x in place of *fix H* and prove it for $H\ x$. If $\nu > len(\Pi_1)$ then also $\nu - 1 > len(\Pi\dagger 1)$, so $term_1\sigma$ is true on both sides, and $\sigma[3] = \zeta$. If $\nu \leqslant len(\Pi_1)$ then we must show

$$\mu'\langle\Pi, \nu+1, \mathbf{I}[\![I]\!]\zeta\rangle \;=\; \mu'\langle\Pi\dagger 1, \nu, \mathbf{I}[\![I]\!]\zeta\rangle \tag{150}$$

where $\mu'\sigma = (x\sigma)[3]$ and $I = \Pi[\nu]$; this follows from the inductive hypothesis. (Note that there are of course other possible strategies for proving this result.)

Remarks.

(1) We have shown the congruence of our compiler with the stack semantics ($\mathbf{E}_1$), not directly with the standard semantics. This illustrates that it is often easier to proceed in stages, with several intermediate definitions, rather than relate an implementation directly with the standard.

(2) In the old days denotational semantics used to be called "mathematical semantics", with the implication that the other kinds were non-mathematical. Note, however, that our machine definition is just as mathematical as the denotational ones, and in fact relies on the same theory of domains and fixed points.

Two More Strategies For This Language

An Interpreter This is another example of an operational semantics.

$$\sigma = \langle \eta, \zeta \rangle \in S = [\mathbf{Exp} + \{ \ulcorner \mathbf{plus} \urcorner \}]^* \times Z$$

ζ — data stack
η — stack of expressions

$$term_2 \langle \eta, \zeta \rangle = (\eta = \langle\rangle) \tag{151}$$
$$step_2 \langle \eta, \zeta \rangle = J(\eta[1]) \langle \eta \dagger 1, \zeta \rangle \tag{152}$$

where

$$J[\![N]\!] \langle \eta, \zeta \rangle = \langle \eta, \langle \mathbf{N}[\![N]\!] \rangle \S \zeta \rangle \tag{153}$$
$$J[\![E_1 + E_2]\!] \langle \eta, \zeta \rangle = \langle \langle E_1, E_2, \ulcorner \mathbf{plus} \urcorner \rangle \S \eta, \zeta \rangle \tag{154}$$
$$J[\![\mathbf{plus}]\!] \langle \eta, \zeta \rangle = \langle \eta, add\, \zeta \rangle \tag{155}$$

$$\mathbf{T}[\![E]\!] \zeta = (machine(step_2, term_2) \langle \langle E \rangle, \zeta \rangle)[2] \tag{156}$$

An Optimising Compiler

$$\Pi \in \mathbf{Prg} = \mathbf{Nml}^*$$
$$\mathbf{C} : \mathbf{Exp} \to \mathbf{Prg}$$
$$\mathbf{C}[\![N]\!] = \langle N \rangle \tag{157}$$
$$\mathbf{C}[\![E_1 + E_2]\!] = \mathbf{C}[\![E_1]\!] \S \mathbf{C}[\![E_2]\!] \tag{158}$$

Its machine

$$\sigma = \langle \Pi, \nu, \alpha \rangle \in S = \mathbf{Prg} \times \mathbf{N} \times \mathbf{N}$$

α — accumulator
ν — program counter
Π — program

$$term_3\langle \Pi, \nu, \alpha \rangle = (\nu > len(\Pi)) \quad (159)$$

$$step_3\langle \Pi, \nu, \alpha \rangle = \langle \Pi, \nu+1, \alpha+\mathbf{N}(\Pi[\nu]) \rangle \quad (160)$$

$$\mathbf{U}[\![\Pi]\!] = (machine(step_3, term_3)\langle \Pi, 1, 0 \rangle)[3] \quad (161)$$

Note that the congruence proof of this is slightly more tricky (it involves associativity of addition, for example); and this strategy makes it much more difficult to extend the programming language (for example by including multiplication).

The Application to Bigger Languages

We have seen how even in our simple language there were advantages in defining intermediate stages between the standard definition and the final implementation. For more realistic languages this is even more a necessity, and in fact it will often be found useful to have several intermediate stages. For example, the proof given in [4] for their rather big language Sal has the following stages.

1. **Standard Semantics.** (with continuations, as outlined in the next chapter).

2. **Store Semantics.** Procedures are now represented by ⟨*code,env*⟩ (where *code* is an environment-independent function). Labels are represented by ⟨*code,env,stack*⟩ (where *stack* is used to hold anonymous partial results of expressions). (So now the pure function represented by *code* is evaluated in the "current" environment, the "current" stack and the "current" store.)

3. **Stack Semantics.** Now denoted values are on the stack too, so environments map names to stack pointers. Continuations for procedure calls are kept on the stack (the germs of return links). So recursive procedures no longer need to use *fix* (though this is still used for the semantics of loops).

4. **Context Conditions.** Since variables now no longer have infinite lifetimes, not all the programs which worked in (1) and (2) above will work in (3). So next is defined the class of programs for which (2) and (3) are congruent, and only these programs are handled from now on.

5. **Compiler (and loader).** into a machine code (Sam).

6. **Denotational Semantics for Sam.** Since this is to be a denotational definition, the meaning of a Sam instruction must be defined in terms of the meaning of its operand fields, for example the jump address fields. So there is a mapping (a "consecution") which maps appropriate field values into continuations.

7. **Pointer Semantics.** An operational semantics for Sam. Everything is now defined in terms of finitary operations on bit patterns.

Notice that the compilers of these examples are defined as functions mapping abstract syntax to abstract syntax. It remains necessary to implement these transformation functions as programs, and to prove the programs correct, using standard program correctness techniques. Similarly, it remains necessary to show that a machine definition is successfully implemented, whether by interpreter, microcode or electronics.

In the present state of the art the congruence proofs for Sal are complicated. Partly this is because they are comparing the small print of two definitions: we must hope for better ways of structuring our definitions, and also for some automated help with the algebraic manipulation.

Reflexive Domains

An extra difficulty arises, however, when one of the definitions which we are relating is a denotational definition which involves reflexive domains, and the other is an operational definition which does not. An example of this might concern a high level functional programming language, where the domain of expressible values might be

$$\mathbf{E} = \mathbf{B} + \mathbf{F}$$

where

$$\mathbf{F} = \mathbf{E} \rightarrow \mathbf{E}.$$

That is to say, **E** contains basic values (**B**) and functions (**F**), and functions map from **E** to **E**. So **E** is reflexive, in that it contains its own function space. On the other hand, we would expect, that an interpreter for this language would *not* involve reflexive domains, but would have a conventional operational definition. (Another example in which the same problem arises is the final stage of the correctness proof for the implementation of Sal mentioned in the previous section.)

The interpreter is a continuous function on trees, with a fixed-point definition: as usual for such definitions, we might employ fixed point induction for its analysis. In essence what we do in this technique is to assume the interpreter "correct" for programs which require, say, n steps of intepretation for their evaluation, and prove that the interpreter is also "correct" for programs which require n+1 steps . In effect we do induction on the duration of the execution. Obviously in this context we cannot expect "correct" to mean "has produced the right answer", since for any particular n there will be some programs that go on for longer that n steps. "Correct" can mean no more than "on the right lines". In more formal terms, all we can prove is that the function defined by the interpreter is an *approximation* (in the sense of Scott's theory) to the function defined by the denotational definition.

On the other hand, when we analyse the denotational definition we use structural induction: we prove that the definitions "agree" for a program assuming they do so for its syntactic subcomponents. That is to say, our induction will be based on the *size* of a program rather than its duration. More important than this, however, is that the predicates asserting "agreement" must be defined circularly: two functions, for example, agree if they give results which agree for all argument which agree. Unfortunately this circular definition of agreement is in no sense monotonic: for example, if no arguments agreed all functions would agree. We can therefore not rely on our usual fixed point theory for the existence of solutions to circular definitions, but instead we have to construct the required predicates explicitly. In effect, what we end up doing is an induction on the order of functionality of our denotational values: we prove that the predicate is well defined for nth order functions assuming that it is well defined for functions of order $n-1$. So, just as before, we cannot expect to be able to assert complete equivalence between the denotational and the operational definitions: for any n, there will always be some programs which manipulate objects of still higher type. The only predicates we can construct will be those which assert that the function defined denotationally is an *approximation* to that defined operationally. This is not to say that the predicates expressing complete congruence do not exist: that is an open question. It is simply that we can not *demonstrate* that they exist, and therefore we cannot use them in our proofs.

Thus in these circumstances we must prove separately that each side is an approximation of the other, and then combine these two results into the correctness property we require. This problem, which is examined in more detail in [7] and [8], and in the discussion on Sal in [4], therefore considerably increases the already tedious length of our correctness proofs.

REFERENCES

1. E.W.Dijkstra: *A Discipline of Programming*; Prentice-Hall International (1976).

2. P.J.Landin: *A Correspondence Between Algol 60 and Church's Lambda-Notation*; Communications of the ACM, **8**, pp. 89-101 and 158-165 (1965).

3. P.J.Landin: *The Mechanical Evaluation of Expressions*; Computer Journal, **6**, pp. 308-320 (1964). See also: *A Lambda-Calculus Approach*; pp. 97-141 of *Advances in Programming and Non-Numerical Computation* (ed. L.Fox), Pergamon Press (1966).

4. R.E.Milne and C.Strachey: *A Theory of Programming Language Semantics*; Chapman and Hall (1976).

5. G.D.Plotkin: *Dijkstra's Predicate Transformers and Smyth's Power Domains*; pp. 527–553 of *Abstract Software Specifications*. Lecture Notes in Computer Science, **86**, Springer-Verlag (1980).

6. D.S.Scott: *Lectures on a Mathematical Theory of Computation*; Technical Monograph PRG-19, Oxford University Computing Laboratory, Programming Research Group (1981).

7. J.E.Stoy: *Denotational Semantics: The Scott-Strachey Approach to Programming Language Theory*; MIT Press (1977).

8. J.E.Stoy: *The Congruence of Two Programming Language Definitions*; Theoretical Computer Science, **13**, pp. 151–174 (1981).

Theme VII.

(J. Stoy)

Starts with a religious allusion in the Oxford accent; concluding with a converging sequence of neighborhoods (see Scott's paper)

Part III
ABSTRACT DATA TYPES

In looking for an implementation-independent way of describing what a program is intended to do, people learned to define their objects and functions in an abstract form. Through this process, a lot of algebraic and categorical ideas found their way into Computer Science. An extensive review of the impact of these abstract data type methods on programming methodology was beyond the scope of this volume. However, the first of the following two articles introduces and motivates the key notions of signature, algebra and theory emphasising their Computer Science connection. In the second article, several type constructions for the specification of a programming language are developed and reviewed: partial abstract types, hierarchical types and parametrized types.

ALGEBRAS, THEORIES AND FREENESS: AN INTRODUCTION FOR COMPUTER SCIENTISTS

R.M. Burstall and J.A. Goguen

Edinburgh University/Xerox PARC and SRI International

INTRODUCTION

In the last ten years or so a lot of algebraic ideas have wormed their way into Computer Science, particularly in work connected with correctness of compilers, with abstract data types and with specification. We have been among those responsible [Burstall and Goguen 1980, 1981]. Most papers begin with a compressed section of definitions, but it is difficult for the well-disposed outsider to make much of these. Reference to books for algebraists, such as Graetzer [1979], or even to those angled towards automata theory [Arbib and Manes 1975] may not be encouraging. So it is perhaps worthwhile to present some of the key algebraic ideas in a leisurely and, we hope, intuitive form, emphasising the Computer Science connection. Some water has flowed under the bridge since one of us was last involved in an attempt to do this [Goguen, Thatcher, Wagner and Wright 1975].

In general we would claim that algebraic and categorial ideas can be a help in understanding computational phenomena. There are of course pitfalls, and we would claim to have fallen into most of them. First you can be overly abstract and fail to understand what you are talking about (if anything). Second you can make the obvious seem impressive and waste everybody's time (including your own). You can impose a lot of unfamiliar general concepts on people without giving adequate reward for their labours in the shape of computational insight. *Caveat lector.*

On the other hand there are advantages .....

M. Broy and G. Schmidt (eds.), Theoretical Foundations of Programming Methodology, 329–349.

The same concepts often come up again and again so that it is worth defining them once and for all, for example 'a collection of sets with some functions defined on them'. Such concepts include 'algebra' and 'homomorphism'. We can learn to think in terms of these concepts instead of in terms of their components such as 'set' or 'n-tuple'; our thoughts can then take bigger strides.

Not only does the same concept come up repeatedly but we often see a family of related concepts. For example the same construction may be carried through using 'sets', 'indexed sets', 'partially ordered sets' and 'sets with error elements'. So it is worth while to treat the general case. (It is even more worth while to treat the simplest special case first so that people can understand what you are talking about - *mea culpa*.) In these matters the general case can often be described using the language of category theory, which is particularly well-adapted for talking about the structure of systems without saying exactly what the components are. As a matter of fact we do that all the time in informal Computer Science. We talk happily about while statements of the form '**while** expression **do** **statement**' without worrying exactly what the expressions and statements are (Do they permit upper and lower case identifiers? Arrays with several subscripts? Mostly we don't care.) But most mathematical treatments of programming languages are very bottom-up and describe these matters in boring detail, or they give a toy language and leave the generalisations to the reader's intuition (not a bad idea, but it is nice to know how to be more precise). So in a way the more abstract categorial concepts correspond more closely to our informal ways of talking.

Although this stuff, particularly the categorial concepts may seem hard to get the hang of, there is hope: we can bring to bear our computing intuition by programming up all the ideas in the mathematics books. One of us has a go at this for some algebraic and categorial concepts [Burstall 1980], with the able assistance of David Rydeheard and Don Sannella. It turns out that many categorial proofs are constructive and hence can give rise to programs; the difficulty in reading the books comes partly from the fact that the resulting programs are very elegant and have little redundancy, they get maximum mileage from a very small number of general procedures.

This paper introduces the ideas of signature, algebra and theory, also initial algebra and free algebra. Our style of treatment motivates, but does not indulge in, the more abstract approach of category theory.

SIGNATURES

We start with the notion of signature. This corresponds in programming terms to a bunch of declarations, declaring types, constants and procedures.

integer, bool, tree: **type**

zero, one: integer
plus, minus, times: **function**(integer, integer) **result** integer
true, false: bool
nil: tree
tip: **function** (integer) **result** tree
node: **function**(tree, tree) **result** tree
isnil, istip: **function**(tree) **result** bool
sum: **function**(tree) **result** integer

Notice that we just name the types; we have not said what they consist of (records, arrays or whatever). Nor do we give the bodies of the procedures. Now we can simplify things a little by regarding the constants such as zero and one as functions with no arguments. Also we will follow mathematical practice and use the word "sort" rather than "type" and "operator" rather than "function name". Thus a signature consists of a set of sorts and a set of operators, each with given input and output sorts, for example

```
  <  {integer, bool, tree},
  {     zero, one:  →integer,  plus, minus, times:  integer ×
                                                    integer →integer,
        nil: →tree,  tip: integer → tree,  node: tree × tree → tree,
        isnil, istip: tree → bool,  sum: tree → integer } >
```

We may think of the second element of this pair as defining for each possible sequence of input sorts and output sort such as <<tree,tree>,tree>, the set of operators having those sorts. More formally we need the notion of indexed family of sets. By an *I-indexed family of sets* we mean a function from I to sets. Let Nat mean {0,1,2,...}. Then for example the function mapping 0 to {a,b}, 1 to {a,c,d}, 2 to the empty set and so on is a Nat-indexed family of sets of letters, written more briefly "$\{a,b\}_0, \{a,c,d\}_1, \emptyset_2, \ldots$".

Now in a signature the operations are a family of sets indexed by their "arity", that is their sequence of input sorts and their output sort.

Defn. A **signature, Σ**, is a pair $<S, \Omega>$, where S is a set (of sorts) and Ω is a family of sets (of operators) indexed by $S^* \times S$.

(Above, for example, the element <integer integer,integer> is the index of the set {plus,minus,times} in Ω).
To save writing we will assume **Σ** is <S,Ω> whenever it is mentioned.

ALGEBRAS

Now signatures are not much fun unless we give some meaning to the names. They are syntax without semantics. If we give these meanings we get an algebra. In a programming language we might say (assuming types integer and bool already defined)

```
representation tree is sexpression

nil: tree = "NIL"
tip: function(n: integer) result tree = cons(n, nil)
node: function(t1,t2: tree) result tree = cons(t1, t2)
isnil: function(t: tree) result bool = null(t)
istip: function(t: tree) result bool = null(cdr(t))
sum: function(t: tree) result integer =  if null(t) then 0 else..
```

This associates some particular set with the sort tree, namely the set of Lisp sexpressions, and some particular function with each operator taking arguments of the appropriate sorts and producing an appropriate result. That is an algebra is a signature together with a function taking sorts to sets and another function taking operators to functions.

A simple example would be the algebra of numbers modulo 4, call it Mod4, with signature

< {number}, {zero,one: → number, plus,times: number × number → number}>

which can be written

number {0, 1, 2, 3}

zero 0

one 1

plus

	0	1	2	3
0	0	1	2	3
1	1	2	3	0
2	2	3	0	1
3	3	0	1	2

times

	0	1	2	3
0	0	0	0	0
1	0	1	2	3
2	0	2	0	2
3	0	3	2	1

The algebra of trees has infinite sets for the sorts integer and tree and the tables for the operations are correspondingly rather too big for our page, but the principle is the same.

We call the family of sets associated with the sorts the *carrier* of the algebra, writing $|A|$ for the carrier of algebra A.

Defn. Let Σ be a signature. A **Σ-algebra** A is an S-indexed family of sets, $|A|$, called the carrier of A, together with an $S^* \times S$-indexed family of functions $\alpha_{us} : \Omega_{us} \to (|A|_u \to |A|_s)$ where $u \in S^*$, $s \in S$ and $|A|_{u1\ldots un} = |A|_{u1} \times \ldots \times |A|_{un}$.

Notation If $\omega \in \Omega_{us}$ and $\langle a_1,\ldots,a_n\rangle \in A_u$, we write $\omega(a_1,\ldots,a_n)$ for $\alpha_{us}(\omega)(a_1,\ldots,a_n)$ where there is no ambiguity.

HOMOMORPHISMS

Now we can, at least informally, understand the idea of evaluating expressions in an algebra. For example "plus(one,one)" in the above algebra gives 2 and hence "times(plus(one,one),one)" also gives 2. We might have two algebras which are rather similar in that evaluating in one gives analogous results to evaluating in the other. Suppose there is a function f from the carrier of the first algebra to that of the second (or if there are several sorts such a function for each sort) with the property that if we evaluate an expression in the first algebra and get answer a, then evaluating it in the second algebra will always give answer f(a). More generally we might ask that if in the first algebra evaluating an expression like plus(x,times(x,y)) with variable x bound to a and y bound to b gives result c, then evaluating the same expression in the second algebra with the corresponding binding, x to f(a) and y to f(b), should give the corresponding result f(c). In this case we say that f is a *homomorphism* from the first algebra to the second one.

For example if we define another algebra, say Mod2, that of numbers modulo 2, with $\{0,1\}$ as the set of numbers and the obvious multiplication, then there is a homomorphism f from Mod4 to Mod2, with

$f(0) = 0 \quad f(1) = 1 \quad f(2) = 0 \quad f(3) = 1$

(Try evaluating some expressions in both these algebras.) We write

f:Mod4 → Mod2

Now consider a "parity" algebra with {even,odd} as the set of numbers and the obvious tables for plus and times, that is plus(even,even) = even and so on. Then there is a homomorphism from the algebra of natural numbers, {0,1,2,3,...} with the usual meaning of plus and times, to Parity.

As a matter of fact all expressions with corresponding bindings give corresponding results if and only if each operator given corresponding arguments gives corresponding results, that is in the example above

f(zero) in Mod4 = zero in Mod2
f(one) in Mod4 = one in Mod2
f(plus(m,n)) in Mod4 = plus(f(m), f(n)) in Mod2
f(times(m,n)) in Mod4 = times(f(m), f(n)) in Mod2

We may as well adopt this property of operators as our official definition of homomorphism, thus

Defn. If $\Sigma = \langle S,\Omega\rangle$ then a Σ-homomorphism from a Σ-algebra $\langle A,\alpha\rangle$ to a Σ-algebra $\langle A',\alpha'\rangle$ is a map $f: |A| \to |A'|$ such that for each $\omega \in \Omega_{us}$ and each $a_1 \in A_{u_1}, \ldots, a_n \in A_{u_n}$, $f_s(\omega(a_1,\ldots,a_n)) = \omega(f_{u_1}(a_1),\ldots,f_{u_n}(a_n))$.

The function f from Mod2 to Parity defined by f(0) = even and f(1) = odd is clearly a homomorphism; so is its inverse g, with g(even) = 0 and g(odd) = 1. We say that f and g constitute an *isomorphism* between Mod2 and Parity. Thus an isomorphism between algebras A and B is a pair of inverse homomorphisms, f: A → B and g: B → A such that $f \circ g = 1_A$ and $g \circ f = 1_B$. This is an important idea, because when we think abstractly about some notion of number we do not care what algebra we are talking about to within an isomorphism. Roman numbers are the same mathematically as Arabic numbers. When we speak of abstract data types we mean that we do not care about the particular set of elements, we are just talking about some algebra to within an isomorphism. We consider two algebras to be "abstractly the same" if they are isomorphic.

WORD ALGEBRAS AND INITIAL ALGEBRAS

But for any signature there is a particularly interesting algebra called the *word algebra*. Its elements for a given sort are all the expressions in the signature denoting elements of that sort. For our number signature the set of "numbers" of the word algebra would be "zero", "one", "plus(zero,zero)", "plus(zero,one)" etc. What are the operations? What should "times" do when given "zero" and "plus(zero,one)"? It should produce "times(zero,plus(zero,one))" of course. The operations simply build larger expressions from smaller ones. To be more definite we may think of the expressions as character strings and the operations as string manipulators, for example, using <> for string concatenation,

```
times: procedure (s1,s2: string) result string =
                  "times(" <> s1 <> "," <> s2 <> ")"
```

Alternatively we could think of the expressions as written in Reverse Polish or represented by trees, using records, or whatever. It doesn't matter because all these expression algebras will be isomorphic. In fact when we speak of "the" word algebra we mean any of these isomorphic algebras.

Is there any more "abstract" way to characterise this algebra (to within an isomorphism) instead of picking a particular representation? The key idea is that given any interpretation of the operators we get a unique value for each expression. But an interpretation of the operators is just another Σ-algebra, A, and the evaluation of the expressions is just a homomorphism from the word algebra to A. Indeed this is the only homomorphism to A. If we know the interpretation of each operator we can work out inductively a value for each term. We call the Σ-algebra having this *unique homomorphism property* the *initial* S-*algebra*.

More formally

Defn. I is an initial Σ-algebra iff for any Σ-algebra, A, there is a unique homomorphism f: I → A.

Making an abstract definition by using a unique homomorphism property, also called a *universal property*, is an important device (used extensively in category theory). Let us try to see what it means. Consider the signature

sorts n
opns z:n   s:n → n

What is the initial algebra for this signature? Algebras of the signature include

Nat	$\|Nat\| = \{0,1,2,...\}$	z = 0 s(x) = x+1
Mod2	$\|Mod2\| = \{0,1\}$	z = 0 s(0) = 1, s(1) = 0

Natnat |Natnat| = {0',1',2',... 0",1",2",...} z = 0'
s(0') = 1', s(1') = 2',...
s(0") = 1", s(1") = 2",...

Now if we try Natnat as the initial algebra we hit trouble. How many homomorphisms are there from Natnat to, say, Nat? Well, z in Natnat must go to z in Nat so 0' goes to 0, and also by the homomorphism property s(0') goes to (s(0), that is 1' goes to 1. But where does 0" go? It could go to 0 or 1 or 2 or anywhere. 0' is there to give a value to z, 1' to give a value to s(z) and so on. But 0" is just junk, a useless piece of baggage. It need not be there and has no idea what to do with itself in a homomorphism. So there are many homomorphisms from Natnat, and it cannot be the initial algebra. No junk, please.

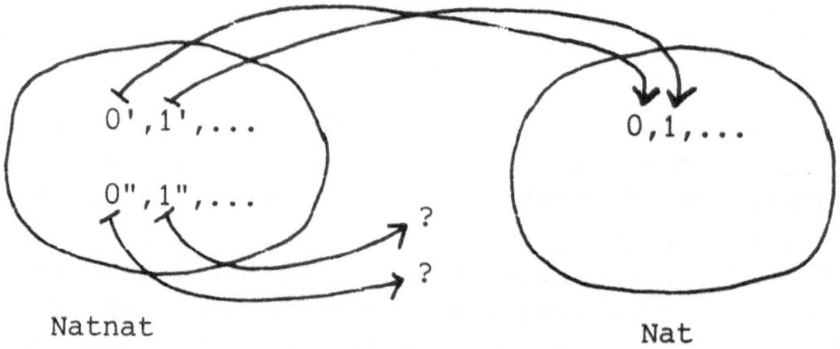

Let us try Mod2 for initial algebra, looking for homomorphisms to, say, Nat. Since z goes to z,0 in Mod2 goes to 0 in Nat, s(0) goes to s(0) so 1 goes to 1, s(1) goes to s(1) so 0 goes to 2. But we already had 0 goes to 1! 0 is schizophrenic. There is no homomorphism from Mod2. In fact 0 in Mod2 is serving as the value of both z and of s(s(2)), introducing this confusion. Mod2 is not the initial algebra. No confusion, please.

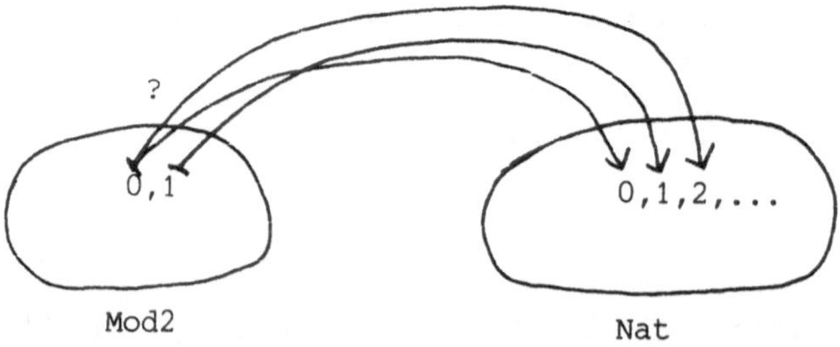

It all ends happily. Nat is the initial algebra because it has

no junk: every element of the carrier is the value of some term

no confusion: different terms get different values.

Thus the unique homomorphism property is a neat way of insisting on these intuitive properties.

Fact. If I and I' are both initial Σ-algebras then they are isomorphic.

Proof. Since I is an initial Σ-algebra we have a homomorphism f: I → I', and similarly one f': I', and similarly one f': I' → I. Now the composition f ∘ f': I → I is a homomorphism. But so is 1_I: I → I, the identity, so by uniqueness f ∘ f' = 1_I. Similarly f' ∘ f = 1_I'. So f and f' are inverses, and I and I' are isomorphic. □

It is not difficult to show by induction that the word algebra is initial.

The fact that there is just one homomorphism from the word algebra to any other algebra is a way of stating the familiar property of expressions that there is just one way of creating a given expression using the operators described. Thus "times(zero,plus(zero,one))" can only be created by applying the operator times to the two expressions "zero" and "plus(zero,one)". That is the operators all have unique inverses.

An example of the application of initial algebras in describing programming concepts is so-called *Abstract Syntax*, McCarthy [*circa* 1776]. By the abstract syntax of a language we mean a collection of rules showing how to construct all phrases of the language in the form of trees rather than character strings (for example S-expressions are the trees generated by the abstract syntax of LISP). Now the abstract syntax of a language is defined by a signature whose sorts are the various syntax classes, e.g. variable, constant, expression, statement, declaration and block. The operators correspond to the rules for building elements of these classes from their components, thus we might have

x,y: variable
0,1: constant
vexpr: variable → expression
cexpr: constant → expression
sumexpr: expression × expression → expression
ifthen: expression × statement × statement → statement
assignment: variable × expression → statement

Now the initial algebra is just the set of elements of the syntax classes (think of them as trees) together with the operations for constructing them. Because we have defined this algebra abstractly, upto isomorphism only, we have not specified any particular concrete syntax. Furthermore a compiler is essentially a homomorphic map from the initial algebra to strings of machine code instructions with suitable combining operations (not quite because this does not take account of environments produced by declarations). For further discussion of this *Initial Algebra Semantics* see Goguen, Thatcher, Wagner and Wright [1977].

In closing we should mention the notion of *final algebra* ('dual' to initial) which has a unique homomorphism from any algebra to it. This is also of use in specification work.

FREE ALGEBRA ON A SET

A slightly more general concept is the word algebra on a set X of variables of given sorts. The set of elements of this algebra is the set of expressions using these variables, for example as well as "plus(zero,zero)" we have "plus(x,zero)". The variables and expressions are of course associated with appropriate sorts. The operators are the expression constructors as before. Call this algebra $W_\Sigma(X)$. Again it enjoys a unique homomorphism property, and to avoid talking of particular representations we may use this property to define a more abstract concept, the free algebra on X.

Defn. By a Σ-algebra on an S-indexed set X we mean a Σ-algebra F(X) with a function $\eta_X: X \to |F(X)|$, such that for any Σ-algebra A, any S-indexed family of functions $f: X \to |A|$ extends uniquely to a homomorphism $f^\#: F(X) \to A$. By $f^\#$ extends f we mean that $\eta_X \cdot |f^\#| = f$.

We will write F_Σ for F when we need to say what the signature is.

Note that we write $|f^\#|$ for the homomorphism $f^\#$ regarded as a function, a subtlety which can be ignored if you wish.

Not surprisingly we can show that the word algebra on X is a free algebra on X. Here η_X is the obvious function taking each variable x in X to the expression "x" in $|F(X)|$. (Compare the LISP function taking x to cons(x,nil)). Now A might be, for example, the usual arithmetic algebra and f is a binding of variables and $f^\#$ is the evaluation function giving a numerical value to each expression. Naturally $f^\#("x") = f(x)$, i.e. $\eta_X \cdot |f^\#| = f$. This is expressed graphically by saying that the following diagram commutes

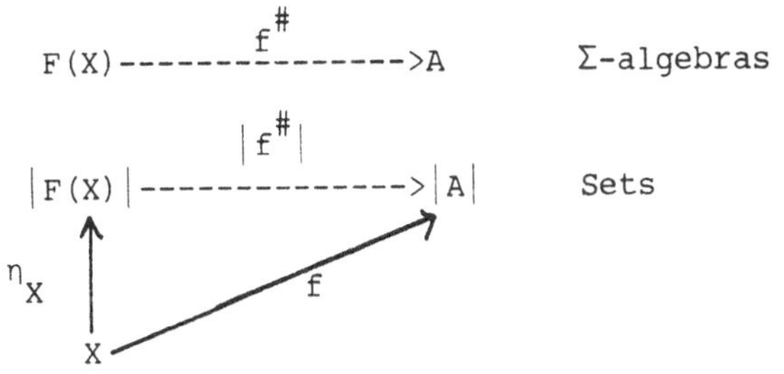

(We say that a diagram commutes if taking the composition of the functions on arrows along any path round a triangle or square or larger polygon gives the same result. Here the path $\eta_X \circ |f^{\#}|$ is equal to the path f. Similarly when we have homomorphisms instead of functions. We use dotted lines for arrows whose existence is being asserted.)

This diagram is mysterious; let us make friends with it. It really says that $f^{\#}$ is an extension of f, i.e. as far as X is concerned $f^{\#}$ is the same as f. But $f^{\#}$ is a homomorphism, so we have to regard it as just a function $|f^{\#}|$ before we can compare it with the function f. Now we would like to say that $|f^{\#}|$ restricted to X is f. That is all right if X is a subset of $|F(X)|$, but the free algebra F(X) is 'abstract'; we can make it with a lot of different carriers. So η_X just tells us which elements in the carrier stand for elements of X, and $\eta_X \bullet |f^{\#}|$ is the nearest thing we can truthfully say to '$f^{\#}$ restricted to X'. This is what we pay for the 'abstractness' of the free algebra notion.

We started with an informal notion of evaluating expressions for motivation. We now have a precise notion of expressions and their evaluation.

A remark on nomenclature: The names "word algebra" and "word algebra on X" apply strictly to the algebras we have described. As we shall see "initial" and "free" refer to more general categorial concepts of which these are particular examples.

THEORIES

Now we may be interested in the collection of algebras with a given signature which satisfy certain laws, such as associativity or commutativity. We can express these laws as equations. An equation in a signature over a set of variables is nothing more than the variables and a pair of expressions in those variables. For example the equation "plus(x,y) = plus(y,x)" over {x,y} can be thought of as the triple <{x,y}, plus(x,y), plus(y,x)>. Formally

Defn. A Σ- **equation, e** , is a triple <X,t1,t2> where X is an S-indexed set (of variables) and t1,t2 ∈ $|W_\Sigma(X)|$ are terms on X of the same sort. (The equation would normally be written "for all X, t1 = t2".)

Now we want to discriminate amongst algebras on the basis of some set of equations. What does it mean for an algebra to satisfy an equation, say "e1(x,y) = e2(x,y)". We understand it to mean that, whatever the values of x and y, if we evaluate the left hand expression in the algebra we get the same result as evaluating the right hand expression with those values. So if A is the algebra and f: {x,y} → |A| is a binding of the variables then we require that its extension to expresssions, $f^\#$, obey the $f^\#(e1(x,y)) = f^\#(e2(x,y))$. In general we have

Defn. A Σ-algebra A satisfies a Σ-equation <X,t1,t2> iff for all maps f: X → |A|, $f^\#(t1) = f^\#(t2)$. We write A ⊨ e for A satisfies e.

Now let E be a set of equations. An algebra is a model of E if it satisfies each of the equations of E. We write E* for the set of all models of E.

Now given a set of equations which describes a set of models we may be interested in what else we can say about these models. Hence we need to speak of all the equations which are satisfied by every model in some set M, call this set of equations M*. (We use the same symbol * again in order to emphasise the duality.) Now E** is the set of all equations which describe the models of E; it is called the *closure* of E. Inference systems aim to generate exactly this set of logical consequences (if they do so they are called consistent and complete). For equations there is a very simple such inference system using reflexivity, symmetry and transitivity of equality plus the substitution properties. But we are concerned here with semantics and models rather than proof theory.

Consider as a very simple example the theory with one sort "person", three nullary operators (constants) and two equations

```
sorts   person
opns    holmes, moriarty, babbage: person
eqns    holmes  = moriarty
        morarity = babbage
```

Now we consider algebras with elements p1,p2,p3,..., say, and various values for the constants. Some of these algebras will satisfy the equations, those in which holmes and moriarty get the same value and moriarty and babbage get the same value; others will not. But if any algebra satisfies the given two equations then it also satisfies the equation "holmes=babbage". So this equation is in the logical closure.

We call a signature with a set of equations a *presentation*. A set of equations is called *closed* if its closure is the set itself. We call a signature with a closed set of equations a *theory*. A theory may well consist of an infinite set of equations.

<u>Defn.</u> A **presentation** is a pair $\langle \Sigma, E \rangle$, where Σ is a signature and E is a set of equations.

<u>Defn.</u> A **theory** is a presentation $\langle \Sigma, E \rangle$, such that E is closed.

When we give a set of equations as above we are usually giving a presentation, but we mean this to denote its closure, a theory. So we will allow ourselves to say "the theory so-and-so" when "so-and-so" is actually a presentation.

Notice that presentations and theories are "syntactic" objects as opposed to algebras which are "semantic" objects. Remember that one theory can have many algebras which satisfy it.

ALGEBRAS FOR THEORIES

Now we can define notions analogous to "Initial Algebra" and "Free Algebra on X" but relative to all the algebras of a given theory. Consider the theory <u>String</u> defined by

```
sorts letter, string
opns    A,B, . . . ,Z: letter
        empty: string
        unit : letter → string
        _ . _: string, string → string
eqns    empty.s = s
        s.empty = s
        s.(t.u) = (s.t).u
```

What is an interesting algebra for string? Well of course A might be the same as C and s.s.s the same as s.s. Also the algebra might have as well as strings like unit(A).unit(B) other strings like πλατο and σοκρατεσ and λιττλεγρεενμεν, but the theory did not say anything about them and the poetic imagination is discouraged in Computer Science education.

So consider the algebra which has <u>no junk</u> (elements are in the algebra only if they are the values of some term) and <u>no confusion</u> (terms denote equal elements only if the equations force them to). This is what we call the initial algebra. A rather neater way to define it is to say that it is the algebra which has a unique homomorphism to every other algebra, where algebra now means algebra which is a model of the given theory.

This leads to

Defn. I is an **initial algebra** of a theory iff for any algebra of the theory, A, there is a unique homomorphism $f\colon I \to A$.

An easy way to get the initial algebra of a theory is to take the word algebra on its signature and then form equivalence classes according to the equations of the theory. We have to ensure that if terms t1 and t2 are in the same equivalence class, then $\omega(t1)$ and $\omega(t2)$ are in the same class, for any unary operator ω, and analogously for operators of several arguments. Such an equivalence is called a *congruence*.

Now of course we can do the same thing with variables. Given a theory T, the free T-algebra on a set X of variables is the one in which any binding of the variables to elements of another algebra determines a unique homomorphism. It can be obtained by taking equivalence classes of elements in the Σ-word algebra on X where Σ is the signature of T. More formally

Defn. By a **free algebra on X** of a theory we mean an algebra F(X) of the theory with a function $\eta_X\colon X \to |F(X)|$, such that for any algebra A of the theory, any function $f\colon X \to |A|$ extends uniquely to a homomorphism $f^{\#}\colon F(X) \to A$. By $f^{\#}$ extends f we mean that $\eta_X \bullet |f^{\#}| = f$.

CHANGING THE SIGNATURE

If we are interested in specifications and their implementations we will be interested in changing signatures, mapping sorts in one signature into sorts in another and operators in one to operators in the other. For example given a signature Integer thus

```
sorts   integer, boolean
opns    zero, one: integer
        plus, minus: integer, integer → integer
        equal: integer, integer → boolean
```

and a signature Entier thus

sorts entier, booleen
opns zero, un: entier
plus, moins: entier, entier → entier
egal: entier, entier → booleen

we can define a function, Translate, thus

integer to entier	boolean to booleen	
zero to zero	one to un	
plus to plus	minus to moins	equal to egal

Note that this function preserves argument and result sorts, thus the input sort of the translation of "plus" is the translation of its input sort and so on. We call such a function a signature morphism. In general a signature morphism may go from one signature to another with extra sorts or operators, and the function need not be one to one, two sorts may map to the same sort in the second signature, and similarly for operators.

Signature morphisms arise in describing programming languages. Suppose that we have some notion of parameterised module and we want to talk about binding the formal types and procedures of the module to actual ones in the external environment. This binding is a signature morphism.

It is more elegant to think of a signature morphism as a pair of functions, one for sorts and one for operators. Formally we have

<u>Defn.</u> A **signature morphism** σ from a signature $\Sigma 1$, say $<S1,\Omega 1>$, to a signature $\Sigma 2$, say $<S2,\Omega 2>$, is a pair $<f,g>$ consisting of a function $f: S1 \to S2$ and a family of functions $g_{us}: \Omega 1_{us} \to \Omega 2_{f^*(u)f(s)}$, where $f^*: S1^* \to S2^*$ is the extension of f to strings. We write $\sigma: \Sigma 1 \to \Sigma 2$.

It is convenient to write $\sigma(s)$ for $f(s)$, $\sigma(u)$ for $f^*(u)$ and $\sigma(\omega)$ for $g_{us}(\omega)$.

What happens to algebras when we change the signature? If we go from $\Sigma 1$ to $\Sigma 2$ does a $\Sigma 1$-algebra turn into an $\Sigma 2$-algebra? No. There is a general and mysterious principle that when you go from syntax to semantics things often go backwards; "contravariant" is the posh word for this. (As in "My husband is in a contravariant mood today.") Given a $\Sigma 2$-algebra and a signature morphism $\sigma: \Sigma 1 \to \Sigma 2$, we can create a $\Sigma 1$-algebra. We get the set for a sort s of the $\Sigma 1$-algebra by taking the set for the sort $\sigma(s)$ in the $\Sigma 2$-algebra, and we get the table for an operator ω of the $\Sigma 1$-algebra by taking the table for $\sigma(\omega)$ in the $\Sigma 2$-algebra. In our example above if we are given an algebra for

Entier we can form one for Integer thus: the set for integer is that for entier, the set for boolean is that for booleen, the table for minus is that for moins and so on. Easy. The contravariance is not too surprising if we reflect that given a Pascal machine and a translater from Ada to Pascal we have an Ada machine.

Given the morphism σ: $\Sigma 1 \rightarrow \Sigma 2$ and the $\Sigma 2$-algebra A, we call the resulting $\Sigma 1$-algebra $U_\sigma(a)$. That is U_σ is the function from the set of all $\Sigma 2$-algebras to the set of all $\Sigma 1$-algebras corresponding to the signature morphism σ: $\Sigma 1 \rightarrow \Sigma 2$.

Defn. If σ: $\Sigma 1 \rightarrow \Sigma 2$ is a signature morphism, and A is a $\Sigma 2$-algebra, say $<|A|,a>$, then we define the function U_σ : $\Sigma 2$-algebras $\rightarrow$ $\Sigma 1$-algebras, by $U_\sigma(A) = A'$, where $|A'|_s = |A|_{\sigma(s)}$, and $\omega(a_1,\ldots,a_n)$ in A' is $\sigma(\omega)(a_1,\ldots,a_n)$ in A.

A useful special case is where σ is an inclusion morphism, that is the sorts of $\Sigma 1$ are a subset of the sorts of $\Sigma 2$, similarly for the operators, and $\sigma(s) = s$, $\sigma(\omega) = \omega$. For example the signature Letters may be included in the signature Strings-of-Letters, so we can get a Letters algebra from a Strings-of-Letters algebra by just forgetting the extra sorts and operators. If we have an inclusion morphism from $\Sigma 1$ to $\Sigma 2$ we call $\Sigma 2$ an *enrichment* of $\Sigma 1$, and talk of the $\Sigma 1$-*reduct* of an $\Sigma 2$-algebra. In this case we write $\Sigma 1 \subseteq \Sigma 2$.

THEORY MORPHISMS

Let us try something a little more ambitious. Given two theories T1 and T2, what would be a reasonable notion of mapping from T1 to T2? Suppose that they have signatures $\Sigma 1$ and $\Sigma 2$ respectively. Then we may consider a signature morphism, σ, from $\Sigma 1$ to $\Sigma 2$. Given an algebra of T2 we can apply U_σ to it. We get a new algebra with signature $\Sigma 1$ but is it a T1 algebra? Not necessarily. In fact we shall say that σ is a theory morphism from T1 to T2 just if U_σ always gives us a T1 algebra. Formally

Defn. If T1 and T2 are theories, say $<\Sigma 1,E1>$ and $<\Sigma 2,E2>$, by a **theory morphism** σ: T1 $\rightarrow$ T2 we mean a signature morphism, σ: $\Sigma 1 \rightarrow \Sigma 2$, such that for each T2-algebra A, $U_\sigma(A)$ is a T1-algebra.

Now since σ translates $\Sigma 1$ to $\Sigma 2$ we can use it to translate $\Sigma 1$ terms to $\Sigma 2$-terms in the obvious way,so we can translate a T1-equation to an equation in signature $\Sigma 2$. Now an important fact is that if σ is a theory morphism then this translation is a T2-equation. (Indeed we previously took this as the definition of theory morphism [Burstall and Goguen 80], but following Reichel [1980] we have learned to prefer the semantic definition.)

Just as we talked about signature inclusions, so we can talk about theory inclusions, where the signature of the first theory is included in that of the second one and the inclusion is indeed a theory morphism. Note that the equations of the second theory then include those of the first theory. For example the theory of monoids, with identity and associative multiplication is included in the theory of groups, which has an inverse operation also.

Notice that we can not only reduce T2-algebras to T1-algebras using σ, but we can reduce T2-homomorphisms to T1-homomorphisms. We simply relabel the maps using the T1-sorts and forgetting any which are not needed. Thus we can define a function, also called U_σ, from T2-homomorphisms to T1-homomorphisms. If $f: A2 \to B2$ is a T2-homomorphism $U_\sigma(f): U_\sigma(A2) \to U_\sigma(B2)$ is a T1-homomorphism.

ENRICHMENTS AND FREE ALGEBRAS

Suppose that signature $\Sigma 2$ is an enrichment of $\Sigma 1$ via the morphism $\sigma: \Sigma 1 \to \Sigma 2$. Given a $\Sigma 2$-algebra, A2, we can reduce it to a $\Sigma 1$ algebra, A1, namely $U_\sigma(A2)$. But can we do the reverse? Is there a 'best' or 'canonical' way of making a $\Sigma 2$-algebra from a $\Sigma 1$-algebra, A1? If σ is just the inclusion $\Sigma 1 \subseteq \Sigma 2$, then we can make the free algebra $F_{\Sigma 2}(|A1|)$. More generally we have to convert the $\Sigma 1$-carrier $|A1|$ to a $\Sigma 2$-carrier by reindexing according to σ, then apply $F_{\Sigma 2}$, then take a congruence on this determined by the operations of A1. But we prefer to define this $\Sigma 2$-algebra by a unique homomorphism property.

Defn. If $\sigma: \Sigma 1 \to \Sigma 2$ is a signature morphism and A1 is a $\Sigma 1$-algebra, then by the **free $\Sigma 2$-algebra on A1**, with respect to σ, we mean a $\Sigma 2$-algebra $F_\sigma(A1)$ and a $\Sigma 1$-homomorphism $\eta_{A1}: A1 \to U_\sigma(F_\sigma(A1))$ such that, for any $\Sigma 2$-algebra A2 and any $\Sigma 1$-homomorphism $f: A1 \to U_\sigma(A2)$, there exists a unique $\Sigma 2$-homomorphism, $f^\#: F_\sigma(A1) \to A2$, such that $\eta_{A1} \circ U_\sigma(F^\#) = f$; that is the following diagram commutes

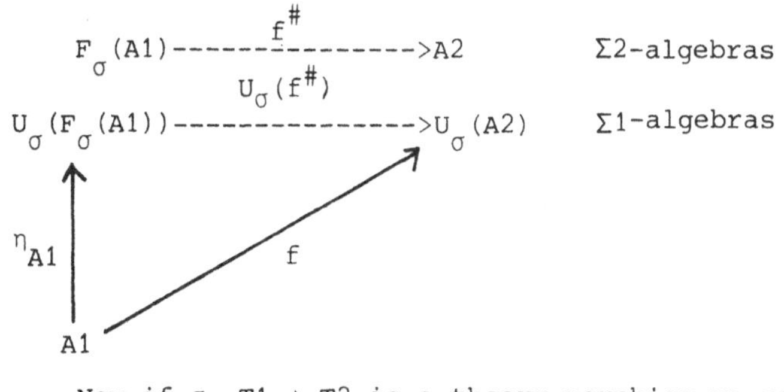

Now if $\sigma: T1 \to T2$ is a theory morphism we can define the **free T2-algebra on a T1-algebra A1**, in just the same way, replacing Σ1 by T1 and Σ2 by T2 in the above definition. The construction is similar too, forming the free T2 algebra on the re-indexed carrier of A1.

Suppose for example that T1 has the signature zero, successor, plus, times with appropriate equations and T2 has a larger signature which also has the string forming operators empty, unitstring and concatenation with equations for identity and associativity. Then for any T1-algebra, say the numbers modulo 3, there is a free extension to a T2-algebra, strings of numbers modulo 3. Note that even though the T1 algebra satisfies extra equations, such as 0 = succ(succ(succ(0))) the new operators get interpreted freely (no junk, no confusion) as string constructors.

This free extension of a theory, gives us a precise algebraic interpretation of the computer scientist's notion of parameterised data type, "string of X". T1 can be used to tell us the *requirements* on the parameter and is usually a subtheory of T2. For strings T1 could be trivial, the theory with one sort and no operators; for more complex data types it could have operators such as a partial order. For a further explanation see Thatcher, Wagner and Wright [1978] or Burstall and Goguen [1981].

Given a T1-homomorphism $f: X \to U_\sigma(Y)$ we get a T2-homomorphism $f^\#: F(X) \to Y$. Without any extra work we can turn a T1-homomorphism $g: X \to X'$ into a T2-homomorphism $F_\sigma(g): F_\sigma(X) \to F_\sigma(X')$. Just form $g \cdot \eta_{X'}: X \to U_\sigma(F_\sigma(X'))$ and put $F_\sigma(g) = (g \cdot \eta_{X'})^\#$. Thus F like U works on both algebras and homomorphisms.

The following table may help to summarise what we have done

Algebras on a signature	Algebras on a theory
Initial Σ-algebra	Initial T-algebra
Free Σ-algebra on an indexed set	Free T-algebra on an indexed set
Free Σ2-algebra on a Σ1-algebra, w.r.t. $\sigma: \Sigma 1 \to \Sigma 2$	Free T2-algebra on a T1-algebra, w.r.t. $\sigma: T1 \to T2$

FREENESS IN GENERAL: THE CATEGORIAL APPROACH

The reader will have noticed the similarity between our definitions of

Free Σ-algebra on a set
Free T-algebra on a set
Free $\Sigma 2$-algebra on a $\Sigma 1$-algebra
Free T2-algebra on a T1-algebra.

They all use the same diagram, although sometimes we write $|\ldots|$ and sometimes U_σ. What is the general definition of freeness? The answer to this and similar questions is the concern of category theory. It deals with arbitrary structures (objects) and structure-preserving functions between them (morphisms). It treats objects (sets, algebras, signatures, theories, etc.) and morphisms (functions, homomorphisms, signature morphisms, theory morphisms, etc.) as 'abstract data types' by concerning itself not with what they are made of ('frogs and snails and puppy dogs' tails'?) but with the operations which may be performed on them, notably composition of morphisms. For a readable introduction to category theory see Arbib and Manes [1975] or the first few chapters of **Goldblatt** [1979].

ACKNOWLEDGEMENTS

RMB would like to thank Peter Landin for educating him in the elements of algebra. We would like to thank the ADJ group (Jim Thatcher, Eric Wagner and Jesse Wright) for much helpful collaboration and our colleagues at Edinburgh and Los Angeles for many algebraic conversations. RMB would like to thank Xerox PARC, where he has been Visiting Scientist, for their support and for their document preparation facilities for a draft (*per ardua ad Alto*), and Eleanor Kerse for typing the final version.

REFERENCES

Arbib, M.A. and Manes, E.G. *Arrows, Structures and Functors: the categorical imperative.* Academic Press: New York, London, 1975.

Burstall, R.M. *Electronic category theory*, in: Proc. Symp. in Math. Foundations of Comp. Science, Rydzyna. Lecture Notes in Computer Science No. 88, 22-40, Springer-Verlag, 1980.

Burstall, R.M. and Goguen, J.A. *The semantics of Clear, a specification language*, in: Proc. of Advanced Course on Software Specifications. Lecture Notes in Computer Science, Berlin: Springer-Verlag, 1980.

Burstall, R.M. and Goguen, J.A. *An informal introduction to specifications using Clear*, in: The Correctness Problem in Computer Science, Academic Press (to appear), 1981.

Goguen, J.A., Thatcher, J., Wagner, E. and Wright, J.B. *An introduction to categories, algebraic theories and algebras.* IBM Technical Report RC 5369, Thos. J. Watson Research Center, Yorktown Heights, N.J., 1975.

Goguen, J.A., Thatcher, J., Wagner, E. and Wright, J.B. *Initial algebra semantics and continuous algebras.* JACM, 24, 1, 68-95, 1977.

Goguen, J.A., Thatcher, J., Wagner, E. and Wright, J.B. *An initial algebra approach to the specification, correctness and implementation of abstract data types*, in: Current Trends in Programming Methodology (ed. R. Yeh), Prentice Hall, N.J., 1978.

Graetzer, G. *Universal Algebra, (Second Edition).* New York: Springer-Verlag, 1979.

Goldblatt, R. *Topoi, the categorial analysis of logic.* Amsterdam: North-Holland, 1979.

Reichel, H. *Initially restricting algebraic theories*, in: Proc. Symp. in Math. Foundations of Comp. Science, Rydzyna. Lecture Notes in Computer Science No. 88, 504-514, Springer-Verlag, 1980.

Thatcher, J.W., Wagner, E.G. and Wright, J. *Data type specification: parameterisation and the power of specification techniques*, in: Proc. of the 10th Annual Symposium on Theory of Computing, ACM, 119-132, 1978.

Theme VIII.

(R.M. Burstall)

A presentation of the Edinburgh approach, from several different angles

AN ANALYSIS OF SEMANTIC MODELS FOR ALGEBRAIC SPECIFICATIONS*

Martin Wirsing and Manfred Broy
Institut für Informatik,
Technische Universität München
Arcisstr. 21
D-8000 München 2

ABSTRACT

Data structures, algorithms and programming languages can be described in a uniform implementation-independent way by axiomatic abstract data types i.e. by algebraic specifications defining abstractly the properties of objects and functions. Different semantic models such as initial and terminal algebras have been proposed in order to specify the meaning of such specifications - often involving a considerable amount of category theory. A more concrete semantics encompassing these different approaches is presented:
Abstract data types are specified in hierarchies, employing "primitive" types on which other types are based. The semantics is defined to be the class of all partial heterogeneous algebras satisfying the axioms and respecting the hierarchy. The interpretation of a specification as its initial or terminal algebra is just a constraint on the underlying data. These constraints can be modified according to the specification goals. E.g. the data can be specified using total functions; for algorithms partial functions with syntactically checkable domains seem appropriate whereas for programming languages the general notion of partiality is needed. Model-theoretic and deduction-oriented conditions are developed which ensure properties leading to criteria for the soundness and complexity of specifications. These conditions are generalized to parameterized types, i.e. type procedures mapping types into types. Syntax and different semantics of parameter are defined and discussed. Criteria

* This research has been partially sponsored by the Sonderforschungsbereich 49, Programmiertechnik, München.

M. Broy and G. Schmidt (eds.), Theoretical Foundations of Programming Methodology, 351–413.

for **proper parameterized specifications are developed. It is** shown that the properties of **proper specifications viz. of snow**balling and impeccable types are preserved under application of parameterized types - finally guaranteeing that the composition of **proper** small specifications always leads to a **proper** large specification.

CONTENTS

1. INTRODUCTION

Since the papers of Liskov, Zilles [29] and Guttag [26] in 1974/1975 abstract data types have been thought of as a tool for designing formal specifications in a modularized and implementation-independent way by abstracting away from the particular representation of the data.
At the same time the ADJ-group [24] introduced the first rigorous though relatively simple approach to the semantics of such specification techniques, the "initial algebra semantics" in which the meaning of a type is the isomorphism class of all initial algebras. But this semantics did not completely meet the original idea of an abstract data type as a "black box" where only the input/output behavior can be observed. For that, Wand [38] proposed in 1977 the "final" or "terminal" algebra semantics. Even earlier (1976), Giarratana, Gimona and Montanari [21] did not consider an isomorphism class but the class of all finitely generated models having the same input/output behavior.
However, partial recursive functions exist which cannot be specified (without hidden sorts) using any of these approaches (cf. Broy, Wirsing [8]). Thus in [7,40] and independently in Reichel [34], abstract types with partial operations (so-called "partial abstract types") are studied which even allow the algebraic specification of the semantics of programming languages [9]. These approaches cope very well with the semantics of small individual data types. But for designing classes of abstract data types and for constructing large specifications from smaller ones while maintaining the independence of the components, additional features are necessary:
"Parameterized types" - these are functions which applied to a type produce a new type (Burstall, Goguen [11], and ADJ [37] which again proposed the first formal semantics); and "hierarchy constraints" (Broy et al [6], also "data constraints" in Burstall, Goguen [12], and "initially restricting theories" in Reichel [34,35]). In the following we study partial abstract types with parameters and hierarchy constraints - parameterized hierarchical partial abstract types.
After an informal introduction to the semantics of this class of types (§2) we define their formal semantics and try to exhibit (syntactic) conditions under which a type T is a proper specification; that means informally:

(1) T admits initial and terminal algebras and (the congruences associated to) its models form a complete lattice (wrt. set inclusion).

(2) The primitive types of T can be implemented independently of those built on top of them.
(3) Every argument type ARG is preserved under type application by a parameterized type T; i.e. every model of ARG can be extended to a model of T(ARG).

In fact, we require these properties only for the "minimally defined models" of a type, which allows us to include the case of total abstract types in a very simple way: we have only to consider types in which the definedness predicates are identically true. Then the class of minimally defined models coincides with the class of all models and all theorems hold for parameterized types with hierarchy constraints (as in CLEAR).

In §3 we develop the theory of partial abstract types without parameters and hierarchy constraints. We define a proof system for partial types and discuss its soundness and completeness (prop. 1,2)- For such types initial algebras always exist (prop. 3) whereas for the existence of terminal algebras, three conditions are needed: "consistency", "weak maximality" and "t-completeness" (prop. 4) which ensure also that the models of T form a complete lattice (prop. 5,6).

Initial and terminal models of nonhierarchical types are almost computable (i.e. semi/cosemicomputable cf. Bergstra, Tucker [4]) whereas nonhierarchical types allow for highly complex models (i.e. hyperarithmetical ones, cf. Bergstra et al [5]) and do not have complete proof systems (prop. 8, cf. Sannella, MacQueen [31]). Therefore in §4 we show first that "weak sufficient completeness" guarantees the equivalence of nonhierarchical and hierarchical models (ensuring a "low" complexity of the latter), and hence the existence of initial algebras of hierarchical types (prop. 9). Together with consistency and weak maximality one obtains weakly terminal algebras and the complete lattice property for hierarchical types (prop. 10). For the independence of the implementation of the primitive types from the overall specification (formally "weak hierarchy persistency") a fourth syntactic condition is introduced: "hierarchy-consistency". Together with the former three conditions it implies that every model of a primitive type can be extended in a persistent way to a model of the overall type and that its extensions form a complete lattice (prop. 12,13).

In §5 the general case of partial abstract types with parameters and hierarchy constraints is studied. A parameterized type is considered as a **function** which applied to a loose type (i.e. a type which possibly allows nonisomorphic models) produces another loose type. A model of a parameterized type is a **function**, too, with a loose (well-behaved) type as argument and a model of the instantiated type as result (§5.3). This notion generalizes the ADJ/Berlin approach [17] in which a parameterized algebra is a functor mapping algebras to algebras. Moreover the abovementioned conditions for hierarchical types carry over (with appropriate changes) to parameterized ones:weak sufficient completeness ensures the existence of initial models for parameterized types (prop. 15). Weakly terminal models exist whenever the type is consistent, weakly maximal and weakly sufficiently complete (prop. 16). Types fulfilling all three criteria are called "snowballing", and "impeccable", if moreover all models of the result type are persistent extensions of the models of the argument type. Impeccability means "persistency + existence of complete lattices" and thus ensures the existence of free constructions (as in [II, I6, I7, 36]) as well as of final ones (as in Ganzinger [I8, I9]. In contrast to ADJ [I6] and Ehrig [I7] we believe in the usefulness of snowballing types e.g. for adding error elements (cf. Ehrich [I5]) or infinite objects (cf. the programming lenguage example PQL). Therefore we establish sufficient conditions for it as well as for impeccability (prop. I8, corr. 2).

In 5.2 we give a mathematical definition of "general parameter passing" where a parameterized type is instantiated by another parameterized type. This is done by simple textual substitution and thus may correspond to a "call-by-name" mechanism in ordinary programming languages. In 5.4 we discuss also three other parameter-passing mechanisms viz. "call-by-theory", by "type" and by "model", and show that the former one is a "call-by-name" mechanism whereas the latter ones are (different) call-by-value parameter-passing mechanisms. The advantage of call-by-name is that it makes specifications easily extensible: If e.g. a parameterized snowballing type is used to add a new language construct to a specification of a programming language by abstract types, then the "semantic equations" are preserved and have not to be rewritten as (often) in denotational semantics (cf. Mosses [42]).

Finally we prove that most of our conditions - in particular those for snowballing and impeccability - are preserved under application of parameterized types. Thus the construction of a large specification from proper small ones always leads to a proper result.

2. INFORMAL INTRODUCTION TO THE SEMANTICS

2.1 *Partial Abstract Types*

A partial abstract type (PAT for short) is written in the form of a "mathematical theory" containing sorts, (**possibly partial**) operations and laws. One of the simplest examples is the theory BOOL of truth-values which might be specified as follows

```
type BOOL ≡ sort bool
            bool true,false
            laws: D(true),D(false) endoftype
```

This means that every data structure M associated with the type BOOL has one carrier set and two constant operations. M is a Σ-algebra where Σ = <{bool},{bool true, bool false}> is its signature. The laws determine the properties of the operations. Here only the definedness of true and false is required but no further equations. E.g. the algebra M1 consisting of 0 and 1 with $true_{M1} \equiv 0$ and $false_{M1} \equiv 1$, or the algebras

$M2 \equiv \langle\{wahr, falsch\}, true_{M2} \equiv wahr, false_{M2} \equiv falsch\rangle$

$M3 \equiv \langle\{2,3,17\}, true_{M3} \equiv 3, false_{M3} \equiv 17\rangle$

$M4 \equiv \langle\{1\}, true_{M4} \equiv 1, false_{M4} \equiv 1\rangle$

$M5 \equiv \langle\{1,21\}, true_{M5} \equiv 1, false_{M5} \equiv undefined\rangle$

are Σ-algebras. But we accept only M1 and M2 as models or equivalently as data structures of type BOOL. The others are rejected for failure to satisfy the following slogans:

(1) "no junk": the algebra should not have unnecessary elements (cf. Burstall,Goguen [12])

(2) "no collapse": trivial algebras where every carrier set has at most one element should be excluded

(3) "no crime": the algebra has to satisfy the laws.

The algebra M3 is disqualified for its "junk" - the number 2 is not the value of any expression; M2 "collapses"; and M4 contains "junk", the number 21, and does not respect the laws since $false_{M5}$ is undefined. However, apart from M1 and M2, the type BOOL contains many other models all of which consist of exactly two elements and are isomorphic to each other. BOOL is therefore called monomorphic. This is not always the **matter**. Consider e.g. the following type

```
type ABC ≡
     sort bool,alpha
     bool true,false
     alpha a,b,c
     funct(alpha)bool isa
     laws: D(true),D(false)
           isa(a) = true
           isa(b) = isa(c) = false    endoftype
```

The strictness of partial functions implies that all operations of type ABC are total. Thus the type ABC has (up to isomorphism) two models, I and Z, which are characterized by the equality in sort alpha:

I satisfies $b \neq c$ whereas Z satisfies $b = c$

(everything else is completely determined by the laws of ABC, e.g. $I \models$ isa(a) = true, isa(a) = false and therefore $a \neq b$). The algebra I is built according to the following slogan (Burstall, Goguen [I2]):

(4) "no confusion":expressions should not be equal unless they are forced to be so by the equations.

Such an algebra is called <u>initial</u>. In contrast to that, Z is a <u>terminal</u> algebra according to

(5) "no apartheid" (Burstall):all expressions should be identified which are not forced to be different by the equations.

A type having nonisomorphic models is called a <u>loose</u> type.

2.2 *Hierarchical and Parameterized Types*

Another example for a loose type are priority queues where on request one may get the least element of a queue. Such queues may be implemented as ordered or unordered sequences or even as binary trees. The abstract type PRIORITY-QUEUE might be specified in the following way:

```
type PRIORITY-QUEUE ≡
PARAMETER:
     type LINORD ⊇
        primitive BOOL
        sort elem
        funct(elem, elem)bool.≤.
        laws ∀ elem x, y, z:
        D(x ≤ y) ∧ x ≤ x
        x ≤ y ∧ y ≤ z ⇒ x ≤ z
        (x ≤ y) = false ⇒ y ≤ x

TARGET:
   primitive BOOL
   sort pqueue
   pqueue empty
   funct(pqueue)bool nonempty
   funct(elem,pqueue)pqueue .o.
   funct(pqueue q :nonempty(q))pqueue remove
   funct(pqueue q :nonempty(q))elem min
   laws:∀ elem x, pqueue q:
   min(x o empty) = x ∧ remove(x o empty) = empty
   nonempty(q) ∧ x ≤ min(q) ⇒ min(x o q) = x
                                ∧ remove(x o q) = q
   nonempty(q) ∧ (x ≤ min(q)) = false ⇒ min(x o q) = min(q)
                          ∧ remove(x o q) = x o remove(q)
```

nonempty(empty) = false
nonempty(x • q) = true

endoftype

The type PRIORITY-QUEUE possesses two additional features: First, it uses the already defined type BOOL as primitive type: PRIORITY-QUEUE is a hierarchical type. The requirement for such types is

(6) "Protection of primitives": New elements for primitives are "junk".

If M is a model of a hierarchical type then its substructure M' which corresponds to the primitive type has to be a model of the primitive type. In particular every element of M' has to be a value of an expression of the primitive type. E.g. for every model M of PRIORITY-QUEUE the carrier set M_{bool} can be written as $\{true_M, false_M\}$. In general, this is a severe condition leading to incompleteness and to (initial) models of hyperarithmetical complexity (cf. Bergstra et al [5] and MacQueen, Sannella [31]).

Second, queues might be built over all data structures which are linearly ordered: PRIORITY-QUEUE is a parameterized type. It can be understood as a function P' taking a type T and forming a new type PRIORITY-QUEUE(T). The only requirement is that T has to be linearly ordered and to contain the primitive type BOOL. The parameter passing mechanism we have chosen (for a discussion see § 5.4) is purely syntactic textual substitution: In the body of PRIORITY-QUEUE the part "LINORD ⊇ .." is substituted by the body of T (where the parameters, primitives, sorts, operations and equations are inserted at the correct place). In general, the models of T are not protected by a parameterized type P: If M is a model of P(T) then its substructure M_T corresponding to the type T has to satisfy the laws of T (since they have been translated) but it may contain elements which are not values of expressions of T (cf. the type of deterministic streams in § 3.7). The slogan is

(7) "selection but no protection of parameters": Types are only allowed as actual parameters, if they satisfy the preconditions for parameters.

3. PARTIAL ABSTRACT TYPES

In this paragraph we define the semantics of partial abstract types without hierarchy-constraints and parameters. We show their relation to total abstract types, establish a sound and complete proof system and give sufficient (syntactic) conditions for the existence of initial and terminal algebras.

3.1 *Partial Heterogeneous Algebras*

A partially defined abstract type has not only total algebras but also partial algebras as models:
A partial heterogeneous algebra $A = \langle \{A_s\}_{s\in S}, \{f_A\}_{f\in F} \rangle$ consists of an indexed family of carrier sets A_s with an indexed family of (possibly partial) operations f_A between those carriers. The indexing system is called (S-sorted) signature and consists of a set S of sorts which indexes the carrier sets and a family F of operation symbols for each of which a functionality $s1 \times \ldots \times sn \to s$ with $s1,\ldots,sn,s \in S$ is defined. An operation symbol $f: s1\times\ldots\times sn \to s$ of F names a (partial) operation $f_A: A_{s1}\times\ldots\times A_{sn} \to A_s$ in a partial algebra with signature Σ.
The class of all partial heterogeneous Σ-algebras is denoted by $PALG(\Sigma)$. Every signature $\Sigma = (S,F)$ defines a set of syntactically correct expressions which can be formed from free variables and the operation symbols of the signature:
Let X be an S-sorted set. For every sort $s \in S$ the set, $W(\Sigma,X)_s$, of terms of sort s (containing elements in X) is the least set containing

(i) every $x \in X_s$ (of sort s) and every nullary operation symbol $f \in F$ with functionality $\to s$ and

(ii) every $f(t1,\ldots,tn)$ where $f: s1\times\ldots\times sn \to s$ is an operation symbol with range s in F and every $ti (i=1,\ldots,n)$ is a term (of sort si) in $W(\Sigma,X)_{si}$.

Terms without elements of X are called ground terms and $W(\Sigma,\emptyset)_s$ is denoted by $W(\Sigma)_s$.

Given a total map $v: X \to A$ (more exactly a family $v_s: X_s \to A_s$ of total maps) we can define the interpretation (wrt. v) $v_A(t)$ of a term t in a partial Σ-algebra A:

(i) $v_A(x) \equiv v(x)$ for $x \in X$ and

(ii) $v_A(f(t1,\ldots,tn)) \equiv f_A(v_A(t1),\ldots,v_A(tn))$ for $f \in F$ provided that every $v_A(ti)$ is defined and the n-tuple $(v_A(t1),\ldots,v_A(tn))$ is in the domain of f_A: otherwise $v_A(f(t1,\ldots,tn))$ is undefined.

If t is a ground term then its interpretation does not depend on the map v; the interpretation is uniquely defined and will be denoted by t_A. Every Σ-algebra A contains a least subalgebra A'. This subalgebra is finitely generated by the constants (i.e. nullary operation symbols) named in its signature; thus every carrier set A'_s of A' can be defined by the interpretations of ground terms in A:

$$A'_s = \{t_A \mid t \in W(\Sigma)_s\}$$

If A does not contain a proper subalgebra (i.e. A = A') then A is called finitely generated or a data structure. The class of all data structures of signature Σ is denoted by $PGEN(\Sigma)$.
The ground terms define a special (total) data structure $W(\Sigma)$, the term algebra of Σ, with

- the carrier set $W(\Sigma)_s$ for $s \in S$ and
- the operations $f_{W(\Sigma)}: W(\Sigma)_{s1} \times \ldots \times W(\Sigma)_{sn} \rightarrow W(\Sigma)_s$, $(t1,\ldots,tn) \mapsto f(t1,\ldots,tn)$, for $f \in F$.

3.2 *Example 1: Priority-Queues over Natural Numbers*

Consider the type PRIORITY-QUEUE where the parameter LINORD is replaced by the type NAT of natural numbers which consists of

```
     primitive Bool
     sort nat
     nat zero
     funct(nat)nat succ
     funct(nat,nat)bool .≤.
laws: ∀ nat x,y:
(1)    zero = succ(x) => true = false
(2)    succ(x) = succ(y) => x = y
(3)    zero ≤ x
(4)    (succ(x) ≤ zero) = false
(5)    (x≤y) = (succ(x) ≤ succ(y))
```

The first law ensures that zero $\neq$ succ(x) whenever true $\neq$ false. The second law asserts the injectivity of the successor operation and the laws (3)-(5) define the ordering "$\leq$" recursively. Then the type PQ(NAT) of priority-queues of natural numbers has the signature

$$S = \{\text{bool},\text{nat},\text{pqueue}\}$$
$$F = \{\text{true},\text{false},\text{zero},\text{succ},\leq,\text{empty},\text{nonempty},\circ,\text{remove},\text{min}\}$$

where the functionalities of the operation symbols have been omitted.
The following algebra Z is in PALG(S,F):
Z has the carrier sets

$$Z_{bool} \equiv \{tt,ff\}$$
$$Z_{nat} \equiv \mathbb{N}$$
$$Z_{pqueue} \equiv \{\langle n1,\ldots,nk\rangle \mid k \geq 0 \text{ and } n1,\ldots,nk \in \mathbb{N} \text{ and } n1 \leq \ldots \leq nk\}$$

For simplicity we denote Z_{bool} by $\mathbb{B}$, Z_{nat} by $\mathbb{N}$, and the set Z_{pqueue} of ordered sequences of natural numbers by $\mathbb{PQ}$.
The operations of Z are defined as follows

$$true_Z \equiv tt, \quad false_Z \equiv ff,$$
$$zero_Z \equiv 0, \quad succ_Z(n) \equiv n + 1,$$
$$n \leq_Z m \iff \exists p \in \mathbb{N}: n + p = m,$$
$$empty_Z \equiv \langle\rangle \text{ "the empty sequence"},$$
$$nonempty_Z(\langle n1,\ldots,nk\rangle) \iff k>0$$
$$n \circ_Z \langle n1,\ldots,nk\rangle \equiv \langle n1,\ldots,n(i-1),n,n(i+1),\ldots,nk\rangle \text{ if } n(i-1) \leq n < n(i+1) \text{ for } 1 \leq i \leq k,$$
$$remove_Z(\langle n1,\ldots,nk\rangle) \equiv \begin{cases} \langle n2,\ldots,nk\rangle & \text{if } k>0 \\ \text{undefined} & \text{otherwise,} \end{cases}$$

$$\min_Z(\langle n1,\ldots,nk\rangle) = \begin{cases} n1 & \text{if } k>0 \\ \text{undefined} & \text{otherwise.} \end{cases}$$

In fact, $\mathbb{B}$ is generated by $true_Z$, $false_Z$, $\mathbb{N}$ by $zero_Z$ and $succ_Z$, and $\mathbb{PQ}$ by $empty_Z$ and $\bullet_Z$. Thus Z is a data structure of (S,F). E.g. true, zero ≤ succ(zero), and nonempty(remove(empty)) are terms of sort bool (where the latter is undefined in Z); succ(succ(zero)) and min(zero • empty) are terms of sort nat and

add(min(empty) • empty)

is a term of sort pqueue (which is undefined in Z). □

3.3 *Equality in Partial Abstract Types*

Given any partial algebra A, any interpretation v_A: X -> A from W(Σ,X) into A induces a congruence $\sim_A$ on W(Σ,X) defined by

$$t \sim_A t' \text{ iff } t_A = t'_A$$

That is, t and t' are congruent wrt A iff either both terms are undefined in A or both terms are defined in A and their interpretations are equal. The congruence $\sim_A$ is called the <u>strong equality</u> of A. This congruence does not allow to express that a term is defined in A. Therefore we introduce a unary semantic predicate D_A (wrt. v_A) on W(Σ,X) which is defined by

$$D_A(t) = \begin{cases} \text{true} & \text{if } v_A(t) \text{ is defined} \\ \text{false} & \text{if } v_A(t) \text{ is } \underline{\text{not}} \text{ defined.} \end{cases}$$

In the examples we omit D(qj) if it is provable from qj = rj.

<u>Remark</u>

Strong equality together with the definedness of the terms gives the socalled <u>existential equality</u> $\overset{e}{\sim}_A$ (cf. Andreka et al. [1]):

$$t \overset{e}{\sim}_A t' \text{ if } D_A(t) \wedge D_A(t') \wedge t \sim_A t'$$

This equality is often used in universal algebra since it leads to enumerable initial models (cf. Bergstra et al. [5]). The reason why we did not choose it is that it is a congruence relation only on defined terms. In general, it is not even an equivalence relation since it is not reflexive.

Now, a positive conditional Σ-formula is an expression of the form

$$(*) \quad \bigwedge_{1 \leq i \leq k} D(pi) \wedge \bigwedge_{1 \leq j \leq l} (D(qj) \wedge qj = rj) \Rightarrow C \quad (k,l \geq 0)$$

where C is either D(t) or t = t' and pi,qj,rj,t,t' are Σ-terms in W(Σ,X) where X is a (finite) set of free variables.

A partial Σ-algebra A satisfies the formula (*) (denoted by $A \models (*)$) iff

for all interpretations v_A
$D_A(pi)$ and $D_A(qj) \wedge qj \sim_A rj$ for i = 1, ...,k and j = 1,...,l
implies $D_A(t)$ (or $t \sim_A t'$, resp.).

That means roughly that for all assignments of variables by elements of the carrier sets of A, if pi is defined in A and qj and rj are defined and equal in A then t is defined in A (or t and t' are strongly equivalent in A, resp.).

Now a partial abstract type T consists of a signature Σ and a (finite) set E of positive conditional formulae. The semantics of T is the class PGEN(Σ,E) of all Σ-data structures satisfying all formulas e of E. An algebra in PGEN(Σ,E) is called a model of T. A congruence $\sim_A$ where A ∈ PGEN(Σ,E) is called a congruence of T.

PGEN(Σ,E) is a subclass of the class PALG(Σ,E) of all partial Σ-algebras satisfying the formulae in E. Clearly the total algebras in PALG(Σ) form a(nother) subclass which is defined by the axioms $E_{TOT} = \{D(f(x1,...,xk)) \mid f \in F\}$.

Example 1 (continued)

The structure Z is a model of type PQ(NAT), since it satisfies the axioms. Consider e.g.

$$\text{nonempty}(q) \wedge x \leq \min(q) \Rightarrow \min(x \circ q) = x$$

Let $q = \langle n1,\dots,nk\rangle$ and $x = n \in N$.
$\text{nonempty}_Z(q)$ implies $k>0$, $x \leq_Z \min(q)$ implies $n \leq n1$, and therefore $x \circ_Z q = \langle n,n1,\dots,nk\rangle$. Thus

$$\min_Z(x \circ_Z q) = \min_Z(\langle n,n1,\dots,nk\rangle) = n = x \qquad \square$$

3.4 *Proofs in Partial Types*

For proof systems of total heterogenous algebras two problems arise:

In general, a proof system is complete for ALG(Σ,E) but not for GEN(Σ,E) since not all algebras satisfying the axioms are considered as models but only those which are quotients of the term algebra (cf. e.g. Wirsing,Broy [39]). On the other hand, the usual proof system for equations is not sound for heterogeneous algebras where some carrier sets might be empty as Goguen and Meseguer [23] have shown.
The latter problem does not exist in the following proof system for partial algebras since in equations the definedness of free variables has to be proved. The former problem is the same for partial algebras, but as in the case for total algebras complete proof systems for PALG(Σ,E) are complete for ground equations of PGEN(Σ,E).

The Proof System (Π)

Let T = (Σ,E) be a partial type. Then Π consists of the axioms

(A) Modified) Axioms

Every axiom $a \Rightarrow b \in E$ is replaced by

$$(\bigwedge_i D_{si}(xi) \wedge a) \Rightarrow b$$

where xi (of sort si) are the free variables occuring in $a \Rightarrow b$;
i.e. quantifiers range only over (defined) objects.

(ST) Strictness

For all operations $f: s1 \times \ldots \times sn \to s \in F$ and variables xi of sort si $(i = 1, \ldots, n)$

$$D_s(f(x1, \ldots, xn)) \Rightarrow D_{si}(xi)$$

i.e. partial functions are strict.

(EQ) Equality Axioms

For all terms $t, t', t'', ti, ti' \in W(\Sigma, X)$

Reflexivity : $t = t$

Comparativity : $t = t' \wedge t = t'' \Rightarrow t' = t''$

Substitution Property :

- For all sort $s \in S$:
  $D_s(t) \wedge t = t' \Rightarrow D_s(t')$;
- For all operations $f: s1 \times \ldots \times sn \to s \in F$
  $t1 = t1' \wedge \ldots \wedge tn = tn' \Rightarrow f(t1, \ldots, tn) = f(t1', \ldots, tn')$

and of two rules

(SV) Substitution of Variables:

$$\frac{D_s(t), G}{G[t/x]}$$

if $t \in W(\Sigma, X)_s$, x is a variable of sort s and G a positive conditional formula

(MP) Modus Ponens

$$\frac{A \Rightarrow B_i, \quad B_1 \wedge \ldots \wedge B_n \Rightarrow C}{A \wedge B_1 \wedge \ldots \wedge B_{i-1} \wedge B_{i+1} \wedge \ldots \wedge B_n \Rightarrow C}$$

where $1 \leq i \leq n$ and $A \Rightarrow B_i$ as well as $\bigwedge_j B_j \Rightarrow C$ are positive conditional formulae.

Because of the restriction to defined terms in the substitution rule (SV) the proof system Π is essentially a proof system for the existential equality (cf. also Reichel [35]). It is complete in the following sense:

Completeness Lemma

For every partial abstract type $T = (\Sigma,E)$ the relation $\sim_T$ on $W(\Sigma;X)$ where $X = (\{x_{is} \mid i \in \mathbb{N}\})_{s\in S}$ and for terms of sort s

$$t \sim_\Pi t' \text{ iff } \Pi \vdash t = t' \text{ or } (\Pi \nvdash D_s(t) \text{ and } \Pi \nvdash D_s(t'))$$

is a congruence relation. The proof follows easily from the axioms of Π.

Proposition 1
For every partial abstract type (Σ,E) the proof system Π is sound and complete, i.e. for all $t,t' \in W(\Sigma;X)$

(1) $\Pi \vdash D_s(t)$ iff $\forall A \in PALG(\Sigma,E): A \models D_s(t)$ and

(2) $\Pi \vdash D(t)$ implies
$\Pi \vdash t = t'$ iff $\forall A \in PALG(\Sigma,E): A \models t = t'$

Proof Apply the completeness lemma.

For the class $PGEN(\Sigma,E)$ of finitely generated models of a type in general every proof system is incomplete (cf. MacQueen, Sanella [31] and Bergstra et al. [5]). Completeness can be guaranteed only for ground equations:

Proposition 2
Let (Σ,E) be a partial abstract type and consider $PGEN(\Sigma,E)$. Then the proof system Π is complete wrt. ground formulas; i.e. for all $t,t' \in W(\Sigma)$

(1) $\Pi \vdash D_s(t)$ iff $\forall A \in PGEN(\Sigma,E): A \models D_s(t)$ and

(2) $\Pi \vdash D_s(t)$ implies
$\Pi \vdash t \overset{s}{=} t'$ iff $\forall A \in PGEN(\Sigma,E): A \models t = t'$

3.5 *Initial and Terminal Algebras*

Homomorphisms for partial algebras have to take the definedness of terms into account: They are partial operations which satisfy the usual homomorphism property on the domain of every function:
For partial heterogenous Σ-algebras A,B a family $(\varphi_s : A_s \to B_s)_{s\in S}$ of (possibly partial) operations is called Σ-homomorphism from A into B (denoted by $\varphi: A \to B$) iff

for all $f: s1 \times \ldots \times sn \to s \in F$ and all
$a1 \in A_{s1}, \ldots, an \in A_{sn}$
$f_A(a1, \ldots, an)$ defined $\Rightarrow$

$$\varphi_s\ (f_A(a1, \ldots, an)) = f_B(\varphi_{s1}(a1), \ldots, \varphi_{sn}(an))$$

Therefore on total algebras this notion of homomorphism coincides with the usual one. We distinguish different kinds of homomorphisms which will be important in the sequel (cf. /Grätzer 68/, /Reichel 79/, /Broy, Wirsing 80a,b/ where a slightly different terminology is used).

Total Σ-homomorphisms are called "Σ-homomorphisms" by Grätzer and preserve the definedness of terms. In particular a Σ-homomorphism $\varphi: A \to B$ is total if for all $f: s_1 \times \ldots \times s_n \to s \in F$ and all $x_1 \in A_{s1}, \ldots x_n \in A_{sn}$
$f_A(x_1, \ldots x_n)$ defined $\Rightarrow f_B(\varphi_{s_1}(x_1), \ldots, \varphi_{s_n}(x_n))$ defined.

A Σ-homomorphism satisfying the converse condition is called weak: A Σ-homomorphism $\varphi: A \to B$ is said to be weak, if for all $f: s_1 \times \ldots \times s_n \to s \in F$ and $x_1 \in A_{s1}, \ldots, x_n \in A_{sn}$
$f_B(\varphi_{s_1}(x_1), \ldots, \varphi_{s_n}(x_n))$ defined $\Rightarrow f_A(x_1, \ldots, x_n)$ defined.

Weak Σ-homomorphisms may be partial operations. On finitely generated algebras they are surjective. They can be characterized analogously to homomorphisms for total algebras:

$\varphi: A \to B$ is a weak Σ-homomorphism iff for all $f: s_1 \times \ldots \times s_n \to s \in F$ and all $x_1 \in A_{s1}, \ldots, x_n \in A_{sn}$

$$\varphi_s(f_A(x_1, \ldots, x_n)) = f_B(\varphi_{s_1}(x_1), \ldots, \varphi_{s_n}(x_n)).$$

Again "=" denotes the strong equality.

Between two finitely generated Σ-algebras there exists at most one total or one weak Σ-homomorphism. If there exists a total one and a weak one they are identical. Such a Σ-homomorphism being both, a total Σ-homomorphism and a weak Σ-homomorphism, is called strong Σ-homomorphism. Therefore $\varphi: A \to B$ is strong if

- $f_B(\varphi_{s1}(a1), \ldots, \varphi_{sn}(an))$ defined $\Leftrightarrow f_A(a1, \ldots, an)$ defined and
- $\varphi_s(f_A(a1, \ldots, an)) = f_B(\varphi_{s1}(a1), \ldots, \varphi_{sn}(an))$.

Now, for a class C of Σ-algebras we define particular elements of C:

A Σ-algebra $I \in C$ is called initial (strongly initial resp.) in C if for all $A \in C$ there exists a unique total (strong resp.) Σ-homomorphism $\varphi: I \to A$.

A Σ-algebra Z $\in$ C is called *terminal* (*strongly terminal* resp.) in C if for all A $\in$ C there exists (at least) a weak (strong resp.) Σ-homomorphism φ: A -> Z.

A Σ-algebra I is said to be *initial in type* T if I is initial in PALG(T).

The terminal algebras of PALG(T) are the trivial algebras whose domains have at most one element if the axioms of T are positive conditional formulas. Instead *terminal algebra semantics* turns to $PALG_0(T)$ where the trivial algebras are removed. We assume (for simplicity) that every type T contains the sort bool, two constants true, false of sort bool and the axioms D(true) and D(false). Then $PALG_0(T)$ is defined to consist of exactly those Σ-algebras A of PALG(T) with A $\models$ true $\neq$ false (cf. Hornung, Raulefs [27]). $PGEN_0(T)$ is defined analogously. Let MDEF(C) = $\{A \in C \mid \forall t \in W(\Sigma,X): A \models D(t) \Leftrightarrow \forall B \in C: B \models D(t)\}$ be the class of the minimally defined algebras of C. Then a Σ-algebra Z is said to be

strongly terminal in type T, if Z is strongly terminal in $PALG_0(T)$, and

weakly terminal in type T, if Z is terminal in $MDEF(PALG_0(T))$

that is, if Z is strongly terminal in the class of minimally defined algebras of T.

Finally, we call a type *monomorphic* (*weakly monomorphic*, resp.) if it possesses strongly initial (initial, resp.) and strongly (weakly, resp.) terminal models which are isomorphic. Otherwise T is called *loose*.

Initial algebras always exist:

Proposition 3
For every partial type T = (Σ,E) there exists an initial algebra I in T which is isomorphic to $W(\Sigma)/\sim_\Pi$ where $\sim_\Pi$ is the congruence defined by the proof system Π.

Proof Broy, Wirsing [7].

Thus an initial algebra I is minimally defined, its equality relation (restricted to defined terms) is recursively enumerable and therefore, roughly speaking, every function in I is partial recursive (cf. Bergstra et al [5]).

Example 1 (continued)

Let I be an initial algebra I in PQ(NAT). NAT and BOOL are

monomorphic. Thus the subalgebras of I corresponding to the types NAT and BOOL are isomorphic to the corresponding subalgebras of Z. But e.g. the equation

x ∘ succ(x) ∘ empty = succ(x) ∘ x ∘ empty

which holds in Z is not provable and therefore does not hold in I. Thus I and Z are not isomorphic: PQ(NAT) is neither monomorphic nor weakly monomorphic.
To ensure the existence of weakly terminal algebras we need three further notions:

A type T = (Σ,E) is called consistent (cf. Hornung, Raulefs [27]) iff Π ⊬ true = false.

A Σ-formula G with at most the free variables x1,...,xk of sorts s1,...sk is called maximal (weakly maximal, resp.;cf. [39]) wrt. ground terms iff for all ground terms t1,...,tn of sort s1,...,sn such that Π ⊢ D(ti) for i = 1,...,n

either A ⊨ G[t1/x1,...,tn/xn] for all A ∈ $PALG_0$(T)
(∈ MDEF($PALG_0$(T)), resp.)
or A ⊨ G[t1/x1,...,tn/xn] for all A ∈ $PALG_0$(T)
(∈ MDEF($PALG_0$(T)), resp.)

A sufficient condition for weak maximality is the following. Let v1,...,vk be those terms of G which do not occur as proper subterms of any other term of G. If the formula

D(v1) ∧ ... ∧ D(vk) ⇒ G

is maximal, then G is weakly maximal.

For example, an equation v1 = v2 is weakly maximal if D(v1) ∧ D(v2) ⇒ v1 = v2 is maximal. Obviously, the definedness predicate D is always weakly maximal.

We call a type T weakly maximal if for any equation qj = rj in the premise of one of the axioms of E the formula qj = x where x is a new variable is weakly maximal wrt. ground terms. T is t(rue)-maximal if for all ground terms t of the sort bool the equation t = true is weakly maximal. Note that Hornung, Raulefs' notion of "completeness" implies t-maximality but not vice-versa.

Example 1 (continued)

The type PQ(NAT) is consistent since its model Z ⊨ true ≠ false and therefore (by prop. 1) Π ⊬ true = false.
It is weakly maximal, too. E.g. nonempty(q) = x is weakly maximal for the following reasons: x has to be true or false. Every term t to be substituted for q is of the form empty or y ∘ q'. Thus a little induction on the length of t shows that if Π ⊢ D(t) then either Π ⊢ nonempty(t) = true or Π ⊢ nonempty(t) = false holds. The weak maximality of nonempty(q) = true and (x ≤ y) = true implies that PQ(NAT) is t-maximal.

Consistency is a necessary condition for the existence of weakly terminal algebras. The need for t- and maximality of the axioms can be seen from the following example:

Example 2 [B. Möller]

<u>type</u> STATUS ≡
    sort bool,status
    bool true,false
    status single,married,widowed
    laws: single = married => true = false
        D(true),D(false),D(single),D(married),D(widowed) <u>endoftype</u>

If we assume true ≠ false in all models, then the first axiom is equivalent to single ≠ married. The type STATUS has up to isomorphism three models A,B,C which are characterized by

A ⊨ single ≠ widowed ≠ married
B ⊨ single = widowed ≠ married
C ⊨ single ≠ widowed = married

A is a (strongly) initial model, but neither B nor C are (weakly or strongly) terminal models since there does not exist any homomorphism between them. Thus STATUS does not have terminal models. On the other hand, the type STATUS is **t-maximal** but not weakly maximal, since in the first law single = x is not weakly maximal (choose widowed for x).

Now, if we drop the first axiom and introduce an operation symbol funct(status)bool f with the axioms f(single) = true and f(married) = false, then the situation with the models A,B,C remains and the new type STATUS does not have any terminal model, either. Since all axioms of STATUS' are equations STATUS' is weakly maximal. But f(widowed) = true is not weakly maximal and thus STATUS' is not t-maximal. □

<u>Proposition 4</u>
Let T = (Σ,E) be a consistent t-maximal and weakly maximal type. Then there exists a weakly terminal algebra Z in T which is characterized as follows:

(1) for all $t \in W(\Sigma)$: $Z \models D(t) \Leftrightarrow \forall A \in PALG_0(T): A \models D(t)$

(2) for all t and $t' \in W(\Sigma)$ such that $Z \models D(t) \wedge D(t')$:
$Z \models t = t' \Leftrightarrow (\Sigma, E \cup \{t = t'\})$ is consistent.

The <u>proof</u> is analogous to the proof of theorem 2 in Broy, Wirsing [7].

Under the conditions of proposition 4 we can establish a lattice structure on the congruences of the minimally defined models:

Proposition 5

Let T be a consistent **t-maximal** and weakly maximal type. Then the set of congruences

$$\{\sim_A \mid A \in \mathrm{MDEF}(\mathrm{PGEN}_0(T))\}$$

forms a complete lattice wrt. the set inclusion $\subset$. Its minimum is the congruence associated to the initial models in T; its maximum is the congruence associated to the weakly terminal models in T.

Proof

Consider the following total type $\overline{T} = (\overline{\Sigma},\overline{E})$ associated to $T = (\Sigma,E)$:

$\overline{\Sigma} \equiv \Sigma \cup \{\text{funct}(s)\ \text{bool}\ \overline{D}_s \mid s \in S\},$
$\cup \{s\ \bot_s \mid s \in S \wedge W(\Sigma)_s \neq \emptyset\}$

$\overline{E}$ consists of the following axioms:

1) $\bigwedge_i \overline{D}_{si'}(xi') \wedge \bigwedge_i \overline{D}_{si}(pi) \wedge \bigwedge_j (\overline{D}_{sj}(qj) \wedge qj = rj) \Rightarrow \overline{C}$
   for every axiom $e \in E$ of the form
   $\bigwedge_i D(pi) \wedge \bigwedge_j (D(qj) \wedge qj = rj) \Rightarrow C$
   with the free variables $x1',\ldots,xm'$ of sort $s1',\ldots,sm'$.
2) $\overline{D}_s(\bot_s) = \text{false}$
   $f(\ldots,\bot_{si},\ldots) = \bot_s$ for all $\text{funct}(s1,\ldots,sn)s \in \Sigma$
3) for all $t \in W(\Sigma)_s$:
   $\overline{D}_s(t) = \text{true}$ if $A \models D(t)$ for all $A \in \mathrm{MDEF}(\mathrm{PGEN}_0(T))$
   $\overline{D}_s(t) = \text{false}$ if $A \not\models D(t)$ for all $A \in \mathrm{MDEF}(\mathrm{PGEN}_0(T))$
4) true $\neq$ false

Then there is a "natural" bijection between the congruences associated to $\overline{T}$ and those associated to $\mathrm{MDEF}(\mathrm{PGEN}_0(T))$ (i.e. associate $\overline{D}_s$ to D and "forget" $\bot_s$). According to prop. 5 of Wirsing, Broy [39] the congruences of $\overline{T}$ form a complete lattice wrt. $\subseteq$. Thus those associated to $\mathrm{MDEF}(\mathrm{PGEN}_0(T))$ form a complete lattice, too.

The initial congruence is the finest congruence of all minimally defined models of T whereas the terminal congruence is the coarsest one.

If, moreover, the definedness predicate is maximal, $PGEN_0(T)$ and $MDEF(PGEN_0(T))$ coincide and therefore strongly initial and strongly terminal models exist:

Proposition 6

Let T be a partial abstract type the definedness predicate of which is maximal. Then the following holds:

(1) T has a strongly initial model I which is initial, too.

(2) If T is consistent, **t-maximal** and weakly maximal, then a strongly terminal model Z exists and the set of congruences

$$\{\sim_A \mid A \in PGEN_0(T)\}$$

forms a complete lattice wrt. $\subseteq$ with $\sim_Z$ as minimum and $\sim_I$ as maximum.

The proof follows directly from proposition 5.

Example 1 (continued)

The type PQ(NAT) is consistent, weakly maximal, **t-maximal** and its definedness predicate is maximal. E.g. D(remove(empty)) = false in all models since

funct(pqueue q: nonempty(q))queue remove

implies the axiom

D(remove(q)) => nonempty(q).

Because of nonempty(empty) = false

D(remove(empty)) = false

must hold, too.
Therefore PQ(NAT) possesses strongly initial and strongly terminal models and its congruences form a complete lattice wrt. $\subseteq$. In particular, the model Z is strongly terminal in PQ(NAT). □

On the other hand, if the definedness predicate is not maximal in a type T, then there exists a term t and two models M1 and M2 such that M1 $\models$ D(t) and M2 $\models \neg$D(t). This implies that any strongly initial algebra of T has to satisfy D(t) as well as $\neg$D(t) which is impossible. Thus we have proved:

Lemma 1
The maximality of the definedness predicate is a necessary condition for the existence of strongly initial and strongly terminal algebras.

This, in fact, gives the key for the importance of strong homomorphisms:

Proposition 7
Let T be a consistent partial abstract type with a strongly initial model. Then the definedness predicate is recursive (i.e. for every ground term t it is decidable whether D(t) is provable or not) and uniformly specified i.e. for every ground term t and every model A of T:

$$A \models D(t) \iff T \vdash D(t).$$

Proof According to lemma 1 the definedness predicate is maximal. Thus

$$\begin{aligned} T \nvdash D(t) &\iff \forall A \in PGEN_0(T):\ A \not\models D(t) \quad \text{[by the maximality of D]} \\ &\iff \forall A \in PGEN_0(T):\ A \models D(t) \Rightarrow \mathbf{true} = \mathbf{false}\ \text{[by the consistenc} \\ &\iff T \vdash D(t) \Rightarrow \mathbf{true} = \mathbf{false}\ \text{[by a slight variation of the completeness lemma]} \end{aligned}$$

Therefore the set $\{t \mid T \nvdash D(t)\}$ is recursively enumerable. Since the set $\{t \mid T \vdash D(t)\}$ is recursively enumerable, too, D is recursive. □
Hence, by types with strongly initial or strongly terminal models one can only specify partial recursive functions with a recursive domain (and, in fact, all such functions by using a recursive definition of the (recursive) predicate D). A programming language normally contains functions with nonrecursive domains and therefore cannot be specified by types with strongly initial models. For this purpose initial models are needed and appropriate as the following example shows:

3.6 *Example 3: A Simple Programming Language for Infinite Streams*

As an example for the algebraic specification of programming languages we define a simple procedural programming language PQL which can be used to write programs which consecutively read (finite) priority queues and write (possibly infinite) priority-queues (cf. Broy, Pair, Wirsing [41]).

The Type PQL

We assume the signature and axioms of PQ(NAT) as given as well as

id, the sort of identifiers for natural numbers together with a uniquely specified equality eq: nat × nat -> bool

exp, the sort of (arithmetic and boolean) expressions built from identifiers of id and the operations of nat and bool together with a substitution function .[./.]: exp × exp × id -> exp

state, the sort of states (i.e. finite sets of pairs of id x (nat ∪ bool)) together with an operation (cf. /Stoy 81/)

$$.+[.\to.]: \text{state} \times \text{id} \times (\text{nat} \cup \text{bool}) \to \text{state}$$

where for $\sigma = \{(x1,n1),\dots,(xk,nk)\}$

$$\sigma + [x \to n] = \begin{cases} \sigma \cup \{(x,n)\} & \text{if } x \notin \{x1,\dots,xk\} \\ \{(x1,n1),\dots,(x(i-1),n(i-1)),(x,n),(x(i+1), \\ n(i+1)),\dots,(xk,nk)\} & \text{if } x = xi \end{cases}$$

and an evaluation function

val: exp × state -> bool ∪ nat

such that val(e,σ) = n is weakly maximal.

Furthermore, the type contains the

sort agent

together with the following operation symbols and axioms:

```
agent nil
funct(id)agent  read
funct(exp)agent  write
funct(id,exp)agent  .:=.
funct(agent,agent)agent  .;.
funct(exp,agent)agent  while
```

As only semantic function we use

```
funct(agent,pqueue,state)pqueue exec
```

The operations "nil" and ";" form a monoid:

(1) a;nil = a = nil;a
(2) a;(a';a") = (a;a');a"

Then exec is inductively defined by

(3) exec(nil,q,σ) = empty
(4) nonempty(q) => exec(read(x);a,q,σ) = exec(a,remove(q), σ +[x → min(q)])
(5) nonempty(q) = false => exec(read(x);a,q,σ) = empty
(6) exec(write(e);a,q,σ) = val(e,σ) • exec(a,q,σ)
(7) exec(x:=e;a,q,σ) = exec(a,q,σ +[x → val(e,σ)])
(8) val(b,σ) = true => exec(while(b,a);a',q,σ) = exec(a;(while(b,a);a'),q,σ)
(9) val(b,σ) = false => exec(while(b,a);a',q,σ) = exec(a',q,σ)

Finally, in order to ensure the syntactic definedness of all agents and of sequences of form exec(..) we require

(19) D(a;a'),D(while(e,a)), and D(exec(a,q,σ))

<u>endoftype</u>

The type PQL satisfies the following properties:

(1) As every partial abstract type it has an initial model which is minimally defined (see prop. 3).

(2) By structural induction one can prove the definedness

D(a) for all terms a containing only operation symbols with range <u>agent</u>
E.g. D(nil) follows from D_{pqueue}(empty) and exec(nil,q,σ) = empty using the strictness of exec; the definedness of read(x),write(e), and x:=e is proved analogously whereas D(a;a') and D(while(e,a)) is implied by the axioms.

(3) The type PQL does not have any strongly initial model nor any strongly terminal model since the definedness predicate is not maximal:
Consider the priority-queue

U ≡ exec(while(true,nil),empty, ∅)

Then the term min(U) is undefined in the initial models of PQL since its definedness cannot be proven. To do this one would have to show that

$$PQL \vdash U = x \bullet q$$

for some nat x and some pqueue q.
But only the axioms (1) and (8) are applicable to U leading to

```
U = exec(while(true,nil);nil,empty,<>)   [by (1)]
  = exec(nil;(while(true,nil);nil),empty,<>)   [by (8)]
  = U                                          [by (1)]
```

But in another model of PQL which is not minimally defined the term min(U) might be defined and equal to some term of the form $succ^n$(zero).

(4) The type PQL has a weakly terminal model. The consistency follows from the nonambiguity of the definition of exec and the weak maximality of the premises of the axioms is implied by the weak maximality of val(b,σ) = b' (as required) and of nonempty(q) = b (see corollary 1, § 4.3). Then because of the weak maximality of $\leq$ we obtain the **t-maximality, too.**

Examples

(1) The following agent FAC computes the **priority queue of all numbers n!, n ∈ ℕ:**

```
FAC ≡   n:=0; res:=1;while(true,write(res);n:=n+1;res:=res*n)
```

where * denotes the multiplication of natural numbers:
Let fac ≡ while(true,write(res);n:=n+1;res:=res*n).
Then

```
exec(FAC,empty, ∅)                          =  [by (7)]
exec(fac,empty,{(n,0),(res,1)})=            [by (8)]
exec(write(res);n:=n+1;res:=res*n;fac,empty,{(n,0),
                          (res,1)})  =     [by (6)]
1 ∘ exec(n:=n+1;res:=res*n;fac,empty,{(n,0),(res,1)})
                                      =     [by (7)]
1 ∘ exec(fac,empty,{(n,1),(res,1*1)}) =   [by (8),(7),(6)]
1 ∘ ...∘ m! ∘ exec(fac,empty,{(n,m),(res,m!)})
```

(2) The following agent ERM erases multiple elements in (finite) priority-queues:

```
ERM ≡ read(x);
       while(true,write(x);read(y);
                  while(eq(x,y),read(y));
                  x:=y                     )
```

where eq denotes an equality relation on the natural numbers.
We have (by axiom (4))

```
exec(read(x);while(...),q, ∅) = exec(while(...),
                           remove(q), {(x, min(q))})
```

iff q ≠ empty and remove(q) and min(q) are defined.

Now, if q is a "finite" nonempty priority-queue, i.e. if q has the form x1 ∘ ... ∘ xm ∘ empty with $m \geq 1$, then min(q) and remove(q) are defined and one can evaluate ERM using the axioms of exec and of PQ(NAT). E.g. exec(ERM,1 ∘ 1 ∘ 2 ∘ 1 ∘ empty,∅) = 1 ∘ 2 is provable in PQL.

But if q is an "infinite" priority-queue - as e.g. exec(FAC,empty,∅) in the previous example - then D(min(q)) is not provable and does not hold in minimally defined models. This fact depends crucially on the precondition $x \leq min(q)$ in the axioms for min and remove. Suppose we would have chosen

(11) min(x ∘ q) = x and
(12) remove(x ∘ q) = q

as axioms for min and remove. Then the new type PQ'(NAT) would simply specify unordered sequences where min(q) is the top of a sequence q and remove(q) is the rest of q. In the corresponding programming language PQL' the operations min and remove are defined for "infinite sequences" as FAC: For i ∈ ℕ the equations (6)-(8), (11),(12) imply min($remove^i$(exec(FAC,empty,∅))) = i!

4. HIERARCHICAL TYPES

The type PQ(NAT) has been designed hierarchically-based on the already defined types BOOL and NAT where NAT itself was based on type BOOL. Such a hierarchical design of data type specifications supports a modular decomposition into structures of manageable size and thereby helps to master the complexity originating from a large number of functions and axioms. But when proceeding in this way it should be made sure that the hierarchy imposed by the design of an abstract data type is reflected within each of its models. This is of particular importance if the specification of a type and the construction of an implementation are done by different people.

In the following we study types with hierarchy-constraints which are similar to the data constraints in CLEAR (cf. Burstall, Goguen [11], Sannella [36]). These constraints allow to specify arbitrarily quantified formulae by means of simple equations and permit therefore the definition of highly complex, noncomputable algebras in the framework of abstract data types (cf. Bergstra et al [5]). In order to distinguish such (unsuitable) specifications we establish necessary and sufficient syntactic conditions ensuring that the initial/terminal algebras of hierarchical types coincide with the initial/terminal algebras of the corresponding types without hierarchy-constraints. By putting these conditions together with our previous propositions we obtain sufficient conditions for the existence of initial and terminal models of hierarchical types. The existence of such models alone does not guarantee the hierarchy-persistency, i.e.

every model of the primitive type can be extended to a model of the hierarchical type. Therefore we show finally that "hierarchy-consistency" is a necessary and sufficient syntactic condition for the hierarchy-persistency of a type.

4.1 *Definitions*

A *hierarchical (partial abstract) type* $T = (\Sigma,E,P)$ is a type in which the type $P = (\Sigma',E')$ is designated as primitive where $\Sigma' \subseteq \Sigma$ and $E' \subseteq E$. The sorts, operation symbols and laws of Σ' and E' are called *primitive* and those of $\Sigma\setminus\Sigma'$ and $E\setminus E'$ *nonprimitive*. Analogously a term $t \in W(\Sigma';x1,\dots,xm)$ is called *primitive*. If $t \in W(\Sigma;x1,\dots,xm)$ where $s' \in S'$ then t is called of *primitive sort*; if $s' \in S\setminus S'$ then t is of *nonprimitive sort*. In general, there exist terms of primitive sort which are not primitive, i.e. $W(\Sigma')_{s'} \subsetneq W(\Sigma)_{s'}$ for $s' \in S'$.

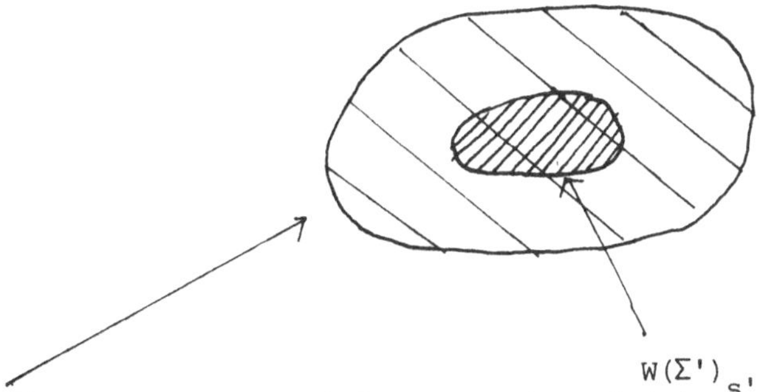

Figure Primitive terms and terms of primitive sort

Example 1 (continued)

NAT and BOOL are the primitive types of PQ(NAT). Thus

{bool,nat} are the primitive sorts and
{true,false,zero,succ,$\leq$} are the primitive operations.

The sort pqueue and the operations empty, .$\circ$., remove,min, nonempty are nonprimitive.

E.g. true and succ(zero)$\leq$zero are primitive terms whereas min(remove(zero $\circ$ empty)) and nonempty(zero $\circ$ empty) are (nonprimitive) terms of primitive sort. □

Now, an algebra A satisfies the hierarchy-constraints if the primitive carrier sets are built only by (interpretations of) primitive ground terms:

$$A_{s'} = \{t_A \mid t \in W(\Sigma')_{s'} \text{ and } t_A \text{ defined}\} \text{ for } s' \in S'$$

Formally, let A be a partial algebra of signature Σ and let Σ' be a subsignature of Σ. We mean by $A|\Sigma'$ the Σ'-reduct of A, i.e. the Σ'-algebra whose domains and operators are those of A named in Σ', and by $\langle A\rangle_{\Sigma'}$ the Σ'-subalgebra of A generated by the constants and operators named in Σ' viz. the smallest Σ'-subalgebra of $A|\Sigma'$.

Then a Σ-algebra A satisfies the hierarchy-constraints wrt. Σ' iff $A|\Sigma' = \langle A\rangle_{\Sigma'}$, i.e. the Σ'-reduct $A|\Sigma'$ of A is a Σ'-data structure. The carrier sets A_s of A where s is a primitive sort are finitely generated by the primitive constants and operations only. The class of all partial Σ-algebras which satisfy the hierarchy-constraints wrt. Σ' and the axioms of E is denoted by HPALG(Σ,E,P) of HPALG(T) of T = (Σ,E,P).
For simplicity we have assumed that T contains exactly one primitive type P which itself is not hierarchical. Types with a finite number P1,...,Pn of primitive types with signatures Σ1,..,Σn can easily be included by considering all hierarchy-constraints wrt. Σ1,..,Σn.

Furthermore, analogously to 3.5 we assume that the type BOOL is among the primitive types of T. Then the class $HPALG_0(T)$ consists of all algebras A of HPALG(T) which satisfy $A \models true \neq false$. HPGEN(T) and $HPGEN_0(T)$ are defined analogously. An algebra is said to be (strongly) initial or (strongly/weakly) terminal in the hierarchical type T if it is so in HPALG(T) or $HPALG_0(T)$, resp.

Examples 1 and 3 (continued)

For the initial and terminal algebras I and Z of PQ(NAT) we have

$$I|\Sigma_{NAT} \cong N \cong Z|\Sigma_{NAT}$$

Thus I and Z are hierarchical algebras in $HPGEN_0(PQ(NAT))$. In fact, all models of PQ(NAT) are hierarchical, i.e. HPGEN(PQ(NAT)) = PGEN(PQ(NAT)).

On the other hand, if we consider PQ(NAT) as primitive for our programming language PQL then PQL does not contain any hierarchical algebra: The term exec(FAC,empty,∅) (see 3.6, example (1)) is defined in all models but because of its infinite behavior it cannot be identified with any of the primitive terms. □

4.2 *Completeness of Hierarchical Types*

Since HPALG(T) is a subclass of PALG(T) every initial model I of the nonhierarchical type which is in HPALG(T) is an initial model of HPALG(T). **But it may happen that I is not a hierarchical model and then either HPALG(T) does not have any initial model or it has an initial model H not isomorphic to I. The situation for terminal algebras is analogous.**

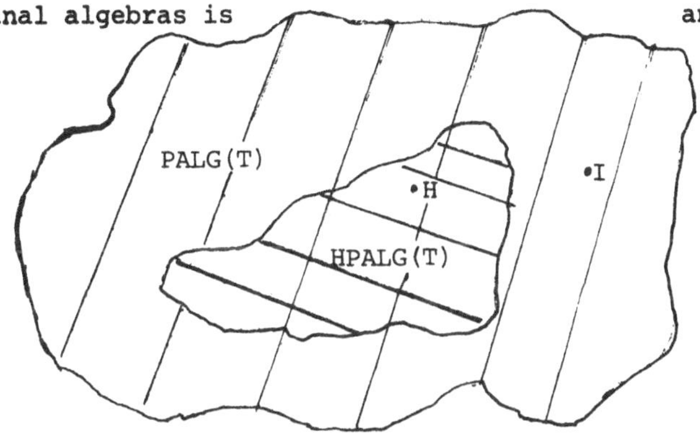

I ≡ initial model in PALG(T)   H ≡ initial model in HPALG(T)

Figure PALG(T) and HPALG(T)

Example 4

Consider the type NAT extended by some uniquely defined function

funct(nat)nat f

Then there exists a conditional specification of the formula $\exists x: f(x) \le z$. That is, we specify the function funct(nat)nat ex, defined by

$$ex(z) = \begin{cases} 1 & \text{if } \exists x \in \mathbb{N}: f(x) \le z \\ 0 & \text{otherwise} \end{cases}$$

To do this, we introduce in analogy to prop. 4 of Bergstra et al. [5] an auxiliary function

funct(nat,nat)nat ex'

by the axioms (where i stands for $succ^i(zero)$)

$$(*)\begin{cases} D(ex'(x,z)) \\ f(x)\le z \Rightarrow ex'(x,z) = 1 \\ (f(x)\le z) = false \Rightarrow ex'(x,z) = 2*ex'(succ(x),z) \end{cases}$$

If $f(x)>z$ for all $x \in \mathbb{N}$ then $ex'(x,z) = 0$ is the only possible solution of the axioms (*) in the natural numbers. Thus the following axioms

$$ex'(0,z) = 0 \Rightarrow ex(z) = 0$$
$$(ex'(0,z)\le 0) = false \Rightarrow ex(z) = 1$$

give the desired specification. All hierarchical models (wrt. NAT) of this type are isomorphic: The hierarchical type is monomorphic.
If e.g. $f(x) \neq 0$ holds for all natural numbers then $ex'(x,0) = 0$ cannot be proven and the carrier set I_{nat} of the nonhierarchical initial model contains the pairwise different elements $ex'(succ^i(zero),zero)_I$ for $i \in \mathbb{N}$ which are different from any "standard" element $succ^j(zero)_I$, too. □

A consequence of this example is the incompleteness of the proof system Π and of any other proof system for hierarchical types even wrt. ground equations and a fortiori wrt. arbitrary equations:

Proposition 8 [MacQueen,Sannella]
There does not exist any proof system ψ which is sound and complete wrt. ground equations of hierarchical types.

Proof We show that there exists a hierarchical type T such that for all sound proof systems ψ there exists a ground equation $t = t'$ of T with

$$\psi \nvdash t = t' \text{ but } \forall A \in HPALG(T): A \models t = t':$$

According to prop. 5 of Bergstra et al [5] there exists a hierarchical type T such that T specifies monomorphically an algebra H the equality $\sim_H$ of which is not recursively enumerable. Hence there exists a ground equation $t = t'$ such that $t \sim_H t'$ but $\psi \nvdash t = t'$. The monomorphicity of T implies

$$\forall A \in HPALG(T): A \models t = t' \qquad □$$

4.3 *Sufficient Completeness*

The equivalence of types with and without hierarchy-constraints is ensured by a slight generalization of Guttag's sufficient completeness condition:
A term t of primitive sort is called reducible (cf. Pair [32])

if there exists a primitive term p such that t = p is provable. **A hierarchical type T is called weakly sufficiently complete if every ground term t of primitive sort the definedness of which is provable is reducible; i.e. if D(t) is provable then there exists a primitive ground term p such that t = p is provable.**

Lemma 2
If T is weakly sufficiently complete, then

MDEF(HPGEN(T)) = MDEF(PGEN(T))

i.e. all minimally defined models of T are hierarchical.
Weak sufficient completeness is also a necessary condition:

Proposition 9
Let T be a hierarchical type. Then weak sufficient completeness is necessary and sufficient to ensure that all minimally defined models of T are hierarchical, i.e.

MDEF(HPGEN(T)) = MDEF(PGEN(T))

Proof The sufficiency follows from the previous lemma. Assume that all minimal models of T are hierarchical. Then the initial model I of PALG(T) is hierarchical; the congruence $\sim_I$ of I is defined by the complete proof system Π. Now let t be a ground term of primitive sort the definedness of which is provable. Then t is defined in I and there exists a primitive term p such that $t \sim_I p$. But then $\Pi \vdash t = p$ and t is reducible. □
Since initial and terminal algebras are always minimally defined data structures, one can apply the results of 3 to weakly sufficiently complete types:

Proposition 10
Let T be a weakly sufficiently complete hierarchical type.

(1.a) T possesses an initial model I which is minimally defined.
(1.b) If T is consistent and weakly maximal then it has a weakly terminal model Z which is minimally defined and the set of congruences

$$\{\sim_A \mid A \in \mathrm{MDEF}(\mathrm{HPGEN}_0(T))\}$$

forms a complete lattice wrt. $\subseteq$ with $\sim_I$ as minimum and $\sim_Z$ as maximum.

(2) If the definedness predicate is maximal, then
a) T has a strongly initial model I which is initial, too.
b) If T is consistent and weakly maximal, then a strongly terminal model Z exists and the set of congruences

$\{\sim_A | A \in HPGEN_0(T)\}$

forms a complete lattice wrt. $\subseteq$. Its minimum is $\sim_I$ and its maximum is $\sim_Z$.

Proof Put props. (4)-(6) together with lemma 2. The **t-maximality** condition is no longer necessary, since it is implied by the weak sufficient completeness together with the fact that BOOL is monomorphic. □

By restricting the form of the axioms one may even drop the requirement of (weak) sufficient completeness:

For a hierarchical type T = (Σ,E,P) where P = (Σ',E',P') we define the set F_0 of nonprimitive output functions by

$$F_0 = \{f \in F \setminus F' \mid f: s1 \times \ldots \times sn \rightarrow s' \text{ and } s' \in S'\}$$

Now, let $\Sigma^- = (S, F \setminus F_0)$. Then a term t is called output-normal (cf. Broy,Pair,Wirsing [41]), if t is in $W(\Sigma^-;x1,\ldots,xm)$ or t is of the form f(q1,...,qk) where $f \in F_0$ and $q1,\ldots,qk \in W(\Sigma^-;x1,\ldots,xm)$. A conditional formula with conclusion t = t' is output-normal, if both terms t and t' are output normal, and a conditional formula with conclusion D(t) is output-normal if t is in $W(\Sigma^-;x1,\ldots,xm)$.

Thus an axiom is output-normal if in its conclusion C either **nonprimitive output function does not occur or in case C has the form t = t' nonprimitive output operation symbols occur at most in outermost position.**

Lemma 3

Every hierarchical type with axioms in output-normal form is weakly sufficiently complete.

Proof Corollary 2 of Broy,Wirsing [7] or for more details Broy,Pair,Wirsing [41]. □

Weak maximality can be replaced by another sufficient condition: A type is called weakly monomorphic, if all its minimally defined models are isomorphic. For weakly monomorphic primitive types every equation between primitive terms is weakly maximal.

Corollary 1

Let T be a consistent hierarchical type with axioms in output normal form, weakly monomorphic primitive types and the terms in the premises of the axioms of primitive sort. Then

(1) If the definedness predicate is maximal, then T has strongly initial and strongly terminal models and the set of congruences $\{\sim_A | A \in HPGEN_0(T)\}$ forms a complete lattice wrt. $\subseteq$.

(2) T has initial and **weakly** terminal models and the set of congruences $\{\sim_A \mid A \in \text{MDEF}(\text{HPGEN}_0(T))\}$ forms a complete lattice wrt. $\subseteq$.

Finally, the hierarchical design of abstract types allows another characterization of weakly terminal models

A Σ-algebra A of a hierarchical type T is called fully abstract (cf. Milner [44]) if for all nonprimitive sorts s and all $x, x' \in A_s$:

$x = x'$ iff for all contexts K of primitive sort
$A \models K[x] = K[x']$, where a context $K[z]$ is an element of $W(\Sigma; z)$.

One can easily see (cf. theorem 3 of Broy,Wirsing [7]) that every minimally defined fully abstract model - if such a model exists - is weakly terminal. But one can also prove the converse.

Proposition 11
Let T be a consistent, weakly sufficiently complete hierarchical type.

(1) If the definedness predicate is maximal then a model of T is strongly terminal iff it is minimally defined and fully abstract.

(2) A model of T is weakly terminal iff it is minimally defined and fully abstract.

Proof Theorem 3 + prop. 5 of Broy,Wirsing [7].

Examples 1 and 3 (continued)

The axioms of PQ(NAT) are in output normal form. The premises in the axioms are of sort bool and BOOL is monomorphic and thus a fortiori weakly monomorphic. Since we have already shown consistency (by defining the model Z) and maximality of the definedness predicate, corollary 1.2 implies that PQ(NAT) has strongly initial and strongly terminal models and that its associated congruences form a complete lattice wrt. $\subseteq$.

The axioms of PQL are also in output normal form and their premises are of sort bool. Thus together with the consistency corollary 1.1 implies that PQL has initial and weakly terminal models and that its minimally defined congruences form a complete lattice wrt. $\subseteq$. □

4.4 *Persistency*

For a hierarchically designed abstract type we would like to implement the primitive type first and then to construct an implementation of the whole type based on this implementation. But the previous theorems guarantee only the existence of initial and terminal models of hierarchical types. They do not ensure that the reducts of these models (to the primitive type) are initial or terminal for the primitive type as well:

Example 5

```
type ABC' ≡
     primitive BOOL
     sort alpha
     alpha a,b,c
     laws: D(a),D(b),D(c)  endoftype,

type ABC" ≡
     primitive ABC'
     funct(alpha)bool isa
     laws: b = c
           isa(a) = true, isa(b) = false  endoftype
```

The type ABC" is monomorphic. Its models are characterized by

$$a \neq b \text{ and } b = c$$

But the (strongly) initial models of ABC' satisfy

$$a \neq b,\ b \neq c, \text{ and } a \neq c$$

and the (strongly) terminal models of ABC' satisfy

$$a = b = c$$

□

Therefore we define the notion of hierarchy-persistency. A type $T = (\Sigma, E, P)$ where $P = (\Sigma', E', P')$ is called hierarchy-persistent wrt. a class $C(P)$ of algebras if for every $A' \in C(P)$ there exists an algebra $A \in C(T)$ such that $A|\Sigma' = A'$. T is called weakly hierarchy-persistent if it is hierarchy-persistent for all minimally defined models of P, and hierarchy-persistent, if it is for all models of P.

On the syntactic level this property is modelled by the following consistency condition: T is called hierarchy-consistent if for all ground positive conditional Σ'-formulae G of the form

$$(**) \quad \bigwedge_i D(pi) \wedge \bigwedge_j D(qj) \wedge qj = rj \Rightarrow t = t' \wedge D(t)$$

the following holds:

$$T \vdash G \Rightarrow P \vdash G$$

where $T \vdash G$ means that $\Pi \vdash \bigwedge D(pi) \wedge \bigwedge (D(qj) \wedge qj = rj) \Rightarrow t = t'$ and $\Pi \vdash D(pi) \wedge (D(qj) \wedge qj = rj \Rightarrow D(t)$ holds.
Roughly speaking this means that T is a conservative extension of the primitive type. It implies the "i-" and "t-completeness" of [27] and is **sufficient and "almost" necessary:**

Proposition 12
Let T be a weakly sufficiently complete type. Then hierarchy-consistency is

(1) a sufficient condition for weak hierarchy-persistency,

(2) a sufficient condition for hierarchy-persistency if the definedness predicates of P are maximal,

(3) a necessary condition for hierarchy-persistency if HPGEN(P) = PGEN(P).

Recall HPGEN(P) = PGEN(P) roughly means that all models of P are hierarchical.

Proof
(1) Let A' be a minimally defined model of P and consider the proof system

$$\Pi' \equiv \Pi \cup \{v = v' \mid A' \models D(v) \wedge v = v' \text{ and } v, v' \in W(\Sigma')\} \cup \{D(v) \mid A' \models D(v) \text{ and } v \in W(\Sigma')\}$$

Then the algebra $A \equiv W(\Sigma)/\sim_{\Pi'}$ satisfies the axioms of T, is minimally defined (since A' is minimally defined) and is hierarchical wrt. Σ' since T is weakly sufficiently complete. Now, let t and t' be ground Σ'-terms and assume $A \models t = t' \wedge D(t)$. Then $t = t'$ and $D(t)$ are provable in Π' and every proof of $t = t'$ and of $D(t)$ contains only a finite number of ground formulas $vi = vi'$ and $D(vj")$ which hold in A'. Thus

$$T \vdash \bigwedge_j D(vj") \wedge \bigwedge_i D(vi) \wedge vi = vi' \Rightarrow t = t' \wedge D(t)$$

The hierarchy-consistency implies

$$P \vdash \bigwedge_j D(vj") \wedge \bigwedge_i D(vi) \wedge vi = vi' \Rightarrow t = t' \wedge D(t)$$

and since the premises of this formula hold in A':

$$A' \models t = t'.$$

By choosing $t = t$ as equation one obtains analogously

$$A \models D(t) \Rightarrow A' \models D(t).$$

Therefore $A|\Sigma'$ is isomorphic to A' and by applying the appropriate isomorphism one obtains an extension of A'.

(2) The maximality of the definedness predicate implies that all models of P are minimally defined. Hence, (1) is applicable.

(3) Let T be hierarchy-persistent and suppose there exists G with $T \vdash G$ and $P \nvdash G$. The completeness of first order logic implies that there exists an algebra $A' \in PALG(P)$ such that $A' \nvDash G$. Since G is a ground formula, $\langle A' \rangle_{\Sigma'} \nvDash G$. But $\langle A' \rangle_{\Sigma'} \in PGEN(P)$ and because of $HPGEN(P) = PGEN(P)$ $\langle A' \rangle_{\Sigma'} \in HPGEN(P)$. Since T is hierarchy-persistent there exists an extension $A \in HPGEN(T)$ of $\langle A' \rangle_{\Sigma}$. Thus $A \nvDash G$ which is a contradiction to $T \vdash G$. □

In many cases, hierarchy-persistent types do not only allow one extension of a model of the primitive type but a complete lattice of extensions:

Proposition 13

Let $T = (\Sigma, E, P)$ be a hierarchy-consistent, weakly sufficiently complete and weakly maximal type (where P is consistent). Then

(1) the minimally defined congruences of T form a complete lattice wrt. $\subseteq$ with the congruence associated to the initial model as minimum and the one of the weakly terminal model as maximum.

(2) the extensions of the congruence of any minimally defined model of P form a complete lattice wrt $\subseteq$. In particular, the congruence of the initial models of T is the minimum in the lattice of the initial models of P and the congruence of the weakly terminal models of T is the maximum in the lattice of the weakly terminal models of P.

Proof

(1) Hierarchy-consistency of T and the consistency of P imply the consistency of T. Thus it is sufficient to apply prop. 10.

(2) Let A' be a minimally defined model of P. Define

$$\begin{aligned}\tilde{E} \equiv\ & \{t = t' \mid A' \models D(t) \wedge t = t' \text{ and } t, t' \in W(\Sigma')\} \\ & \cup \{t = t' \Rightarrow true = false \mid A' \nvDash t = t' \text{ and } t, t' \in W(\Sigma')\} \\ & \cup \{D(t) \mid A' \models D(t) \text{ and } t \in W(\Sigma')\} \\ & \cup \{D(t) \Rightarrow true = false \mid A' \nvDash D(t) \text{ and } t \in W(\Sigma')\}\end{aligned}$$

and consider the type $\tilde{T} = (\Sigma, E \cup \tilde{E}, \tilde{P})$ where $\tilde{P} = (\Sigma', E' \cup \tilde{E}, P')$. The weak hierarchy-persistency implies that $\tilde{T}$ is consistent. From the definition of $\tilde{E}$ and the weak maximality of T one obtains the weak maximality of $\tilde{T}$ and the weak sufficient completeness of T together with the weak hierarchy-persistency imply the weak sufficient completeness of $\tilde{T}$. Now we can apply prop. 10. □

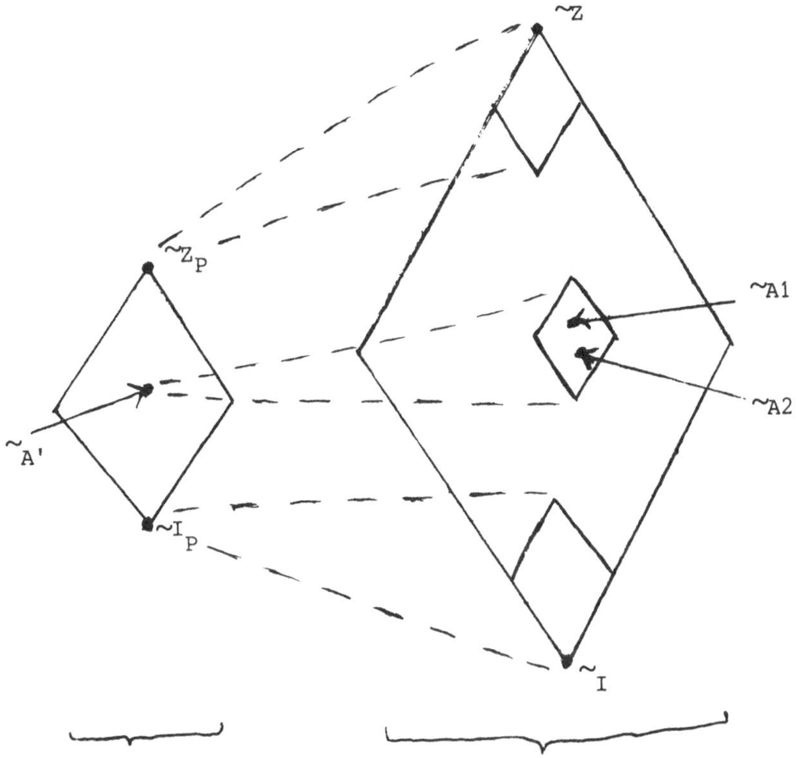

lattice of min. defined congruences of P	lattice of min. defined congruences of T
$I_P \equiv$ initial model of P	$I \equiv$ initial model of T
$Z_P \equiv$ weakly terminal model of P	$Z \equiv$ weakly terminal model of P
$A' \equiv$ some min. def. model of P	$A1, A2 \equiv$ some min. def. models of T with $A1\,|\,\Sigma' \cong A' \cong A2\,|\,\Sigma'$

Figure Minimally defined congruences of a hierarchy-persistent type

5. PARAMETERIZED TYPES

The examples PRIORITY-QUEUE and PQL do not describe a single type but they are type schemata which for any given type (satisfying the parameter requirements) specify a new type: A parameterized type T is a function from the class of algebraic types to the class of algebraic types. Parameterized types define classes of data structures and are therefore particularly well-suited for supporting the development of classes of algorithms being one of the main goals of program development by transformations (cf. Bauer, Wössner [2], CIP [13]).

In the following we study loose parameterized types with hierarchy-constraints. Thus the full power of CLEAR is included in this approach and even a bit more by dealing with terminal algebras, partial functions and conditional formulas. In order to manage this complexity we do not only give definitions of syntax and semantics of such types, but we also make a first attempt to develop a theory of parameterized types by studying conditions for the existence of particular models, for checking whether a type preserves hierarchy and parameters and whether it is persistent. As before we use as little of category theory as possible in order to extend and apply our previous results for nonparameterized types. Several parameter passing mechanisms for the compositions of parameterized types are discussed and it is shown that properties as e.g. sufficient completeness, output normal form, hierarchy-consistency are preserved under composition and type application guaranteeing the existence of initial and terminal algebras for types built of a large number of components.

5.1 *Syntax of Parameterized Types*

A <u>parameterized type</u> T = (PA,Σ,E,PR) consists of a signature Σ = (S,F), laws E, a (possibly empty) finite list PA of hierarchical types PARi = (Σ.pari,E.pari,PR.pari) and a (possibly empty) finite list PR of primitive parameterized types PRIMj = (PA.primj, Σ.primj,E.primj,PR.primj). The types PARi (i.e. i=1,...,n) are called <u>parameters</u> and their signature and laws have to be contained in Σ and E, i.e. S.pari $\subseteq$ S,F.pari $\subseteq$ F and E.pari $\subseteq$ E. The primitive types PRIMj (j=1,...,m) have to be subtypes of T, i.e. Σ.primj $\subseteq$ Σ,E.primj $\subseteq$ E, every parameter of (the parameter list) PA.primj has to be a subtype of one of the PARi and every primitive type of (the list) PR.primj has to be a (parameterized) subtype of PRIMj.
We do not require that the primitive parameter types are subtypes of the primitive types PRIMj of T. One reason is that for the existence of initial models weak sufficient completeness is needed only for the primitive types in PR and often not true for those of PA.

In the following we write

$R.p$ for $\bigcup_i R.pi$

where $R \in \{PA, \Sigma, E, PR\}$ and $q \in \{prim, par\}$, i.e. e.g.

$\Sigma.par$ for $\bigcup_{1\leq i\leq n} \Sigma.pari$ and

$E.prim$ for $\bigcup_{1\leq i\leq m} E.primi$.

A term $t \in W(\Sigma;x1,\ldots,xk)_s$ where $s \in \Sigma.par$ is of parameter sort; if, moreover, $t \in W(\Sigma.par;x1,\ldots,xk)$, then t is called parameter term. A nonprimitive parameter sort is a sort in $S.nppar \equiv S.par \setminus S.PR.par$ The constants, operation symbols and laws which are neither parameters nor primitive are called new. Otherwise they are called old. We define
$\Sigma.old = \Sigma.par \cup \Sigma.prim$, $E.old \equiv E.par \cup E.prim$

$\Sigma.new \equiv \Sigma \setminus \Sigma.old, E.new \equiv E \setminus E.old$

Example 1 (continued)

The type PRIORITY-QUEUE has one parameter, the type LINORD which is based on the primitive type BOOL. Therefore bool and elem are the two parameter sorts, elem is the only nonprimitive parameter sort. Σ.new is exactly the signature mentioned in the target of PRIORITY-QUEUE: the sort elem and the operations empty,nonempty,.•.,remove and min. □

Example 6

A standard example is the parameterized type of finite sets which is not specifiable without inequalities in any approach without hierarchy constraints (cf. Ganzinger [18] where an equivalence relation is specified instead of equality and ADJ [17]). In our approach the inequality true ≠ false is hidden by the exclusion of trivial algebras

```
type SET ≡
PARAMETER:
     type ELEM ⊇
          primitive BOOL
          sort elem
          funct(elem,elem)bool eq
          laws ∀ elem x,y:
             (1)  D(eq(x,y))
             (2)  eq(x,y) = true => x = y
             (3)  x = y => eq(x,y) = true
```

```
TARGET:
   primitive ELEM
   sort set
   set Ø
   funct(set,elem)set .U.
   funct(elem,set)bool .ε.
   laws ∀set s, elem x,y:
   x ε Ø = false
   x ε s U x = true
   eq(x,y) = false => x ε s U y = x ε s    endoftype
```

The parameter laws imply

eq(x,y) = true <=> x = y [by (2),(3)]

and also

eq(x,y) = false <=> x ≠ y

since eq(x,y) is defined for all x,y [by (1)] and its value has to be true or false because of the hierarchy-constraint for BOOL (cf. 4.1) □

5.2 *Syntax of Parameter Passing*

In order to apply a parameterized type to an actual parameter we have to specify the renaming of the sorts and functions of the formal parameter by the actual ones and to ask whether the actual parameter satisfies the "preconditions" i.e. the axioms and hierarchy-constraints of the formal parameter. The renaming will be described by a signature morphism and the latter by a theory morphism. As ADJ [17] we define a "generalized parameter passing" where the actual type may be a parameterized type itself; that is we describe the composition of parameterized types.
A signature morphism (cf. Sannellla [36]) maps the sorts and operations of one type into sorts and operations of another such that the functionality is preserved. A theory morphism (cf. Burstall, Goguen [11]) is a signature morphism preserving laws and hierarchy:
Let T = (PA,Σ,E,PR) and T' = (PA',Σ',E',PR') where PA and PR are lists of types of the form (Σ.par,E.par,Pr.par) and (PA.prim, Σ.prim,E.prim,PR.prim). PA' and PR' are analogously defined.
A <u>signature morphism</u> h: T → T' consists of two functions hS: S → S' and hF: F → F' such that for all f: s1 × ... × sn → s ∈ F the functionality of hF(f) is hS(s1) × ... × hS(sn) → hS(s). A <u>theory morphism</u> h: T → T' is a signature morphism such that

(1) E' ⊢ h(e) for all e ∈ E and

(2) for all types prim ∈ PR there exists a type prim' ∈ PR'.par ∪ PR' such that h is surjective, i.e. hS(S.prim) = S'.prim and hF(F.prim) = F'.prim', and (the restriction h.prim of) h is a theory morphism from prim to prim'.

Thus every theory morphism is a specification morphism (cf. ADJ [17]) which preserves the hierarchy.
Now, given a parameterized type T with parameters PA = (PAR1,...,PARn), a list ARG = (ARG1,...,ARGn) of actual (parameterized) types and n signature morphisms hi : PARi -> ARGi what should be the result T(ARG1,...,ARGn) after applying the "procedure" T to the "expressions" ARG1,...,ARGn?
Certainly, T(ARG1,...,ARGn) should be defined only if ARG satisfies the "preconditions"; or equivalently if all hi are **theory morphisms from PARi to ARGi. If this happens, the** procedure should work as follows:
Rename all PARi appropriately according to hi and produce the result signature Σ", the result laws E", the result parameters PA" **and the result primitive types PR".**

Formally, let $T = (PA,\Sigma,E,PR)$, $ARG = (PA.arg,\Sigma.arg,PR.arg)$ and let h be a theory morphism from PA into ARG. Then $T\bullet_h ARG = (PA'',\Sigma'',E'',PR'')$ (or T(ARG) for short) denotes the application of T to ARG via h and is defined as follows:
Let $h'': \Sigma \to \Sigma''$ be a signature morphism from Σ into some signature Σ'' such that h'' extends $hi: \Sigma.pari \to \Sigma.argi$, i.e. $h''(s) = hi(s)$, if $s \in S.pari$, $h''(f) = hi(f)$, if $f \in F.pari$: furthermore h'' introduces new names if this is necessary because of name clashes with the actual parameters. (For a formal treatment of this using tags see Sannella [36].)

The signature and laws of the result type $T\bullet_h ARG$ consist of signature and laws of the actual parameters together with the translation of signature and laws of T:

$$\Sigma'' \equiv h''(\Sigma) \cup \bigcup_i \Sigma.argi$$

$$E'' \equiv h''(E) \cup \bigcup_i E.argi$$

The parameters are those of ARG:

$$PA'' = \langle PA.argi \rangle$$

The primitive types consist of the primitive types of ARG and the type applications of the primitive types of T.
Let PRIMj have the parameter PARj1,...,PARjk. Then
$PR'' \equiv \langle PR.argi \rangle \circ \langle PRIMj \circ_h (ARGj1,\ldots,ARGjk) \rangle$

We call this parameter passing mechanism "call by Specification" (for a comparison with other call mechanisms see 5.4). A first consequence is the associativity of the composition of parameterized types:

<u>Corollary</u>
Let T0(PAR0), T (Par1) and T2 be parameterized types such that $T0\bullet_h T1$ and $T1\bullet_{h'} T2$ is defined. Then $(T0\bullet_h T1)\bullet_{h'} T2$ is defined and

$$(T0\bullet_h T1)\bullet_{h'} T2 = T0\bullet_{h'\bullet h} (T1\bullet_{h'} T2)$$

The <u>proof</u> is obvious from the construction of the type instantiation.

5.3 *Semantics of Parameterized Types*

The semantics of a parameterized type T(PA) is a function $\mathcal{T}$ which associates to every hierarchical type ARG (without parameters) and every theory morphism $h: PA \to ARG$ the result type $T\bullet_h ARG$:

$\mathcal{T}$: {hierarchical types} × {theory morphisms} → {hierarchical types}

$$(\text{ARG},h) \mapsto \begin{cases} \text{T}\bullet_h\text{ARG} & \text{if } h: \text{PA} \to \text{ARG} \text{ is a theory morphism} \\ \text{undefined} & \text{otherwise} \end{cases}$$

Or from our model-theoretic point of view the semantics of $\mathcal{T}$ is the function

$\mathcal{T}'$: {HPGEN(H) | H ∈ {hier. types}} × {theory morphisms}

→ {HPGEN(H) | hier. types},

$$(\text{HPGEN(ARG)},h) \mapsto \begin{cases} \text{HPGEN}(\text{T}\bullet_h\text{ARG}) & \text{if } h: \text{PA} \to \text{ARG} \text{ is theory morphism} \\ \text{undefined} & \text{otherwise} \end{cases}$$

where a "model-theoretic" theory morphism h: T → T' satisfies (1') for all A ∈ HPALG(T) : h(A) ∈ HPALG(T) instead of (1).

Remark:

(1) From a category theory point of view a parameterized type would be a functor between categories such thst the objects are classes of algebras and the morphisms are theory morphisms.

(2) For the development and implementation of specifications a notion of equivalence is used which abstracts from the particular names of sorts and operations (see Wirsing [43] and cf. the notion of institution in Burstall and Goguen [12]): T and T' are called equivalent of there exists a theory morphism from T to T' and vice versa.

Given a parameterized type - say PRIORITY-QUEUE - and an argument type - say NAT - we are interested in the structure of the model class of the result type PRIORITY-QUEUE(NAT); e.g. we would like to know whether there exist initial or terminal models of PRIORITY-QUEUE(NAT) or not. More generally, we would like to know for a parameterized type T under which conditions the result type T(ARG) has initial or terminal algebras for every well-behaved argument type ARG. We cannot require this for every type ARG since if ARG does not have initial or terminal algebras then in most cases T(ARG) does not admit such algebras either.
Therefore, according to the results in §3 we say that an argument type ARG is <u>well-behaved</u> if it is weakly sufficiently complete and if the set of congruences associated to the class $\text{MDEF}(\text{HPGEN}_0(\text{ARG}))$ of its minimally defined models is nonempty and forms a complete lattice wrt. set inclusion ⊆.

Then we say thet a parameterized type T(PA) has (or admits) initial models if for every well-behaved nonparameterized (hierarchical) type ARG and every theory morphism $h: PA \rightarrow ARG$ $T\bullet_h ARG$ has initial models. The notions of admitting strongly initial, strongly terminal and weakly terminal models are analogously defined.

Remark In our approach we do not need "parameterized models" in the sense of ADJ[17]. Such parameterized models are model schemata which often allow a simple description of e.g. the class of initial models of a parameterized type (as free construction) and thus are useful for special implementations of types.

Example 1 (continued)

For every well-behaved argument type M of PRIORITY-QUEUE a weakly terminal model A(M) of PRIORITY-QUEUE(M) can be defined as follows:

Let V be a weakly terminal model of M

$$A(M)_{bool} \equiv V_{bool}, A(M)_{elem} \equiv V_{elem}$$

$$A(M)_{pqueue} \equiv \{\langle v1,\ldots,vk\rangle \mid k\geq 0 \text{ and } v1,\ldots,vk \in V_{elem} \text{ and } v1\leq\ldots\leq vk\}$$

The parameter operations of A(M) are defined as follows:

$$true_{A(M)} \equiv true_V, false_{A(M)} \equiv false_V$$

$$zero_{A(M)} \equiv zero_V, succ_{A(M)}(v) \equiv succ_V(v)$$

$$v \leq_{A(M)} w \Leftrightarrow v \leq_V w$$

The nonprimitive operations are defined exactly as for the model Z of 3.2, e.g.

$$empty_{A(M)} \equiv \langle\rangle, nonempty(\langle v1,\ldots,vk\rangle) \Leftrightarrow k>0.$$

In fact, Z is a special instance of A(M) i.e. Z = A(Nat) where Nat denotes the standard model of the natural numbers.

The construction of A(M) is similar to those in Ganzinger [19] and Hornung, Raulefs [27]. It can be considered as a function A which for every argument type M produces a model of $T \circ_h M$. Thus A can be seen as "parameterized model" (cf. the remark above). □

5.4 *Other Parameter Passing Mechanisms*

The parameter passing mechanism we described so far is defined by textual substitution which is related to the syntax and not to the semantics of a type. But as in programming languages there exist other call-mechanisms: E.g. Burstall and Goguen [11] consider the theory of ARG instead of the axioms and ADJ [17] define the semantics by a free functor which is equivalent to consider all ground equations which are provable in ARG.
The signatures of the result type are the same for all approaches. The laws of the result type may always be defined by

$$E''_{call} \equiv h''(E) \cup CALL(ARG)$$

where only the (possibly infinite) set CALL(ARG) of laws varies. We characterize the different call-mechanisms according to the set CALL(ARG):

(1) Call by Specification
SPEC(ARG) is the set of laws of ARG:

$$SPEC(ARG) \equiv E.arg$$

Thus "call by specification" is the call mechanism of 5.2. It is a "call by name" mechanism considering the expression ARG (without "evaluating" the type).
All other mechanisms are variations of "call by value" considering the set of those formulas which determine the semantics of a type:

(2) Call by Theory
THE(ARG) is the set of all positive conditional formulas which hold in all hierarchical algebras:

THE(ARG) = {c | A ⊨ c for all A ∈ HPALG(ARG)
and c is a pos. cond. Σ.arg-**formula**}

(3) Call by Type

The only difference to call by theory is that now all formulas c are considered which hold in all models:

TY(ARG) ≡ {c | A ⊨ c for all A ∈ HPGEN(ARG)
and c is a pos. cond. Σ.arg-**formula**}

(4) Call by Algebra

Here only ground formulas are considered:

AL(ARG) ≡ {c | A ⊨ c for all A ∈ HPALG(ARG)
and c is a ground pos. cond. Σ.arg-**formula**}

Let us denote the resulting types by T"spec, T"the, T"by, and T"al. Then T"the corresponds to the parameter passing in CLEAR [36], T"spec to the syntax of parameter passing of ADJ/Ehrig [17] and T"al to their semantical construction.

Example 7

To see the differences between the parameter mechanisms consider the following trivial parameterized type:

```
type BL ≡
  PARAMETER:type BL' ⊇
            sort bl,
            bl tt,ff

TARGET: bl er
  laws: tt = er => tt = ff
        ff = er => tt = ff   endoftype
```

and the following actual parameters:

```
type BOOL' ≡
     sort bool
     bool true,false
     funct(bool)bool not,id
laws: ∀bool x
      id(x) = x
      not(true) = false
      not(false) = true    endoftype
```

Consider the **theory morphism** h **defined by** h(tt) = **true** and h(ff) = false. Then, "call by specification" leads to the type

```
type BOOL"spec   ≡
     sort bool
     bool true,false,er
     funct(bool)bool not,id
laws: ∀bool x.
  (1) not(true) = false, not(false) = true
  (2) id(x) = x
  (3) true = er => true = false
      false = er => true = false    endoftype
```

Using "call by type" one obtains (at least) one further law

(4) not(not(x)) = x

whereas "call by algebra" allows only to prove (1) and (3), but not (4) and instead of (2)

(2') id(true) = true, id(false) = false

That is, e.g.

BOOL"spec satisfies "id(er) = er" but not "not(not(er)) = er
BOOL"ty satisfies both "id(er) = er" and "not(not(er)) = er
BOOL"al satisfies neither "id(er) = er" nor "not(not(er))=er"
□

Therefore, "call by specification", "call by type" and "call by algebra" may lead to different result types, whereas call by specification and call by theory coincide.

FACT
(1) call by specification = call by theory
(2) If T and ARG are weakly persistent, then all call mechanisms coincide.

Proof
(1) SPEC(ARG) ⊂ THE(ARG) implies $HPALG(T_{THE}) \subseteq HPALG)T_{SPEC})$. Since all laws of THE(ARG) hold in all algebras of $HPALG(T_{THE})$ the converse is true, too.
(2) For weakly persistent types THE(ARG) = TY(ARG) = AL(ARG) holds. Thus we can apply (1) □

Hence, we can concentrate our discussions on the calls by specification, type and algebra.
"Call by algebra" has been introduced since in the initial algebra approach it corresponds to the construction of the free algebra over the (initial) argument algebras. The problem is that if the free algebra construction is not persistent

(cf. e.g. Ehrich [16]) the result algebra may not satisfy the axioms of the argument algebras. Only the relationships between ground argument types are preserved. This does not fit in the "extensibility"-philosophy for types (cf. Mosses [42]) where we would like new objects of a sort to be treated exactly as the old ones. E.g. in our type PQ'(NAT) (cf. end of 3.7) we need to apply the axioms (11) and (12) to "infinite" terms of form exec(...) which is not possible using "call by algebra".
The two other parameter passing mechanisms, call by specification and call by type, have the extensibility property. Call by type would correspond exactly to the "no junk" slogan by taking only (finitely generated) models into consideration. Moreover, it is independent from the particular axioms of a type, i.e. if E and E' are axiomatisations of the same class of algebras $C \subseteq$ HPGEN(Σ) and we apply a parameterized type T to (Σ,E) and (Σ,E') then HPGEN(T(Σ,E)) = HPGEN(T(Σ,E')). But on the other hand we may be able to prove too many formulas; for instance, if T(Σ,E) adds new constructors to (Σ,E) all Σ-formulas provable by structural induction in T(Σ,E) are provable without taking the new constructors into consideration (e.g. in the type BOOL"ty of example 7 not(not(x)) = x is provable by induction without using the constructor er).
In contrast to that, call by specification translates exactly those formulas which hold in all extensions of the argument types. Furthermore it can simulate call by type (but without the above mentioned drawback concerning induction) by **a general construction:**
One only has to extend the list PR **of primitive types of a parameterized type T by its list of parameter types PA. E.g. in example 7 we have to join**

primitive BL'

to the target of BL. Then in BOOLPAR".spec the formula not(not(x)) = x is provable, but now er = true $\vee$ er = false holds in all models of BOOL"spec. Call by specification is independent of the axiomatisation of HPALG but in general it is not independent of the axiomatisation of HPGEN. E.g. in example 7 the axiom

not(not(x)) = x instead not(true) = false $\wedge$ not(false) = true

for type BOOL' leads to a different BOOL'spec. But this dependence may happen only if the parameters are not considered as primitive or equivalently if the specification of the nonprimitive part of T is not "finished".
Therefore, we have chosen call by specification as parameter passing mechanism.

5.5 *Snowballing and Impeccable Types*

As for hierarchical types for parameterized ones we are interested to find conditions telling us how proper a specification is - e.g. whether a type admits initial and terminal models or whether the parameters are preserved under type application. To see which conditions might be interesting let us first analyze how the models of a parameterized type look after a type application to a "well-behaved" type:
A Σ-algebra A is called an extension of the Σ'-algebra A' if the Σ'-reduct $A|\Sigma'$ of A contains A' as a subalgebra, and persistent extension if $A|\Sigma' = A'$.

Now, consider a well-behaved argument type M, its image denoted by T(M), and its restriction $T(M)\upharpoonright M$ to M and, furthermore, a model Q of M, an extension Q" of Q in T(M) and its restriction $Q''\upharpoonright Q$ to M.

In general, T(M) does not form a complete lattice. Every model of $T(M)\upharpoonright M$ satisfies the axioms of M, but it might not be finitely generated by the constants and functions named in M. Thus it might happen that Q does not have any subalgebra of $Q''\upharpoonright M$.

If T(M) is consistent, weakly sufficiently complete and weakly maximal then T(M) forms a complete lattice but still extensions of Q might not exist or they might not be persistent. On the other hand, even if every model has a (persistent) extension, then T(M) might not form a complete lattice. What we would like is that

(1) T and therefore T(M) admits initial and weakly terminal models.
(2) (The minimally defined models of) T(M) form a complete lattice.
(3) Every (minimally defined) model Q of M admits a complete lattice of extensions.
(4) All extensions of minimally defined models of M are persistent.
(5) Implementation of primitive types can be constructed independently of the nonprimitive part i.e. hierarchy consistency (see 4.4).

Types which achieve all these properties except (4) will be called snowballing and types satisfying all properties will be called impeccable.

For hierarchical types the first step in order to obtain (semi)computable initial models was to ensure the equivalence between nonhierarchical and hierarchical minimally defined models. For this, weak sufficient completeness was a necessary and sufficient condition. For parameterized types we first give a semantic description of this situation:

A parameterized type T is called hierarchy-preserving if for all nonparameterized wellbehaved (cf. 5.3) types ARG and all theory morphisms h: PA → ARG $HPGEN_0(T\bullet_h ARG) = PGEN_0(T\bullet_h ARG)$, and T is called weakly hierarchy-preserving, if for all ARG and h as above $MDEF(HPGEN_0(T\bullet_h ARG)) = MDEF(PGEN_0(T\bullet_h ARG))$, i.e. if the result type $T\bullet_h ARG$ is weakly sufficiently complete.

It is parameter-preserving, if furthermore T does not produce terms of (nonprimitive) parameter sort which are not reducible to parameter terms:

T = (PA,Σ,E,PR) is called weakly parameter-preserving if the type T^+=(PA,Σ,E,(PR+PA)) is weakly hierarchy-preserving. Then the type T^+ is the type T where the list of primitive types has been extended by the parameter types of T.

To generalize the notion of weak sufficient completeness to parameterized types we introduce terms the only free variables of which are of nonprimitive parameter sort:

A term t is called quasi-ground, if all its free variables (if any) are of a sort in S.nppar (cf. 5.1).

Then an obvious generalization of the notions for hierarchical types would be to call type T "weakly sufficiently complete" if every quasiground term of primitive sort the definedness of which is provable is reducible and "weakly parameter complete" if the type T^+ is "weakly sufficiently complete".

For types without primitive types this characterizes weak parameter- and weak hierarchy-preserving (cf. Ganzinger [19]).

Proposition 14

Let T be a parameterized type without primitive types. Then T is weakly parameter-preserving iff T^+ is "weakly parameter complete". T is weakly hierarchy-preserving iff T is "weakly sufficiently complete".

Proof

"⇐" Put the weak sufficient completeness of the actual parameter ARG together with the one of T or the one of T^+ resp.

"⇒" Consider the type T augmented by an infinite set of constants for every nonprimitive parameter sort and apply to this hierarchical type prop. 9.

□

For types with hierarchy-constraints the situation is more intricate as we see from the priority-queue example.

Example 1 (continued)

Consider the term

t ≡ min(add(x1,add(x2,empty)))

of parameter sort elem. Then

$$(x1 \le x2) = true \Rightarrow t = x1$$
and $$(x1 \le x2) = false \Rightarrow t = x2$$

is provable, but there does not exist a quasi-ground parameter term p in $W(LINORD,X)_{elem}$ such that t = p is provable. □

Therefore we define a quasiground term t of old/primitive sort (cf. 5.1) to be <u>reducible</u> if there exists a finite number of positive conditional formulae containing exactly the variables occurring in t of the form Gi ⇒ t = pi where pi is a quasiground old/primitive term and Gi a old/primitive formula such that all Gi ⇒ t = pi are provable in T and D(t) ⇒ VGi is valid in T. Then a parameterized type T is called <u>weakly parameter complete/</u> <u>weakly sufficient complete</u> if every quasiground term t of an old sort/a primitive sort is reducible.

T is called <u>weakly structurally complete</u> if T is both weakly parameter complete and weakly sufficiently complete.

These properties are preserved under composition

<u>Fact</u>

Let T and ARG be parameterized types such that $T\circ_h ARG$ is well defined and T is weakly parameter complete.

(1) If ARG is weakly parameter complete then $T\circ_h ARG$ is weakly parameter complete

(2) If T and ARG are weakly sufficiently complete then $T\circ_h ARG$ is weakly sufficiently complete.

(3) If T and ARG are weakly structurally complete then $T\circ_h ARG$ is weakly structurally complete.

In the following let us always assume that the primitive types types of T are weakly hierarchy preserving and, if we prove something about parameter completeness, weakly parameter preserving. Then weak sufficient/parameter completeness is obviously a sufficient condition for weak hierarchy/parameter preserving:

Fact

Every weakly sufficiently complete type is weakly hierarchy-preserving and every weakly parameter complete type is weakly parameter-preserving.

It is not obvious and possibly false that weak sufficient completeness and weak parameter completeness are also necessary conditions as the following example indicates:

Example 8

We assume the type NAT of natural numbers to be given together with a boolean operation odd: nat -> bool, odd(n) = true, if n is odd, and odd(n) = false, otherwise.

```
type CARD1 ≡
PARAMETER:
     type CARD ⊇
       primitive NAT,BOOL
       sort copy
       funct(copy)nat f
       laws: ∀copy x,y
             D(f(x))
             f(x) = f(y) => x = y

TARGET:
     funct(copy)copy div
     laws: ∀copy x,y
           D(div(x))
           f(x) = 0 => div(x) = x
           f(x) = succ(f(y)) ∧ odd(f(x)) => div(x) = div(y)
           f(x) = succ(f(y)) ∧ odd(f(x)) = false =>
                                  f(div(x)) = succ(f(div(y)))
                                                    endoftype
```

For the parameter type CARD only one function f from copy into the natural numbers is required. This function has to be injective and therefore can be seen as a numbering of the elements of copy. W.l.o.g. we can assume that the carrier set B_{copy} of every model B of CARD has the form $\{a_i \mid 0 \leq i < |B_{copy}|\}$ where $|Y|$ denotes the cardinality of a set Y. Then $\bar{f}$ can be assumed to be the numbering function $f(b_i) = i$. Then one can prove that $div(b_i) = b_{i/2}$ where ./. denotes integer division.

Therefore CARD1 is parameter-preserving and persistent (see below) but to prove weak parameter completeness we need the function ./2 in NAT (which is not available) or an infinite number of formulae: For every $n \in N$

$$f(x) = succ^{2n}(zero) \wedge f(y) = succ^{n}(zero) \Rightarrow t(x) = y$$

$$f(x) = succ^{2n+1}(zero) \wedge f(y) = succ^{n}(zero) \Rightarrow t(x) = y \quad \square$$

Output normality carries easily over to parameterized types by specifying a "primitive output function" as an $f \in F.new$ with primitive range and a "parameter output function" as $f \in F.new$ with parameter range. Then T is <u>output-normal</u> if it is wrt. all primitive output functions, and <u>parameter-normal</u> if it is output normal wrt. all parameter output functions (cf. 4.3).

<u>Lemma 4</u>
Every parameterized type T with axioms in output-normal form is weakly sufficiently complete. If furthermore the axioms are in parameter normal form then T is weakly parameter complete.
<u>Proof</u> As lemma 3 □

Applying proposition 10.1 we get

<u>Proposition 15</u>
Every weakly sufficiently complete parameterized type T possesses initial models which are minimally defined. If T is weakly parameter complete, then for every well-behaved actual parameter ARG of T the reduct $I|\Sigma.arg$ of any initial model I of $T \circ_h ARG$ to the signature of ARG is a quotient of an initial model of ARG.

<u>Proof</u> The weak sufficient completeness of T implies the one of $T \circ_h ARG$ for all well-behaved actual parameters ARG of T. According to prop. 10.1 the initial model I of $T \circ_h ARG$ exists. If T is weakly parameter complete then obviously $I|\Sigma.arg$ is a minimally defined model of ARG. Therefore $T \circ_h ARG$ is a quotient of the initial model of ARG. □

In order to be able to ensure the existence of weakly terminal parameterized models the notions of consistency and maximality have to be generalized.

A parameterized type $T = (PA, \Sigma, E, PR)$ is called <u>weakly maximal</u> if all $e \in E$ are weakly maximal wrt. <u>quasiground</u> terms. The definedness predicate D is called maximal if D is maximal wrt. quasiground terms. T is called <u>consistent</u> if $T \bullet_h ARG$ is consistent for all well-behaved actual parameters ARG of T and all theory morphisms $h: PA \to ARG$.

<u>Proposition 16</u>
Let T be a consistent parameterized type which is weakly sufficiently complete and weakly maximal:

(1) T possesses weakly terminal models which are minimally defined.
If T is weakly parameter complete, then for every wellbehaved actual parameter ARG of T the reduct $Z|\Sigma.arg$ of any weakly terminal model Z of $T \bullet_h ARG$ to the signature of ARG is a weakly terminal model of ARG.
(2) If the definedness predicate of T is maximal, then T has strongly initial and strongly terminal models.

<u>Proof</u> The consistency, weak sufficient completeness, weak maximality and maximality of T imply the corresponding properties for all types $T \bullet_h ARG$ which are nonparameterized hierarchical types. Hence we can apply proposition 10. □

Weak maximality is a fairly strong condition.

It can be guaranteed as in corollary 1 by the weak monomorphicity of the primitive types:
Let PR^- be the set of all primitive types P of T and PA such that P does <u>not</u> contain any nonprimitive parameter sort.

<u>Proposition 17</u>
A parameterized weakly sufficiently complete type T is weakly maximal, if all premises of the axioms are of a sort of a weakly monomorphic type in PR^-.

<u>Proof</u> Every well-behaved type ARG is weakly sufficiently complete. Therefore $T \bullet_h ARG$ is weakly sufficiently complete wrt. PR^- and all minimally defined models are indistinguishable wrt. PR^-. Thus the weak monomorphicity of the primitive types occuring in the premises of the axioms implies the weak maximality. □
E.g. the types PRIORITY-QUEUE, SET and CARD1 (see examples 1, 6, 8) can be shown to be weakly maximal according to the above

proposition.
For nonparameterized types consistency follows from the consistency of the primitive types and hierarchy-consistency which on the other hand implies persistency. For parameterized types one obtains analogous results wrt. the parameters.
Let T = (PA,Σ,E,PR) be a parameterized types where PA = (Σ0,E0,PR0) and PR = (PA1,Σ1,E1,PR1). We denote by $T\restriction Ei \vdash H$ the fact that H is provable in T by using only parameter laws $e \in Ei$.

Now T is called <u>parameter/hierarchy-consistent</u> if for every <u>quasiground</u> positive conditional formula $G \Rightarrow t = t' \wedge D(t)$ with terms in W(Σ0,X)/W(Σ1,X) which is provable in T is already provable in $T\restriction E0$/$T\restriction E1$.

If T is both hierarchy-consistent and parameter-consistent then T is called <u>structurally consistent</u>.

Therefore, parameter consistency ensures that all quasiground equations between parameter terms which under certain assumptions are provable in T are already provable in the parameter type PA. Similarly hierarchy consistency guarantees that all proofs of quasiground equations between primitive equations can be carried out within the primitive type PR.

Obviously consistency is preserved under composition of types.

<u>Fact</u>

Let T and ARG be parameterized types such that $T\circ_h ARG$ is well-defined.

If T and ARG are parameter consistent / hierarchy-consistent / structurally consistent then $T\circ_h ARG$ is parameter consistent / hierarchy consistent / structurally consistent. □

For weakly structurally complete types parameter-consistency guarantees that all the result models of a parameterized type are extensions of its domain algebras whereas hierarchy-consistency implies that the result models are persistent extensions of their primitive subalgebras:

Proposition 18

Let T be a parameter-consistent, weakly structurally complete and weakly maximal type with a consistent parameter type PA.

(1) T has initial and weakly terminal models.

(2) Every minimally defined model Q of an actual parameter ARG of T has a persistent extension to a model of type T(ARG) and the congruences of its minimally defined extensions form a complete lattice wrt. $\subseteq$.

(3) If T is hierarchy-consistent, then every minimally defined model of a primitive type PR of T(ARG) has a persistent extension to a model of T(ARG) and the congruences of its minimally defined extensions form a complete lattice wrt. $\subseteq$.

Proof of proposition 18

(1) follows directly from (2).

(2) and (3): For every (nonparameterized) actual parameter ARG the type T(ARG) satisfies the assumption of proposition 13. □

The type PRIORITY-QUEUE satisfies the assumption of proposition 18. Hence for every actual parameter ARG, PRIORITY-QUEUE (ARG) preserves all minimally defined models (and in fact all models) of its primitive types and of ARG in a persistent way.

In order to obtain similar properties for types which are not weakly parameter complete we define a stronger notion of consistency:

A parameterized type T with parameter (ΣO,EO,PRO) is called extension-complete if for all universally quantified formulas G containing only quasiground parameter terms (i. e. terms in W(ΣO,X))

$$T \vdash G \Rightarrow T \upharpoonright EO \vdash G$$

holds.

Proposition 19

Let T be a extension-consistent, weakly sufficiently complete and weakly maximal type with a consistent parameter type PA.

(1) T has initial and weakly terminal models.
(2) Every minimally defined model Q of an actual parameter ARG of T can be extended to a model of type T(ARG) and the congruences of its minimally defined extensions form a complete lattice wrt. $\subseteq$.
(3) If T is hierarchy-consistent, then every minimally defined model of a primitive type PR of T(ARG) has a persistent extension to a model T(ARG) and the congruences of its minimally defined extensions form a complete lattice wrt. $\subseteq$.

Proposition I8 und I9 provide sufficient conditions for the existence of such types:
A type T(ARG) satisfies the snowballing conditions if the following five conditions hold: PAR is consistent, T is hierarchy- and extension-consistent, all axioms are in primitive output-normal form and the premises of the axioms are sorts of weakly monomorphic primitive types in PR^-.
T(PAR) satisfies the impeccability conditions if PAR is consistent, T is structurally consistent, weakly sufficiently complete and weakly maximal.
The conditions for snowballing are stronger than those for impeccability since in general, if T(PAR) is not parameter-complete, the result T(ARG) of an application of T to a type ARG is neither weakly sufficiently complete nor weakly maximal. Now, we are able to define snowballing and impeccable types:
A type T(PAR) is called snowballing if for every actual parameter ARG the type T is weakly hierarchy-persistent and every minimally

defined model Q of ARG has a nonempty complete lattice of congruences associated to its minimally defined extensions. A snowballing type is called impeccable if all minimally defined extensions of minimally defined models are persistent. T(ARG) is called totally snowballing/continually impeccable if the above conditions hold for all models of ARG and all extensions (not only for minimally defined ones).

Corollary 2

(1) If a type satisfies the snowballing conditions then it is snowballing and, if its definedness predicate is maximal, then it is continually snowballing.

(2) If a type satisfies the impeccability conditions then it is impeccable and if its definedness predicate is maximal, then it is continually impeccable.

E.G. the type PQL satisfies the snowballing conditions but it is not impeccable. The types PRIORITY-QUEUE and SET satisfy the impeccability conditions. CARD1 is impeccable, but it does not satisfy the corresponding conditions.

6. CONCLUDING REMARKS

Parameterized hierarchical partial abstract types provide a powerful framework for studying and specifying computational structures, relevant in computer science, be it for theoretical investigations or for practical software design. In particular, they allow for the formal specification of the semantics of programming languages (cf.[9], [20], [32]) and abstract description of hardware systems (which is the same from an abstract point of view) as well as for the structured design of complex software systems. Moreover the important steps in software development such as problem analysis and specification, stepwise implementation by given primitives, interpretation and compilation of programming languages and system integration can be very precisely described and analysed within this framework (see also [42]).

In the theory of abstract types nowadays a first consolidation of notions and definitions may be observed. Most of the theoretical foundations are at least roughly explored. Now it is the work of the future to apply this theory to areas of computer science, where other methods up to now have been cumbersome, such as the semantics of unconventional programming languages, and the comparison of semantic concepts and semantic models. Practical tools have to be invented including a morphology of types [2], type schemes and their implementation [15, 16, 28], a means of analysing and developing abstract types [14, 33], and tools for simulating them by term rewriting systems ("testing a specification") together with a support for verification [3].

Concentration on a deduction-oriented approach to algebraic types omitting the more complex model-theoretic or category-theoretic mathematical background should lead to practically useful "syntactic" principles for the design and the analysis of algebraic specifications (cf. [28, 30]).

ACKNOWLEDGEMENT

Thanks go to Don Sannella for interesting discussions and for reading a draft and to Eleanor Kerse for excellent typing.

REFERENCES

1 Andreka, H., Burmeister, B. and Nemeti, I.: "Quasivarieties of partial algebras - a unifying approach towards a two-valued model theory for partial algebras". TH Darmstadt, FB Math. Preprint Nr. 557.

2 Bauer, F.L. and Wössner, H.: "Algorithmische Sprache und Programmentwicklung". Berlin: Springer 1981

3 Bergman, M. and Deransart, P.: "Abstract data types and rewriting systems: application to the programming of algebraic abstract data types in Prolog". 6th CAAP, Genova, March 1981. LNCS 112, 1981.

4 Bergstra, J. and Tucker, J.: "Initial and final algebra semantics for data type specifications: two characterisation theorems". Math. Centre, Dept. of Comp. Science Report IW 142, Amsterdam, 1980.

5 Bergstra, J.A., Broy, M., Tucker, J.V. and Wirsing, M.: "On the power of algebraic specifications". 10th MFCS, 1981, LNCS 118, 1981.

6 Broy, M., Dosch, W., Partsch, H., Pepper, P. and Wirsing, M.: "Existential quantifiers in abstract data types". 6th ICALP, LNCS 71, 73-87, 1979.

7 Broy, M. and Wirsing, M.: "Partial abstract types". to appear in Acta Informatica. Preliminary version in: TU München, TUM-I8018, 1980.

8 Broy, M. and Wirsing, M.: "Partial-recursive functions and abstract data types". Bull. EATCS 11, June 1980.

9 Broy, M. and Wirsing, M.: "On the algebraic specification of nondeterministic programming languages". 6th CAAP, Genova, March 1981, LNCS 112, 1981.

10 Burstall, R.M. and Goguen, J.A.: "Putting theories together to make specifications". Proc. IJCAI, MIT, Cambridge, Mass. 1045-1058, 1977.

11 Burstall, R.M. and Goguen, J.A.: "The semantics of CLEAR: a specification language". Proc. Copenhagen Winter School on Abstract Software Specifications, 1980.

12 Burstall, R.M. and Goguen, J.A.: An informal introduction to specifications using CLEAR". In Boyer, R. and Moore, J. eds. The Correctness Problem in Computer Science. Academic Press, 1981.

13 CIP: Report on a wide spectrum language for program specification and development. TUM-I8104, May 1981.

14 Dosch, W., Wirsing, M., Ausiello, G., and Mascari, G.T.: "Polynomials - the specification, analysis and development of an abstract data type". 10. GI-Jahrestagung, Saarbrücken, Informatik-Fachberichte 33, 306-320, 1980.

15 Ehrich, H.D.: "On the theory of specification implementation and parameterization of abstract data types". Res. rep. Universität Dortmund, 1978. Short version: LNCS 64, 155-164, 1978.

16 Ehrig, H.: "Algebraic theory of parameterized specifications with requirements". 6th CAAP, Genova, March 1981,LNCS 112,1981.

17 Ehrig, H., Kreowski, H.J., Thatcher, J.W., Wagner, E.G. and Wright, J.B.: "Parameterized data types in algebraic specification languages". 7th ICALP, LNCS 85, 157-168, 1980.

18 Ganzinger, H.: "Parameterized specifications: parameter passing and implementation". To appear in TOPLAS.

19 Ganzinger, H.: "A final algebra semantics for parameterized specifications". Draft version, UC Berkeley, November 1980.

20 Gaudel, M.C.: Generation et preuve de compilateurs basées sur une sémantique formelle des langages de programmation. These d'Etat, Nancy, 1980.

21 Giarratana, V., Gimona, F. and Montanari, U.: "Observability concepts in abstract data type specifications". 5th MFCS, LNCS 45, 576-587, 1976.

22 Goguen, J.A.: "Abstract errors for abstract data types". IFIP Working Conference on Formal Description of Programming Concepts",St. Andrews, New Brunswick, 21.1 - 21.32, 1977.

23 Goguen, J.A. and Meseguer, J.: "Completeness of many-sorted equational logic". Manuscript, SRI International, Menlo Pk., 1980.

24 Goguen, J.A., Thatcher, J.W., Wagner, E.G. and Wright, J.B.: "Initial algebra semantics and continuous algebras". IBM-RC5701, 1975. JACM 24, pp. 68-95, 1977.

25 Goguen, J.A., Thatcher, J.W. and Wagner, E.W.: "An initial algebra approach to the specification, correctness and implementation of abstract data types". In: Current Trends in Programming Methodology IV. Prentice Hall, 80-144, 1978.

26 Guttag, J.V.: "The specification and application to programming of abstract data types". Ph.D. thesis, Univ. of Toronto, 1975.

27 Hornung, G. and Raulefs, P.: "Terminal algebra semantics and retractions for abstract data types". 7th ICALP, LNCS 85, 310-323, 1980.

28 Klaeren, H.A.: "On parameterized abstract software modules using inductively specified operations". Res. Rep. RWTH Aachen 66, 1980.

29 Liskov, B. and Zilles, S.: "Programming with abstract data types". Proc. ACM Sigplan Conference on Very High Level Languages, Sigplan Notices 9:4, pp. 55-59, 1974.

30 Loeckx, J.: "Algorithmic specifications of abstract data types". Univ. Saarbrücken, Rep. A 80/12/

31 MacQueen, D.B. and Sannella, D.T.: "Completeness of proof systems for equational specifications". In preparation.

32 Pair, C.: "Types abstraits et semantique algebrique des langages de programmation". Centre Recherche Informatique de Nancy, 80-R-011, 1980.

33 Partsch, H. and Broy, M.: "Examples for change of types and object structures". Proc. Summer School on Program Construction, Marktoberdorf 1978, LNCS 69, 1979.

34 Reichel, H.: "Theorie der Aequoide",Dissertation B. Humboldt-Universität Berlin, 1979.

35 Reichel, H.: "Initially restricting algebraic theories". 9th MFCS, Rydzyna, LNCS 88, 504-514, 1980.

36 Sannella, D.T.: "A new semantics for CLEAR". Dept. of Computer Science, University of Edinburgh, Rep. CSR-78-81, March 1981.

37 Thatcher, J.W., Wagner, E.G. and Wright, J.B.: "Data type specification: parameterization and the power of specification techniques". SIGACT 10th Annual Symposium on the Theory of Computing, 1978.

38 Wand, M.: "Final algebra semantics and data type extensions". Indiana University TR65, 1978.

39 Wirsing, M. and Broy, M.: "Abstract data types as lattices of finitely generated models". 9th MFCS, LNCS 88, 673-685, 1980.

40 Wirsing, M., Pepper, P., Partsch, H., Dosch, W. and Broy, M.: "On hierarchies of abstract data types". To appear in Acta Informatica. Preliminary version in: TU München, TUM-I8007,1980.

41 Broy, M., Pair, C. and Wirsing, M.: "A systematic study of models of abstract data types". In preparation.

42 Mosses, P.: "A semantic algebra for binding constructs". Proc. Formalization of Programming Concepts, LNCS 107, 408-419, 1981.

43 Wirsing, M.: "Structured algebraic specifications". Proc. AFCET Symposium on Math. for Computer Science, Paris March 1982.

44 Milner, R.: "Fully abstract models of typed λ-calculi". TCS 4, 1-22, 1977.

Theme IX.

(M. Wirsing)

When he stops thinking about abstract models and turns to more concrete matters, he decidedly prefers the French style; so he is represented by an adaptation of a theme by Jean Francaix

Part IV

INFINITE STRUCTURES

Semantics is in one way or the other the study of infinite trees: e.g., the tree of execution paths of an imperative sequential program or the expression tree of an applicative program. Therefore, programming methodology in heavily concerned with infinite trees. The first paper of this section is devoted to a detailed study of such trees with respect to algebraic and combinatorial questions. The methods employed are topological and order-theoretic. Substitutions are first-order and second-order. The second paper presents a theory of processes and synchronized systems of processes in terms of finite and infinite behaviors over an alphabet of actions.

FUNDAMENTAL PROPERTIES OF INFINITE TREES

To the memory of C.C.ELGOT

Bruno COURCELLE

University of Bordeaux I, FRANCE

ABSTRACT.- Infinite trees naturally arise in the formalization and the study of the semantics of programming languages. This paper investigates some of their combinatorial and algebraic properties that are especially relevant to semantics.

This paper is concerned in particular with regular and algebraic infinite trees, not with regular or algebraic sets of infinite trees. For this reason most of the properties stated in this work become trivial when restricted either to finite trees or to infinite words.

It presents a synthesis of various aspects of infinite trees, investigated by different authors in different contexts and hopes to be a first step towards a theory of infinite trees that could take place near the theory of formal languages and the combinatorics of the free monoid.

M. Broy and G. Schmidt (eds.), Theoretical Foundations of Programming Methodology, 417–471.

Introduction.

Infinite trees naturally arise in mathematical investigations on the semantics of programming languages. They arise in essentially one way : when one unloops or unfolds a program undefinitely. One obtains then either a tree of execution paths (infinite in general) in the case of a program written in an imperative language like FORTRAN or an expression tree in the case of a program written in an applicative language like LISP. In the latter case, the expression tree is usually infinite although its value can be finitely computed in each case ; this is possible by using if-then-else as a base function (like the addition of integers) and not as a piece of control structure. Once again, the infiniteness of the tree corresponds to the infiniteness of the set of possible computations.

In both cases, the semantics of the program is completely defined by the associated tree. Hence two programs are equivalent if the associated trees are the same (the converse being not true). Roughly speaking, this allows to distinguish between the equivalence of programs which is due to the control structure (loops, recursive calls, etc...) from the equivalence which is due to the properties of the domains of computation and the given functions on these domains.

It should be noted that these infinite trees are finitely defined. Hence we are lead to trying to decide whether two infinite trees defined in some finitary way are equal. Two types of infinite trees will be considered : the regular trees which are defined by unlooping FORTRAN-like program or flowcharts and the algebraic trees which are defined by unfolding recursive program schemes more or less derived from LISP programs.

We shall also introduce operations on trees : the first-order substitution which corresponds (roughly) to the sequential composition of flowcharts (by the operator ; of ALGOL) or to functional application (in the case of an applicative language). We shall also introduce the second-order substitution which corresponds to the replacement of a function symbol in an expression tree by some expression tree (intented to denote the same function). The theory of regular and algebraic trees will be developed for itself. The relevance to semantics will be shown with examples only in subsections (1.7) and (1.8).

Here is a brief survey of the content of the paper which is intented to be a synthesis of several aspects of infinite trees usually defined and studied separately for different purposes :

(1) Topological (i.e. metric) and order-theoretical properties of infinite trees are investigated in parallel in order to enlighten similarities and differences.

(2) First- and second-order substitutions are investigated in the two above frameworks.

(3) Regular trees, rational expressions defining them are studied. The concept of an iterative theory, due to C.C. Elgot, is one of the possible algebraic frameworks where to study infinite trees ; the set of regular trees forms the free iterative theory. Regular trees also arise as most general first-order unifiers in a generalized sense.

(6) Algebraic trees play a similar role with respect to second-order substitutions as regular trees with respect to first-order ones. Their combinatorial properties as sufficiently complicated to yield an open problem equivalent to the DPDA equivalence problem.

The following text will give precise, unifying definitions of all the above subjects. Two kinds of proofs will be omitted : those which are simple (but possibly long and tedious) checkings from the definitions, and those which are long and complex. References will be given as much as possible in the former case and always in the latter. Some of the given proofs are simpler than the original ones. Some of the stated results are "new" in the sense that they had never been published before (to the author's knowledge) but are not really difficult to establish and were probably known from the specialists. Other ones can be really claimed as new : for instance proposition (3.5.4) concerning second-order substitution and proposition (5.9.1) saying that if the most general unifier of two algebraic trees exists, then it is algebraic.

Acknowledgements : This paper is dedicated to the memory of C.C. Elgot whose contribution to the subject is essential as the text below will show.

I thank G. Cousineau, R. Tindell, S. Bloom and some of the participants of the Marktoberdorf summer school for comments which have been useful for the preparation of the final version of the paper. I thank Mrs. Polzin for the beautiful typing.

Due to a lack of space, large portions of the original text have been omitted at places indicated by .... The complete text is available from the author as report AAI-8105 and will be submitted for publication.

Author's mailing address :

Université Bordeaux I
U.E.R. de Mathématiques et Informatique
351, Cours de la Libération
33405 TALENCE CEDEX (FRANCE)

1. Basic definitions and examples.

In this section we make precise some mathematical notations, we define finite and infinite trees over a ranked alphabet and we show informally how infinite trees can be associated with program schemes of various types.

(1.1) Mathematical notations.

We denote by $\mathbb{N}_+$ ($\mathbb{N}$) the set of positive (non-negative) integers, by $[n]$ the interval $\{1,2,3,\ldots,n\}$ for $n \geq 0$ (with $[0] = \emptyset$).

The domain of a partial mapping $f : A \to B$ will be denoted by Dom(f). The restriction of f to a subset A' of A will be denoted by $f\restriction A'$.

(1.2) Definitions.

In order to define trees, we shall use ranked alphabets. A ranked alphabet is a pair (F,ρ) consisting of a set F , not necessarily finite, and a mapping $\rho : F \to \mathbb{N}$ which defines the rank of any symbol f in F .

For such a set F , we denote by F_i the set $\{f\in F/\rho(f)=i\}$, for $i\geq 0$.

In many cases the symbols in F will be considered as function symbols ; the rank of a function symbol is called its arity and a symbol of arity 0 is called a constant symbol (or a constant).

Following [31, 46, 30] we define a tree over a ranked alphabet F (the rank function will always be ρ) as a partial mapping $t : \mathbb{N}_+^* \to F$ such that :

(1.2.1) Dom(t) is prefix-closed i.e. if $\alpha,\beta\in\mathbb{N}_+^*$, $\alpha\beta\in$ Dom(t) then $\alpha\in$ Dom(t) ,

(1.2.2) if $\alpha\in\mathbb{N}_+^*$, $i,j\in\mathbb{N}$, $1\leq i\leq j$ and $\alpha j\in$ Dom(t) then $\alpha i\in$ Dom(t) ,

(1.2.3) if $t(\alpha) = f$ of arity $k \geq 0$ then, for $i \in \mathbb{N}_+$, $\alpha i \in \underline{\mathrm{Dom}}(t)$ if and only if $1 \leq i \leq k$.

We shall use the following terminology and notations :

$M^{\infty}(F)$ for the set of all trees over F,

$M(F)$ for the set of finite trees over F, i.e. of trees t having a finite set of nodes $\underline{\mathrm{Dom}}(t)$,

$\underline{\mathrm{First}}(t)$ for $t(\varepsilon)$, the label of the root of t,

$\underline{\mathrm{Occ}}(f,t)$ for $\{\alpha \in \underline{\mathrm{Dom}}(t) \;/\; t(\alpha) = f\}$, the set of occurrences of f in t,

t/α for the subtree of t issued from node α i.e. the tree $t' = \lambda \beta \in \mathbb{N}_+^{*}.t(\alpha\beta)$,

$\underline{\mathrm{Subtree}}(t) = \{t/\alpha \;/\; \alpha \in \underline{\mathrm{Dom}}(t)\}$,

$|t| = \underline{\mathrm{Card}}(\underline{\mathrm{Dom}}(t))$ (an element of $\mathbb{N}_+ \cup \{\infty\}$) is the size of a tree t.

Remark : Condition (1.2.2) follows from (1.2.3)

(1.3) Examples.

Let $F = \{c,f,g,h,k,a,b,v_1,v_2\}$ with $\rho(c)=3$, $\rho(f)=\rho(g)=2$, $\rho(h)=\rho(k)=1$, $\rho(a)=\rho(b)=\rho(v_1)=\rho(v_2)=0$.

(1) Let s be defined as follows :

$$s(\varepsilon) = f$$
$$s(1) = s(11) = k$$
$$s(111) = a$$
$$s(2) = h$$
$$s(21) = b\ .$$

It is depicted in figure 1 (see below).

(2) Let now t be defined by

$$t(\varepsilon) = f$$
$$t(1) = a$$
$$t(2^n) = g$$
$$t(2^n 1) = b \quad \text{for all} \quad n \geq 1 \; .$$

This infinite tree is shown on figure 2. The subtree $t/2$ consists of all the g's and b's of t .

(3) The tree u such that $u(1^n) = h$ for all $n \geq 0$ consists of one infinite branch. We identify it with the infinite word h^ω. See Nivat [40] and section (5.10).

(4) Our last example will be the tree w of figure 3. It can be defined as follows :

$$w(3^n) = c$$
$$w(3^n 1) = v_1$$
$$w(3^n 2 1^n) = v_2$$
$$w(3^{n+1} 2 1^m) = h \quad \text{for all} \quad n \geq m \geq 0 \; . \; \square$$

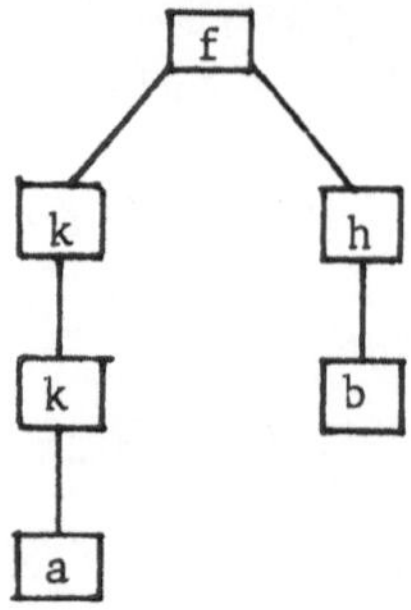

Figure 1

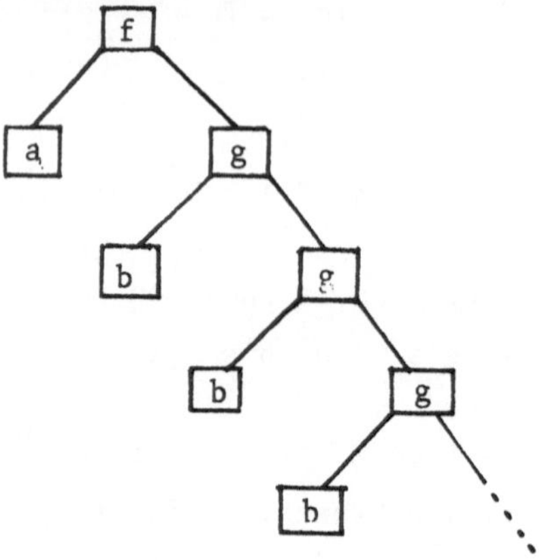

Figure 2

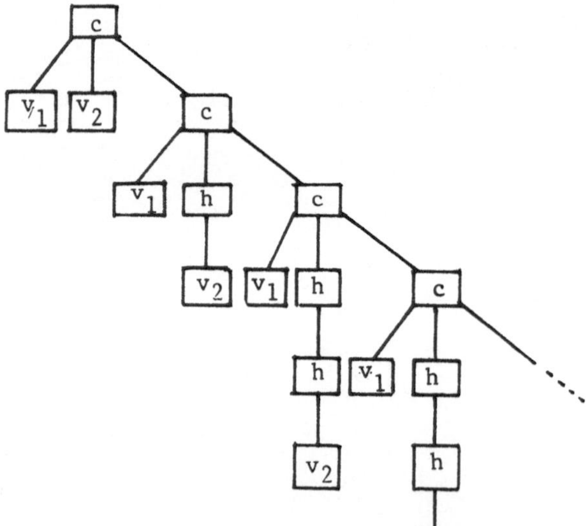

Figure 3

If we consider F as a set of function symbols, the finite trees over F can be identified with well-formed terms over F and written linearly with commas and parentheses. For instance, the tree t of example (1.3) can be written

$$f(k(k(a)),h(b))$$

From such a notation, one can infer the arities of the symbols f,k,a,h,b .

We shall also omit the parentheses surrounding the arguments of monadic (i.e. of arity 1) function symbols ; the above tree can also be written

$$f(kka,hb)$$

$$\text{or} \quad f(k^2a,hb) \; .$$

Within a proof or a theorem, we shall only write down well-formed trees and terms ; hence when declaring "let t be of the form $f(t_1,t_2,\ldots,t_n)\ldots$" we also declare that f is of arity n. And this allows n to be 0 (in this case $(t_1,\ldots,t_n)$ is the empty sequence i.e. $t = f$) .

(1.4) F-magmas, F-algebras.

The standard operation defined by a symbol f of F_k , $k\geq 0$ is the mapping $\bar{f} : M^\infty(F)^k \to M^\infty(F)$ such that $\bar{f}(t_1,\ldots,t_k) = t'$ where :

$$t'(\varepsilon) = f \; ,$$

$$t'(i\alpha) = t_i(\alpha) \text{ if } 1\leq i\leq k \; ,$$

$$t'(\alpha) \text{ is undefined otherwise.}$$

The mapping $\bar{f}$ maps $M(F)^k$ into $M(F)$. The notation of finite trees with commas and parentheses allows us to write

$$\bar{f}(t_1,\ldots,t_k) = f(t_1,\ldots,t_k)$$

i.e. to identify $\bar{f}$ with f.

We shall do the same for infinite trees and specify finite or infinite trees by "let $t = f(t_1,\ldots,t_k)$ for $t_1,\ldots,t_k$ in $M^{\infty}(F)$... ".

The definition of the $\bar{f}$'s makes $M(F)$ and $M^{\infty}(F)$ into F-magmas (equivalently F-algebras [30] but we prefer the former terminology since the terms algebra, algebraic are overused in mathematics). We shall not distinguish between the sets of trees $M(F)$ and $M^{\infty}(F)$ and the associated magmas, and we shall frequently do the same for arbitrary F-magmas.

It is well-known that $M(F)$ is the (isomorphic to any) initial F-magma. We shall denote by $eval_A$ the unique morphism of $M(F)$ into an F-magma A (we consider t as a syntactic object denoting the value $eval_A(t)$ of A).

We shall refer to the $\bar{f}$'s as the F-operations on $M(F)$ (and on $M^{\infty}(F)$) to contrast them with other operations to be introduced below.

....

(1.6) Representations of trees by languages.

The definitions given in (1.2) show that a tree t can be represented i.e. completely defined by an indexed family of languages $(\underline{Occ}(f,t))_{f\in F}$ which reduces to a tuple if the number of symbols of F occuring in t is finite, or by the single language $L(t) = \{\alpha f/\alpha\in \underline{Occ}(f,t),\ f\in F\} \subseteq \mathbb{N}_+^* F$.

Another representation has be defined by Rosen [47], and investigated in depth by Courcelle [12]. One represents a tree t by the language $\underline{Brch}(t)$ of its branches which is defined as follows.

An alphabet $\bar{F}$ is associated with a ranked alphabet F by :

$$\bar{F} = \{ [f,i] \ / \ f\in F,\ \ 1\leq i\leq \rho(f)\}\cup F_o \ .$$

Let t be a tree in $M^{\infty}(F)$.

For every $a\in F_o$, every $\alpha\in \underline{Occ}(a,t)$, let $\bar{\alpha}$ be the word

$$\bar{\alpha} = [f_1,i_1]\,[f_2,i_2]\ldots[f_n,i_n]a$$

where $\alpha = i_1 i_2 \ldots i_n$

$$f_j = t(i_1 i_2 \ldots i_{j-1}) \quad \text{for} \quad 1\leq j\leq n.$$

Then we define $\underline{Brch}(t)$ as the set of all such words $\bar{\alpha}$.

Note in particular that $\underline{Brch}(t) = \emptyset$ if t has no occurrence of any symbol in F_o . Hence $\underline{Brch}(t)$ does not represent all infinite trees, but only the locally finite ones (in the sense of [12]), defined as follows :

$$M^{loc}(F) = \{t\in M^{\alpha}(F)/ \text{ for all } \alpha \text{ in } \underline{Dom}(t), \text{ there exists } \beta \text{ such that } t(\alpha\beta)\in F_o\}$$

is the set of locally finite trees.

Note that $M(F) \subsetneq M^{loc}(F)$.

(1.6.1) Examples :

Let us use the trees of example (1.3).

(1) $L(s) = \{f, 1k, 11k, 111a, 2h, 21b\}$

$\underline{Brch}(s) = \{f_1kka,\ f_2hb\}$

(we use f_i for $[f,i]$ when $\rho(f) \geq 2$ and f for $[f,1]$ when $\rho(f)=1$ in our examples for more clarity).

(2) $\underline{Brch}(t) = \{f_1a, f_2g_2^n g_1 b \ /\ n \geq 0\}$.

Note that t is locally finite. Its language of branches is regular. We shall see that t is a regular tree.

(3) $L(u) = \{1^n h \ /\ n \geq 0\}$

$\underline{Brch}(u) = \emptyset$.

This tree is not locally finite. It is also a regular tree.

(4) $\underline{Brch}(w) = \{c_3^n c_1 v_1, c_3^n c_2 h^n v_2 \ /\ n \geq 0\}$.

This tree is locally finite. Its language of branches is context-free. We shall see that w is an algebraic tree.

(1.6.2) Proposition : For t, t' in $M^{\infty}(F)$:

(1) $t = t'$ if and only if $L(t) = L(t')$

(2) if $t \in M^{loc}(F)$, then $t = t'$ if and only if $\underline{Brch}(t) = \underline{Brch}(t')$.

It follows that each family of languages C naturally defines two families of trees :

$$C_L = \{t \in M^{\infty}(F) \ /\ L(t) \in C\}$$

$$C_B = \{t \in M^{loc}(F) / \underline{Brch}(t) \in C\}.$$

Certain classes of trees can be characterized in this way (see (4.11), (5.5) and Damm [23]).

(1.7) Flowchart schemes and infinite trees.

Consider the flowchart scheme S of figure 4 (see below). Its infinite unlooping yields the tree of figure 2.

An interpretation **I** for S is an object $\mathbf{I} = < D_I, f_I, g_I, a_I, b_I >$ consisting of

- a nonempty set D_I,
- partial mappings $a_I, b_I : D_I \to D_I$,
- partial mappings $f_I, g_I : D_I \to D_I \times \{1,2\}$ (the 1 and 2 correspond respectively to the left and right exits of actions f and g).

There corresponds to S and I a partial mapping $S_I : D_I \to D_I$ in an obvious way ($S_I(d) = d'$ if and only if there exists a sequence $d_o, d_1, d_2, \ldots, d_n$ of elements of D_I with $d_o = d$, $d_n = d'$, which corresponds to a "computation" of S in I ; note that f_I and g_I are not only "tests" since they can modify their data).
Such a model of computation is due to Elgot [25].

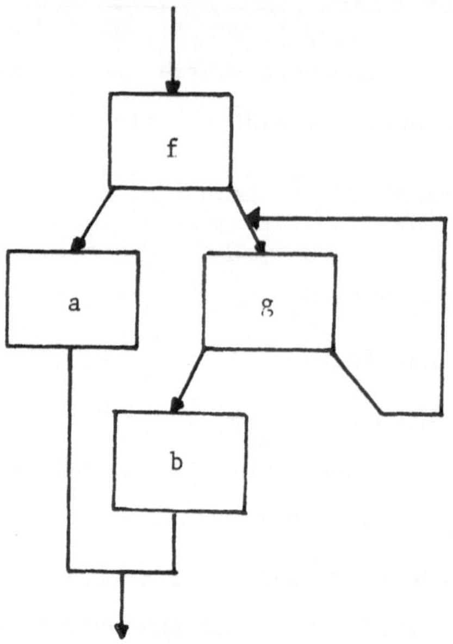

Figure 4

Let P be the following program (with integer variables) :

```
begin
x ← x+3 ;
y ← 2x+7 ;
if  x ≤ y  then  y ← 27x+3  else
    begin
    do  x ← x-8 ; y ← 10x  until  x≤0  od ;
    x ← 82y ;
    y ← 0 ;
    end ;
end
```

We do not claim that P computes anything interesting, but we chose it only as an example.

It "corresponds" to the pair (S,I) consisting of flowchart scheme S and interpretation I defined as follows :

$D_I = \mathbb{Z}^2$

$a_I((x,y)) = (x',y')$ with $y' = 27x+3$ and $x' = x$

$b_I((x,y)) = (x',y')$ with $x' = 82y$ and $y' = 0$

$f_I((x,y)) = ((x',y'),i)$ with $x' = x+3$, $y' = 2x+13$, and
$i = \underline{\text{if}}$ $x+3 \leq 2x+13$ $\underline{\text{then}}$ 1 $\underline{\text{else}}$ 2

$g_I((x,y)) = ((x',y'),i)$ with $x' = x-8$, $y' = 10x-80$ and
$i = \underline{\text{if}}$ $x \leq 8$ $\underline{\text{then}}$ 1 $\underline{\text{else}}$ 2.

By "corresponds" we mean in particular that the function computed by P is S_I .

The tree of figure 2, let us call it $t(S)$, can also be seen as an "infinite" flowchart scheme from which a partial function $t(S)_I$ can be defined as for finite schemes.

The two main facts are the following ones :

(1) $S_I = t(S)_I$ for all interpretation I

(2) for any two schemes S and S' :
$S_I = S'_I$ for all interpretation I i.e. S and S' are <u>equivalent</u> if and only if $t(S) = t(S')$.

(1.8) Recursive program schemes and infinite trees.

Let us consider the following (fancy) recursive definition :

$$K(x_1,x_2) = \underline{\text{if}}\ x_1 = x_2\ \underline{\text{then}}\ x_1+3\ \underline{\text{else}}\ 18.K(x_1,x_2-1)$$

It can be considered as an instance of the following recursive program scheme (see Guessarian [33]) :

$$\varphi(v_1,v_2) = c(v_1,v_2,\varphi(v_1,h(v_2)))$$

where the domain of computation is $D = \mathbb{Z}\cup\{\perp\}$ ($\perp$ means undefined) and :

$$h(x) = \lambda x\in D.[x-1]$$

$$c(x,y,z) = \lambda x,y,z \in D.[\text{if } x = y\ \underline{\text{then}}\ x+3\ \underline{\text{else}}\ 18.z]$$

A formal computation of φ i.e. a infinite unfolding of the recursion leaving c and h unevaluated yields the infinite tree of figure 3 as an expression tree $t(\varphi(v_1,v_2))$ denoting the semantics of $\varphi(x_1,x_2)$. As in the case of flowcharts :

(1) $K = \varphi_D$ and this function can be defined from $t(\varphi(v_1,v_2))$

(2) for any two recursive program schemes $\varphi(v_1,v_2)$ and $\varphi'(v_1,v_2)$, the functions φ_D and φ'_D are the same for all interpretation D i.e. φ and φ' are equivalent if and only if $t(\varphi(v_1,v_2)) = t(\varphi'(v_1,v_2))$.

These facts are investigated in depth in the book by I. Guessarian [33].

******

2. Topological and order-theoretical properties of trees.

We show that the set of infinite trees can be considered as a compact metric space and also as a complete partial order. In both cases an infinite tree can be considered as the limit (in some sense) of a sequence of infinite trees and this allows to extend "continuous" mappings from finite trees to infinite ones.

Hence a double theory of infinite trees can be developped either in the framework of topology or in that of the theory of ordered sets. In particular two universal characterizations of the F-magma of infinite trees can be given.

Since the introduction of infinite trees has been motivated by studies of semantics of programming languages (via program schemes) the order-theoretical approach has been developped first. It seems better suited for semantics (but this was not the opinion of Elgot since his theory of monadic computations [25] avoids orderings ; Arnold and Nivat also avoid orderings in [2]).

In the present paper where we investigate syntactical properties of trees, both of them are useful.

(2.1) Contracting magmas.

We shall not recall the basic definitions concerning metric spaces.

Let (E,d) and (E',d') be two metric spaces. We shall say that a mapping $f : E \to E'$ is contracting if there exists a real number c, $0 \leq c < 1$ such that, for all x, x' in E :

$$d'(f(x),f(x')) \leq c.d(x,x')$$

and we shall let $\|f\|$ be the greatest lower bound of all such numbers.

Hence a contracting mapping is uniformly continuous. Let $k \geq 2$; we shall also denote by d the distance on E^k which is defined by :

$$d((t_1,\dots,t_k),(t'_1,\dots,t'_k))=\underline{\mathrm{Max}}\ \{d(t_i,t'_i)/1\le i\le k\}.$$

An F-magma $A = \langle A,(f_A)_{f\in F}\rangle$ is contracting if

(1) A is a complete metric space with distance d_A,
(2) $d_A(x,y)\le 1$ for all $x,y\in A$,
(3) f_A is contracting for all f in F and
(4) $\|A\| = \underline{\mathrm{Sup}}\{\|f_A\| \,/\, f\in F\} < 1$.

A morphism of contracting F-magmas is a morphism of F-magmas which is uniformly continuous.

We state a useful lemma which justifies our interest in contracting mappings :

(2.1.1) Unique fix-point lemma : *Let E be a complete metric space. Every contracting mapping $f : E \to E$ has a unique fix-point.*

Proof : We are to show the existence and unicity of x in E such that $f(x) = x$.

Assume we have two such fix-points x and x'. Then $d(x,x') = d(f(x),f(x'))$ since x and x' are fix-points and $d(f(x),f(x')) \le c.d(x,x')$ since f is contracting. Hence $d(x,x') = 0$ and $x = x'$. For the existence, let x_o be any element of E. Let $x_n = f^n(x_o)$. We have

$$d(x_{n+1},x_n) \le c^n.d(x_1,x_o)$$
$$d(x_{m+n},x_n) \le c^n(c^{m-1}+c^{m-2}+\dots+1)\ d(x_1,x_o)$$
$$\le c^n(1-c)^{-1}.d(x_1,x_o)$$

hence $(x_n)_{n\ge 0}$ is a Cauchy sequence. It has a limit x and $x=f(x)$, by continuity. ¤

(2.2) $M^{\infty}(F)$ as a metric space.

Let t and t' be two elements of $M^{\infty}(F)$.

Let us define

$$\delta(t,t') = \infty \quad \text{if} \quad t = t'$$
$$= \underline{\mathrm{Min}}\{|\alpha| \,/\, \alpha \in \underline{\mathrm{Dom}}(t) \cap \underline{\mathrm{Dom}}(t'), t(\alpha) \neq t'(\alpha)\} \text{ if } t \neq t' .$$

(We omit the proof of the existence of $\delta(t,t')$ when $t \neq t'$).

Finally we let

$$d(t,t') = 0 \qquad \text{if} \quad t = t'$$
$$= 2^{-\delta(t,t')} \quad \text{if} \quad t \neq t' .$$

It is easy to show that d is a distance on $M^{\infty}(F)$ making it into a complete metric space [3 , 8 , 5]. This distance is even ultrametric. It is essentially the same as the distance that one puts on the ring of formal power series.

Note that $d(t,t') \leq 1$ for all t,t' in $M^{\infty}(F)$.

It can be shown that $M^{\infty}(F)$ is compact if and only if F is finite [3], that $M(F)$, the set of finite trees is a dense subset of $M^{\infty}(F)$ and that $M^{\infty}(F)$ is the topological completion of $M(F)$.

It follows in particular that a uniformly continuous mapping : $M(F)^k \to E$ where E is complete extends uniquely into a uniformly continuous mapping : $M^{\infty}(F)^k \to E$.

(2.2.1) Remark :

This also applies to a property $P(t)$ for t in $M^{\infty}(F)$ such that :

(1) $\forall t \in M(F) . P(t)$

(see (1.5.2)) and which is continuous in the sense that :

(2) if $t = \underline{\mathrm{Lim}}_{n\to\infty} t_n$ where $t, t_o, \ldots, t_n, \ldots, \in M^\infty(F)$ and $\forall n \in \mathbb{N}$. $P(t_n)$ holds then $P(t)$ holds too,

(i.e. which defines a continuous mapping from $M^\infty(F)$ into the discrete space {true, false}).

From (1) and (2) one can conclude that $P(t)$ holds for all t in $M^\infty(F)$. ¤

Let us now consider the F-operations on $M^\infty(F)$.

The mappings $\bar{f} : M^\infty(F)^k \to M^\infty(F)$ are contracting (with $\|f\| = 1/2$ if $\rho(f) \geq 1$, $\|f\| = 0$ if $\rho(f) = 0$) hence :

(2.2.2) Proposition : $M^\infty(F)$ is a contracting F-magma.

Let us answer to the natural question :

(2.2.3) Proposition : $M^\infty(F)$ is the initial contracting F-magma.

Proof : A very similar result is proved in Bloom and Patterson [8].¤

(2.2.4) Remark : The hypothesis that $\underline{\mathrm{Sup}}\{\|f_A\| \ / \ f \in F\} < 1$ that we made in the definition of a contracting F-magma is essential to insure proposition (2.2.3).

This hypothesis is not made in [8]. It follows that $M^\infty(F)$ is initial (in the corresponding category) if and only if $F - F_o$ is finite.

It is shown that there is no initial object if $F - F_o$ is infinite. ¤

Proposition (2.2.3) says that a tree t in $M^{\infty}(F)$ can be seen as a syntactic object denoting an element of A, where A is a contracting F-magma. We denote it by $\underline{eval}_A(t)$. Hence we also denote by $\underline{eval}_A$ the unique uniformly continuous extension to $M^{\infty}(F)$ of the mapping $\underline{eval}_A : M(F) \to A$ defined in (1.4)).

(2.3) Complete magmas.

An ω-complete F-magma A is an F-magma equipped with a partial order $\leq_A$ such that :

(1) A has a least element,

(2) every countable directed subset B (equivalently every increasing sequence) has a least upper bound $\underline{Sup}(B)$ and,

(3) the functions f_A's are monotone and ω-continuous, (i.e. preserve the Sup's of countable directed subsets).

Hence this concept coincides with that of ω-continuous F-algebra introduced in [30].

We call A complete if the least upper bounds are taken with respect to arbitrary directed sets in (2) and (3).

All properties we shall state below hold for both completeness concepts. The ω-completeness will be sufficient for dealing with trees.

We refer the reader to [21,51] for more details about partial orders and other possible concepts of completeness.

It is certainly not necessary to give the proof of the following well-known lemma :

(2.3.1) Least fix-point lemma : Let E be an ω-complete partial order with least element e ; let $f : E \to E$ be ω-continuous. The element $u_0 = \mathrm{Sup}\{f^n(e)/n \geq 0\}$ of E is the least fix-point of f in E and also, the least solution in E of the inequation $f(u) \leq u$.

We shall denote u_0 by $\mu x.f(x)$.

This applies to systems of equations since a system $S = < x_i = f_i(x_1,\ldots,x_n)\ ;\ 1 \leq i \leq n >$ where $x_i \in E_i$ for $1 \leq i \leq n$ can be considered as a single equation $x = f(x)$ to be solved in $E_1 \times E_2 \times \ldots \times E_n$ with $f((d_1,\ldots,d_n)) = (f_1(d_1,\ldots,d_n),\ldots,f_n(d_1,\ldots,d_n))$ for all d_1 in $E_1,\ldots,d_n$ in E_n .

Another fundamental lemma is the following one

(2.3.2) Lemma (Bekič [4]) : Let E and E' be two ω-complete partial orders, let $f : E \times E' \to E$ and $g : E \times E' \to E'$ be ω-continuous.

(1) The mapping $h : E' \to E$ defined by $h(y) = \mu x.f(x,y)$ is ω-continuous.

(2) The two systems $S = < x = f(x,y),\ y = g(x,y) >$ and $S' = < x = h(y),\ y = g(h(y),y) >$ have the same least solution in $E \times E'$.

In other words the least solution (x_0,y_0) of S can be defined by $y_0 = \mu y.g(h(y),y)$ and $x_0 = h(y_0) = \mu x.f(x,y_0)$.

Proof : Part (1) is easy. Let us sketch the proof of part (2).

Let $y_1 = \mu y.g(h(y),y)$. It is easy to verify that $(h(y_1),y_1)$ is a solution of S , hence $(x_o,y_o) \leq (h(y_1),y_1)$.

Since (x_o,y_o) is a solution of S , $x_o = f(x_o,y_o)$ hence $h(y_o) \leq x_o$ by definition of h . Hence $g(h(y_o),y_o) \leq g(x_o,y_o) = y_o$. Hence $y_1 = \mu y.g(h(y),y) \leq y_o$. Since h is monotone, $h(y_1) \leq h(y_o) \leq x_o$. Hence $(h(y_1),y_1) \leq (x_o,y_o)$.

Hence we have shown that $(x_o,y_o) = (h(y_1),y_1)$. ¤

(2.4) $M_\Omega^\infty(F)$ as an ω-complete partial order.

Let F be a ranked alphabet. Let Ω be a new symbol of arity 0 that we add to F .

For any complete F-magma A we shall define the value Ω_A of Ω as the least element of A .

Since Ω will play a special role, we shall use the notations

$M_\Omega(F)$ for $M(F\cup\{\Omega\})$ and

$M_\Omega^\infty(F)$ for $M^\infty(F\cup\{\Omega\})$.

We define a partial order on $M_\Omega^\infty(F)$ denoted by $\leq$ as follows :

$t \leq t'$ if and only if $\underline{Dom}(t) \subseteq \underline{Dom}(t')$ and for all α in $Dom(t)$, if $t(\alpha) \neq \Omega$ then $t'(\alpha) = t(\alpha)$.

It is fairly easy to show that $\leq$ is a partial order, that Ω is the least element of $M_\Omega^\infty(F)$ with respect to $\leq$.

Every directed subset A of $M_\Omega^\infty(F)$ has a least upper bound, $a = \underline{Sup}(A)$ in $M_\Omega^\infty(F)$ defined by :

$\underline{Dom}(a) = \cup \{\underline{Dom}(t) \ / \ t \in A\}$,

for all α in $\underline{Dom}(a)$:

$a(\alpha) = f \in F$ if $t(\alpha) = f$ for some t in A

$= \Omega$ if $t(\alpha) = \Omega$ for all t in A such that $\alpha \in \underline{Dom}(t)$.

The mappings $\bar{f}$ are continuous hence we can conclude that $M^{\infty}_{\Omega}(F)$ is an ω-complete F-magma (in fact a complete F-magma as well).

The following proposition is analogous to (2.2.3) :

(2.4.1) Proposition [30] : $M^{\infty}_{\Omega}(F)$ is the initial ω-complete F-magma.

(2.4.2) Proposition : Let E be an ω-complete partial order. Every monotone mapping $h : M_{\Omega}(F)^k \rightarrow E$ can be uniquely extended into an ω-continuous mapping : $M^{\infty}_{\Omega}(F)^k \rightarrow E$.

Hence, if A is an ω-complete F-magma the monotone mapping $\underline{eval}_A : M_{\Omega}(F) \rightarrow A$ extends uniquely to $M^{\infty}_{\Omega}(F)$. We also denote by $\underline{eval}_A$ its extension. This means that a tree t in $M^{\infty}_{\Omega}(F)$ denotes an element $(\underline{eval}_A(t))$ of A .

(2.4.3) Remark : Let us call a property $P(t)$ of trees in $M^{\infty}_{\Omega}(F)$ ω-continuous if $P(t)$ is true whenever it is true for all element of an increasing sequence t_n in $M^{\infty}_{\Omega}(F)$ with least upper bound t .

If P is ω-continuous, in order to establish $\forall t \in M^{\infty}_{\Omega}(F).P(t)$, it suffices to establish the validity of

$$\forall t \in M_{\Omega}(F).P(t)$$

for instance by means of structural induction (1.5.2).

3. Substitutions.

By introducing variables we shall make trees denote functions and not only values as we did up to now.

Then we shall define the first-order-substitution i.e. the substitution of trees for variables in another tree as a syntactic counterpart of the composition of functions and we shall state its basic properties.

We shall also introduce the second-order substitution i.e. the substitution of trees for function symbols in trees. This corresponds to replacing a function name by its definition everywhere its occurs in some tree.

When reducing trees to words (if $\rho(f)=1$ for all f in F) the first-order substitution reduces to the concatenation of words whereas the second-order one reduces to the homomorphism.

We shall use (possibily infinite) ranked alphabets F and G, not necessarly disjoint or distinct.

(3.1) Trees with variables.

Let V be a set of variables i.e. of symbols of arity 0. By using them together with F we can define the following sets of trees :

$$M(F \cup V) \text{ also denoted by } M(F,V)$$

$$M^{\infty}(F \cup V) \text{ also denoted by } M^{\infty}(F,V)$$

and similarly for $M_{\Omega}(F,V)$ and $M_{\Omega}^{\infty}(F,V)$.

When using the notation $M^{\infty}(F \cup V)$ we use the elements of V as constants, whereas we emphasize their special role (see below) when we use the notation $M^{\infty}(F,V)$.

If we need an enumeration of V we shall take $V=\{v_1,v_2,v_3,\dots,v_n,\dots\}$, $V_k=\{v_1,\dots,v_k\}$ and $V_o=\emptyset$.

Alternative sets of variables will be W, X, Y.

For t in $M^\infty(F,V)$ we define Var(t), the set of variables from V occuring in t i.e. $\underline{Var}(t)=\{v\in V\ /\ \underline{Occ}(v,t)\neq\emptyset\}$.

(3.2) Derived operators.

Let A be an F-magma.

It is clear that a tree t in $M(F,V_k)$ can be seen as denoting a mapping : $A^k\to A$. Such a mapping is called a derived operator (derived from the F-operators) and is denoted by $\underline{derop}_A(t)$.

The mapping $\underline{derop}_A(t)$ can be defined "point-wise" as follows :

(3.2.1)
$$\underline{derop}_A(t)\ (a_1,\dots,a_k)=\underline{eval}_{A'}(t)$$
$$A'=\langle A,(f_A)_{f\in F},(a_i)_{v_i\in V_k}\rangle$$

or "globally" by primitive recursion :

(3.2.2) $\underline{derop}_A(v_i)$ is the ith projection : $A^k\to A$,

$\underline{derop}_A(f)$ is the constant function equal to f_A for $f\in F_o$

$\underline{derop}_A(f(t_1,\dots,t_n))=\lambda a_1,\dots,a_k\in A.f_A(\alpha_1(a_1,\dots,a_k),\dots,\alpha_n(a_1,\dots,a_k))$ where $\alpha_i=\underline{derop}_A(t_i)$ for $i=1,\dots,n$.

Definition (3.2.1) says that $M(F,V_k)$ is the free F-magma generated by V_k. The same holds for V instead of V_k and for $M^\infty(F,V)$ when A is contracting and for $M^\infty_\Omega(F,V)$ when A is ω-complete. More precisely :

(3.2.3) Proposition : (1) $M(F,V)$ is the free F-magma generated by V.
(2) $M^{\infty}(F,V)$ is the free contracting F-magma generated by V.
(3) $M^{\infty}_{\Omega}(F,V)$ is the free ω-complete F-magma generated by V.

Definition (3.2.2) uses an obvious $(F\cup V_k)$-magma structure on $(A^k \to A)$, the set of total mappings : $A^k \to A$, with v_i denoting the ith projection. Hence it is based on the fact that $M(F,V_k)=M(F\cup V_k)$ the initial $(F\cup V_k)$-magma.

By defining $d'(\alpha,\alpha') = \mathrm{Sup}\{d(\alpha(a_1,\ldots,a_k),\alpha'(a_1,\ldots,a_k))/ a_1,\ldots,a_k \in A\}$ we make $(A^k \to A)$ into a contracting $(F\cup V_k)$-magma (if A is contracting).

By defining $\alpha \leq \alpha'$ iff $\alpha(a_1,\ldots,a_k) \leq \alpha'(a_1,\ldots,a_k)$ for all $a_1,\ldots,a_k$ we make it into an ω-complete F-magma (if A is ω-complete).

We can now state :

(3.2.4) Proposition : (1) If A is an F-magma, derop_A is the unique morphism : $M(F,V_k) \to (A^k \to A)$.
(2) If A is a contracting F-magma, derop_A is the unique morphism (of contracting F-magmas) : $M^{\infty}(F,V_k) \to (A^k \to A)$.
(3) If A is an ω-complete F-magma, derop_A is the unique ω-continuous morphism : $M^{\infty}_{\Omega}(F,V_k) \to (A^k \to A)$.

Proofs : (2) follows from the fact that $\mathrm{derop}_A : M(F,V_k) \to (A^k \to A)$ is uniformly continuous.

(3) follows from the fact that $\mathrm{derop}_A : M_{\Omega}(F,V_k) \to (A^k \to A)$ is monotone. □

(3.3) First-order substitutions.

Definition :

Let $t \in M^{\infty}(F,V)$, let $V' \subseteq V$ and let $\sigma(v)$ be a tree in $M^{\infty}(G)$ for all v in V' .

The result of the substitution of $\sigma(v)$ for $v \in V'$ in t is the tree t' defined as follows :

For all α in $\mathbb{N}_+^*$, $t'(\alpha)$ is defined if and only if :
either $\alpha \in \underline{Dom}(t)$, $t(\alpha) \notin V'$ and then $t'(\alpha) = t(\alpha)$
or $\alpha = \beta\alpha'$ for some $\beta \in \underline{Occ}(v,t)$ and some v in V for some α' in $\underline{Dom}(\sigma(v))$ and then $t(\alpha) = \sigma(v)(\alpha')$.

It can be checked that t' is a perfectly well defined tree in $M^{\infty}(F \cup G,(V-V'))$. We denote it by $\sigma(t)$ hence we consider σ as extended from V' to $M^{\infty}(F,V)$.

Such a mapping σ is called a first-order substitution.

We shall also use the notation $t\,[\sigma(v)/v \; ; v \in V']$.

In many cases, V' will be $\{v_{i_1},\ldots,v_{i_k}\}$ and we shall use the notation $t\,[u_1/v_{i_1},\ldots,u_k/v_{i_k}]$ with $u_j = \sigma(v_{i_j})$ for $1 \leq j \leq k$.

We shall also use the notation $t\,[u_1,\ldots,u_k]$ when $V' = \{v_1,\ldots,v_k\}$ is known from the context.

(3.3.1) Remark :

The restriction of a first-order substitution σ to $M(F,V)$ satisfies the following properties :

$\sigma(x) = x$ if $x \in F_o \cup V - V'$,

$\sigma(f(t_1,\ldots,t_k)) = f(\sigma(t_1),\ldots,\sigma(t_k))$ for $k \geq 0$ and $t_1,\ldots,t_k$ in $M(F,V)$.

This is easy to check from the definition.

......

(3.5) Second-order substitutions.

The second-order substitution consists in substituting trees for function symbols in trees.

Every first-order substitution can be viewed as a second-order one, but second-order substitutions are more difficult to study than first-order ones for the following reasons :

(1) The result of a second-order substitution cannot be easily defined as in the case of a first-order one (cf.(3.3)). Hence we shall not define them "directly" on all trees, but only on finite trees first, and this by primitive recursion.

(2) The extension to infinite trees is not at all obvious, due to a lack of continuity for certain erasing substitutions.

Our two approaches, namely the topological and the order-theoretical ones differ on this point.

Let us note that this very problem occurs when one wants to define homomorphisms of infinite words.

Definitions :

Let $t \in M(F,V)$, let F' be a subset of F ; for each f in F' , let $\nu(f)$ be an element of $M^{\infty}(G,V_{\rho(f)})$.

In a context where we use the constant Ω , we always assume that $\Omega \notin F'$.

The result of the substitution of $\nu(f)$ for $f \in F'$ in t is the tree $\theta(t)$ also denoted by $t\{\nu(f)/f \; ; \; f \in F'\}$ and defined as follows by induction on the structure of t :

if $t \in V$ then $\theta(t) = t$

if $t = f(t_1,\ldots,t_n)$ with $f \notin F'$ then $\theta(t)=f(\theta(t_1),\ldots,\theta(t_n))$

if $t = f(t_1,\ldots,t_n)$ then $\theta(t) = \nu(f)\,[\theta(t_1),\ldots,\theta(t_n)]$.

Such a mapping $\theta : M(F,V) \to M^{\infty}(F \cup G,V)$ is called a <u>second-order substitution on finite trees</u>. (If we say "let θ be a second-order substitution : $M(F,V) \to M^{\infty}(F \cup G,V)$...", we need not specify "on finite trees").

It is <u>erasing</u> if $\nu(f) \in V$ for some f in F' and nonerasing otherwise.

It $F' = \{f_1,\ldots,f_k\}$ we shall also use the notation $t\{\nu(f_1)/f_1,\ldots,\nu(f_k)/f_k\}$ for $t\{\nu(f)/f \; ; \; f \in F'\}$ and the notation $t\{\nu(f_1),\ldots,\nu(f_k)\}$ if the sequence $f_1,\ldots,f_k$ is known from the context.

(3.5.1) <u>Lemma</u> : <u>Let</u> θ <u>be a second-order substitution</u> : $M(F,V) \to M^{\infty}(F \cup G,V)$.

(1) θ <u>is nonerasing if and only if</u> $\|\dot{\theta}\| \leq 1$.

(2) <u>If we also assume that</u> $\Omega \in F$ <u>then</u> θ <u>is monotone</u>.

......

4. Regular trees

This section is devoted to regular trees. Such trees naturally arise in the process of "unlooping" flowcharts. They also appear as results of first-order unification in the generalized sense of Huet [37].

We shall characterize regular trees as solutions in $M^{\infty}(F)$ of certain systems of equations. Solving such systems equation by equation will allow us to denote regular trees by some kind of rational expressions (Cousineau [10]). We shall also characterize them as forming the free iterative theory generated by F (Elgot et al. [26]). Finally we shall characterize them in terms of their language of branches (Courcelle [12], section (1.6)) or their languages of occurrences (Ginali [29], section (1.6)).

All definitions of this section will given with respect to a fixed finite ranked alphabet F . The extension of all definitions and results to a sorted alphabet does not raise any difficulty except perhaps for notations. It will not be done.

(4.1) Definitions

A tree t is regular if the set Subtree (t) of all its subtrees is finite. We shall denote by R(F) the set of regular trees over F .

It is not difficult to establish the following properties :

(4.1.1) $M(F) \subsetneq R(F) \subsetneq M^{\infty}(F)$

(provided F contains at least two symbols of arity ≥ 1) ,

(4.1.2) R(F) is closed under the F-operations.

(4.1.3) Any subtree of a regular tree is regular.

We shall also use regular trees with variables. As for M(F,V) and $M^{\infty}(F,V)$ we shall use the notation R(F,V) for

$R(F \cup V)$ in order to specify which symbols are variables i.e. are subject to substitution. The following fact is a straightforward consequence of lemma (3.4.1) :

(4.1.4) $R(F,V)$ <u>is closed under substitution</u>.

(4.2) <u>Systems of regular equations</u>.

A <u>system of regular equations</u>, (we shall also say a <u>regular system</u>) is a finite system of the form $S=<x_1= u_1,\ldots,x_n= u_n>$ where $x_1,\ldots,x_n$ are the unknowns and $u_1,\ldots,u_n$ are elements of $F(\{x_1,\ldots,x_n\})$ i.e. are all of the form f for f in F_o or $f(x_{i_1},\ldots,x_{i_k})$ for f in F_k , $k\geq 1$.

A <u>solution</u> of S is an n-tuple $(t_1,\ldots,t_n)$ of trees in $M^\infty(F)$ <u>satisfying</u> the equations where each f in F has its standard meaning on $M^\infty(F)$ (see section 1).

(4.2.1) <u>Theorem</u> : <u>A regular system has a unique solution in $M^\infty(F)$. All components of this solution are regular trees. Every regular tree is a component of the unique solution of some regular system.</u>

<u>Proof</u> : The first two assertions will be proved later for more general systems of equations (theorem (4.3.1)).

Let t be a regular tree. For each element u of <u>Subtree</u>(t) let us introduce an unknown x_u . Let X be this set of unknowns. For each x_u in X we have to define an equation of the form $x_u= s_u$ for some s_u in $F(X)$. If $u= f(u_1,\ldots,u_k)$ for $u_1,\ldots,u_k$ in <u>Subtree</u>(t) we take $s_u= f(x_{u_1},\ldots, x_{u_k})$. The system of all these equations is regular, the family of trees $(u)_{x_u \in X}$ is a solution of this system, hence its (unique) solution. The component of this solution corresponding to x_t is clearly t .□

(4.2.2) Example :

Let $F=\{f,g,a,b\}$ with $\rho(f)=\rho(g)=2$, $\rho(a)=\rho(b)=0$.

The tree $t=f(a,g(b,g(b,g(b,\ldots)))$ is regular since Subtree$(t)=\{t,t_1,a,b\}$ where $t_1= g(b,g(b,g(b,\ldots)))$. It is the first component of the unique solution in $M^{\infty}(F)$ of the regular system $S= <x_o= f(x_2,x_1),\ x_1= g(x_3,x_1),\ x_2= a,\ x_1= b>$. ¤

(4.3) More general systems of regular equations.

A generalized system of regular equations (or a generalized regular system) is a finite system of the form $S=<x_1= u_1,\ldots, x_n= u_n>$ where $u_1,\ldots,u_n$ are elements of $M^{\infty}(F,X_n)$ and $X_n= \{x_1,\ldots,x_n\}$ is the set of unknowns.

It is in Greibach normal form if $u_i \notin X_n$ for all $i=1,\ldots,n$. Except in (4.10) below, all systems will be assumed in Greibach normal form.

A solution of S is an n-tuple of trees $(t_1,\ldots,t_n)$ in $M^{\infty}(F)$ such that $t_i= u_i[t_1,\ldots,t_n]$.

If a system S as above has regular right-hand sides i.e. if $u_1,\ldots,u_n$ belong to $R(F,X)$ we say that S is a system of extended regular equations or an extended regular system.

(4.3.1) Theorem : A generalized regular system has a unique solution in $M^{\infty}(F)$. All components of the solution of an extended regular system are regular.

Proof : Let $E = M^{\infty}(F)^n$ considered as a metric space. Then E is compact since $M^{\infty}(F)$ is since F is finite ; see (2-2)).

Let us associate with S the mapping $|S| : E\to E$ such that $|S|(t_1,\ldots,t_n)= (t'_1,\ldots,t'_n)$ with $t'_i= u_i[t_1,\ldots,t_n]$ for $i=1,\ldots,n$. Since none of the u_i's belongs to $\{x_1,\ldots,x_n\}$, the mapping $|S|$ is contracting i.e. there exists $0\leq c <1$ such that

$$d(|S|(s),\ |S|(t)) \leq c.d(s,t)$$

for all s,t in E . It follows that $|S|$ has a unique fixpoint i.e. that S has a unique solution in $M^{\infty}(F)$ (see(2.1.1)).

Let us now assume that the u_i's are regular. Let $(t_1,\ldots,t_n)$ be the solution of S . By lemma (3.4.1) we have :

$$\underline{\text{Subtree}}(t_i) = \underline{\text{Subtree}}(t_1) \cup \ldots \cup \underline{\text{Subtree}}(t_n) \cup \{t'[t_1,\ldots,t_n] \ / \ t' \in \underline{\text{Subtree}}(u_i)\}.$$

Furthermore, for all $i=1,\ldots,n$ we have :

(*) $\underline{\text{Subtree}}(t_i) \subseteq \{t'[t_1,\ldots,t_n] / \ \ j \in [n],\ t' \in \underline{\text{Subtree}}(u_j)\} = A$.

We show that t_i/w belongs to A for all $i \in [n]$, all $w \in \mathbb{N}^*$ by induction on the length of w .

If $w \in \underline{\text{Occ}}(u_i)$ then $t_i/w = (u_i/w)\ [t_1,\ldots,t_n]$ and belongs to A with $t' = u_i/w$ and $j=i$.

Otherwise, $w=w'w''$ for some occurrence w' of $v_{i'}$ in u_i and some occurrence w'' in $t_{i'}$, (since $t_i = u_i[t_1,\ldots,t_n]$). Since $u_i \notin \{x_1,\ldots,x_n\}$ we have $|w'|>0$ and $|w''|<|w|$. Hence $t_i/w = t_{i'}/w''$ which belongs to A by the induction hypothesis.

Since $\underline{\text{Subtree}}(u_j)$ has been assumed finite for all j , (*) shows that $\underline{\text{Subtree}}(t_i)$ is finite for all i . ∎

(4.3.2) Remark : Let us define a generalized system as proper if it has a unique solution.

In order to characterize the proper systems let us say that an unknown x_i of S (as above) is singular if $|S|^p(x_1,\ldots,x_n) = (s_1,\ldots,s_n)$ with $s_i = x_i$ for some $p \geq 1$ (such a p can be taken less than n if it exists).

It has been shown by Bloom et al. [6] that a system is proper if and only if it has no singular unknown.

Two singular unknowns x_i and x_j are <u>independent</u> if for all $p\geq 1$, $s_i \neq x_j$ where as above, $|S|^p(x_1,\ldots,x_n)= (s_1,\ldots,s_n)$.

All solutions of a nonproper system can be defined parametrically in a unique way in terms of arbitrary values given to independent singular unknowns. More details will be given in remark (4.9.8). ¤

(4.4) <u>Solving regular systems equation by equation.</u>

Let $S=<x_1= u_1,\ldots,x_n= u_n>$ be a proper generalized regular system. Let us single out its first equation. It can be solved in $M^\infty(F,\{x_2,\ldots,x_n\})$ by considering the unknowns $x_2,\ldots,x_n$ as constants. Let t_1 denote its solution i.e. the unique tree such that $t_1= u_1[t_1/x_1]$.

Let now S' be the system $< x_2= u'_2,\ldots,x_n= u'_n>$ where $u'_i= u_i[t_1/x_1]$ for $i=2,3,\ldots,n$. Let $(t'_2,\ldots,t'_n)$ be its solution.

(4.4.1) <u>Claim</u> : <u>The solution of</u> S <u>is the</u> n-<u>tuple</u> $(t'_1,\ldots,t'_n)$ <u>where</u> $(t'_2,\ldots,t'_n)$ <u>is the solution of</u> S' <u>and</u> $t'_1= t_1[t'_2/x_2,\ldots,t'_n/x_n]$.

The proof is very simple since we are dealing with systems having a unique solution. One just has to verify that $(t'_1,\ldots,t'_n)$ is a solution of S :

$$\begin{aligned} t'_1 &= t_1[t'_2/x_2,\ldots,t'_n/x_n] \\ &= u_1[t_1/x_1]\,[t'_2/x_2,\ldots,t'_n/x_n] \\ &= u_1[t_1[t'_2/x_2,\ldots,t'_n/x_n]/x_1\ ,\ t'_2/x_2,\ldots,t'_n/x_n] \\ &= u_1[t'_1/x_1\ ,\ t'_2/x_2,\ldots,t'_n/x_n]\ . \end{aligned}$$

For $i=2,\ldots,n$:

$$\begin{aligned} t'_i &= u'_i[t'_2/x_2,\ldots,t'_n/x_n] \\ &= u_i[t_1/x_1]\,[t'_2/x_2,\ldots,t'_n/x_n] \end{aligned}$$

$= u_i[t'_1/x_1,\ldots,t'_n/x_n]$.

with similar computations as for t'_1 . □

Arguing by induction on the number of equations, one can show that solving a system of n equations reduces to solving n single equations and composing appropriately their solutions.

This method is fully similar to the one used in language theory to solve regular systems of equations in terms of rational expressions. This suggests to do the same for regular trees.

(4.5) Rational expressions denoting regular trees.

Cousineau has defined in [10] a class of "rational expressions" in order to denote regular tree obtained as solutions of regular systems. Our presentation of his results differs substantially from his.

(4.5.1) Definition

We introduce on $M^\infty(F,V_k)$ a new operation named Star. For t in $M^\infty(F,V_k)-\{v_1\}$ we define Star(t) as the unique tree in $M^\infty(F,V_{k-1})$ such that :

$$\underline{\text{Star}}(t)= t[\underline{\text{Star}}(t)/v_1,v_1/v_2,\ldots,v_{k-1}/v_k]$$

The existence and unicity of Star(t) follows from theorem (4.3.1).

If $t\in M^\infty(F,\{v_1\})$ then $\underline{\text{Star}}(t)\in M^\infty(F)$ $(=M^\infty(F,V_o))$ and $t\in M^\infty(F)$ if and only if $\underline{\text{Star}}(t)=t$.

Note that Star(v_1) is not defined. We could define it as Ω , the "bottom" tree (see section 3) and this would be useful for expressing least solutions of possibly nonproper regular system (Ω is clearly the least solution of the equation $x_1= x_1$). We shall discuss this later (see (4.10)) but we restrict here our attention to regular systems in Greibach normal form which have unique solutions.

Remark finally that the star operation depends on a precise set of variables, here $V=\{v_1,v_2,\ldots,v_n,\ldots\}$ which will be kept fixed in this section.

(4.5.2) Lemma : Let $u\in M^{\infty}(F,V_{k+1})-\{v_1\}$ and $t_1,\ldots,t_k\in M^{\infty}(F,V_{\ell})$. Then $\underline{Star}(u)[t_1,\ldots,t_k]=\underline{Star}(u[t'_1/v_2,\ldots,t'_k/v_{k+1}])$ where $t'_i=t_i[v_2/v_1,\ldots,v_{\ell+1}/v_{\ell}]$ for $i=1,\ldots,k$.

Proof : We have $\underline{Star}(u)=u[\underline{Star}(u)/v_1,v_1/v_2,\ldots,v_k/v_{k+1}]$. Let $u'=\underline{Star}(u)[t_1,\ldots,t_k]$. Lemma (3.4.2) gives us :

$$(1)\qquad u'=u[u'/v_1,t_1/v_2,\ldots,t_k/v_{k+1}].$$

On the other hand, $\underline{Star}(u[t'_1/v_2,\ldots,t'_k/v_{k+1}])$ is the unique tree w in $M^{\infty}(F,V_{\ell})$ such that :

$$(2)\qquad w=u[t'_1/v_2,\ldots,t'_k/v_{k+1}][w/v_1,v_1/v_2,\ldots,v_{\ell}/v_{\ell+1}].$$

Since $t'_1,\ldots,t'_k$ have no occurrence of v_1 , (2) can be written :

$$(3)\qquad w=u[w/v_1,t''_1/v_2,\ldots,t''_k/v_k]$$

where

$$t''_i=t'_i[v_1/v_2,\ldots,v_{\ell}/v_{\ell+1}]\quad\text{for}\quad i=1,\ldots,k\,.$$

By definition of t'_i :

$$(4)\qquad \begin{aligned} t''_i&=t_i[v_2/v_1,\ldots,v_{\ell+1}/v_{\ell}][v_1/v_2,\ldots,v_{\ell}/v_{\ell+1}]\\ &=t_i\,. \end{aligned}$$

Since $u\neq v_1$, the same holds for $u[t_1/v_2,\ldots,t_k/v_{k+1}]$ hence the equation

$$x=u[x/v_1,t_1/v_2,\ldots,t_k/v_{k+1}]$$

has a unique solution in $M^{\infty}(F,V_{\ell})$. This shows together with (1), (3) and (4) that $u'=w$. ◻

(4.5.3) Definition.

A rational expression is (in this paper) an element e (or $e_1, e' \ldots$) of $M(F \cup \{*\}, V)$. The tree $\underline{Val}(e)$ it possibly denotes can be inductively defined as follows :

if $e = v_i$ then $\underline{Val}(e) = v_i$,

if $e = f(e_1,\ldots,e_k)$ then $\underline{Val}(e)=f(\underline{Val}(e_1),\ldots,\underline{Val}(e_k))$

if each of $\underline{Val}(e_1),\ldots,\underline{Val}(e_k)$ is defined and $\underline{Val}(e)$ is undefined otherwise,

if $e = *(e')$ then $\underline{Val}(e) = \underline{Star}(\underline{Val}(e'))$ if $\underline{Val}(e')$ is defined and is not v_1 and $\underline{Val}(e)$ is undefined otherwise.

We say that e is defined (undefined) if $\underline{Val}(e)$ is.

It is easy to check that e is undefined if and only if it contains a subexpression of the form $*(*(\ldots*(v_k))\ldots))$ with k occurrences of $*$.

We can already solve some regular systems : for instance let $S=\langle x_1= f(x_1,g), x_2= h(x_1,x_2)\rangle$. Its solution is (e_1,e_2) i.e. more precisely the pair $(\underline{Val}(e_1),\underline{Val}(e_2))$ where $e_1= *(f(v_1,g))$ and $e_2= *(h(*(f(v_1,g)),v_1))$.

In order to apply the method of (4.4) and solve arbitrary systems we need a way to form a rational expression $\underline{Comp}(e,e_1,\ldots,e_k)$ having the value $\underline{Val}(e)[\underline{Val}(e_1),\ldots,\underline{Val}(e_k)]$. It is easy to check that taking $e[e_1/v_1,\ldots,e_k/v_k]$ would be incorrect.

(4.5.4) Definitions.

Let $\underline{Comp}(e,e_1,\ldots,e_k)$ be the rational expression defined as follows by induction on the structure of e :

if $e=v_i$ and $1 \le i \le k$ then $\underline{Comp}(e,e_1,\ldots,e_k)= e_i$

if $e=v_i$ and $i>k$ then $\underline{Comp}(e,e_1,\ldots,e_k)= v_i$

if $e=f(e'_1,\ldots,e'_\ell)$ then $\underline{Comp}(e,e_1,\ldots,e_k)= f(e''_1,\ldots,e''_\ell)$

where $e_i''=\underline{Comp}(e_i',e_1,\dots,e_k)$ for $i=1,\dots,\ell$

if $e=*(e')$ then $\underline{Comp}(e,e_1,\dots,e_k)= *(\underline{Comp}(e',v_1,e_1'',\dots,e_k''))$ where $e_i''= \underline{Comp}(e_i,v_2,v_3,\dots,v_{\ell+1})$ for all $i=1,\dots,k$ and ℓ is large enough such that $\underline{Val}(e_i)\in M^{\infty}(F,V_\ell)$ for all $i=1,\dots,k$.

In the last clause above, we shall put $\underline{Shift}(e_i)$ instead of e_i'' where $\underline{Shift}$ is a mapping on regular expressions acting as $\lambda e.\underline{Comp}(e,v_2,v_3,\dots,v_{\ell+1})$ but which can be defined directly.

Actually, we shall define $\underline{Shift}(k,e)$ for $k\geq 1$ and we shall take $\underline{Shift}(e) = \underline{Shift}(1,e)$:

$$\underline{Shift}(k,v_i) = v_i \quad \text{if} \quad i<k$$
$$= v_{i+1} \quad \text{if} \quad i\geq k$$

$$\underline{Shift}(k,f(e_1,\dots,e_\ell)) = f(\underline{Shift}(k,e_1),\dots,\underline{Shift}(k,e_\ell))$$

$$\underline{Shift}(k,*(e)) = *(\underline{Shift}(k+1,e)) .$$

(4.5.5) Claim : <u>For all defined rational expression</u> e <u>and all integer</u> $k\geq 1$, $\underline{Val}(\underline{Shift}(k,e))=\underline{Val}(e)[v_{k+1}/v_k,v_{k+2}/v_{k+1},\dots,v_{\ell+1}/v_\ell]$ <u>where</u> ℓ <u>is such that</u> $\underline{Val}(e)\in M^{\infty}(F,V_\ell)$.

<u>Proof</u> : By induction on the structure of e . We only consider the case $e = *(e')$ where $\underline{Val}(e')\in M^{\infty}(F,V_{\ell+1})$. Then :

$$\underline{Val}(e)[v_{k+1}/v_k,\dots,v_{\ell+1}/v_\ell]$$

$$= \underline{Star}(\underline{Val}(e'))[v_{k+1}/v_k,\dots,v_{\ell+1}/v_\ell]$$

$$= \underline{Star}(\underline{Val}(e'))[v_1,v_2,\dots,v_{k-1},v_{k+1},\dots,v_{\ell+1}]$$

(1) $$= \underline{Star}(\underline{Val}(e')[v_2/v_2,\dots,v_k/v_k,v_{k+2}/v_{k+1},\dots,v_{\ell+2}/v_{\ell+1}]$$

$$= \underline{Star}(\underline{Val}(e')[v_{k+2}/v_{k+1},\dots,v_{\ell+2}/v_{\ell+1}])$$

(2) $$= \underline{Val}(*(\underline{Shift}(k+1,e')))$$

$$= \underline{Val}(\underline{Shift}(k,e)).$$

We have used lemma (4.5.2) to obtain (1) and the induction hypothesis to obtain (2). ¤

(4.5.6) Claim : If $e, e_1, \ldots, e_k$ are rational expressions having values then $\underline{Val}(\underline{Comp}(e,e_1,\ldots,e_k)) = \underline{Val}(e)[\underline{Val}(e_1),\ldots,\underline{Val}(e_k)]$.

Proof : By induction on the structure of e. Once again the only interesting case is $e = *(e')$. Then :

$$\underline{Val}(\underline{Comp}(e,e_1,\ldots,e_k)) = \underline{Val}(*(\underline{Comp}(e',v_1,\underline{Shift}(e_1),\ldots,\underline{Shift}(e_k))))$$
$$= \underline{Star}(\underline{Val}(e')\,[t_1'/v_2,\ldots,t_k'/v_{k+1}])$$

by induction and (4.5.5), with $t_i' = \underline{Val}(e_i)[v_2/v_1,\ldots,v_{\ell+1}/v_\ell]$.
Hence, by lemma (4.5.2) :

$$\underline{Val}(\underline{Comp}(e,e_1,\ldots,e_k)) = \underline{Star}(\underline{Val}(e'))\,[\underline{Val}(e_1),\ldots,\underline{Val}(e_k)]$$
$$= \underline{Val}(e)\,[\underline{Val}(e_1),\ldots,\underline{Val}(e_k)]. ¤$$

(4.5.7) Theorem : $R(F,V)$ is the set of values of rational expressions. For any extended regular system of equations, one can find rational expressions defining its solution.

Proof : The value of a rational expression is a regular tree : this is an easy consequence of theorem (4.3.1) and the definition of Star.

Conversely, let $S = \langle x_1 = u_1, \ldots, x_n = u_n \rangle$ be a regular system. We shall construct an n-tuple of rational expressions denoting its solution.

Actually, we shall do the construction in a more general case, where S is an extended regular system where $u_i = \underline{Val}(e_i)[x_1/v_1,\ldots, x_n/v_n]$ for some rational expressions $e_1,\ldots,e_n$. (It is useful not to identify x_i and v_i as it will appear soon).

If $n=1$ then the solution of S is $t_1 = \underline{Val}(*(e_1))$.

Otherwise, we solve S by following the method of (4.4) (and using the same notations). It is clear that

$t_1 = \underline{\text{Val}}(*(e_1))\,[x_2/v_1,\ldots,x_n/v_{n-1}]$.

We let $e_i' = \underline{\text{Comp}}(e_i, *(e_1), v_1, \ldots, v_{n-1})$ for $i=1,\ldots,n$ so that the system S' of (4.5) is exactly $\langle x_2 = u_2', \ldots, x_n = u_n' \rangle$ with $u_i' = \underline{\text{Val}}(e_i')\,[x_2/v_1,\ldots,x_n/v_{n-1}]$ for $i=2,\ldots,n$.

By induction, we can assume that we know rational expressions $e_2'',\ldots,e_n''$ defining the solution of S' and we need only compute $e_1'' = \underline{\text{Comp}}(e_1, e_2'', \ldots, e_n'')$ to obtain an n-tuple $(e_1'', e_2'', \ldots, e_n'')$ defining the solution of S . That $t_i' = \underline{\text{Val}}(e_i'')$ follows from claim (4.4.1) and the induction hypothesis for $i=2,\ldots,n$ and from claims (4.4.1) and (4.5.6) for $i=1$. ¤

(4.5.8) Example : Let S be the system :

$$x = f(x)$$
$$y = g(x,y,z)$$
$$z = h(y,z) \quad .$$

Solving the first equation gives us

$$x = *(f(v_1))$$

(for simplicity, we identify a rational expression with its value). Then the system S reduces to the following two equations :

$$y = g(*(f(v_1)),\ v_1, v_2)\ [y/v_1, z/v_2]$$
$$z = h(v_1,v_2)\ [y/v_1, z/v_2] \quad .$$

By defining e as $g(*(f(v_1)), v_1, v_2)$, we get

$$y = *(e)\ [z/v_1]$$

and we are reduced to solve

$$z = h(*(e), v_1)\ [z/v_1] \quad .$$

We now obtain the final expressions for z and y :

$$z = *(h(*(g(*(f(v_1)), v_1, v_2)), v_1))$$

$y = \underline{\text{Comp}}(*(e), e')$ where e' is the rational exp. defining z,

$y = *(g(\ *(f(v_1)), v_1,\ *(h(\ *(g(\ *(f(v_1)), v_1, v_2)), v_1)))$.

......

5. Algebraic trees.

This section investigates *algebraic trees*. Such trees are interesting for at least two reasons. They naturally arise in the study of recursive program schemes (modeled after systems of mutually recursive functional (i.e. applicative) procedures), when one "unfolds" the recursion *ad infinitum* in order to characterize by means of a unique infinite tree what in the function defined by a recursive program scheme depends on the interpretation. Another reason is their deep connection with deterministic languages through their branch languages. Whereas "all properties" of regular trees are decidable (see (4.17)), many problems on algebraic trees are undecidable and others are decidable (in particular the equality problem), if and only if the equivalence problem for DPDA's is.

As in section 4, F will denote a fixed ranked alphabet. The extension to a many sorted alphabet is immediate and needs not be done formally.

(5.1) Systems of algebraic equations.

In order to define systems of algebraic equations, we shall use the operations of composition and tupling introduced in (4.6). Moreover, since we shall only use 1-trees we shall use the notations T_k for $T_{1,k} = M^{\infty}(F,V_k)$ and $T_{\Omega,k}$ $M^{\infty}_{\Omega}(F,V_k)$. The symbol Φ with always denote a finite ranked alphabet the elements of which will be used as unknowns in algebraic systems.

The set of *scalar monomials of type* p *over* a ranked alphabet G is the set $SM_p(G)$ of expressions inductively defined as follows :

$e \in SM_p(G)$ if and only if :

either $e = \pi_{i,p}$ for some i in $[p]$

or $e = f.(e_1,\ldots,e_k)$ for some f in G_k and some $e_1,e_2,\ldots,e_k$ in $SM_p(G)$.

Hence, for every algebraic theory A, if for all $k \geq 0$, every f in G_k denotes an element $\nu(f)$ of $A_{1,k}$, then for all $p \geq 0$, every scalar monomial e in $SM_p(G)$ denotes an element $e_{A,\nu}$ of $A_{1,k}$, inductively defined in an obvious way.

Let Φ be the ranked alphabet $\{\phi_1,\ldots,\phi_n\}$ (the same notation will be kept in this chapter). A system of algebraic equations (or an algebraic system) over F in the set of unknowns Φ is a system of the form $\Sigma = \langle \phi_1 = e_1,\ldots,\phi_n = e_n \rangle$ where $e_i \in SM_{k_i}(F \cup \Phi)$ and k_i denotes $\rho(\phi_i)$ for all $i=1,\ldots,n$.

A solution of Σ is an n-tuple $(t_1,t_2,\ldots,t_n)$ in $T_{k_1} \times T_{k_2} \times \ldots \times T_{k_n}$ such that $t_i = e_{iT,\nu}$ where naturally, $\nu(f) = f(v_1,\ldots,v_k)$ and $\nu(\phi_j) = t_j$ for all f in F_k and j in $[n]$.

Such a system is in Greibach normal form if the left-most symbol of each e_i is in F.

An alternative way of writing a system Σ as above (used for instance in many works on recursive program schemes [22,39,15,32,33]) is

$$\langle \phi_1(v_1,\ldots,v_{k_1}) = u_1,\ldots,\phi_n(v_1,\ldots,v_{k_n}) = u_n \rangle$$

where u_i is the element of $M(F\cup\Phi, V_{k_i})$ denoted by e_i in the algebraic theory $M_{F\cup\Phi}$. In that case, a solution of Σ is defined as an n-tuple $(t_1,\ldots,t_n)$ in $M^{\infty}(F,V_{k_1}) \times \ldots \times M^{\infty}(F,V_{k_n})$ such that $t_i = u_i\{t_1/\phi_1,\ldots,t_n/\phi_n\}$ where we denote by $u\{t_1/\phi_1,\ldots,t_n/\phi_n\}$ the result of the second-order substitution of t_1 for $\phi_1,\ldots,t_n$ for ϕ_n; we also denote it more simply by $\theta(u)$. Since u is finite we can define it as follows :

$$\theta(v_i) = v_i$$

$$\theta(f(u_1,\ldots,u_k)) = f(\theta(u_1),\ldots,\theta(u_k))$$

$$\theta(\phi_i(u_1,\ldots,u_{k_i})) = t_i[\theta(u_1),\ldots,\theta(u_{k_i})] \; .$$

(5.1.1) Example

Here are two notations for the same algebraic equation.

$$\phi = c.(\pi_1, \pi_2, \phi.(\pi_1, h.\pi_2))$$

$$\phi(v_1,v_2) = c(v_1,v_2,\phi(v_1,h(v_2))) .$$

The solution t (it is actually unique) is depicted in figure 3.

(5.1.2) Theorem : An algebraic system Σ as in (5.1) has a least solution in $T_{\Omega,k_1} \times \ldots \times T_{\Omega,k_n}$. If Σ is in Greibach normal form, it has a unique solution which belongs to $T_{k_1} \times \ldots \times T_{k_n}$.

Proof : We sketch the well-known proof in order to show how the theory develops.

Let E_Ω be the ω-complete partial order $T_{\Omega,k_1} \times \ldots \times T_{\Omega,k_n}$ (its least element is $(\Omega,\Omega,\ldots,\Omega))$ and $|\Sigma|: E_\Omega \to E_\Omega$ be the mapping such that $|\Sigma|(w_1,\ldots,w_n) = (w'_1,\ldots,w'_n)$ with $w'_i = u_i\{w_1/\phi_1,\ldots,w_n/\phi_n\}$ for $i=1,\ldots,n$. This mapping is monotone and ω-continuous hence it has a least fixpoint $(t_1,\ldots,t_n)$ in E_Ω which is the least solution of Σ .

Let now E be the complete metric space $T_{k_1} \times \ldots \times T_{k_n}$. The restriction of $|\Sigma|$ to E is a contracting mapping : $E \to E$ (since Σ has been assumed in Greibach normal form). This mapping has a unique fixpoint $(t'_1,\ldots,t'_n)$ in E which is the unique solution of Σ (this proof technique is used in Bloom [5] and in a more general situation in Arnold and Nivat [2]).

Remark now that if we consider Ω as an ordinary constant then E_Ω is also a complete metric space and $|\Sigma|$ has a unique solution in E_Ω . Since $E \subsetneq E_\Omega$ the solutions of $|\Sigma|$ in E and E_Ω are the same and $(t_1,\ldots,t_n) = (t'_1,\ldots,t'_n)$. ¤

......

(5.3) Algebraic trees and schematic tree languages.

We state that the components of the least solution $(t_1,\ldots,t_n)$ of an algebraic system Σ can be defined as the least upper bounds of directed sets of trees $L_1,\ldots,L_n$ generated from a context-free tree grammar Σ_Ω associated with Σ.

Let $\Sigma = < \phi_1(v_1,\ldots,v_{k_1}) = u_1,\ldots,\phi_n(v_1,\ldots,v_{k_n}) = u_n>$. Let Σ_Ω be the set of pairs $(\phi_i(v_1,\ldots,v_{k_i}), u_i)$ and $(\phi_i(v_1,\ldots,v_{k_i}),\Omega)$ and let $\xrightarrow[\Sigma_\Omega]{*}$ be the semi-Thue relation associated with Σ_Ω.

The triple (F,Φ,Σ_Ω) is in fact a context-free tree-grammar (see Engelfriet and Schmidt [27] for a detailed study) of a special form : we call it a schematic tree-grammar as in Courcelle [13] (since it comes from a recursive program scheme).

For every u in $M(F\cup\Phi,V_k)$, the schematic grammar Σ_Ω generates a tree-language $L(u,\Sigma_\Omega) = \{w\in M_\Omega(F,V_k)/\ u \xrightarrow[\Sigma_\Omega]{*} w\}$. Such a tree-language is called a schematic tree-language.

(5.3.1) Lemma (Nivat [39]) : $L(u,\Sigma_\Omega)$ is directed with respect to $\leq$.

Hence $L(u,\Sigma_\Omega)$ has a least upper bound that we shall denote by $\tau(u)$.

Let us denote by θ the second order substitution of t_1 to $\phi_1,\ldots,t_n$ to ϕ_n where $(t_1,\ldots,t_n)$ is here the least solution of Σ in $T_{\Omega,k_1}\times\ldots\times T_{\Omega,k_n}$.

The following result is often refered to as Schützenberger's theorem (by reference to a similar result of [48]) :

(5.3.2) Theorem : (1) The n-tuple of trees $(\tau(\phi_1(v_1,\ldots,v_{k_1})),\ldots,\tau(\phi_n(v_1,\ldots,v_{k_n})))$ is the least solution of Σ.

(2) For all u in $M(F\cup\Phi,V_k)$, $\tau(u)=\theta(u)$.

......

(5.5) Algebraic trees and deterministic languages.

The following theorem draws a bridge between algebraic trees and deterministic context-free languages. Let us recall that these languages can be defined by deterministic pushdown automata (DPDA's) or equivalently, by LR(k) or strict deterministic grammars (see Harrison [34]).

The equivalence problem for DPDA's i.e. the problem of deciding whether two DPDA's A_1 and A_2 define the same language is open. Many decidable subcases have been discovered (Valiant [49], Oyamaguchi et al.[42,43] in particular).

By using the notations of (1.6) :

(5.5.1) Theorem : (1) A tree in $M^{loc}(F)$ is algebraic if and only if Brch(t) is a deterministic language.

(2) A tree t in $M^{\infty}(F)$ is algebraic if and only if L(t) is a deterministic language.

(3) If a tree t in $M^{\infty}(F)$ is algebraic then Occ(f,t) is a deterministic language for all f in F .

Part (1) is proved in Courcelle [12] and part (2) follows easily and part (3) is proved in Gallier [28]. The proofs are much too technical to be even sketched here.

This result is fully similar to theorem (4.17.1) concerning regular trees, except that the converse to (3) yields an open problem.

(5.5.2) Open problem.

Is it true that a tree t in $M^{\infty}(F)$ such that F is finite and Occ(f,t) is a deterministic language for all f in F is algebraic ?

The answer is yes if F consists of two symbols. This is due to the fact that the complement of a deterministic language L can be recognized by the same automaton as L except for accepting modes.

We do not make any conjecture concerning the general case but we give an equivalent formulation :

Is it true that if $(L_1, L_2, \ldots, L_n)$ is a partition of X^* into n deterministic languages then the language $L_1 1 \cup L_2 2 \cup \ldots \cup L_n n$ over $X \cup [n]$ is deterministic ?

(5.5.3) Theorem : The equivalence problem for DPDA's and the equality problem for algebraic trees are interreducible.

The reduction from algebraic trees to DPDA's follows from the remark that $t = t'$ if and only if Occ(f,t) = Occ(f,t') for all f in F and part (2) of theorem (5.5.1). It can also be established by means of Brch(t) (Courcelle [12]).

The other reductions are even more technical ; they are proved in Courcelle [12] and Gallier [28].

(5.5.4) Consequences.

The above cited constructions yield the following facts :

(1) Every decidable case of the equivalence problem for DPDA's yields decidable cases of the equality problem for algebraic trees. (Not just one case because there exist several reductions of the equality problem for algebraic trees to the equivalence problem for DPDA's : two by Courcelle and one by Gallier [11,12,28]. Actually it is not at all easy to have handy characterizations of the corresponding classes of algebraic systems. But this is a direction for future research.

(2) Every decidable case of the equality problem for algebraic trees yields decidable cases of the equivalence problem for DPDA's. This is also a largely open research direction.

......

(5.7) Two congruences associated with an algebraic system.

Let Σ , $(t_1,\ldots,t_n)$, θ be as in (5.6). In particular Σ is trim.

Since Σ is in Greibach normal form, the second-order substitution θ is continuous and extends uniquely to $M^{\infty}(F\cup\Phi,V)$ by proposition (3.5.4).

There corresponds to Σ a congruence $\overset{*}{\underset{\Sigma}{\longleftrightarrow}}$ on $M(F\cup\Phi,V)$ generated by Σ considered as a set of pairs of terms, and a congruence on $M^{\infty}(F\cup\Phi,V)$ defined by :

$$t \equiv_{\Sigma} t' \text{ if and only if } \theta(t) = \theta(t') .$$

In the following theorem, we compare $\overset{*}{\underset{\Sigma}{\longleftrightarrow}}$ with the restriction of $\equiv_{\Sigma}$ to $M(F\cup\Phi,V)$, also denoted by $\equiv_{\Sigma}$.

(5.7.1) Theorem : (1) $\overset{*}{\underset{\Sigma}{\longleftrightarrow}}$ and $\equiv_{\Sigma}$ are stable congruences on $M(F\cup\Phi,V)$ and $\overset{*}{\underset{\Sigma}{\longleftrightarrow}} \subset \equiv_{\Sigma}$.

(2) $\equiv_{\Sigma}$ is simplifiable.

(3) $\overset{*}{\underset{\Sigma}{\longleftrightarrow}}$ is semi-decidable and $\equiv_{\Sigma}$ is semi-refutable.

(4) $t \equiv_{\Sigma} t'$ if and only if for all m in $\mathbb{N}$ there exist u, u' in $M(F\cup\Phi,V)$ such that $t \overset{*}{\underset{\Sigma}{\longleftrightarrow}} u$, $t' \overset{*}{\underset{\Sigma}{\longleftrightarrow}} u'$ and $\delta(u,u') \geq m$.

(5.7.2) Remarks : Proving that $\equiv_{\Sigma}$ is semi-decidable is equivalent to proving that the equivalence problem for DPDA's is solvable.

Whether $\overset{*}{\underset{\Sigma}{\longleftrightarrow}}$ is decidable in general is an open question raised by R. Milner.

......

6. Conclusion.

The present work has developed several aspects of finite and infinite trees which are especially relevant to the theory of computing. To summarize :

(1) A double theory of infinite trees, by topological or order-theoretical methods can be developed ;

(2) First-order and second-order substitutions are two important concepts ; some of their basic combinatorial properties have been stated ; their continuity properties have been investigated in detail.

(3) Regular trees and their relations with first-order unification have been studied ; rational expressions denoting regular trees have been introduced.

(4) Algebraic trees have been studied ; their combinatorial properties are complex enough to yield an open problem which is interreducible with the equivalence problem for DPDA's ; decidable special cases have been stated.

Many other interesting aspects (raising open problems) could have been treated as well (except for the author's time availability):

(5) Higher-order algebraic trees corresponding to higher-order recursive program schemes (Indermark et al. [24], Damm [23]),

(6) frontiers of infinite trees as generalized infinite words (Courcelle [13], Heilbrunner [36]) and even more important :

(7) extensions of congruences from finite trees to infinite ones : whereas the theory is rather neat in the approach with partial orders, (Courcelle [15]), it is much more difficult in the topological approach (Courcelle [14]).

And finally, the theory of tree languages. But we think that it constitutes a theory by itself that should be distinguished from the theory of infinite trees, which the present work hopes to contributes to.

References

LNCS means : Lecture Notes in Computer Science, Springer-Verlag, Berlin, Heidelberg, New-York.

[1] A. Arnold, M. Dauchet, *Théorie des magmoîdes*, RAIRO Informatique Théorique 12 (1978) pp.235-257.

[2] A. Arnold, M. Nivat, *Metric interpretations of infinite trees and semantics of non deterministic recursive programs*, Theor. Comput. Sci.11 (1980) pp.181-205.

[3] A. Arnold, M. Nivat, *The metric space of infinite trees. Algebraic and topological properties*, Fundamenta Informaticae III.4 (1980) pp.445-476.

[4] H. Bekič, *Definable operations in general algebras, and the theory of automata and flowcharts*, Unpublished work, IBM Laboratory, Vienna, 1969.

[5] S. Bloom, *All solutions of a system of recursion equations in infinite trees and other contraction theories*, J. Comput. System Sci.

[6] S. Bloom, C. Elgot, J. Wright, *Solutions of the iteration equation and extensions of the scalar iteration operation*, SIAM J. Comput. 9 (1980) pp.25-45.

[7] S. Bloom, S. Ginali, J. Rutledge, *Scalar and vector iteration*, J. Comput. System Sci. 14 (1977) pp.251-256.

[8] S. Bloom, D. Patterson, *Easy solutions are hard to find*, Proceedings of the 6th Colloquium on trees in algebra and programming Geno a 1981, (to appear in L.N.C.S).

[9] G. Cousineau, *La programmation en EXEL*, Revue Technique de Thomson-CSF 10 (1978) pp.209-234 and 11 (1979) pp.13-35.

[10] G. Cousineau, *An algebraic definition for control structures*, Theor. Comput. Sci. 12 (1980) pp.175-192.

[11] B. Courcelle, *On jump-deterministic pushdown automata*, Math. Systems Theory 11 (1977) pp.87-109.

[12] B. Courcelle, *A representation of trees by languages*, Theor. Comput. Sci. 6 (1978) pp.255-279 and 7 (1978) pp.25-55.

[13] B. Courcelle, *Frontiers of infinite trees*, RAIRO Informatique Théorique 12 (1978) pp.319-337.

[14] B. Courcelle, *Arbres infinis et systèmes d'équations*, RAIRO Informatique Théorique 13 (1979) pp.31-48.

[15] B. Courcelle, *Infinite trees in normal form and recursive equations having a unique solution*, Math. Systems Theory 13 (1979) pp.131-180.

[16] B. Courcelle, *An axiomatic approach to the Korenjak-Hopcroft algorithms*, Report AAI-8018, University of Bordeaux I and communication to the 8th Intern. Conf. on Automata, Languages and Programming, Acre,Israël, 1981 (to appear in L.N.C.S).

[17] B. Courcelle, Work in preparation.

[18] B. Courcelle, P. Franchi-Zannettacci, *Attribute grammars and recursive program schemes*, to appear in Theor. Comput. Sci.

[19] B. Courcelle, G. Kahn, J. Vuillemin, *Algorithmes d'équivalence et de réduction à des expressions minimales, dans une classe d'équations récursives simples*, 2nd Int. Coll. on Automata, Languages and Programming, Saarbrücken 1974, L.N.C.S. 14, pp.200-213.

[20] B. Courcelle, M. Leprévost, Unpublished work.

[21] B. Courcelle, J.C. Raoult, *Completions of ordered magmas*, Fundamenta Informaticae III.1 (1980) pp.105-116.

[22] B. Courcelle, J. Vuillemin, *Completeness results for the equivalence of recursive schemes*, J. Comput. System Sci. 12 (1976) pp.179-197.

[23] W. Damm, *The IO - and OI - hierarchies*, Report 41, RWTH Aachen, October 1980.

[24] W. Damm, E. Fehr, K. Indermark, *Higher type recursion and self-application as as control structures*, in Formal descriptions of programming concepts, E. Neuhold ed., North-Holland 1978, pp.461-487.

[25] C. Elgot, *Monadic computation and iterative algebraic theories*, Proc. Logic Colloq. 73, North-Holland Pub. Co., Amsterdam (1975) pp.175-230.

[26] C. Elgot, S. Bloom, R. Tindell, *The algebraic structure of rooted trees*, J. Comput. System Sci. 16 (1978) pp.362-399.

[27] J. Engelfriet, E. Schmidt, *IO and OI*, J. Comput. System Sci. 15 (1977) pp.328-353 and 16(1978) pp.67-99.

[28] J. Gallier, *DPDA's in "atomic" form and applications to the equivalence problems*, Theor. Comput. Sci. 14 (1981) pp.155-186.

[29] S. Ginali, *Regular trees and the free iterative theory*, J. Comput. System Sci. 18 (1979) pp.228-242.

[30] J. Goguen, J. Thatcher, E. Wagner, J. Wright, *Initial algebra semantics and continuous algebras*, J. Assoc. Comput. Mach. 24 (1977) pp.68-95.

[31] S. Gorn, *Explicit definitions and linguistic dominoes*, Systems and Computer Science, J. Hart and S. Takasu ed., Sept.1965.

[32] I. Guessarian, *Program transformations and algebraic semantics*, Theor. Comput. Sci. 9 (1979) pp.39-65.

[33] I. Guessarian, *Algebraic Semantics*, L.N.C.S. 99 (1981).

[34] M. Harrison, *Introduction to formal language theory*, Addison-Wesley, 1978.

[35] M. Harrison, I. Havel, A. Yehudai, *On equivalence of grammars through transformation trees*, Theor. Comput. Sci. 9 (1979), pp.173-205.

[36] S. Heilbrunner, *An algorithm for the solution of fixed-point equations for infinite words*, RAIRO Informatique Théorique, 13 (1979) pp.131-141.

[37] G. Huet, *Résolution d'équations dans les langages d'ordre 1,2,...,ω*, Doctoral dissertation, Univ. Paris 7, Paris, Sept. 1976.

[38] G. Markowsky, B. Rosen, *Bases for Chain-Complete posets*, IBM Journ. of Res. and Dev. 20 (1976) pp.138-147.

[39] M. Nivat, *On the interpretation of recursive polyadic program schemes*, Symposia Mathematica 15, Academic Press, (1975) pp.255-281.

[40] M. Nivat, *Mots infinis engendrés par une grammaire algébrique*, RAIRO Informatique Théorique 11 (1977) pp.311-327.

[41] M. Nivat, Private communication.

[42] M. Oyamaguchi, N. Honda, *The decidability of the equivalence for deterministic stateless pushdown automata*, Information and Control 38 (1978) pp.367-376.

[43] M. Oyamaguchi, N. Honda, Y. Inagaki, *The equivalence problem for real-time strict deterministic languages*, Information and Control 45 (1980) pp.90-115.

[44] M. Paterson, M. Wegman, *Linear Unification*, J. Comput. System Sci. 16 (1978) pp.158-167.

[45] J. Robinson, *A machine-oriented logic based on the resolution principle*, J. Assoc. Comput. Mach. 12 (1965) pp.23-41.

[46] B. Rosen, *Tree-manipulating systems and Church-Rosser theorems*, J. Assoc. Comput. Mach. 20 (1973) pp.160-187.

[47] B. Rosen, *Program equivalence and context-free grammars*, J. Comput. System Sci. 11 (1975) pp.358-374.

[48] M. Schützenberger, *On context-free languages and push-down automata*, Information and Control 6 (1963) pp.246-264.

[49] L. Valiant, *The equivalence problem for deterministic finite-turn push-down automata*, Information and Control 25 (1974) pp.123-133.

[50] J. Wright, J. Thatcher, E. Wagner, J. Goguen, *Rational algebraic theories and fixed-point solutions*, 17th Symp. on Foundations of Computer Science, Houston, Texas, 1976, pp.147-158.

[51] J. Wright, E. Wagner, J. Thatcher, *A uniform approach to inductive posets and inductive closure*, Theor. Comput. Sci. 7 (1978) pp.57-77.

References added at revision

[52] S. Bloom, C. Elgot, *The existence and construction of free iterative theories*, J. Comput. Sci. 12 (1976), 305-318.

53 S. Bloom, R. Tindell, *Compatible orderings on the metric theory of trees*, SIAM J. Comput. 9 (1980), 683-691.

54 C.C. Elgot, *Structured programming with and without GOTO statements*, IEEE Trans. on Software Eng. Vol. SE-2 (1976) pp. 41-54.

[55] J. Gallier, *Recursion-closed algebraic theories*, to appear in J. Comput. System Sci.

[56] B. Courcelle, F. Lavandier, *Définitions récursives par cas*, to appear.

****

Theme X.

(B. Courcelle)

The theme of IX turned into an infinite tree (only a finite approximation can be implemented)

BEHAVIORS OF PROCESSES AND SYNCHRONIZED SYSTEMS OF PROCESSES

Maurice Nivat
University Paris VII

I. INTRODUCTION

In these notes we consider processes as given by their sets of behaviors, including the infinite behaviors. The set $HR^{\infty}(p)$ of a process p is what we call an infinitary language, i.e. a subset of the set $A^{\infty} = A^{*} \cup A^{\omega}$ of all finite words (A^{*}) and infinite words (A^{ω}) on an alphabet of actions A. In a first part we study infinitary languages and then recognizability by transition systems with a special emphasis on infinitary rational languages which are defined as the family of those languages which are recognizable by finite transition systems. The definition of recognizability of a finite word by a transition system is the standard one which goes back to S. Kleene: the transition system $\mathcal{S}$ recognizes the word $f \in A^{*}$ iff there exists a computation sequence of $\mathcal{S}$ reading f which starts in an initial configuration and terminates in a final configuration (both sets of initial and final configurations are given as parts of the definition of $\mathcal{S}$). The notion of recognizability of an infinite word we use is also standard, if less well known: the word $u \in A^{\omega}$ is recognized by $\mathcal{S}$ iff there exists an infinite computation sequence of $\mathcal{S}$ reading u which starts in an initial configuration and passes infinitely

M. Broy and G. Schmidt (eds.), Theoretical Foundations of Programming Methodology, 473–551.

many times in a given set of "infinite" configurations (the subset of infinite configurations is also given as part of the definition of S). Buchi [8] may be the first author to introduce this notion of recognizability for infinite words which has been used later on by Mac Naughton [18], L. Landweber [15], Eilenberg [11]. We shall use in a crucial way results of these four authors: the major difference between infinite words and finite ones is that given a finite transition system S one can easily build a deterministic finite transition system S_d which recognizes the same set of finite words. Whereas there exist purely infinitary languages which can be recognized by non-deterministic finite transition systems which cannot be recognized by any deterministic finite transition system. This fact does not prevent the equivalence of two finite transition systems to be decidable (by equivalence we mean equality of the infinitary languages which they recognize): the decision procedure is, however, more complicated than in the finitary case. Our originality in this first part is to deal with infinitary languages rather than finitary languages (subsets of A^*) whose theory is widely taught under the name of automata theory or purely infinitary ones (subsets of A^ω) as this is the case in the works of Buchi, Mac Naughton, Robin, Eilenberg and their followers. If one thinks of an infinitary language L as being the set of behaviors of a process p some natural properties of the process p are reflected by properties of L which link the finitary part $L^{fin} = L \cap A^*$ and the purely infinitary part $L^{inf} = L \cap A^\omega$ of L.

Such properties, defined below, are the properties of being closed, normal, ideal and central: the last one which is defined by the condition that "every finite behavior can be extended into an infinite one" expresses the fact that p whose set of behaviors is equal to L has no deadlocks. As everybody knows one of the major problems when dealing with processes and systems of processes is to detect and avoid deadlocks:

we thus believe that the notion of center we introduce, the center of $L \subseteq A^{\infty}$ being the largest normal and central language contained in L, is quite important. The center of a rational language L is rational and computable: in our eyes the best solution to the general problem of avoiding deadlocks is precisely the computation of the center.

In a second part of these notes we study synchronized systems of processes. We assume that h processes $p_1, \ldots, p_h$ are given by their sets of behaviors $HR^{\infty}(p_1) \subseteq A_1^{\infty}, \ldots, HR^{\infty}(p_h) \subseteq A_h^{\infty}$. These processes will have to work simultaneously in accordance with a certain synchronization condition Syn ; such a synchronization condition specifies rules of the following type:

"p_i cannot perform a without p_j performing b at the same time"

"p_i cannot perform a when p_j is performing b"

"p_i must wait to perform a until p_j has performed b"

If one thinks about a programming system involving several components i.e. CPU's, reading and printing devices sharing a common memory it is easy to find practical examples of conditions of the three above types (a value cannot be used in future computations, nor printed, before it has been computed, no two CPU's can use the same peripheral equipment at the same time, when a unit sends a value to another unit then this unit receives it and so on).

Synchronization conditions may be expressed in several ways. We claim that a very general form one can give to them is the following: Syn is formed of a synchronization mechanism $\mathcal{S}$, which will be in fact a transition system, performing actions in a set A and the mapping $\vec{\theta}$ maps A into $A_1 \times \ldots \times A_h$ (we assume that each A_i contains the special action e which is the action of waiting or empty action).

The mapping $\vec{\Theta}$ specifies which vectors of actions may be performed by the components $p_1, \ldots, p_h$ of the system when S is performing an action $a \in A$. For such a system $<\vec{p}, \text{Syn}>$ where $\text{Syn} = <S, \vec{\Theta}>$ we define a <u>synchronized multibehavior</u> as a h-uple of words

$$\vec{\alpha} = <\alpha_1, \ldots, \alpha_h> \quad \text{such that}$$

$$\forall\, j \in [h] \quad \alpha_j \in HR^{\infty}(p_j)$$
$$\exists\, \beta \in L^{\infty}(S) \quad \vec{\alpha} \in \vec{\Theta}(\beta)$$

In other words call $HR^{\infty}(\vec{p})$ the cartesian product

$$HR^{\infty}(\vec{p}) = HR^{\infty}(p_1) \times \ldots \times HR^{\infty}(p_h)$$

This is the set of mulibehaviors of the system $\vec{p}$ without any synchronization condition (the p_i's act independently from each other). Then the set of synchronized multibehaviors of $<\vec{p}, \text{Syn}>$ is $HR^{\infty}(\vec{p}, \text{Syn}) = HR^{\infty}(\vec{p}) \cap \vec{\Theta}(L^{\infty}(S))$.
This set is a set of <u>multiwords</u> $\vec{\alpha} \in A_1^{\infty} \times \ldots \times A_h^{\infty}$ that is an infinitary relation (or infinitary <u>multilanguage</u>).

The programming language COSY, designed in Newcastle upon Tyne by P. Lauer and others [16] explicitly uses synchronization conditions of this type. We give several other examples which show that this form is a very general one by which one can describe the synchronization conditions which are expressible in the various languages proposed to deal with systems. This leads us to the systematic study of infinitary relations which extends the study of infinitary languages which is done in the first part, with also a special emphasis on the rational case relations are recognized by multitransition systems (or h-tapes automata) and rational relations are defined as those relations which can be recognized by finite multitransition systems.

Finitary rational relations subsets of $A_1^* \times \ldots \times A_h^*$ were first studied by Elgot and Mezei [12] and a characterization of them was given by Nivat [23] as the set of images $\vec{\Theta}(L)$ of a rational finitary language under a multimorphism $\vec{\Theta} = < \Theta_1, \ldots, \Theta_h >$ where $\Theta_i : A \to A_i^*$. As will be seen, the extension of this result to the infinitary case is not an entirely trivial matter.

Now if one thinks of an infinitary relation R as the set of multibehaviors of a system several properties of R are naturally introduced: some are more or less the same as the properties of infinitary languages, closedness, normality, ideality, centrality. There is one more which also plays a crucial role in the analysis and design of actual systems, this is fairness. Asserting that $R = HR^\infty(\vec{p}, \text{Syn})$ is fair amounts to assert that after any length of time during which $< \vec{p}, \text{Syn} >$ has been following an allowed initial behavior, it is possible to activate each component of the process. This leads to a notion of fair center of a relation: in the rational case the fair center is proved to be rational and computable. In our eyes this "solves" the problem of fairness in this case.

Remarks and acknowledgements

These notes are lecture notes, sketches of the main proofs and constructions will be given, but many details are left to the reader. In many cases we shall refer to several other papers of the author.

Even if we mainly restrict our study to the rational case (finite alphabets, finite transition systems) we try to give as general definition as we can. Our theory is very far from being complete and we do hope that it will be completed, why not by participants in the 1981 Marktoberdorf NATO Summer School?

The author has benefitted from many fruitful discussions and co-operations, let me mention A. Arnold, L. Boasson, B. Courcelle, F. Gire, J. van Leeuwen, G. Roucairol, G. Ruggiu and A. Salwicki.

The author wishes to thank heartly the organizers of the School in Marktoberdorf and especially Prof. F. L. Bauer, M. Broy and M. Wirsing for giving him the opportunity of expressing his ideas orally first in front of a large and interesting audience and in these pages which tend to reflect faithfully the content of his lectures.

For making his stay in Marktoberdorf as pleasant and fruitful as it was (most of these lines were written there), Herr Kuss and Herr Hönle were extremely helpful: let they be thanked too and also NATO, without which this course would not have been given.

II. INFINITARY LANGUAGES

Infinitary languages are introduced in order to describe the set of behaviors of a process.

We need first recall some facts about infinite words on an alphabet A. Let A denote any set of symbols which we call actions. As usual we denote by A^* the set of finite words on A: a finite word $f \in A^*$ has a length $|f| \in \mathbb{N}$. The only word of length 0 is the empty word denoted ε. A word f of length $n > 0$ is a sequence of n letters in A, which we write from left to right $f = f(1) \dots f(n)$.
The set of infinite words on A is denoted A^ω: every $u \in A^\omega$ has length $|u| = \omega = \mathrm{card}(\mathbb{N})$ and is a countable sequence of letters in A which we also write from left to right $u = u(1)\ u(2) \dots u(n) \dots$

Remark: This is not the only notion of infinite words which one may introduce and study:

- infinite words in both directions are studied in [11]
- infinte words with "holes" meant to describe the yield of an infinite tree are defined in [10].

The one way infinite words are exactly what we need to describe the infinite behavior of a process which starts performing actions at a given time and goes on for ever.

We denote by A^∞ the set of finite and infinite words i.e. the union $A^\infty = A^* \cup A^\omega$

For the sake of clarity we usually denote by $f, g, h \dots$ the finite words we deal with, by $u, v, w \dots$ the infinite words and by $\alpha, \beta, \gamma \dots$ the words whose length is either finite of infinite.

II. a. Left factors

The main tool to deal with infinite words is the relation "is a left factor of" which we now define.

For all $\alpha \in A^\infty$ and $n \in \mathbb{N}$ we define $\alpha[n]$ by

$$\alpha[n] = \alpha(1)\ \alpha(2) \ldots \alpha(n) \quad \text{if} \quad n \leq |\alpha|$$

$$\alpha[n] = \alpha \quad \text{if} \quad n \geq |\alpha|$$

The finite word $\alpha[n]$ is called the left factor of length n of α (though the length of $\alpha[n]$ may be strictly less than n)
We then define the order relation $\leq$ on A^∞ :

$$\alpha \leq \beta \Leftrightarrow \quad \text{either} \quad |\alpha| < \omega \quad \text{and} \quad \alpha = \beta[|\alpha|]$$

$$\text{or} \quad |\alpha| = |\beta| = \omega \quad \text{and} \quad \alpha = \beta$$

We say that α is a left factor of β iff $\alpha \leq \beta$.
We shall mainly use the set of finite left factors of a word β denoted $FG(\beta)$ and defined as

$$FG(\beta) = \{ \beta[n] \mid n \in \mathbb{N} \}$$

Obviously $\operatorname{card}(FG(\beta)) < \omega \Leftrightarrow \beta \in A^*$
It is clear that

$$\alpha \leq \beta \Leftrightarrow FG(\alpha) \subseteq FG(\beta)$$

and $$\alpha = \beta \Leftrightarrow FG(\alpha) = FG(\beta)$$

and also that $\leq$ is an order relation.
When we refer to A^∞ as an ordered set, we always refer to A^∞ ordered by $\leq$ unless otherwise specified.
For example we say that the sequence α_n, $n \in \mathbb{N}_+$ of words in A^∞ is increasing iff for all $n \in \mathbb{N}_+$ $\alpha_n \leq \alpha_{n+1}$.
And the major property is the following :

Property II. a. 1.

Every increasing sequence α_n, $n \in \mathbb{N}_+$ of elements of A^∞ has a least upper bound denoted $\operatorname{Sup}(\alpha_n)$.

Proof

Either the sequence is stationary, i.e.

there exists $n_o \in \mathbb{N}_+$ such that $n \geq n_o$ implies $\alpha_n = \alpha_{n_o}$
and then $\mathrm{Sup}(\alpha_n) = \alpha_{n_o}$

Or α_n is not stationary, i.e. $|\alpha_n| \xrightarrow[n\to\infty]{} \infty$ and there exists a unique infinite word $u \in A^\omega$ such that $\alpha_n \leq u$ for all n.
Then $\mathrm{Sup}(\alpha_n) = u$. □

Remark

One can thus prove that A^∞ is a complete partial order [24]. The set A^∞ has also the structure of a complete metric space.
One defines on A^∞ the following distance $d : A^\infty \times A^\infty \to \mathbb{R}$ where $\mathbb{R}$ is the set of real numbers

$$d(\alpha, \beta) = 2^{-\min\{n \mid \alpha[n] = \beta[n]\}} \quad \text{if} \quad \alpha \neq \beta$$
$$= 0 \quad \text{if} \quad \alpha = \beta$$

Then $d(\alpha, \beta) \geq 0$ for all α, β
$d(\alpha, \beta) = 0$ iff $\alpha = \beta$
$d(\alpha, \beta) = d(\beta, \alpha)$ for all α, β
$d(\alpha, \beta) \leq \max(d(\alpha, \gamma), d(\gamma, \beta))$ for all α, β, γ

The distance d is thus an ultrametric distance [5].
The topology on A^∞ induced by d is referred to as the d-topology. It is easy to show that A^∞ equipped with d is a complete metric space. Indeed a sequence α_n, $n \in \mathbb{N}$ is d-Cauchy
iff $\forall\, p \in \mathbb{N} \quad \exists\, n_p \in \mathbb{N}_+ : n, n' \geq N_p \Rightarrow \alpha_n[p] = \alpha_{n'}[p]$

And every d-Cauchy sequence converges towards

$$\lim(\alpha_n) = \mathrm{Sup}(\alpha_{n_p})$$

For more details on this topology see [4] . □

II. b. Multiplication of A^∞

The set of finite words A^* has the structure of a monoid.
If $f, g \in A^*$ the product fg is defined by

$$|fg| = |f| + |g|$$

$$\forall\ n \quad 1 \le n \le |f| \quad (fg)(n) = f(n)$$

$$\text{and} \quad 1 \le n \le |g| \quad (fg)(|f| + n) = g(n)$$

We extend this multiplication to A^∞ by the rules

If $f \in A^*$ and $u \in A^\omega$ the product $fu \in A^\omega$ is

$$(fu)(n) = f(n) \quad \text{for all} \quad 1 \le n \le |f|$$

$$(fu)(n) = u(n - |f|) \quad \text{for all} \quad n > |f|$$

If $u \in A^\omega$ and $\alpha \in A^\infty$ we set

$$u\alpha = u \quad .$$

This product is associative and ε is a neutral element, i.e. satisfies $\varepsilon\alpha = \alpha\varepsilon = \alpha$ for all $\alpha \in A^\infty$
Whence A^∞ is a monoid.

The following properties hold

Property II. b. 1.

For all $f, g \in A^*$, $u, v \in A^\omega$

$fu = gv \Rightarrow \exists\ h \in A^*$ such that

either $f = gh$ and $v = hu$

or $g = fh$ and $u = hv$

Remark: this is an extension of the well known Levi's lemma

Property II. b. 2.

For all $\alpha, \beta \in A^\infty$

$$\alpha \le \beta \Leftrightarrow \exists\ \gamma \in A^\infty : \alpha\gamma = \beta$$

Remark

We say that α is extendable into β iff there exists γ such that $\alpha\gamma = \beta$ and thus

α is a left factor of β iff α is extendable into β

Contrary to what happens for finite words

$\beta\alpha^{-1} = \{\gamma \mid \alpha\gamma = \beta\}$ may be different from a singleton. In fact

either $\beta\alpha^{-1} = A^{\infty}$ when $\alpha = \beta \in A^{\omega}$

or $\mathrm{card}(\beta\alpha^{-1}) = 1$ when $\alpha \in FG(\beta)$

or $\mathrm{card}(\beta\alpha^{-1}) = 0$

II. c. Infinitary languages and behaviors of processes

We call infinitary language any subset of A^{∞}. If $L \subseteq A^{\infty}$ we denote by L^{fin} and call the finitary part of L the set $L^{fin} = L \cap A^{*}$ of all finite words in L. And we denote by L^{inf} and call the purely infinitary part of L the set $L^{inf} = L \cap A^{\omega}$ of all infinite words in L. The language L is said to be finitary (resp. purely infinitary) iff $L = L^{fin}$ (resp. $L = L^{inf}$). Obviously $L^{fin} \cap L^{inf} = \emptyset$.

Now a process is any mechanism which performs actions. We assume that all the actions which p may perform are instantaneous, i.e. are performed in one unit of time. Let A be this set of actions. The process p is entirely defined by its set of finite and infinite behaviors $HR^{\infty}(p)$.

We denote $HR^{*}(p) = HR^{\infty}(p) \cap A^{*}$

$HR^{\omega}(p) = HR^{\infty}(p) \cap A^{\omega}$

and for all $n \in \mathbb{N}$ $HR^{n}(p) = HR^{*}(p) \cap A^{n}$

Then to say that $f \in HR^{n}(p)$ means that the process p can perform $f(1)$ in the interval of time between the instant 0 and the instant 1 (we say that p performs $f(1)$ at time 0) then p can perform $f(2)$ at time 1 after performing $f(1)$ at time (0) and so on, up to

p can perform f(n) at time (n - 1) after performing f(1) at time 0, f(2) at time 1 , ... , f(n - 1) at time (n - 2) and p can stop at time n .
To say that $u = u(1) \ldots u(n) \ldots$ belongs to $HR^{\omega}(p)$ means that one possible behavior of p is to perform f(1) at time 0, f(2) at time 1 , ..., f(n) at time (n - 1), and so on for ever. Immediately some properties of infinitary languages appear which are interesting if one thinks of an infinitary language as of the set of behaviors of a process.

II. c. 1. Closedness

Define the adherence of the infinitary language L as the set of infinite words

$$Adh(L) = \{u \in A^{\omega} \mid FG(u) \subseteq FG(L)\}$$

In a standard way FG(L) denotes the set

$$FG(L) = \cup \{FG(\alpha) \mid \alpha \in L\}$$

Adherences have been extensively studied in [4] .

In fact, Adh(L) is the set of cluster points of L in the d-topology (and thus improperly called adherence which according to Bourbaki should be $L \cup Adh(L)$: we use the word adherence to designate Adh(L) , however, for it has proved to be convenient). The one has

Property II. c. 1.

The infinitary language L is closed in the d-topology iff one of the three following conditions is satisfied:

$$Adh(L) \subseteq L$$
$$L = \overline{L} = L \cup Adh(L)$$
$$L^{inf} = Adh(L)$$

The language $\overline{L} = L \cup Adh(L)$ is called the closure of L and is indeed the topological closure of L in A^{∞} with its d-topology.

A process p is said to be closed iff $HR^{\infty}(p)$ is closed. Intuitively p is closed iff it satisfies the condition "if p behaves well from time 0 up to time n for all n then it behaves well for ever". By p behaves well from time 0 up to time n , we mean that p performs f(1) at time 0 , f(2) at time 1 , up to f(n) at time (n - 1) where

$$f = f(1) \ldots f(n) \in FG(HR^{\infty}(p))$$

(f is also called an <u>initial behavior</u> of p).

We shall see that the property for a process of being closed is a very important one. We give now an example of a typical non closed process.

- Assume that $HR^{*}(p) = (a \cup b)^{*}$

we can think of p as being a process which serves two users a meaning serves the first and b meaning serves the second.

- Take $HR^{\omega}(p) = \{ u \in (a \cup b)^{\omega} \mid |u|_a = |u|_b = \omega \}$

By $|u|_a$ we denote $card\{n \mid u(n) = a\}$. Thus setting $u \in HR^{\omega}(p) \Leftrightarrow |u|_a = |u|_b = \omega$ amounts to require that each user will be served infinitely many times along every allowed infinite behavior of p . This is equivalent to say that "none of the users will wait for ever before being served again".

This is a <u>fairness</u> requirement (see below a definition of fairness).

Clearly now if $HR^{\infty}(p) = HR^{*}(p) \cup HR^{\omega}(p)$ one has

$$FG(HR^{\infty}(p)) = (a \cup b)^{*}$$

and

$$Adh(HR'^{\infty}(p)) = (a \cup b)^{\omega}$$

Whence p is <u>not closed</u> □

<u>II. c. 2. Normality and ideality</u>

We define L is <u>normal</u> iff $FG(L) \subseteq FG(L^{fin})$

L is <u>ideal</u> iff $L^{fin} = FG(L)$

Clearly L ideal implies L normal . In fact, ideality is the strongest possible form of normality. Intuitively, a process p , which is said to be normal iff $HR^{\infty}(p)$ is normal, is a process which can be stopped after having performed any allowed initial behavior.

And a process is ideal iff it can be stopped immediately after having performed an allowed initial behavior. Certainly all real processes will be normal or at least one will try to keep them normal.

II. c. 3. Centrality

L is central iff $L^{fin} \subseteq FG(L^{inf})$

A process p is central iff $HR^{\infty}(p)$ is central, i.e. iff every finite behavior (or initial behavior) of p can be extended into an infinite behavior.

This is certainly a property which one will require of a large number of actual processes which are meant to work for ever: reservation system, communication exchange system, query and updating system for a data base and so on.

It is usual to introduce at that point the notion of deadlock : we call deadlock of p any initial behavior $f \in FG(HR^{\infty}(p))$ such that for all a in A fa is not an initial behavior of p . Or else f is a deadlock if f is an allowed (initial) behavior and if after performing f the process p cannot perform any action. The following property is clear:

Property II. c. 3.

p is central $\Leftrightarrow$ p has no deadlock . (We say then that p is deadlock free) . The converse of this property is not usually true unless p is closed.

Property II. c. 4. If p is closed, then

p is central $\Leftrightarrow$ p is deadlock free .

Proof

p central $\Rightarrow$ p deadlock free follows from II. c. 3.
Assume now p is deadlock free and consider $f \in HR^*(p)$. There exists an $a_1 \in A$ such that $fa_1 \in FG(HR^\infty(p))$. There exists an $a_2 \in A$ such that $fa_1a_2 \in FG(HR^\infty(p))$. We can repeat the argument to find an infinite sequence

$a_1, a_2, \ldots, a_n, \ldots$ of letters in a such that

$$\forall\ n \in \mathbb{N}_+ \quad fa_1a_2 \ldots a_n \in FG(HR^\infty(p)) .$$

This implies, since p is closed, that if $u = a_1a_2 \ldots a_n \ldots$

$fu \in HR^\omega(p)$ □

II. d. Operations on infinitary languages

The theory of finitary languages uses in a crucial way some operations on languages, especially

- the boolean operations of union, intersection, complementation
- the product $LL' = \{ ff' \mid f \in L ,\ f' \in L' \}$
- Kleene's star $L^* = \{ \varepsilon \} \cup \{ f_1f_2 \ldots f_n \mid n \in \mathbb{N}_+$ and $f_i \in L$ for all $1 \leq i \leq n \}$
- morphisms and inverse morphisms: if φ is a mapping of the alphabet B into A^* there exists a unique (monoid) morphism, which we also denote φ, of B^* into A^*

  $\varphi(\varepsilon) = \varepsilon$

  $\forall\ f = f(1) \ldots f(n) \in B^n \qquad \varphi(f) = \varphi(f(1)) \ldots \varphi(f(n))$.

  The morphism φ is said to be continuous (resp. alphabetic, resp. strictly alphabetic) iff $\varphi(B) \subseteq A^+ = AA^* = A^* \setminus \{\varepsilon\}$ (resp. $\varphi(B) \subseteq A \cup \{\varepsilon\}$, resp. $\varphi(B) \subseteq A$)

  If L is a language in B^* $\varphi(L) = \{ \varphi(f) \mid f \in L \}$ and if L is a language in A^* $\varphi^-(L) = \{ f \in B^* \mid \varphi(f) \in L \}$

We shall use similar operations on infinitary lnaguages and a few others which are quite natural to build infinitary languages from finitary ones.

The set theoretic operations raise no problem. One has

$$(L_1 \cup L_2)^{fin} = L_1^{fin} \cup L_2^{fin}$$
$$(L_1 \cup L_2)^{inf} = L_1^{inf} \cup L_2^{inf}$$

and similar relations for the intersection $L_1 \cap L_2$ and the complement $L_1 \setminus L_2 = \{ f \in A^\infty \mid f \in L_1 \text{ and } f \notin L_2 \}$.

The product of two infinitary languages L_1 and L_2 is defined as $L_1 L_2 = \{ \alpha\beta \mid \alpha \in L ,\ \text{and}\ \beta \in L_2 \}$.

Using the fact that for all $u \in A^\omega$ and $\alpha \in A^\infty$ $u\alpha = u$ we get

$$(L_1 L_2)^{fin} = L_1^{fin} L_2^{fin}$$
$$(L_1 L_2)^{inf} = L_1^{inf} \cup L_1^{fin} L_2^{inf}$$

The Kleene's star or finite star of a language L is then given by $L^* = \{\varepsilon\} \cup \{\alpha_1\alpha_2 \dots \alpha_n \mid n \in \mathbb{N}_+ \text{ and } \alpha_i \in L\}$

and we may write

$$(L^*)^{fin} = (L^{fin})^*$$
$$(L^*)^{inf} = (L^{fin})^* L^{inf}$$

Then we have the infinite power L defined as

$$L^\omega = (L^{fin})^\omega = \{f_1 f_2 \dots f_n \dots \mid f_n \in L^{fin} \setminus \{\varepsilon\} \text{ for all } n \in \mathbb{N}_+\}$$

(clearly $L^\omega \subseteq A^\omega$) .

The infinite star L^∞ is simply the union

$$L^\infty = L^* \cup L^\omega .$$

Among the other operations we shall use

- the adherence $Adh(L) = Adh(FG(L)) = \{u \in A \mid FG(u) \subseteq FG(L)\}$
- the closure $\bar{L} = L \cup Adh(L)$
- the limit in Eilenberg's sense

$$Elim(L) = Elim(L^{fin}) = \{u \in A^\omega \mid card\{n \mid u[n] \in L^{fin}\} = \omega\}$$

There are some obvious and some less obvious relations among all the languages thus defined. We shall not list them all, but rather take examples borrowed from various papers.

Property II. d. 1.

$$Adh(L) = Adh(FG(L)) = Elim(FG(L)) .$$

Property II. d. 2. [4]

$$\overline{L_1 \cup L_2} = \overline{L}_1 \cup \overline{L}_2$$
$$\overline{L_1 L_2} = \overline{L}_1\, \overline{L}_2$$
$$\overline{L^*} = (\overline{L})^*$$

Property II. d. 3.

$$\mathrm{Elim}(L_1 \cup L_2) = \mathrm{Elim}(L_1) \cup \mathrm{Elim}(L_2)$$
$$\mathrm{Elim}(L_1 L_2) = \mathrm{Elim}(L_1) \cup L_1^{fin}\, \mathrm{Elim}(L_2)$$
$$\mathrm{Elim}(L^*) = (L^{fin})^*\, \mathrm{Elim}(L) \cup L^\omega$$

Proof The first identity is trivial for

$\mathrm{card}\{n \mid u[n] \in L_1 \cup L_2 = \omega$ obviously implies either $\mathrm{card}\{n \mid u[n] \in L_1\} = \omega$ or $\mathrm{card}\{n \mid u[n] \in L_2\} = \omega$

The second identity comes from the fact that if

$\mathrm{card}\{n \mid u[n] \in L_1 L_2\} = \omega$ then for infinitely many n's there exists $f_n \in L_1$ and $g_n \in L_2$ such that

$u[n] = f_n g_n$.

Then two cases arise:

- $|f_n|$ is bounded whence there exists one $f_o \in \{f_n \mid n \in \mathbb{N}_+\}$ such that $f_n = f_o$ for infinitely many n's .
  And one has $u[n] = f_o g_n$, when $g_n \in L_2$ for infinitely many n's which implies $u \in L_1^{fin}\, \mathrm{Elim}(L_2)$.
- $|f_n|$ is not bounded whence $u[n] \in L_1$ for infinitely many n's and $u \in \mathrm{Elim}(L_1)$.

The third identity is proved in a way which is very similar to this one □

Property II. d. 4.

The union of any number of normal (resp. central) languages is normal (resp. central) .

Proof

Obviously if $L = \cup \{L_i \mid i \in I\}$ then

$$FG(L) = \cup\ FG(L_i) \mid i \in I\}$$

Thus $FG(L_i) \subseteq FG(L_i^{fin})$ for all $i \in I$ implies

$$FG(L) = \cup \{FG(L_i) \mid i \in I\} \subseteq \cup \{FG(L_i^{fin}) \mid i \in I\}$$

and thus $FG(L) \subseteq FG(L^{fin})$.

Similarly $FG(L_i) \subseteq FG(L_i^{inf})$ for all $i \in I$ implies

$$FG(L) \subseteq FG(L^{inf}) \quad \square$$

Ideal languages will play an important role in the sequel, for we have:

Property II. d. 5.

The intersection of any number of ideal languages is an ideal language.

Proof

Consider $L = \cap \{L_i \mid i \in I\}$ where for all $i \in I$

$$FG(L_i) = L_i^{fin}$$

Then $FG(L) = \cap \{FG(L_i) \mid i \in I\}$

For $f \in FG(L_i) = L_i^{fin}$ for all $i \in I$ implies

$$f \in \cap \{L_i \mid i \in I\} \subseteq FG(L)$$

Conversely $f \in FG(L)$ implies for all $i \in I$ $f \in FG(L_i)$ $\square$

Closed languages have a similar property.

Property II d. 6.

The intersection of any number of closed languages is closed.

Proof

This is a well-known property of topologically closed sets. We can also argue in the following way:

Assume $\quad Adh(L_i) \subseteq L_i \quad$ for all $\quad i \in I$

$$Adh(\cap \{L_i \mid i \in I\}) = \{u \in A^\omega \mid FG(u) \subseteq FG(L_i) \quad \text{for all } i\}$$

$$\subseteq \cap \{Adh(L_i) \mid i \in I\} \subseteq \cap \{L_i \mid i \in I\} \qquad \square$$

For further properties we shall refer to [4].

But we need say a few words about "morphisms":

If $\quad \varphi : B \to A^* \quad$ we define for all $\quad u \in B^\omega$

$$\varphi(u) = \varphi(u(1)) \ \varphi(u(2)) \ \dots \ \varphi(u(n)) \ \dots$$

There is no difficulty in defining the infinite product

$$\varphi(u(1)) \ \dots \ \varphi(u(n)) \ \dots \ .$$

We thus get an application of B^∞ into A^∞, denoted φ.

Unfortunately it is not a morphism: Clearly

$$|\varphi(u)| = \omega \iff \mathrm{card}\{n \mid \varphi(u(n)) \neq \varepsilon\} = \omega \ .$$

If φ is such that there exist

$b_0 \in B \quad$ satisfying $\quad \varphi(b_0) = \varepsilon$

$b_1 \in B \quad$ satisfying $\quad \varphi(b_1) \neq \varepsilon$

one has $\quad b_0^\omega b_1 = b_0^\omega \quad$ and

$$\varphi(b_0^\omega) = \varepsilon \ , \quad \varphi(b_0^\omega)\varphi(b_1) = \varepsilon\varphi(b_1) \neq \varepsilon$$

Also $\quad \varphi(b_0^\omega b_1) = \varphi(b_0^\omega) = \varepsilon \quad$ and thus

$$\varphi(b_0^\omega b_1) \neq \varphi(b_0^\omega)\varphi(b_1)$$

The mapping φ is <u>not</u> a morphism of B^ω into A^ω. We shall however call morphism of B^∞ into A^∞ any application of the previous form, i.e. $\quad \varphi : B^\omega \to A^\infty \quad$ such that

$$\varphi(\varepsilon) = \varepsilon$$

$$\varphi(f(1) \ \dots \ f(n)) = \varphi(f(1)) \ \dots \ \varphi(f(n))$$

$$\varphi(u(1) \ \dots \ u(n) \ \dots \ ) = \varphi(u(1)) \ \dots \ \varphi(u(n)) \ \dots$$

And we define consequently the inverse morphic image of a language in A^∞ :

$$\varphi^{-1}(L) = \{\beta \in B^\infty \mid \varphi(\beta) \in L\}$$

The properties of images of a closed, normal, ideal or central language L under a morphism or inverse morphism have been studied in [4] and we shall use them freely.

III. TRANSITION SYSTEMS

They are also called automata.

III. a. A transition system S is composed of

- a set of configurations C
- an alphabet A
- a set of transitions $T \subseteq C \times A \times C$
- a set $D \subseteq C$ of initial configurations
- a set $C_{fin} \subseteq C$ of final configurations
- a set $C_{inf} \subseteq C$ of infinite configurations

Transition systems are used to recognize languages and to realize processes. Intuitively, the transition system is in a configuration at each instant of time and can move in one unit of time from configuration c to configuration c' by performing a iff $(c, a, c') \in T$. Define for all $t = (c, a, c') \in T$

$$\pi_1(t) = c \qquad \pi_2(t) = a \qquad \pi_3(t) = c'$$

We say that t' can follow immediately t and we write t (> t' iff $\pi_1(t') = \pi_3(t)$.

Then a computation of S is a sequence δ of transitions which may be finite or infinite and satisfies

$$\forall\, n < |\delta| \qquad \delta(n) \text{ (> } \delta(n+1) .$$

We denote $Cal^*(S)$ (resp. $Cal^\omega(S)$) the set of finite (resp. infinite) computations of S .

We shall use the following notations

- for all $f \in A^n$ and $c, c' \in C$ we write
  $c \overset{f}{\rightarrow} c' \Leftrightarrow \exists\, \delta \in Cal^n(S): \pi_1(\delta(1)) = c,\ \pi_2(\delta) = f$ and $\pi_3(\delta(n)) = c'$
  We also write $c \overset{\varepsilon}{\rightarrow} c$ for all $c \in C$.

- for all $u \in A^\omega$, $c \in C$ and $c' \subseteq C$ we write
  $c \overset{u}{\twoheadrightarrow} c' \Leftrightarrow \exists\, \delta \in Cal^\omega(S): \pi_1(\delta(1)) = c,\ \pi_2(\delta) = u$
  and $\mathit{card}\{n \mid \pi_3(\delta(n)) \in c'\} = \omega$.

We say that the infinite computation δ passes infinitely many times in c' iff the last condition is satisfied.

A few obvious lemmas will be used continuously

Lemma III-a-1 $c \xrightarrow{f} c'$ and $c' \xrightarrow{g} c'' \Rightarrow c \xrightarrow{fg} c''$

Lemma III-a-2 $c \xrightarrow{fg} c'' \Rightarrow \exists c' : c \xrightarrow{f} c'$ and $c' \xrightarrow{g} c''$

Lemma III-a-3 $c \xrightarrow{f} c'$ and $c' \overset{u}{\twoheadrightarrow} C' \Rightarrow c \overset{fu}{\twoheadrightarrow} C'$

Lemma III-a-4 $c \overset{fu}{\twoheadrightarrow} C' \Rightarrow \exists c' : c \xrightarrow{f} c'$ and $c' \overset{u}{\twoheadrightarrow} C'$

We can now define the languages recognized by S.

The finitary language recognized by S is :

$$L^*(S) = \{f \in A^* \mid \exists c_o \in D , c \in C_{fin} \quad c_o \xrightarrow{f} c$$

(We also write $L^*(S) = \{f \in A^* \mid D \xrightarrow{f} C_{fin}\}$)

The purely infinitary language recognized by S is

$$L^{\omega}(S) = \{u \in A^{\omega} \mid D \overset{u}{\twoheadrightarrow} C_{inf}\}$$

The infinitary language recognized by S is

$$L^{\infty}(S) = L^*(S) \cup L^{\omega}(S)$$

And a process p is said to be directly realized by the transition system S iff $HR^{\infty}(p) = L^{\infty}(S)$. The process p, with A as set of actions, is realized by S, with B as set of actions, iff there exists an alphabetic morphism $\varphi : B \to A \cup \{\varepsilon\}$ such that

$$HR'^{\infty}(p) = \varphi(L^{\infty}(S)) \; ; \quad (\varphi \text{ is an "observation" morphism}).$$

III. b. Rational infinitary languages and finite transition systems

One now has the following general problems

- given a property $P \in \{\text{closed, normal, ideal, central}\}$ or, in the set of boolean combinations of those, can one describe a family of transition systems $TS(P)$ such that

  L has the property $P \Leftrightarrow \exists S \in TS(P) : L = L^{\infty}(S)$
- can one decide whether a given language $L = L^{\infty}(S)$ is closed, normal, ideal or central?

It happens that in the general case of transition systems, where both A and C may be infinite, there is not much to say.

We shall thus now on restrict ourselves to the consideration of finite transition systems, i.e. the family of all S's such that $card(A) < \omega$ and $card(C) < \omega$.

We define then the family of rational infinitary languages on a finite alphabet A as

$$\mathrm{Rat}(A^\infty) = \{L \subseteq A^\infty \mid \text{there exists a finite transition systems } S \text{ such that } L = L^\infty(S)\} .$$

Remark

This definition is improper, since, if one follows Eilenberg's definitions [11] any singleton $\{m\}$ should be a rational language in Rat(M), the family of rational subsets of the monoid M.

This ist not satisfied here. Indeed, one proves easily that for all $u \in A^\omega$ one has $\{u\} \in \mathrm{Rat}(A^\infty) \Leftrightarrow u$ is periodic where u is periodic $\Leftrightarrow \exists f, g \in A^* \quad u = fg^\omega$.

Assume $\{u\} = L^\omega(S)$ for some finite transition system S.

Then $D \overset{v}{\twoheadrightarrow} C_{inf} \Leftrightarrow v = u$.

Since C is finite and consequently $C_{inf} \subseteq C$ is also finite, one has for all $\delta \in \mathrm{Cal}^\omega(S)$, $card\{n \mid \pi_3(\delta(n)) \in C_{inf}\} = \omega \Rightarrow \exists c' \in C_{inf} \quad card\{n \mid \pi_3(\delta(n)) = c'\} = \omega$.

Whence if $D \overset{u}{\twoheadrightarrow} C_{inf}$ there exists certainly $c' \in C_{inf}$ such that $D \overset{f}{\rightarrow} c'$ for some $f \in A^*$ and

$c' \overset{g}{\rightarrow} c'$ for some $g \in A^*$.

Then one can consider $c' \in C_{inf}$ such that

$\exists f \in A^* \quad D \overset{f}{\rightarrow} c'$

and $\mathrm{Loop}(c') = \{g \in AA^* \mid c' \overset{g}{\rightarrow} c'\}$ is non empty.

And one has $\{u\} = f(\mathrm{Loop}(c'))^\omega$. The result follows from

Lemma III. b. 1.

For all $L_1, L_2 \subseteq A^*$

$$card\, L_1(L_2^\omega) = 1 \Rightarrow \exists f, g \in A^*$$

$$L_1(L_2^\omega) = \{fg^\omega\}$$

Proof

Assume $\{u\} = L_1(L_2)^{\omega}$. Then both L_1 and L_2 are non empty and for any $f \in L_1$ and $g \in L_2$

$$fg^{\omega} \in L_1(L_2^{\omega}) = \{u\} \Rightarrow u = fg^{\omega} \quad \square$$

We recall some theorems.

Theorem III. b. 1. [11]

The language $L \subseteq A^{\infty}$ is rational iff its finitary part L^{fin} is finitary rational and its purely infinitary part L^{inf} is a finite union

$$L^{inf} = \cup \{K_i(K'_i)^{\omega} \mid i \in [p]\}$$

where $K_i, K'_i \in Rat(A^*)$

Proof

Assume $L = L^{\infty}(S) = L^*(S) \quad L^{\omega}(S)$. Clearly $L^*(S) \in Rat(A^*)$.

Call $L^*(D, S, c)$ the language $L^*(D, S, c) = \{f \in A^* \mid D \overset{f}{\to} c\}$

and $Loop(S, c) = L^+(c, S, c) = \{f \in A^+ \mid c \overset{f}{\to} c$.

Then $L^{inf} = \cup \{L^*(D, S, c) (Loop(S, c))^{\omega} \mid c \in C_{inf}\}$.

One way the inclusion is clear.

Conversely we just remark that $\delta \in Cal^{\omega}(S)$ and

$\omega = card\{n \mid \pi_3(\delta(n)) \in C_{inf}\}$ implies, if C is finite

$\exists\, c \in C_{inf}$ such that $card\{n \mid \pi_3(\delta(n)) = c\} = \omega$.

If now $L = L^{fin} \cup \{K_i(K'_i)^{\omega} \mid i \in [p]\}$ we easily build such that $L = L^{\infty}(S)$: first we build S''_i such that

$K_i(K'_i)^{\omega} = L^{\omega}(S_i)$.

We take transition systems S_i and S'_i such that

$L\ (S_i) = K_i$ and $L\ (S'_i) = K'_i$.

We assume the sets of configurations are disjoint. The set of configurations of S''_i is $C_i \cup C'_i$. The set of transitions is

$T''_i = \{(c, a, c') \mid (c, a, c') \in T_i$ or $(c, a, c') \in T_i$ or $c' \in D'_i$ and $\exists\, c'' \in C_{ifin} : (c, a, c'') \in T_i\}$

The sets of initial configurations and final configurations are:

$$D''_i = D_i (\cup D'_i \quad \text{if} \quad K_i \ni \varepsilon)$$

$$D''_{iinf} = D'_{ifin}$$

We build a transition system recognized $L_1 \cup L_2$ in the standard way (with $C_1 \times C_2$ as a set of configurations) □

Theorem III. b. 2. [13]

$Rat(A^\infty)$ is the smallest family of subsets of A^∞ which contains the finite finitary subsets of A^∞ (i.e. subsets of A^*) and is closed under union, product, finite star and infinite power. We shall mainly use the following

Theorem III. b. 3. [8]

$Rat(A^\infty)$ is closed under insertion.

Proof

Let $L_1 = L^\infty(S_1)$ and $L_2 = L^\infty(S_2)$

Take $C = C_1 \times C_2$ and

$$T = \{((c_1, c_2), a, (c'_1, c'_2)) \mid (c_1, a, c'_1) \in T_1 \text{ and } (c_2, a, c'_2) \in T_2\}$$

Clearly $(c_1, c_2) \overset{f}{\to} (c'_1, c'_2)$ in T iff $c_1 \overset{f}{\to} c'_1$ in T_1 and $c_2 \overset{f}{\to} c'_2$ in T_2

And one has $L_1^{fin} \cap L_2^{fin} = \{f \in A^* \mid D_1 \times D_2 \overset{f}{\to} C_{1fin} \times C_{2fin}\}$

The situation is not quite so simple for infinite words.

Indeed $u \in L_1^{inf} \cap L_2^{inf}$ iff there exists an infinite computation in T :

$$\delta = (c_1^{(0)}, c_2^{(0)}), u(1), (c_1^{(1)}, c_2^{(2)}), u(2), \ldots, c_1^{(n-1)}, c_2^{(n-1)}), u(n), (c_1^{(n)}, c_2^{(n)})$$

with $(c_1^{(0)}, c_2^{(0)}) \in D_1 \times D_2$ and the two following conditions are satisfied:

$$card\{n \mid c_1^{(n)} \in C_{1inf}\} = \omega$$
$$card\{n \mid c_2^{(n)} \in C_{2inf}\} = \omega$$

If δ is such a computation there exists a pair (c_1, c_2) such that $c_1 \in C_{1inf}$, $(c_1^{(n)}, c_2^{(n)}) = (c_1, c_2)$ for all $n \in N$, where N is an infite subset of $\mathbb{N}$ and for $n_1, n_2 \in N$ there exists n_3 such that $n_1 < n_3 < n_2$ and $c_2^{(n_3)} \in C_{2inf}$.

Whence we can define for all $(c_1, c_2) \in C_{1inf} \times C_2$ the languages

$$H_{c_1,c_2} = \{f \mid D_1 \times D_2 \xrightarrow{f} (c_1, c_2)\}$$

and $H'_{c_1,c_2} = \{ff' \mid \exists\ c_3 \in D_{2inf} \quad c_2 \xrightarrow{f} c_3,\ c_3 \xrightarrow{f'} c_2 \text{ and } c_1 \xrightarrow{ff'} c_1\}$

These languages are obviously finitary and rational and one has

$$L_1^{inf} \cap L_2^{inf} = \cup\ \{H_{c_1,c_2}(H'_{c_1c_2})^{\omega} \mid (c_1, c_2) \in C_{1inf} \times C_2\} \quad .$$

We have proved one inclusion, the reverse inclusion is obvious. We apply theorem III. b. 1. to end the proof and if needed build a finite transition system which recognizes $L_1^{inf} \cap L_2^{inf}$ □

Example

Consider $\quad L_1 = \{u \in (a \cup b)^{\omega} \mid |u|_a = \omega\}$

and $\quad L_2 = \{u \in (a \cup b)^{\omega} \mid |u|_b = \omega\}$

Obviously $\quad L_1^{inf} \cap L_2^{inf} = \{u \in (a \cup b)^{\omega} \mid |u|_a = |u|_b = \omega\}$

We give two finite transition systems recognizing L_1 and L_2

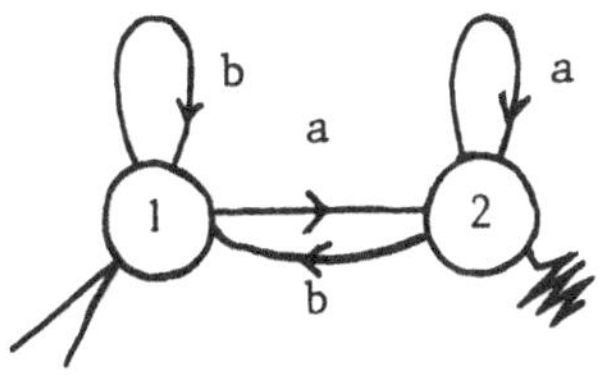

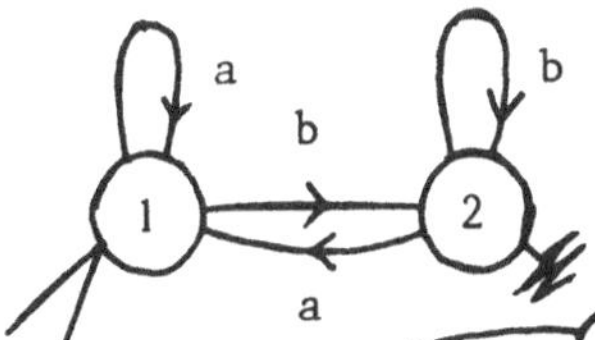

The product of S_1 and S_2 is:

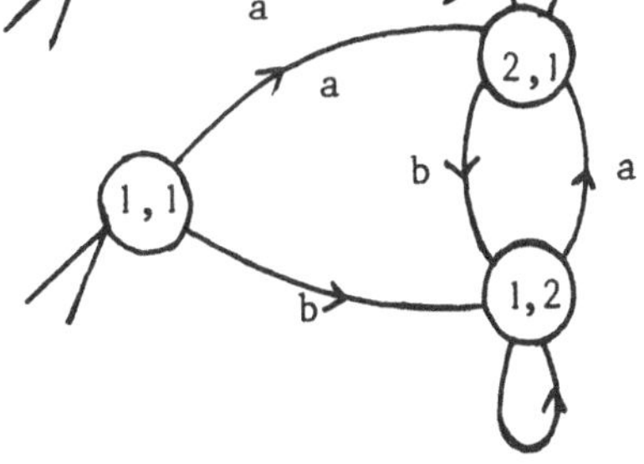

and one cannot find a subset of $\{(1,1), (2,1), (1,2)\}$ such that $L_1^{inf} \cap L_2^{inf}$ is equal to the set of infinite words for which there exists an infinite computation passing infinitely many times in this subset. But one can write

$$L_1^{inf} \cap L_2^{inf} = (a \cup b)^* a \; (bb^* a^* a)^\omega$$

and thus recognize $L_1^{inf} \cap L_2^{inf}$ by the following transition system

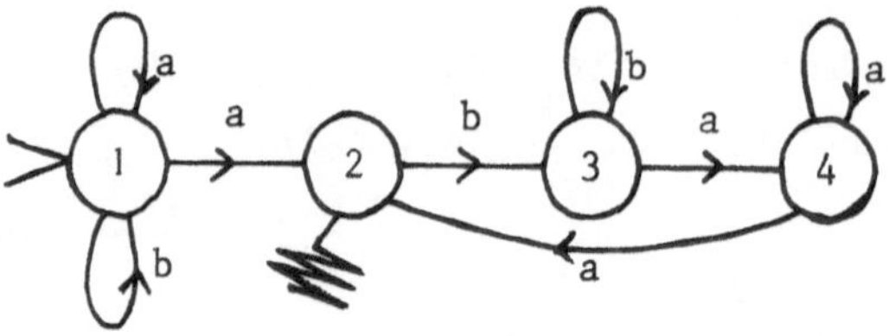

The "simplest" transition system recognizing $L_1^{inf} \cap L_2^{inf}$ is

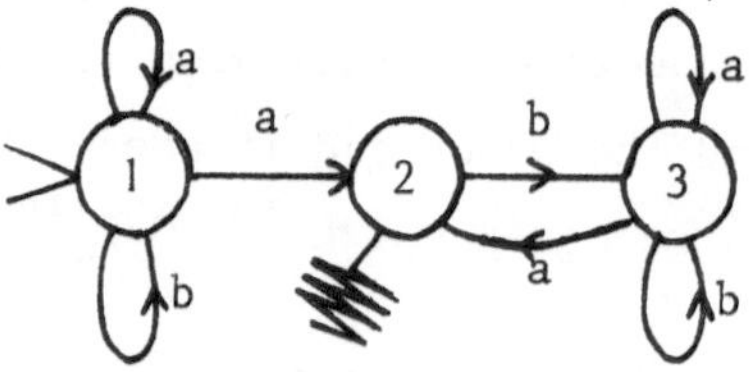

it corresponds to the factorization

$$\{ u \mid |u|_a = |u|_b = \omega \} = (a \cup b)^* a \; [\; b(a \cup b)^* a \;]^\omega \qquad \square$$

III. c. Deterministic rational languages

A transition system S is said to be <u>deterministic</u> iff

$$card(D) < \omega$$

and $\forall\, c \in C,\ a \in A$ the set $\lambda(c, a) = \{c' \mid (c, a, c') \in T\}$ contains at most one element.
We denote by $DRat(A^\infty)$ the family of subsets of A^∞ which are recognizable by a deterministic finite transition system. As above $DRat(A^\omega)$ and $DRat(A^*)$ are defined as the families of purely infinitary and finitary deterministic rational languages. The important fact here is that, whereas $DRat(A^*) = Rat(A^*)$, the inclusion $DRat(A^\omega) \subseteq Rat(A^\omega)$ is strict.

A typical example of rational infinitary language which is not deterministic rational is

$$L = (a \cup b)^* b^\omega = \{ u \in (a \cup b)^\omega \mid |u|_a < \omega \}$$

We shall prove it using the following

Theorem III. c. 1. [11]

The three following conditions are equivalent

1 - $L \in DRat(A^\omega)$

2 - $L = Elim(K)$ for some rational finitary language $K \in Rat(A^*)$

3 - L is a finite union of the form

$$L = \bigcup_{i=1}^{i=p} K_i (K'_i)^\omega \quad \text{where } K_i, K'_i \in Rat(A^*) \text{ and}$$

are prefix.

(a language K is prefix iff $f \in K$ and $g \in K \Rightarrow g = \varepsilon$)

Example

Assume $L = (a \cup b)^* b^\omega = Elim(K)$ for some finitary rational language K. Take $n_o > 0$. From $a^{n_o} b^\omega \in Elim(K)$ one can derive the existence of $n'_o > 0$ such that $a^{n_o} b^{n'_o} \in K$. Consider $n_1 > 0$. From $a^{n_o} b^{n'_o} a^{n_1} b^\omega \in K$ one can find $n_1 > 0$ such that $a^{n_o} b^{n'_o} a^{n_1} b^{n_1} \in K$.

Repeating the argument one can find

$u = a^{n_o} b^{n'_o} a^{n_1} b^{n'_1} \ldots a^{n_i} b^{n'_i} \ldots$ such that

$\forall i \in \mathbb{N} \quad a^{n_o} b^{n'_o} a^{n_1} b^{n'_1} \ldots a^{n_i} b^{n'_i} \in K$

But this implies $u \in Elim(K)$ and a contradiction since $|u|_a = \omega$ □

We recall next the important result whose proof is difficult.

Theorem III. c. 2. (Mac Naughton [18])

$Rat(A^\omega) = (DRat(A^\omega))^B$

where L is a family of subsets L^B denotes the boolean closure of L i.e. the set of subsets which can be obtained from elements in L by a finite number of boolean operations (union, intersection, complementation).

Example

$$L = \{u \in (a \cup b)^{\omega} \mid |u|_a < \omega\} \quad \text{is equal to}$$
$$(a \cup b)^{\omega} \setminus \{u \in (a \cup b)^{\omega} \mid |u|_a = \omega\}$$

and since $\{u \in (a \cup b)^{\omega} \mid |u|_a = \omega\} = ((a \cup b)^* a)^{\omega}$
$= \mathrm{Elim}((a \cup b)^* a)$

one gets L as the difference of two deterministic rational languages in $(a \cup b)^{\omega}$.

For proofs of III. c. 2. see [11] .

III. d. Reduced transition systems

III. d. 1. Accessibility

Let S be a transition system. We write $c \overset{*}{\rightarrow} c'$ iff there exists $f \in A^*$ such that $c \overset{f}{\rightarrow} c'$ and we say that c' is accessible from c iff $c \overset{*}{\rightarrow} c'$. The set $\mathrm{Acc}(c)$ is defined as

$$\mathrm{Acc}(c) = \{c' \mid c \overset{*}{\rightarrow} c'\} .$$

We say that c is coaccessible from c' and write $c \in \mathrm{Coacc}(c')$ iff $c \overset{*}{\rightarrow} c'$.

III. d. 2. Liveness

Let S be a transition system. We write $c \overset{\omega}{\twoheadrightarrow} C_{inf}$ iff there exists $u \in A^{\omega}$ such that $c \overset{u}{\twoheadrightarrow} C_{inf}$. And we say that c is alive with respect to C_{inf} iff $c \overset{\omega}{\twoheadrightarrow} C_{inf}$ defining

$$\mathrm{Viv}(C_{inf}) = \{c \in C \mid c \overset{\omega}{\twoheadrightarrow} C_{inf}\}$$

And the subset C_{inf} is said to be alive iff every $c \in C_{inf}$ is alive with respect to C_{inf} .

III. d. 3. Reducibility

Let S be a transition system.

Define $C_R = Acc(D) \cap Coacc(C_{fin} \cup Viv(C_{inf}))$

$T_R = C_R \times A \times C_R \cap T$:

Assume $c \in C_R$: there exists $c_o \in D$ and $f \in A^*$ such that $c_o \overset{f}{\to} c$ in T and either

$c' \in C_{fin}$ and $g \in A^*$ such that $c \overset{g}{\to} c'$ in T

or $c' \in Viv(C_{inf})$ and g such that $c \overset{g}{\to} c'$ in T .

In this last case there exists also $u \in A$ such that

$c' \overset{u}{\twoheadrightarrow} C_{inf}$ in T .

We claim that $c_o \in D_R = D \cap C_R$ and that $c_o \overset{f}{\to} c$ in T_R : this is fairly obvious since all the configurations along a computation going from c_o to c reading f are accessible from D by definition and coaccessible from $C_{fin} \cup Viv(C_{inf})$. Thus all these configurations are in C_R and all the transitions are in T_R .

The set $Viv(C_{inf}) \cap C_{inf} \cap C_R = C_{Rinf}$ is alive. For if $c' \in C_{Rinf}$ then there exists $u \in A^\omega$ such that $c' \overset{u}{\twoheadrightarrow} C_{inf}$ and c' is accessible from D_R . By an argument similar to the previous one $c' \overset{u}{\twoheadrightarrow} C_{Rinf}$ in T_R . Then it is clear that c in C_R is coaccessible (in T_R) from $C_{Rfin} \cup C_{Rinf}$ where $C_{Rfin} = C_{fin} \cap C_R$.

We say that a transition system S is <u>reduced</u> iff the three following conditions are satisfied

1 - every $c \in C$ is accessible from D

2 - the set of infinite configurations is alive

3 - every $c \in C$ is coaccessible from $C_{fin} \cup C_{inf}$

And we state

Property III. d. 3.

For every transition system S there exists a reduced transition system S_R which recognizes the same infinitary language.

Proof

It is almost entirely written above.

We take S_R given by C_R, T_R, D_R, C_{Rfin}, C_{Rinf} .

We know that S_R is reduced.

Then $L^*(S_R) = L^*(S)$: the inclusion $L^*(S_R) \subseteq L^*(S)$ is obvious and the reverse inclusion comes from the fact that

$$c \in C_{fin},\ c_o \in D \quad \text{and} \quad c_o \xrightarrow{f} c \ \text{in}\ T \quad \text{imply}$$
$$c \in C_{Rfin},\ c_o \in D_R \quad \text{and} \quad c_o \xrightarrow{f} c \ \text{in}\ T_R \ .$$

One also has $L^\omega(S_R) = L^\omega(S)$. One inclusion is also obvious, namely $L^\omega(S_R) \subseteq L^\omega(S)$: it comes immediately from the inclusions $C_R \subseteq C$, $D_R \subseteq D$, $T_R \subseteq T$, $C_{Rinf} \subseteq C_{inf}$.

The reverse inclusion comes from the fact that for all $c_o \in D$ $c_o \overset{u}{\twoheadrightarrow} C_{inf}$ in T implies $c_o \overset{u}{\twoheadrightarrow} C_{Rinf}$ in T_R for each configuration in C_{inf} which is passed through along an infinite computation starting in c_o and reading u is obviously alive with respect to C_{inf} i.e. belongs to $Viv(C_{inf}) \cap C_{inf}$ and is accessible from D . □

The last property is quite general and holds for every transition system whichever are the cardinalities of C and A . In general of course, one cannot compute S_R .

Below we give a slightly different definition of a reduced system which is quite useful in the case of finite transition systems.

Let S be a transition system and $c \in C$.

A loop of c is a non empty word $f \in A^+$ such that $c \xrightarrow{f} c$.

And the configuration c is said to be looping iff it has a loop.

If S is finite then one has

$$c \overset{u}{\twoheadrightarrow} C_{inf} \iff c \overset{u}{\twoheadrightarrow} \{c' \in C_{inf} \mid c' \text{ is looping}\} \ .$$

One can take

$$C_{Rinf} = \{c' \in C_{inf} \cap Acc(D) \mid c' \text{ is looping}\}$$
$$C_{Rfin} = C_{fin} \cap Acc(D)$$
$$D_R = D \cap Coacc(C_{Rinf} \cup C_{Rfin})$$
$$C_R = Acc(D_R) \cap Coacc(C_{Rinf} \cup C_{Rfin})$$
$$T_R = T \cap C_R \times A \times C_R$$

And S_R is defined as $C_R, T_R, D_R, C_{Rfin}, C_{Rinf}$. Clearly $L^\infty(S_R) = L^\infty(S)$ and S_R is reduced in the following sense:

- every configuration is accessible from D_R
- every infinite configuration is looping
- every configuration is coaccessible form $C_{Rfin} \cup C_{Rinf}$

And one has

Porperty III. d. 4.

Given any finite transition system S one can compute a reduced transition system S_R such that $L^\infty(S) = L^\infty(S_R)$.

Proof

Clearly if S is finite, i.e. $card(C) = N < \omega$ for all $c, c' \in C$ $c \overset{f}{\to} c' \iff \exists\, g, |g| \leq N$, such that $c \overset{g}{\to} c'$ (by the standard pumping lemma of Bar-Hillel-Perles-Shamir, recalled in [11]) and thus if $card(A) < \omega$ one can decide whether $c \overset{*}{\to} c'$ □

We have a straightforward consequence of the previous construction which solves part of the problem stated in § III. b.

Property III. d. 5.

If S is a reduced finite transition system and $L = L^\infty(S)$ is closed , then

$$L^{inf} = \{u \in A^\omega \mid D \overset{u}{\twoheadrightarrow} C\} \ (= L^{t\omega}(S)) .$$

Proof

$L^{inf} = L^{\omega}(S) = \{u \in A^{\omega} \mid D \overset{u}{\twoheadrightarrow} C_{inf}\}$

The inclusion $C_{inf} \subseteq C$ obviously implies $L^{inf} \subseteq L^{t\omega}(S)$.
Conversely we prove $L^{t\omega}(S) \subseteq L^{inf}$: $D \overset{u}{\twoheadrightarrow} C \Rightarrow \forall n \quad D \xrightarrow{u[n]} C$
$D \xrightarrow{u[n]} C \Rightarrow u[n] \in FG(L)$ since S is reduced.
Whence $D \overset{u}{\twoheadrightarrow} C \Rightarrow FG(u) \subseteq FG(L) \Rightarrow u \in L^{inf}$ □

Theorem III. d. 6.

The rational infinitary language L is closed iff $L = L^{\infty}(S)$ for some finite transition system S such that $C_{inf} = C$.

Proof

We need prove that $C_{inf} = C$ implies $Adh(L^{\infty}(S)) \subseteq L$.
We apply Koenig's lemma. If $u \in Adh(L^{\infty}(S))$ $\forall n \ \exists c \in C \quad D \xrightarrow{u[n]} c$.
Call E_n the set of pairs (c, c') such that $D \xrightarrow{u[n]} c$ and $c \xrightarrow{u(n+1)} c'$. For all n E_n is non empty and finite (as a subset of $C \times C$). For all n and $(c, c') \in E_{n+1}$ there exists c'' such that $(c'', c) \in E_n$. Whence there exists an infinite computation of S reading u ,

$$c_0, u(1), c_1, u(2), \ldots, c_n, u(n+1), c_{n+1}, \ldots$$

and $u \in L^{\omega}$ □

Remarks

1. The theorem is false if C is infinite and does not have the property of bounded non-determinism (see [26]) .

2. The theorem shows the necessity of introducing $C_{inf} \subseteq C$ in general to deal with not closed languages and processes. As we shall see later "fair" processes are usually not closed.

III. e. Determinization of a transition system

The construction is very well known.

Let S be a transition system.

We have defined $\lambda(\hat{c},a) = \{c' \mid (c,a,c') \in T\}$ for all $c \in C$, $a \in A$

Define $\lambda(\hat{c}, a) = \cup \{\lambda(a, c) \mid c \in \hat{c}\}$ for all $\hat{c} \in \iota^{c}$, $a \in A$

Then take S given by

$$\hat{C} = P(C)$$

$$\hat{T} = \{(\hat{c}, a, \hat{c}') \mid \hat{c}' = \lambda(\hat{c}, a)\}$$

$$\hat{D} = \{D\} \quad \text{and} \quad \hat{C}_{fin} = \{\hat{c} \in \hat{C} \mid \hat{c} \cap C_{fin} = \emptyset\}$$

This system is deterministic and

for all $f \in A^*$ $\hat{D} \overset{f}{\to} \hat{C}_{fin} \Leftrightarrow (D, f) \cap C_{fin} \neq \emptyset$

and this happens iff $\exists\, c_o \in D$, $c \in C_{fin}$ $c_o \overset{f}{\to} c$.

Thus $L^*(\hat{S}) = L^*(S)$.

In fact we shall use the reduced version of $\hat{S}$ obtained by considering only the elements in $\hat{c}$ which are accessible from $\hat{D}$ and accessible from $\hat{C}_{fin}$ in $\hat{T}$. This gives us the system

$$S_d = \langle C_d, T_d, D_d, C_{dfin} \rangle$$

Remark

We have noted before that in general one cannot define $\hat{C}_{inf}$ such that $\hat{D} \overset{u}{\twoheadrightarrow} \hat{C}_{inf}$ (in $\hat{T}$) $\Leftrightarrow D \overset{u}{\to} C_{inf}$ (in T) .

A special case where this is possible gives rise to :

Property III. e. 1.

$L \in Rat(A^\infty)$ and L closed imply $L \in DRat(A^\infty)$.

Proof

In S_d one has $D_d \overset{f}{\to} C_d \Leftrightarrow D \overset{f}{\to} C$ in S (in other words,

$$\lambda(D_d, f) \neq \emptyset \Leftrightarrow \exists\, c_o \in D,\ c \in D \quad c_o \overset{f}{\to} c) \ .$$

Whence the sequence of equivalences

$$D \overset{u}{\twoheadrightarrow} C_{inf} \Leftrightarrow \forall n \quad D \xrightarrow{u[n]} C \quad (\text{taking } S \text{ such that } C_{inf} = C)$$
$$\Leftrightarrow \forall n \quad D_d \xrightarrow{u[n]} C_d$$
$$\Leftrightarrow \quad D_d \overset{u}{\twoheadrightarrow} C_d \quad \text{in } S_d$$

And if we take $C_{dinf} = C_d$ then $L^{\omega}(S_d) = L^{\omega}(S)$ □

III. f. Normal transition systems

Theorem III. f. 1.

The rational infinitary language is normal iff $L = L^{\infty}(S)$ for some finite transition system S such that $C \in \text{Coacc}(C_{fin})$. Such a transition system is said to be normal .

Proof

The fact that $C \in \text{Coacc}(C_{fin})$ implies

$$FG(L) \subseteq \{f \in A^* \mid D \overset{f}{\rightarrow} C\} \subseteq FG(L^{fin})$$

is quite obvious (and independent of the finiteness of C). To prove the converse we first take a reduced transition system S such that $L = L^{\infty}(S)$. And we build $\underline{S}$ given by

$$\underline{C} = \text{Coacc}(C_{inf})$$
$$\underline{T} = T \cap \underline{C} \times A \times \underline{C}$$
$$\underline{D} = D \cap \underline{C}, \quad \underline{C}_{fin} = \underline{C} \quad \text{and} \quad \underline{C}_{inf} = C_{inf} \cap \underline{C}$$

Then we have

$$L^*(\underline{S}) = FG(L^{inf}) \quad \text{and} \quad L^{\omega}(\underline{S}) = L^{inf} \quad .$$

Afterwards, using $\underline{S}$ and the deterministic transition system S_d previously built, we construct S' as follows:

$$C' = \underline{C} \cup \{\emptyset\} \times C_d$$
$$T' = \{((\underline{c}, c_d), a, (\underline{c}' c_d')) \mid (\underline{c}, a, c_d') \in \underline{T} \quad \text{and} \quad \lambda(c_d, a) = c_d'\}$$
$$\cup \{((\underline{c}, c_d), a, (\emptyset, c_d')) \mid \lambda(\underline{c}, a) = \emptyset \quad \text{and} \quad \lambda(c_d, a) = c_d'\}$$
$$\cup \{((\emptyset, c_d), a, (\emptyset, c_d')) \mid \lambda(c_d, a) = c_d'\}$$

$$D' = \underline{D} \times D_d$$
$$C'_{fin} = \underline{C} \cup \{\emptyset\} \times C_{dfin}$$
$$C'_{inf} = \underline{C} \times C_d$$

We check easily that every configuration in C' is coaccessible from C'_{fin} since $(\underline{c} \cup \emptyset, c_d) \xrightarrow{f} (\underline{c}' \cup \emptyset, c'_d) \Leftrightarrow c_d \xrightarrow{f} c'_d$ and in S_d every configuration c_d is coaccessible from C_{dfin}. This also proves that $L^*(S') = L^*(S_d) = L^*(S)$.

Assume now that L is normal:

$$D \xrightarrow{u} C_{inf} \Rightarrow \forall n \quad \underline{D} \xrightarrow{u[n]} \underline{C}$$
$$\Rightarrow \forall n \quad D_d \xrightarrow{u[n]} C_d$$

Whence for all n $\underline{D} \times D_d \xrightarrow{u[n]} \underline{C} \times C_D$ and $D' \xrightarrow{u} C'_{inf}$. This shows $L^{\omega}(S) \subseteq L^{\omega}(S')$. The reverse inclusion is obvious □

This proof does not depend on the cardinality of C (but for the fact that $\underline{S}$ and S_d are not usually computable if $|C| = \omega$)

<u>Example</u>

S :

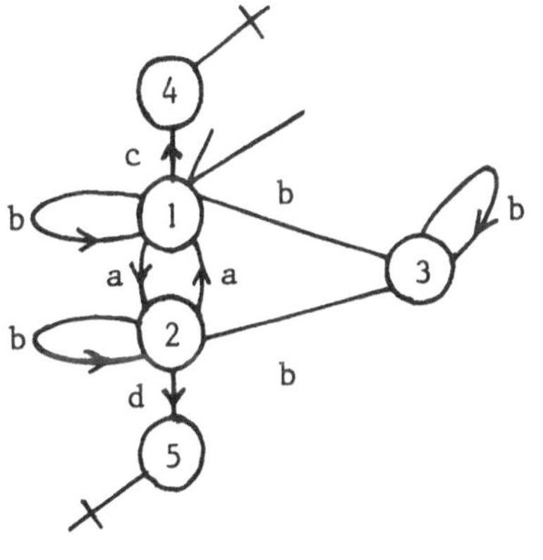

$D = \{1\}$

$C_{fin} = \{4, 5\}$

$C_{inf} = \{3\}$

We have $L^{\omega}(S, C_{inf}) = (a \cup b)^* b^{\omega}$

$$L^*(S, C_{fin}) = \{fc \mid f \in (a \cup b)^* \text{ and } |f|_a \text{ is even}\}$$
$$\cup \{fd \mid f \in (a \cup b)^* \text{ and } |f|_a \text{ is odd }\}$$

$L = L^{\infty}(S)$ is clearly normal but in S 3 is not accessible from C_{fin} .

S_d :

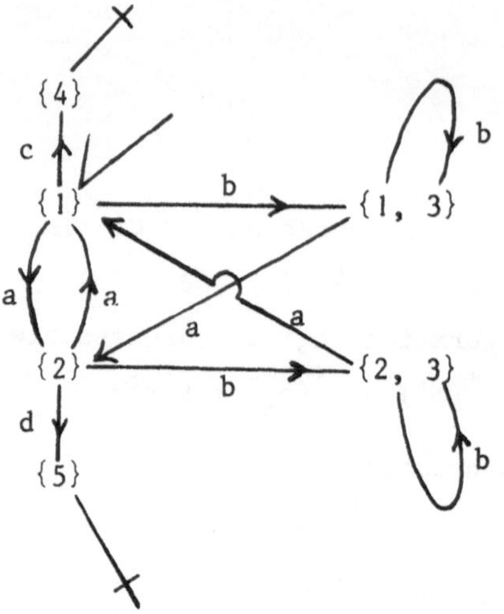

$D_d = \{\{1\}\}$

$C_{dfin} = \{\{4\}\},\{\{5\}\}$

We remark that for no subset of $C_d = \{\{1\},\{2\},\{4\},\{5\},\{1,3\},\{2,3\}\}$ does one have $L^{\omega}(S_d) = L^{\omega}(S)$.

$\underline{S}$:

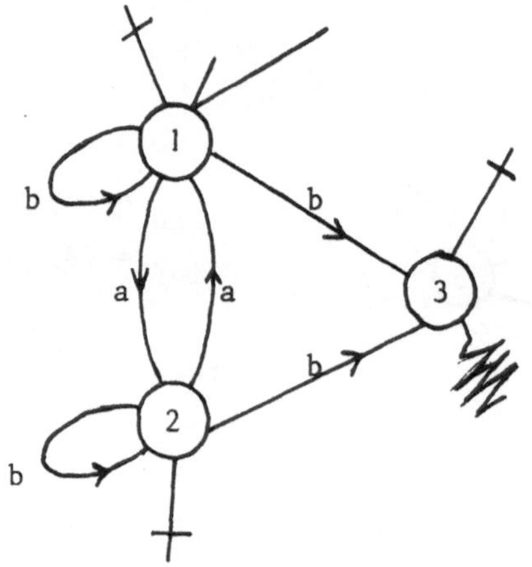

$\underline{D} = \{1\}$

$\underline{C}_{fin} = \{1, 2, 3\}$

$\underline{C}_{inf} = \{3\}$

Eventually we build S' which has the desired property :

S' :

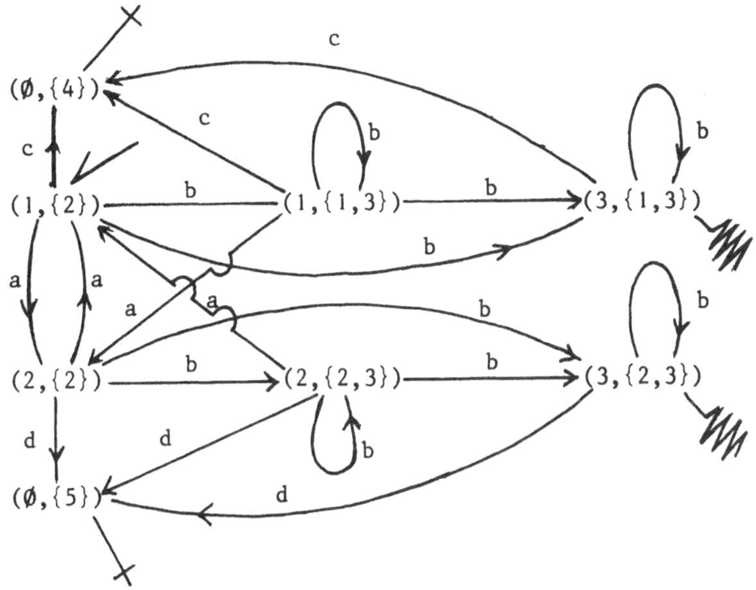

$$D' = \{1,\{1\}\},\ C'_{fin} = \{(\emptyset,\{4\}),\ (\emptyset,\{5\})\}$$

$$C'_{inf} = \{(3,\{1,3\}),\ (3,\{2,3\})\}$$ □

The two next statements are immediate consequences of the two previous theorems.

Property III. f. 2.

The rational language L is closed and normal iff $L = L^{\infty}(S)$ for some finite transition system S such that

$$C_{inf} = C \qquad \text{and} \qquad C \subseteq Coacc(C_{fin})$$

Property III. f. 3.

The rational language is ideal and closed iff $L = L^{\infty}(S)$ for some finite transition system such that $C_{fin} = C_{inf} = C$.

Remark: Our previous work on permanent transition systems [27] was in fact restricted to this case.

III. g. Permanent systems and centers

We say that S is permanent iff

$$\forall c \in C \quad \exists u \in A^{\omega} \qquad c \overset{u}{\twoheadrightarrow} C_{inf} \quad .$$

Theorem III. g. 1.

L is central iff $L = L^{\infty}(S)$ for some permanent system S .

Proof

S permanent obviously implies $L^{\infty}(S)$ central .

Conversely assume $L = L^{\infty}(S)$ with S reduced and L central.

The Permanent system $\underline{\underline{S}}$ is the "product" of $\underline{S}$ by S_d , i.e.

$$\underline{\underline{C}} = \underline{C} \times C_d$$
$$\underline{\underline{T}} = \{((\underline{c},c_d),\ a,\ (\underline{c}',c_d')) \mid (\underline{c},a,\underline{c}') \in \underline{T} \quad \text{and} \quad \lambda(c_d,a) = c_d'\}$$
$$\underline{\underline{D}} = \underline{D} \times D_d$$
$$\underline{\underline{C}}_{fin} = \underline{C} \times C_{dfin}$$
$$\underline{\underline{C}}_{inf} = \underline{C}_{inf} \times C_d$$

The system $\underline{\underline{S}}$ is permanent, for $\underline{S}$ is permanent :

$\underline{S}$ is the transition system which recognizes $FG(L^{inf}) \cup L^{inf}$.

It can be defined in the most general case as

$$\underline{C} = \{c \in C \mid \exists u : \ c \overset{u}{\twoheadrightarrow} C_{inf}\}$$
$$\underline{T} = T \cap \underline{C} \times A \times \underline{C}$$
$$\underline{D} = D \cap \underline{C},\ \underline{C}_{fin} = \underline{C},\ \underline{C}_{inf} = C_{inf} \cap \underline{C}$$

And $\underline{\underline{S}}$ is permanent for if

$$\underline{c}_o,\ u(1),\ \underline{c}_1,\ \ldots,\ \underline{c}(n-1),\ u(n),\ \underline{c}_n,\ \ldots$$

is an infinite computation of $\underline{S}$ reading u from $\underline{c}_o$ and $\underline{c}_o \in c_d$ then $\lambda(c_d, u[n])$ contains $\underline{c}_n$ for all n .

Thus

$$(\underline{c}_o,\ c_d),\ u(1),\ (\underline{c}_1,\ \lambda(c_d,\ u(1))),\ \ldots,$$
$$(\underline{c}_{(n-1)},\ \lambda(c_d,\ u[n-1])),\ u(n),\ (\underline{c}_n,\ \lambda(c_d,\ u[n])),\ \ldots$$

is an infinite computation of $\underline{\underline{S}}$ reading u .

Now we use the fact that L is central: $L^{fin} \subseteq FG(L^{inf})$ implies that for all $f \in A^*$ and $c_d \in C_d$

$$\lambda(c_d, f) \neq \emptyset \Rightarrow \lambda(c_d, f) \cap \underline{C} = \emptyset .$$

The verification that $L = L^{\infty}(\underline{\underline{S}})$ is immediate. We have proved immediately above $L^{inf} = L^{\omega}(\underline{\underline{S}})$. Now if $\lambda(D_d, f) \in C_{cfin}$ and $f \in L^{fin} \subseteq FG(L^{inf})$ is the left factor $u[n]$ of some $u \in L^{inf}$ there exists an infinite computation of $\underline{S}$ reading u

$$\underline{c}_o, u(1), \underline{c}_1, \ldots, \underline{c}_{(n-1)}, u(n), \underline{c}_n, \ldots$$

with $\underline{c}_o \in D = D_d$. Clearly $(\underline{c}_o, D) \in \underline{\underline{D}}$ and $(\underline{c}_o, D) \overset{f}{\rightarrow} (\underline{c}_n, \lambda(D_d, f))$ whence $f \in L^*(\underline{\underline{S}})$.

Note that $\underline{\underline{S}}$ is also reduced □

Particular cases

The theorem just proved does not depend on *card*(C) .

1. If we assume that S is finite and reduced then S is permanent iff $C \subseteq Coacc(C_{inf})$.

2. Let us assume that L is closed and central. Then we can always find S such that $L = L^{\infty}(S)$ and $C_{inf} = C$ since L is closed. If we consider S_d we then have

   $u \in L^{inf} \Leftrightarrow \lambda(D_d, u[n]) \neq \emptyset$ for all n .

   And L central $\Rightarrow S_d$ permanent.

We can thus state:

Property III. g. 2.

The language L is rational, closed and central, iff $L = L^{\infty}(S)$ for some deterministic permanent finite transition system such that $C_{inf} = C$.

3. It is tempting to prove that a language L is rational, normal and central, iff there exists a finite reduced transition system S such that $L = L^{\infty}(S)$ and $C \subseteq Coacc(C_{fin}) \cap Coacc(C_{inf})$

The language $L = (a \cup b)^*(a \cup b^\omega)$ is a counter example due to A. Arnold (the proof is not difficult).
We can thus prove the corresponding property iff L is deterministic ans this gives rise to

Property III. g. 3.

L is rational, closed, normal and central, iff $L = L^\infty(S)$ for some deterministic finite reduced transition system S such that $C_{inf} = C$ and $C \subseteq Coacc(C_{fin})$.

Property III. g. 4.

L is rational, closed, ideal and central, iff $L = L^\infty(S)$ for some deterministic finite reduced transition system S such that $C_{inf} = C_{fin} = C$.

We call center of an infinitary language L the largest language $\underline{\underline{L}}$ contained in L . Since the union of any number of central languages is central, the center $\underline{\underline{L}}$ always exists. One has obviously:

$$\underline{\underline{L}}^{fin} = L^{fin} \cap FG(L^{inf})$$
$$\underline{\underline{L}}^{inf} = L^{inf}$$

For $\underline{\underline{L}}$ thus defined is central, and if L' is central contained in L , $L'^{fin} \subseteq L^{fin}$ and $L'^{fin} \subseteq FG(L'^{inf}) \subseteq FG(L^{inf})$ imply $L'^{fin} \subseteq \underline{\underline{L}}^{fin}$.

The best construction we know of a transition system recognizing $\underline{\underline{L}}$ is to take the reduced version of $\underline{\underline{S}}$, denote it $\underline{\underline{S}}_R$: we restrict $\underline{\underline{S}}$ to the set of configurations $(\underline{c}, c_d)$ which are accessible from $\underline{C} \times D_d$. By the argument given above $\underline{\underline{S}}_R$ is permanent and $L^\omega(\underline{\underline{S}}_R) = L^\omega(S) = L^{inf}$. The inclusion $L^*(\underline{\underline{S}}_R) \subseteq L^*(S) = L^{fin}$ is clear, since $\underline{D} \times D_d \overset{f}{\rightarrow} \underline{C} \times C_{dfin}$ in $\underline{\underline{S}}_R$ implies $\lambda(D_d, f) \in C_{dfin}$ and thus $f \in L^{fin}$ and also $\underline{D} \overset{f}{\rightarrow} \underline{C}$ whence $f \in FG(L^{inf})$.
The reverse inclusion is obvious from what precedes □

Centers have been studied in [3, 4, 25]: in these papers the definition was slightly different, there was only defined the center L^{c} of a finitary language L as $L^{c} = FG(adh(L))$

Our actual definition is more general, one has

$$L^{c} = (\overline{\underline{\underline{L}}})^{fin} \quad \text{for all} \quad L \subseteq A^{*}$$

and $\underline{\underline{L}} = (L^{fin})^{c} \cup L^{inf}$ for all closed ideal language $L \subseteq A^{\infty}$.

A construction of the center of a closed rational language which is much simpler than the previous one has been given in [27] .

We may then assume that $L = L^{\infty}(S)$ for some finite transition system S such that $C_{inf} = C$ (theorem III. d. 6). The one determines $Viv(c) = \underline{C}$ and builds $\underline{S}$ as follows:

$$\underline{C} = Viv(c), \quad \underline{T} = T \cap \underline{c} \times A \times \underline{C}, \quad \underline{D} = D \cap \underline{C},$$

$$\underline{C}_{fin} = C_{fin} \cap \underline{C} \quad \text{and} \quad \underline{C}_{inf} = \underline{C}$$

and the following equality holds $\underline{\underline{L}} = L^{\infty}(\underline{S})$.

III. h. Conclusion

We conclude this chapter by stressing the importance of the notion of center and raising some questions.

Given a process, described by any mean, i.e. its set of behaviors or any transition system which realizes it, one certainly wishes first to take its "largest deadlock free part". The process will be used as a component of some synchronized system of processes (see below), and most of these systems are intended to work for ever, in other words, to be also deadlock free. Deadlocks of the components of a system certainly induce either a deadlock of the whole system or at least some unfairness.

At the same time, the larger is the set of behaviors of each component process of a synchronized system, the larger will be the set of behaviors of the whole system. This is shere common sense: a large set of behaviors means a lot of freedom in choosing at each time the action which the component p_i may perform.

Thus we believe that the first thing to do is to make sure that each process to be considered for inclusion in a system is central, if it is not central as given then replace it by its center. We know how to compute the center of a rational process. A general question is, how to compute a transition system $\underline{S}$ recognizing the center $\underline{L}$ of $L = L^{\infty}(S)$ when S is not finite. We know how to do it if S is a push down transition system, but we do not know what happens when S is a multi-counter transition system (describing a producer - consumer process) and a fortiori when S is a Petri net (see [31]).

Another wish that one may have is to ensure the normality of each process. This gives rise to the notion of normal center , i.e. the largest normal and central language $\underline{\underline{L}}_n$ contained in L (it exists by Prop. II.d.4.). How can it be computed? (The question is raised even in the rational case: an example given in § V. 2.2. shows that the problem is not easy).

IV. SYNCHRONIZATION OF PROCESSES

IV. 1. We begin with a very typical example which is also a well-known one: this is the celebrated problem of the philosophers eating noodles due to E. W. Dijkstra:

Assume that 6 philosophers named $p_1, \ldots, p_6$ are eating noodles in a dining room where a round table is set with 4 seats, 4 plates and 4 forks as shown by the following figure:

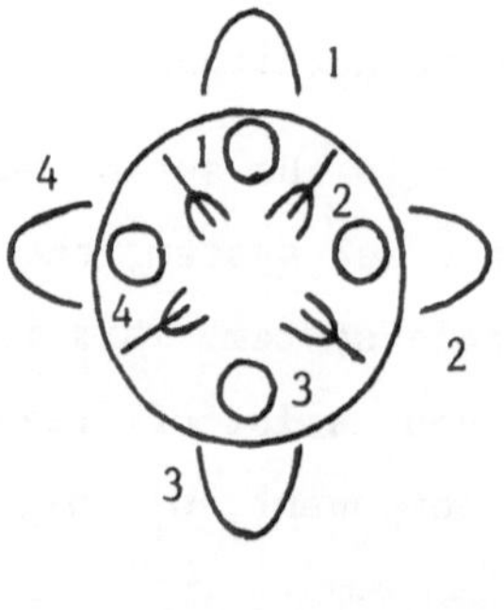

The problem lies in the fact that in order to eat, a philosopher needs two forks (sometimes in the literature noodles are replaced by rice, and forks by chopsticks). Precisely, we describe the set of possible behaviors of a philosopher by a finite transition system, the same for the six of them:

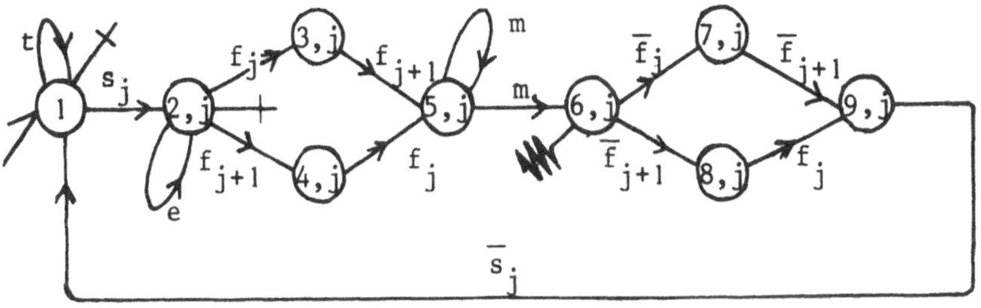

We show only one fourth of it: from (1) are starting four arrows $s_1, \ldots, s_4$ leading to four states (2,1), ..., (2,4) etc.
The actions have the following intuitive meaning:

t : think

s_j: sit down in seat number j

f_j: take the fork number j
(when seated in seat number j a philosopher can take its left fork, i.e. perform the action f_j or take its right fork, i.e. perform the action f_{j+1} where $j+1$ is defined by $j+1$ = if $j < 4$ then $j+1$ else 1)

$\overline{f}_j$: replace the fork number j on the table

m : eat

$\overline{s}_j$: get up from seat number j

e : do nothing

Each action in this set A is instantaneous: but clearly a philosopher can think for a while when in state (1) , do nothing (or wait) for a while when in one of the states (2,1) , ..., (2,4) , and eat for a while when in one of the states (5,1) , ..., (5,4)

The set of behaviors of each philosopher is the language recognized by the transition system :

Write for all $j \in [4]$

$$K_j = s_j \, e^*(f_j f_{j+1} \cup f_{j+1} f_j) m^* m(\overline{f}_j \overline{f}_{j+1} \cup \overline{f}_{j+1} \overline{f}_j) \overline{s}_j$$

Then $HR^*(p_i) = (t^*(K_1 \cup K_2 \cup K_3 \cup K_4))^{**} t^*$

and $HR^\omega(p_i) = (t^*(K_1 \cup K_2 \cup K_3 \cup K_4))^\omega$.

The system we are considering is also composed of forks and seats. Let us denote by p_7, p_8, p_9, p_{10} the forks number 1, 2, 3, 4 . The set of behaviors of the fork p_{6+j} , for all $j \in [4]$, is the language recognized by the transition system.

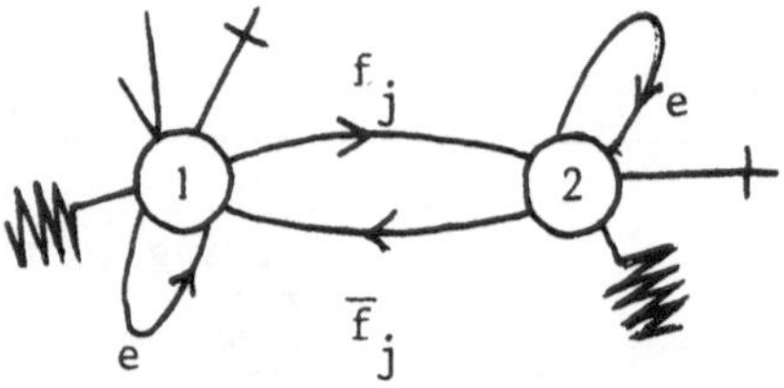

We have $HR^\infty(p_{6+j}) = (e^* f_j \, e^* \overline{f}_j)^* (e^\infty \cup e^* f_j \, e^\infty) \cup (e^* f_j \, e^* \overline{f}_j)^\omega$

The behavior of the seats denoted $p_{11}, p_{12}, p_{13}, p_{14}$ are very similar given by the transition system

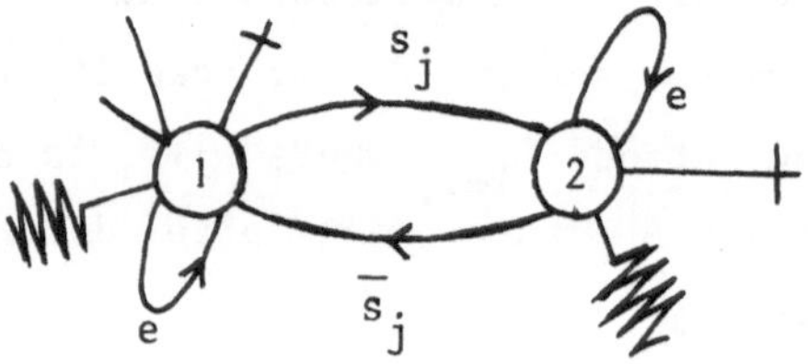

In the configuration (1) the seat number j is empty, it is occupied in configuration (2) . We can formulate the intuitive synchronization condition:

- No two philosophers can take the same fork at the same time. At each instant of time each philosopher p_i performs an action in A . Thus the vector of philosophers $\vec{p} = \langle p_1, \ldots, p_6 \rangle$ performs a vector of actions $\vec{a} = \langle \pi_1(\vec{a}), \ldots, \pi_6(\vec{a}) \rangle$ in $A \times A \times A \times A \times A \times A = A^{(6)}$.

The synchronization condition says that

- at most one $\pi_i(\vec{a})$ belongs to $\{f_j, \bar{f}_j\}$ for all $j \in [4]$
- at most one $\pi_i(\vec{a})$ belongs to $\{s_j, \bar{s}_j\}$ for all $j \in [4]$

Call S the subset of $A^{(6)}$ formed of all the vectors satisfying these two conditions. We can define for all $j \in [4]$ the morphisms of S^∞ :

$$\begin{aligned}\varphi_{6+j}(\vec{a}) &= f_j \quad \text{if there exists } i \text{ such that } \pi_i(\vec{a}) = f_j\\ &= \bar{f}_j \quad \text{if there exists } i \text{ such that } \pi_i(\vec{a}) = \bar{f}_j\\ &= e \quad \text{otherwise}\end{aligned}$$

$$\begin{aligned}\varphi_{10+j}(\vec{a}) &= s_j \quad \text{if there exists } i \text{ such that } \pi_i(\vec{a}) = s_j\\ &= \bar{s}_j \quad \text{if there exists } i \text{ such that } \pi_i(\vec{a}) = \bar{s}_j\\ &= e \quad \text{otherwise}\end{aligned}$$

These morphisms are well defined on S .

Now a vector of words of length n $\vec{f} = \langle\pi_1(\vec{f}), \ldots, \pi_6(\vec{f})\rangle \in (A^n)^{(6)}$ is a vector of six initial behaviors which the six philosophers may perform simultaneously if and only if

$$\begin{aligned}&\forall\, i \in [6] && \pi_i(\vec{f}) \in FG(HR^\infty(p_i))\\ \text{and}\quad &\forall\, n' \in [n] && \vec{f}(n') = \langle\pi_1(\vec{f})(n'), \ldots, \pi_1(\vec{f})(n')\rangle \in S\\ \text{and}\quad &\forall\, j \in [8] && \varphi_{6+1}(\vec{f}) \in FG(HR^\infty(p_{6+j}))\ .\end{aligned}$$

The second condition is needed in order to define $\varphi_{6+j}(\vec{f})$.

It is crucial to remark that a vector $\vec{f}$ of words all of whose components have the same length can be considered as a word in $(A^{(6)})^\infty$: we shall say that a vector of words $\vec{f} \in (A^\infty)^{(6)}$ is <u>uniform</u> iff $|\pi_1(\vec{f})| = |\pi_2(\vec{f})| = \ldots = |\pi_6(\vec{f})|$.

The set of uniform vectors in $(A^\infty)^{(6)}$ denoted $Un((A^\infty)^{(6)})$ is isomorphic to $(A^{(6)})^\infty$.

Let us denote P the system we are considering and $IR(P)$ the set of initial behaviors of P just defined.

The set of finite behaviors of P can only be described as the set of uniform vectors $\vec{f} \in (A^n)^{(6)}$ satisfying

$$\begin{array}{lll} & \forall\ i \in [6] & \pi_i(\vec{f}) \in HR^*(p_i) \\ \text{and} & \forall\ n' \in [n] & \vec{f}(n') \in S \\ \text{and} & \forall\ j \in [8] & \varphi_{6+j}(\vec{f}) \in FG(HR^\infty(p_{6+j})) = HR^*(p_{6+j}) \ . \end{array}$$

And the set of infinite behaviors of P is the set of all uniform vectors $\vec{u} \in (A^\omega)^{(6)}$ satisfying

$$\begin{array}{lll} & \forall\ i \in [6] & \pi_i(\vec{u}) \in HR^\omega(p_i) \\ \text{and} & \forall\ n \in \mathbb{N} & \vec{u}(n) \in S \\ \text{and} & \forall\ j \in [8] & \varphi_{6+j}(\vec{u}) \in HR^\omega(p_{6+j}) \ . \end{array}$$

Let us give examples of initial, finite and infinite behaviors of the system P we consider :

$$\vec{f} = \left\{ \begin{array}{ccccccccc} s_1 & f_1 & f_2 & m & m & \overline{f}_1 & \overline{f}_2 & \overline{s}_1 & t \\ s_3 & f_3 & f_4 & m & \overline{f}_4 & \overline{f}_3 & \overline{s}_3 & t & t \\ s_2 & e & e & e & e & e & e & f_2 & f_3 \\ s_4 & e & e & e & e & f_4 & f_1 & m & m \\ t & t & t & t & t & t & t & t & t \\ t & t & t & t & t & t & t & t & t \end{array} \right.$$

is an initial behavior : one checks easily that each column of this matrix is in S , each row is in $FG(HR^\infty(p_i))$ and if one computes $\varphi_{6+j}(\vec{f})$ for all $j \in [8]$ one gets the 8 words $f_1 \overline{f}_1 f_1$, $f_2 \overline{f}_2 f_2$, $f_3 \overline{f}_3 f_3$, $f_4 \overline{f}_4 f_4$, $s_1 \overline{s}_1$, s_2, $s_3 \overline{s}_3$, s_4 which belong to $FG(HR^\infty(p_{6+1}))$, ..., $FG(HR^\infty(p_{6+8}))$ respectively.

$$\vec{f} = \left\{ \begin{array}{ccccccccc} s_1 & f_1 & f_2 & m & m & \overline{f}_1 & \overline{f}_2 & \overline{s}_1 & t \\ s_3 & f_3 & f_4 & m & f_4 & f_3 & s_3 & t & t \\ s_2 & e & e & e & e & e & e & e & e \\ s_4 & e & e & e & e & e & e & e & e \\ t & t & t & t & t & t & t & s_3 & e \\ t & t & t & t & t & t & t & t & s_1 \end{array} \right.$$

is a finite behavior. Each row is in $HR^*(p)$ and each column is in S and for all $j \in [8]$ $\varphi_{6+j}(\vec{f}) \in HR^*(p_{6+j})$. (The 8 words $\varphi_{j+1}(\vec{f}), \ldots, \varphi_{6+8}(\vec{f})$ are respectively $f_1 \bar{f}_1, f_2 \bar{f}_2, f_3 \bar{f}_3, f_4 \bar{f}_4, s_1 \bar{s}_1 s_1, s_2, s_3 \bar{s}_3 s_3, s_4$ which belong to $HR^*(p_{j+1}), \ldots, HR^*(p_{j+8}))$.

$$\vec{u} = \begin{array}{l} ((s_1 f_1 f_2 m \bar{f}_1 \bar{f}_2 \bar{s}_1) t^8 (s_2 f_3 f_2 m \bar{f}_3 \bar{f}_2) t^8)^\omega \\ ((s_3 f_3 f_4 m \bar{f}_3 \bar{f}_4 \bar{s}_3) t^8 (s_4 f_1 f_4 m \bar{f}_1 \bar{f}_4) t^8)^\omega \\ t^4 ((s_2 f_3 f_2 m \bar{f}_3 \bar{f}_2) t^8 (s_1 f_1 f_2 m \bar{f}_1 \bar{f}_2 \bar{s}_1) t^8)^\omega \\ t^4 ((s_4 f_1 f_4 m \bar{f}_1 \bar{f}_4) t^8 (s_3 f_3 f_4 m \bar{f}_3 \bar{f}_4 \bar{s}_3) t^8)^\omega \\ (t^8 (s_1 f_1 f_2 m \bar{f}_1 \bar{f}_2) t^8 (s_2 f_2 f_3 m \bar{f}_3 \bar{f}_2))^\omega \\ (t^8 (s_3 f_3 f_4 m \bar{f}_3 \bar{f}_4) t^8 (s_4 f_1 f_4 m \bar{f}_4 \bar{f}_1))^\omega \end{array}$$

This vector is in $HR^\omega(P)$: check it !

IV. 2. Systems of synchronized processes

The dining philosophers were a typical example of the general situation we describe below.

We consider a vector $\vec{p} = \langle p_1, \ldots, p_h \rangle$ of h processes. For all $i \in [h]$ p_i is given by its set of finite and infinite real behaviors $HR^\infty(p_i) \subseteq A_i^\infty$.

We denote by A the cartesian product $A = A_1 \times \ldots \times A_h$ and by $\mathcal{A}$ the cartesian product $\mathcal{A} = A_1^\infty \times \ldots \times A_h^\infty$. A vector of words $\vec{\alpha} \in \mathcal{A}$ is said to be uniform iff

$$|\pi_1(\vec{\alpha})| = |\pi_2(\vec{\alpha})| = \ldots = |\pi_h(\vec{\alpha})| \in \mathbb{N} \cup \{\omega\} .$$

The set of uniform vectors in $\mathcal{A}$ is denoted

$$Un(\mathcal{A}) = A_1 \circledast A_2 \circledast \ldots \circledast A_h .$$

More generally if for all $i \in [h]$ $L_i \subseteq A_i^\infty$
$L_1 \circledast L_2 \circledast \ldots \circledast L_h$ is the set of uniform vectors in the cartesian product $L_1 \times \ldots \times L_h$, i.e. $Un(L_1 \times \ldots \times L_h) = L_1 \circledast \ldots \circledast L_h$ and this set is called the uniform cartesian product of the L_i's .

In the absence of synchronization condition, the p_i's may behave independently from each other, i.e.:
any uniform vector $\vec{f} \in FG(HR^{\infty}(p_1)) \circledast \ldots \circledast FG(HR^{\infty}(p_h))$ is an initial behavior of $\vec{p}$. We thus define

$IR(\vec{p}) = IR(p_1) \circledast \ldots \circledast IR(p_h)$, where for all

$i \in [h] \quad IR(p_i) = FG(HR^{\infty}(p_i))$.

A vector $\vec{f} \in IR(\vec{p})$ is a finite behavior of $\vec{p}$ iff for all

$i \in [h] \quad \pi_i(\vec{f}) \in HR^{*}(p_i)$.

Infinite behaviors of $\vec{p}$ are defined as

$$HR^{\omega}(\vec{p}) = HR^{\omega}(p_1) \times \ldots \times HR^{\omega}(p_h) .$$

A major difference between single processes and systems immediately appears. One has certainly

$FG(HR^{\infty}(\vec{p})) \subseteq IR(\vec{p})$, but usually the reverse inclusion is false.

Example IV. 2. 1.

Take $HR^{\infty}(p_1) = (ab)^{\infty}$

and $HR^{\infty}(p_2) = (ab)^{*}a$.

Let $\vec{p}$ be the vector $\vec{p} = \langle p_1, p_2\rangle$

Then $IR(\vec{p}) = \langle ab, ab\rangle^{*}$

$HR^{*}(\vec{p}) = \emptyset$ since for all $\vec{f} \in \{a, b\}^{*} \circledast \{a, b\}^{*}$

$|\pi_1(\vec{f})| = |\pi_2(\vec{f})|$

But $\pi_1(\vec{f}) \in HR^{*}(p_1) \Rightarrow |\pi_1(\vec{f})|$ is even ,

$\pi_2(\vec{f}) \in HR^{*}(p_2) \Rightarrow |\pi_2(\vec{f})|$ is odd

and the two conditions $\pi_1(\vec{f}) \in HR^{*}(p_1)$ and $\pi_2(\vec{f}) \in HR^{*}(p_2)$ cannot be simultaneously satisfied.

Clearly too $HR^{\omega}(\vec{p})$ is empty since $HR^{\omega}(p_2)$ is empty.

Thus $HR^{\infty}(\vec{p}) = \emptyset$ and $FG(HR^{\infty}(\vec{p})) = \emptyset$.

Obviously $IR(\vec{p}) \not\subseteq FG(HR^{\infty}(\vec{p}))$ □

Remark

The inclusion $IR(\vec{p}) \subseteq FG(HR^{\infty}(\vec{p}))$ is satisfied, if for all $i \in [h]$, $HR^{\infty}(p_i)$ is ideal.

For then $FG(HR^{\infty}(p_i)) = IR(p_i) = HR^{*}(p_i)$

And thus $IR(\vec{p}) = IR(p_1) \circledast \ldots \circledast IR(p_h)$ is equal to

$HR^{*}(p_1) \circledast \ldots \circledast HR^{*}(p_h) = HR^{*}(\vec{p})$.

We have even proved that if p_i is ideal for all i , $\vec{p}$ is ideal □

A synchronization condition is a condition of the form

$$Syn = \{\vec{\alpha} \in S^{\infty} \mid \varphi(\vec{\alpha}) \in L_o\}$$

where S is a subset of A , L_o is a language in A_o^{∞} , and φ is a morphism of S^{∞} into A_o^{∞} .

And we define if P is the system $\langle\vec{p}, Syn\rangle$ composed of the h processes $p_1, \ldots, p_h$ synchronized by Syn

$$IR(P) = IR(\vec{p}) \cap FG(Syn)$$

$$HR^{\infty}(P) = HR^{\infty}(\vec{p}) \cap Syn$$

<u>Example IV. 2. 2.</u>

That is exactly what we have done in the example of the dining philosophers : S is the set of vectors in $A^{(6)}$ satisfying

- for all $j \in [4]$ there exists at most one $i \in [6]$ such that $\pi_i(\vec{a}) \in \{f_j, \bar{f}_j\}$
- and for all $j \in [4]$ there exists at most one $i \in [6]$ such that $\pi_i(\vec{a}) \in \{s_j, \bar{s}_j\}$.

A_o is the cartesian product

$$\{f_1, \bar{f}_1, e\} \times \ldots \times \{f_4, \bar{f}_4, e\} \times \{s_1, \bar{s}_1, e\} \times \ldots \times \{s_4, \bar{s}_4, e\}$$

The morphism φ is the morphism of S^{∞} into A_o^{∞} defined by

$$\varphi(\vec{a}) = \langle\varphi_7(\vec{a}), \ldots, \varphi_{14}(\vec{a})\rangle \in A_o .$$

Then L_o is the language A_o^{∞} defined by

$$L_o = HR^{\infty}(p_7) \circledast \ldots \circledast HR^{\infty}(p_{14}) .$$

The definitions given above are equivalent to

$IR(P) = IR(\vec{p}) \cap FG(Syn)$ and

$HR^{\infty}(P) = HR^{\infty}(\vec{p}) \cap Syn$ □

Very often the synchronization condition Syn will be ideal and closed as it is in the example of dining philosophers. Intuitively the constraint imposed by the synchronization condition is an instantaneous one, as is for example the use by a vector of processes of restricted resources or shared variables. If one does not care about the behavior of these resources and then state at the end of a computation, then certainly $FG(Syn) \subseteq Syn$ and $Adh(Syn) \subseteq Syn$.

IV. 3. Properties of systems of synchronized processes

Form the previous definitions it follows that if $P = \langle \vec{p}, Syn \rangle$ is a system of synchronized processes, the sets $IR(P)$ and $HR^{\infty}(P)$ are sets of uniform vectors which can be considered as languages on the alphabet $A = A_1 \times \ldots \times A_h$.
If $\vec{\alpha} \in A^{\infty} = A_1^{\infty} \otimes \ldots \otimes A_h^{\infty}$ is such a uniform vector, one defines easily

$$FG(\vec{\alpha}) = \{\vec{\alpha}[n] \mid n \in \mathbb{N}\}$$

where $\vec{\alpha}[n] = \langle \pi_1(\vec{\alpha})[n], \ldots, \pi_h(\vec{\alpha})[n] \rangle$

The set $FG(\vec{\alpha})$ is also a set of uniform vectors. We then say that

- P is <u>safe</u> iff $IR(P) = FG(HR^{\infty}(P))$. Intuitively P is safe iff each initial behavior of P can be extended into a finite or infinite havior of P.
- P is <u>closed</u> iff $Adh(HR^{\infty}(P)) \subseteq HR^{\omega}(P)$. Intuitively the safe system P is closed iff it behaves well from time 0 up to time n for all n then it behaves well forever.
- P is <u>normal</u> iff $FG(HR^{\omega}(P)) \subseteq FG(HR^{*}(P))$
- P is <u>ideal</u> iff $FG(HR^{\infty}(P)) = HR^{*}(P)$
- P is <u>central</u> iff $FG(HR^{*}(P)) \subseteq FG(HR^{\omega}(P))$

These conditions have the same meaning for safe systems of processes as the corresponding conditions for a single process (we have assumed that every single process is safe).

IV. 3. 1. Visible behaviors and fairness

Assume for all $i \in [h]$ the set of actions A_i contains the empty action e.

We consider the morphism $\sigma : A_i^\infty$ onto $(A_i \setminus \{e\})^\infty$ defined by $\sigma(e) = \varepsilon$ and $\sigma(a) = a$ for all $a \in A_i \setminus \{e\}$.

Clearly $\sigma(A_i) = A_i \setminus \{e\}$ and

$$\sigma(A_i^\infty) = \sigma(A_i)^\infty .$$

We need use the shuffle product on A_i^∞ (see [28])

- if $\alpha, \beta \in A_i^*$ the shuffle of α and β is the ordinary shuffle product (see for example [11]) given by
  $$\alpha \sqcup\!\sqcup \beta = \{f_1 g_1 \dots f_n g_n \mid f_i, g_i \in A_i^*, \ \alpha = f_1 \dots f_n \text{ and } = g_1 \dots g_n\}$$
- if $\alpha \in A_i^*$ and $\beta \in A_i^\omega$ we take
  $$\alpha \sqcup\!\sqcup \beta = \{f_1 g_1 \dots f_{n-1} g_{n-1} f_n u \mid f_i, g_i \in A_i^*, \ u \in A_i^\omega \ \alpha = f_1 \dots f_n, \ \beta = g_1 \dots g_{n-1} u\}$$
- a symetric definition holds for $\alpha \sqcup\!\sqcup \beta$ where $\alpha \in A_i^\omega$ and $\beta \in A_i^*$
- if $\alpha, \beta \in A_i^\omega$ then
  $$\alpha \sqcup\!\sqcup \beta = \{f_1 g_1 \dots f_n g_n \mid f_i, g_i \in A_i^*, \ \alpha = f_1 \dots f_n \dots \text{ and } \beta = g_1 \dots g_n \dots\}$$

The shuffle product thus defined is associative and commutative.

We shall use the shuffle product of two languages

$$L_1 \sqcup\!\sqcup L_2 = \cup \{\alpha \sqcup\!\sqcup \beta \mid \alpha \in L_1, \ \beta \in L_2\}$$

Property IV. 3. 1. 1.

For all $f \in (\sigma(A_i))^*$, the set $\sigma^{-1}(f) = \{\alpha \in A_i^\infty \mid \sigma(\alpha) = f\}$ is equal to $f \sqcup\!\sqcup e^\infty$ and is thus the union of

$$(\sigma^{-1}(f))^{fin} = f \sqcup\!\sqcup e^* \subseteq A_i^*$$
$$(\sigma^{-1}(f))^{inf} = f \sqcup\!\sqcup e^\omega \subseteq A_i^\omega$$

For all $u \in (\sigma(A_i))^\omega$ the set $\sigma^{-1}(u)$ is purely infinitary and equal to $u \sqcup\!\sqcup e^\infty$.

Let us then define for every process p_i such that $HR^\infty(p_i) \subseteq A_i^\infty$ the set of visible behaviors of p_i as the set

$$HV^\infty(p_i) = \{\alpha \in (\sigma(A_i))^\infty \mid \exists\, \beta \in HR^\infty(p_i) : \sigma(\beta) = \alpha\}$$

We also write $HV^\infty(p_i) = \sigma(HR^\infty(p_i))$.

We define $HV^*(p_i) = HV^*(p_i) \cap A_i^*$

and $HV^\omega(p_i) = HV^\omega(p_i) \cap A_i^\omega$.

One has immediately from property IV. 3. 1. 1. :

$$HR^*(p_i) \subseteq HV^*(p_i) \sqcup\!\sqcup\, e^*$$

$$HR^\omega(p_i) \subseteq HV^*(p_i) \sqcup\!\sqcup\, e^\omega \cup HV^\omega(p_i) \sqcup\!\sqcup\, e^\infty .$$

We shall concentrate our attention on the proccesses p_i's such that both equalities hold. We define

p_i is delayable iff $HR^*(p_i) = HV\ (p_i) \sqcup\!\sqcup\, e^*$

and $HR^\omega(p_i) = HV^*(p_i) \sqcup\!\sqcup\, e^\omega \cup HV^\omega(p_i) \sqcup\!\sqcup\, e^\infty$

Let $P = \langle\vec{p}, Syn\rangle$ be a system of synchronized processes whose set of real behaviors is $HR^\infty(P) = HR^\infty(\vec{p}) \cap Syn$. A visible, a synchronizable, a asynchronous behavior of P is the image

$$\sigma(\vec{\alpha}) = \langle\sigma(\pi_1(\vec{\alpha})), \ldots, \sigma(\pi_h(\vec{\alpha}))\rangle$$

of a real behavior of P . We denote by $HV^\infty(P)$ this set

$$HV^\infty(P) = \sigma(HR^\infty(P)) .$$

We can write $\vec{\beta} \in HV^\infty(P) \Leftrightarrow \exists\, \vec{\alpha} \in HR^\infty(P) : \sigma(\vec{\alpha}) = \vec{\beta}$

Clearly then $\vec{\alpha}$ is a uniform vector in $\sigma^{-1}(\vec{\beta})$:

For all $\vec{\beta} \in \sigma(A_1)^\infty \times \ldots \times \sigma(A_h)^\infty$ we can write

$$\sigma^{-1}(\vec{\beta}) = \pi_1(\vec{\beta}) \sqcup\!\sqcup\, e^\infty \times \ldots \times \pi_h(\vec{\beta}) \sqcup\!\sqcup\, e^\infty$$

$= \vec{\beta} \sqcup\!\sqcup\, (\vec{e})^\infty$ if we extend the shuffle product componentwise. The set of uniform vectors in $\sigma^{-1}(\vec{\beta})$ is

$$\sigma^{-1}(\vec{\beta}) \cap A^\infty = \pi_1(\vec{\beta}) \sqcup\!\sqcup\, e^\infty \otimes \ldots \otimes \pi_h(\vec{\beta}) \sqcup\!\sqcup\, e^\infty .$$

We denote it by $\vec{\beta} \sqcup\!\sqcup_{un} (\vec{e})^\infty$.

These notations are extended to relations :

$$\sigma^{-1}(R) = \cup\, \{\sigma^{-1}(\vec{\beta}) \mid \vec{\beta} \in R\} = R \sqcup\!\sqcup\, (\vec{e})^\infty$$

$$\sigma^{-1}(R) \cap A^\infty = R \sqcup\!\sqcup_{un} (\vec{e})^\infty .$$

Now a real infinite behavior of p_i is said to be fair iff its image by σ is infinite :

Intuitively we say that the process p_i is <u>activated</u> when it performs a non empty action. An infinite behavior u is fair iff the process is activated infinitely many times, when p_i performs u. We can write

$$\alpha \text{ is fair} \Leftrightarrow |\sigma(\alpha)| = \omega$$
$$\Leftrightarrow \alpha \in HV^{\omega}(p_i) ш e^{\infty}$$

and define
$$FR^{\omega}(p_i) = \{u \in HR^{\infty}(p_i) \mid \sigma(u) = \omega\}$$
$$= (HV^{\omega}(p_i) ш e^{\infty}) \cap HR^{\omega}(p_i) \quad .$$

Note that if p_i is delayable, one has the equality
$$FR^{\omega}(p_i) = HV^{\omega}(p_i) ш e^{\infty} \quad .$$

We say that the process p_i is <u>fair</u> iff
$$FG(HR^{\infty}(p_i)) \subseteq FG(FR^{\omega}(p_i)) \quad .$$

That is p_i is fair iff all its initial behaviors can be extended into a fair infinite behavior. We say that p_i is <u>completely fair</u> iff it is fair and
$$HR^{\omega}(p_i) = FR^{\omega}(p_i) \; .$$

Let us now consider a system $P = \langle\vec{p}, Syn\rangle$, a real infinite behavior $\vec{u}$ of P is fair iff all its components are fair, i.e. $\vec{u} \in FR^{\omega}(p_1) \times \ldots \times FR^{\omega}(p_h) = FR^{\omega}(p)$.

The set of fair infinite behaviors is
$$FR^{\omega}(P) = FR^{\omega}(\vec{p}) \cap Syn$$

And the safe system P is said to be fair iff
$$FG(HR^{\infty}(P)) \subseteq FG(FR^{\omega}(\vec{p})) \; .$$

This means, that every initial behavior can be extended into an infinite fair behavior or else, it is possible to activate each process p_i, $i \in [h]$, after having performed any initial behavior. We also say that P is completely fair, iff
$$HR^{\omega}(P) = FR^{\omega}(P) \quad .$$

Examples

1. $HR^{\infty}(p_1) = (ab \cup e)^{\infty}$
   One has $FG(HR^{\infty}(p_1)) = (ab \cup e)*(a \cup \varepsilon)$
   $FR^{\omega}(p_1) = (e^* ab)^{\omega}$
   Whence $FG(HR^{\infty}(p_1)) \subseteq (FG(FR^{\omega}(p_1))$ and p_1 is fair
   clearly p_1 is not completely fair since $e^{\omega} \notin HR^{\omega}(p_1)$.

2. Similarly p_2 such that $HR^{\infty}(p_2) = (cd \cup e)^{\infty}$ is fair but not completely fair.

3. Take now $\vec{p} = \langle p_1, p_2 \rangle$, $Syn = (\langle a,e \rangle \cup \langle b,c \rangle \cup \langle a,d \rangle)^{\infty}$
   and $P = \langle \vec{p}, Syn \rangle$
   One computes easily $HR^{\infty}(P)$:
   $HR^*(P)$ is empty, for if $\vec{f} \in (a \cup b \cup e)^* \circledast (c \cup d \cup e)^*$
   $\vec{f} \in Syn$ and $|\vec{f}| = n \Rightarrow \vec{f}(n) \in \langle a,e \rangle \cup \langle b,c \rangle \cup \langle a,d \rangle$
   and $\vec{f} \in HR^*(p_1) \circledast HR^*(p_2) \Rightarrow$
   $\vec{f}(n) \in \{\langle b,e \rangle , \langle b,d \rangle , \langle e,d \rangle , \langle e,e \rangle\}$
   And $HR^{\omega}(P) = \langle (ab)^{\omega}, e(cd)^{\omega} \rangle$

   This is an interesting example of the bad phenomena which may happen: indeed

$$FG(HR^{\infty}(p_1)) \circledast FG(HR^{\infty}(p_2)) = (ab \cup e)^*(a \cup \varepsilon) \circledast (cd \cup e)^*(c \cup \varepsilon)$$

   And $IR(P) = FG(HR^{\infty}(\vec{p})) \cap FG(Syn)$
   $= \langle a,e \rangle (\langle b,c \rangle \langle a,d \rangle)^*$
   is contained in $FG(HR^{\omega}(P))$ whence P is <u>safe</u> . But obviously P is not normal though both p_1 and p_2 are normal □

IV. 3. 2. Delayable processes

Some of the unpleasant properties of systems are prevented if we assume that all the p_i's are delayable, i.e. satisfy

$$HR^{\infty}(p_i) = \sigma^{-1}(HV^{\infty}(p_i)) \quad .$$

Then $HV^{\infty}(\vec{p}) = \sigma(HR^{\infty}(\vec{p}))$ is equal to

$$HV^{\infty}(p_1) \times \ldots \times HV^{\infty}(p_h) .$$

The inclusion $HV^{\infty}(\vec{p}) \subseteq HV^{\infty}(p_1) \times \ldots \times HV^{\infty}(p_h)$ always holds.
Assume then that all the p_i's are delayable and

$$\vec{\beta} \in HV^{\infty}(p_1) \times \ldots \times HV^{\infty}(p_h) .$$

For all $i \in [h]$: $\pi_i(\vec{\beta})e^{\omega} \in HR^{\omega}(p_i)$ since

$$\sigma(\pi_i(\vec{\beta})e^{\omega}) = \pi_i(\vec{\beta}) .$$

We also have

$$HR^{\infty}(\vec{p}) = \sigma^{-1}(HV^{\infty}(\vec{p})) \cap A^{\infty} .$$

One inclusion has hust been proved.
The reverse inclusion is fairly obvious : if $\vec{\alpha}$ is a uniform vector in $\sigma^{-1}(HV^{\infty}(\vec{p}))$ then for all $i \in [h]$

$$\pi_i(\vec{\alpha}) \in \sigma^{-1}(HV^{\infty}(p_i)) = HR^{\infty}(p_i)$$

And $\vec{\alpha} \in HR^{\infty}(p_1) \otimes \ldots \otimes HR^{\infty}(p_h) = HR^{\infty}(\vec{p})$.

We can then prove:

If all the p_i's are delayable

$$HV^{\infty}(P) = HV^{\infty}(\vec{p}) \cap \sigma(Syn)$$

One certainly has $\sigma(HR^{\infty}(\vec{p}) \cap Syn) \subseteq \sigma(HR^{\infty}(\vec{p})) \cap \sigma(Syn)$.
Consider $\vec{\beta} \in \sigma(HR^{\infty}(\vec{p})) \cap \sigma(Syn)$. There exists

$\vec{\alpha} \in Syn \cap \sigma^{-1}(\beta)$ and since Syn is uniform

$$\vec{\alpha} \in \sigma^{-1}(\beta) \cap A^{\infty} \subseteq HR^{\infty}(\vec{p}) .$$

All this means that if the p_i's are delayable, the uniform synchronization condition Syn can be replaced by the non uniform or asynchronous synchronization condition $\sigma(Syn)$ so that

$$HV^{\infty}(P) = HV^{\infty}(p_1) \times \ldots \times HV^{\infty}(p_h) \cap \sigma(Syn) .$$

We wish to explain in more details, what asynchronous means.

Let $p_1, \ldots, p_h$ be delayable processes.
Let Syn be a synchronization condition of the form :

$$\mathrm{Syn} = \{\vec{\alpha} \in S^{\infty} \mid \varphi(\vec{\alpha}) \in HR^{\infty}(p_o)\}$$

where S is a subset of $A_1 \times \ldots \times A_h$, φ maps S into A_{h+1} and p_o is the synchronization mechanism, i.e. a process whose set of actions is A_{h+1}, which we also assume to be delayable.

Let $B \subseteq A_o \times \ldots \times A_h = \{\langle\alpha_o, \alpha_1, \ldots, \alpha_h\rangle \mid \langle\alpha_1, \ldots, \alpha_h\rangle \in S$ and $\varphi\langle\alpha_1, \ldots, \alpha_h\rangle = \alpha_o\}$

One has $HR^{\infty}(\vec{p}, \mathrm{Syn}) = HR^{\infty}(\vec{p}) \cap \mathrm{Syn}$
and we can rewrite this equality as

$$HR^{\infty}(\vec{p}, \mathrm{Syn}) = \{\pi_{[h]}(\vec{\beta}) \mid \vec{\beta} \in B^{\infty},\ \forall i = 0, \ldots, k : \pi_i(\vec{\beta}) \in HR^{\infty}(p_i)\} .$$

We denote $\pi_{[h]}(\vec{\beta}) = \langle\pi_1(\vec{\beta}), \ldots, \pi_h[\vec{\beta}]\rangle$
Then denote for all $i = 0, \ldots, h$

$$A'_i = (A_i \setminus \{e\}) \cup \{\varepsilon\}$$

Define $B' = A'_o \times \ldots \times A'_h$

One has $HV^{\infty}(p_i) = \sigma(HR^{\infty}(p_i))$
and $HR^{\infty}(p_i) = \sigma^{-1}(HV^{\infty}(p_i))$ for all $i = 0, \ldots, h$

$$HV^{\infty}(\vec{p}, \mathrm{Syn}) = \{\pi_{[h]}(\vec{\alpha}) \mid \vec{\alpha} \in (B')^{\infty} \cap HV^{\infty}(p_o) \times \ldots \times HV^{\infty}(p_h)\}$$
$$= \{\langle\alpha_1, \ldots, \alpha_h\rangle \mid \exists\, \alpha_o \in B'_o : \langle\alpha_o, \alpha_1, \ldots, \alpha_h\rangle \in (B')^{\infty} \text{ and } \langle\alpha_o, \alpha_1, \ldots, \alpha_h\rangle \in HV^{\infty}(p_o) \times \ldots \times HV^{\infty}(p_h)\}$$

Assuming all the $p_o, p_1, \ldots, p_h$ are rational, this is a rational relation recognizable by a multitransition system (see below). We then know the set of asynchronous behaviors of P i.e., the set of all $\vec{\alpha} \in A$ such that

$$\exists\, \vec{\beta} \in A^{\infty} \text{ satisfying } \vec{\beta} \in HR^{\infty}(P)$$

A major problem is then to find an algorithm which given an asynchronous behavior $\vec{\alpha} \in HV^{\infty}(P)$ determines one (or the shortest) synchronous behavior $\vec{\beta} \in HR^{\infty}(P)$ such that $\sigma(\vec{\beta}) = \vec{\alpha}$ (this may be called the serialization problem, see [30]).

V. MULTITRANSITION SYSTEMS

We have described above how one process p can be realized by a transition system.
In this paragraph we define multitransition systems intended to realize systems of processes.

A multitransition system (abbreviated mts) is a structure of the form

$$S = \langle C, A_1, \ldots, A_h, T, D, C_{fin}, C_{inf} \rangle$$

where C is a set of configurations .
$A_1, \ldots, A_h$ are finite alphabets of actions
$T \subseteq C \times (A_1 \cup \{\varepsilon\} \times \ldots \times A_h \cup \{\varepsilon\}) \times C$ is a set of multitransitions
D, C_{fin}, C_{inf} are subsets of C .

Denote by $\hat{A}$ a set of symbols in one to one correspondence with $A_1 \cup \{\varepsilon\} \times \ldots \times A_h \cup \{\varepsilon\} = \sigma(\hat{A})$ (one can take $\hat{A} = A_1 \cup \{e\} \times \ldots \times A_h \cup \{e\}$ in which case σ is defined as above by $\sigma(e) = \varepsilon$ and $\sigma(a) = a$ for all $a \neq \varepsilon$) .
$\hat{T} = \{(c, \hat{a}, \sigma') \mid (c, (\hat{a}), c') \in T\}$.

The mts S is said to be proper iff

$$C \times \{\vec{\varepsilon}\} \times C \cap T = \emptyset \quad .$$

Now for all $\vec{f} \in (\sigma(\hat{A}))^* = A_1^* \times \ldots \times A_h^*$ and for all $c, c' \in C$
we define $c \overset{\vec{g}}{\Rightarrow} c' \iff \exists\, \vec{g} \in (\hat{A})^* \; c \overset{\vec{g}}{\Rightarrow} c'$ in $\hat{T}$
and for all $\vec{u} \in (\sigma(\hat{A}))^\omega$, $c \in C$

$$c \overset{\vec{u}}{\Rightarrow} C_{inf} \iff \forall\, \vec{v} \in (\hat{A})^\omega \quad c \overset{\vec{v}}{\rightarrow} C_{inf} \text{ in } \hat{T} \ .$$

It is clear that if S is proper, then

$$c \overset{\vec{v}}{\rightarrow} C_{inf} \text{ in } \hat{T} \Rightarrow \sigma(\vec{v}) \in A^{inf}$$

(since for each n $\exists\, i \in [h]$ $\pi_i(\sigma(\vec{v}))(n) \neq \varepsilon$ whence $\exists\, i \in [h]$ $\pi_i(\sigma(\vec{v}))(n) \neq \varepsilon$ for infinitely many n's and this implies $|\pi_i(\sigma(\vec{v}))| = \omega$) .

A multitransition system is said to be finite iff cond(C) is finite and uniform iff $T \subseteq C \times A \times C$.
As we remarked previously, A freely generates $A^\infty = A_1 \otimes \ldots \otimes A_h$, and a uniform relation $R \subseteq A^\infty$ can be considered as a language on A and a uniform mts as a simple transition system with A as alphabet (we then write $c \overset{\vec{f}}{\rightarrow} c'$ and $c \overset{\vec{u}}{\twoheadrightarrow} c'$ rather than $c \overset{\vec{f}}{\Rightarrow} c'$ and $c \overset{\vec{u}}{\twoheadrightarrow} c'$).

We define for any mts S

$$R^*(S) = \{\vec{f} \in A^{fin} \mid D \overset{\vec{g}}{\Rightarrow} C_{fin}\}$$
$$R^\omega(S) = \{\vec{u} \in A^{inf} \mid D \overset{\vec{u}}{\twoheadrightarrow} C_{inf}\}$$
$$R^\infty(S) = R^*(S) \cup R^\omega(S)$$ is the relation recognized or realized by S.

V. 1. Rational relations

The family of <u>rational infinitary relations</u> Rat(A) is defined as the family of all relations $R \subseteq A$ such that $R = R^\infty(S)$ for some finite mts S.
Rat(A) has been studied mainly in [13, 26].
We shall use the following results:

Theorem V. 1.1.

Rat(A) is the smallest family of subsets of A which contains the finite subsets of A^{fin} and is closed by union, product, finite star and infinite power.
The operations of product, finite star and infinite power are defined componentwise

$$R_1R_2 = \{\vec{\alpha}\vec{\beta} \mid \vec{\alpha} \in R_1, \vec{\beta} \in R_2\}$$

where $\vec{\alpha}\vec{\beta}$ is the vector $<\pi_1(\vec{\alpha})\pi_1(\vec{\beta}), \ldots, \pi_h(\vec{\alpha})\pi_h(\vec{\beta})>$

$R^* = \cup \{R^n \mid n \in \mathbb{N}$ where $R_o = \{\vec{\varepsilon}\}$ and for all n $R^{n+1} = RR^n$

$R^\omega = \{\vec{f}_1\vec{f}_2 \ldots \vec{f}_n \ldots \mid \forall n \ \ \vec{f}_n \in R^{fin} \setminus \{\vec{\varepsilon}\}\}$.

Theorem V. 1. 2.

A relation $A \subseteq \mathcal{A}$ is rational iff there exists an alphabet B, a rational language $L \subseteq B^{\infty}$ and a multimorphism $\vec{\varphi} = \langle \varphi_1, \ldots, \varphi_h \rangle$ where for all i φ_i is a morphism of B^{∞} into A_i^{∞} such that $R = \vec{\varphi}(L)$.

This is a rephrasing of the definition when $\vec{\varphi}$ is an alphabetic multimorphism, i.e. satisfies $\forall\, i \quad \varphi_i(B) \subseteq A_i \cup \{\ \}$.

We come now to a major difference between infinitary rational relations and finitary rational relations. In our notations one has the following result of S. Eilenberg (theorem IX. 6. 1. in [11]).

Theorem V. 1. 3.

If $R \subseteq \mathcal{A}^*$ and $R \subseteq \mathrm{Rat}(\mathcal{A})$ then $R \subseteq \mathrm{Rat}(\mathcal{A}^*)$.

This means that a uniform and rational infinitary relation is uniformly rational, i.e. satisfies one of the equivalent conditions

- $R = R^*(S)$ for some uniform multitransition system
- $R = \vec{\varphi}(L)$ for some finitary rational language L and

strictly alphabetic multimorphism $\vec{\varphi}(\varphi_i(B) \subseteq A_i$ for all i).

The following example shows that the same property is not true for a purely infinitary R.

Example

$$R = \langle aa, b\rangle^* \langle d^{\omega}, c^{\omega}\rangle = \{\langle a^{2n} d^{\omega}, b^n d^{\omega}\rangle \mid n \in \mathbb{N}\}$$

is rational as the product of the two rational relations

$$R_1 = \langle aa, b\rangle^* \quad \text{and} \quad R_2 = \langle d, d\rangle^{\omega}$$

R cannot be recognized by a uniform finite transition system.

Assume $R = R^{\omega}(S)$ for some finite S with $\mathrm{cond}(C) = N$.

There exists a sequence $c_o, c_1, \ldots, c_n, c_{n+1}, \ldots, c_{2n}, c_{2n+1}, \ldots$ such that $c_o \in D$, for all $j \in [n]$ $(c_{j-1}, \langle a, b \rangle, c_j) \in T$ for all $j \in \{n+1, \ldots, 2n\}$ $(c_{j-1}, \langle a, d \rangle, c_j) \in T$ for all $j \geq 2n+1$ $(c_{j-1}, \langle d, d \rangle, c_j) \in T$ and $\text{cond}\{j \mid c_j \in C_{inf}\} = \omega$.

If $n \in N$ there exists j and $j' \in [n]$ such that $j < j'$ and $c_j = c_j'$.
One can then delete in the sequence $c_o, \ldots, c_n, \ldots$ the elements $c_{j+1}, \ldots, c_j,$ and still get a computation sequence of S : the infinite words it recognizes is not in R □

V. 2. Realization of a system of synchronized processes

Assume we are given the following objects

- $S_1, \ldots, S_h$ are reduced transition systems which realize the processes $p_1, \ldots, p_h$, i.e. $HR^\infty(p_i) = L^\infty(S_i) \subseteq A_i^\infty$ for all $i \in [h]$
- S_o is a reduced transition system which recognizes a language $L_o = L^\infty(S_o) \subseteq A_o^\infty$
- S is a subset of $A = A_1 \times \ldots \times A_h$
- φ maps S into A_o : as usual we denote also φ the morphism of S^∞ into A^∞ which extends φ
- $\text{Syn} = \{\vec{\beta} \in S^\infty \mid \varphi(\vec{\beta}) \in L_o\}$

All this defines completely a system $P = \langle \vec{p}, \text{Syn} \rangle$
We have $HR^\infty(\vec{p}) = HR^\infty(p_1) \circledast \ldots \circledast HR^\infty(p_h)$

$$HR^\infty(P) = HR^\infty(\vec{p}) \cap \text{Syn} .$$

We can also write since the S_i's are reduced

$$IR(p_i) = FG(HR^\infty(p_i)) = \{f \in A_i^* \mid D_i \xrightarrow{f} C_i \text{ in } S_i\}$$
$$IR(\vec{p}) = IR(p_1) \circledast \ldots \circledast IR(p_h)$$
$$IR(P) = IR(\vec{p}) \cap FG(\text{Syn})$$

Let us denote for every transition system

$$S = <C, A, T, D, C_{fin}, C_{inf}>$$

$$L^{**}(S) = \{f \in A^* \mid D \overset{f}{\to} C\}$$

this set being equal to $FG(L^{\infty}(S))$ if S is reduced.

Our aim in this chapter is to build a uniform multitransition system S such that

$$R^{\infty}(S) = HR^{\infty}(P)$$

and $$R^{**}(S) = IR(P)$$

V. 2. 1. Uniform product

We can build the product $S = S_1 \otimes \ldots \otimes S_h$ defined by

$$C = C_1 \times \ldots \times C_h$$

$$A = A_1 \times \ldots \times A_h$$

$$T = \{((c_1, \ldots, c_h), (a_1, \ldots, a_h), (c'_1, \ldots, c'_h)) \mid \forall i \in [h] \quad (c_i, a_i, c'_i) \in T_i\}$$

$$D = D_1 \times \ldots \times D_h$$

$$C_{fin} = C_{1fin} \quad \ldots \quad C_{hfin}$$

This is a uniform mts, and it is very easy to prove that

$$\forall \vec{c}, \vec{c}' \in C \quad , \quad \vec{f} \in A^*$$

$$\vec{c} \overset{\vec{f}}{\to} \vec{c}' \iff \forall i \in [h] \quad \pi_i(\vec{c}) \xrightarrow{\pi_i(\vec{f})} \pi_i(\vec{c}')$$

Whence $R^{**}(S) = IR(p_1) \otimes \ldots \otimes IR(p_h)$, $R^*(S) = HR^*(p_1) \otimes \ldots \otimes HR^*(p_h)$ and one also has

$$R^{\omega\omega}(S) = \{\vec{u} \in A^{\omega} \mid D \overset{\vec{u}}{\twoheadrightarrow} C\}$$

$$= L^{\omega\omega}(S_1) \times \ldots \times L^{\omega\omega}(S_h)$$

where for all $i \in [h]$ $L^{\omega\omega}(S_i) = \{u \in A_i^{\omega} \mid D_i \overset{u}{\twoheadrightarrow} C_i\}$

Applying this to the case where all the S_i's are finite we get

Property V. 2. 1. 1. The uniform product of h finitary rational languages $L_1, \ldots, L_h$ is a uniformely rational relation.

The unform product of h closed infinitary languages is a closed uniformly rational infinitary relation.

Proof

This comes from the obvious fact that $S_1 \otimes \ldots \otimes S_h$ is finite if all the S_i's are finite.
The second statement follows from the fact that if S_i is finite and reduced, $L^{\omega\omega}(S_i) = \mathrm{Adh}(L^{\omega}(S_i))$
and if all the S_i's are finite

$$R^{\omega\omega}(S) = \mathrm{Adh}(R^{\omega}(S)) = L^{\omega\omega}(S_1) \times \ldots \times L^{\omega}(S_h) .$$

Whence $L^{\omega\omega}(S_i) = L^{\omega}(S_i)$ for all i implies

$$\mathrm{Adh}(R^{\infty}(S)) = L^{\omega}(S_1) \times \ldots \times L^{\omega}(S_h) \subseteq L_1 \otimes \ldots \otimes L_h \quad \square$$

There is an obvious difficulty to build S such that

$$R^{\omega}(S) = L_1{}^{\omega} \times \ldots \times L_h{}^{\omega}$$

if the L_i's are not closed.
If we build the same product $S = S_1 \otimes \ldots \otimes S_h$ as above then $\vec{u} \in L_1{}^{\omega} \times \ldots \times L_h{}^{\omega}$ iff there exists an infinite sequence $\vec{c}_o, \ldots, \vec{c}_n, \ldots$ of elements of C such that

$$\vec{c}_o \in D, \quad \forall\, n \in \mathbb{N} \quad (\vec{c}_{n-1}, \vec{u}(n), \vec{c}_n) \in T$$

and for all $i \in [h]$ $\mathrm{card}\{n \mid \pi_i(\vec{c}_n) \in C_{i\inf}\} = \omega$

To get around this difficulty, one uses the same method as to prove Buchi's theorem (III. b. 3.) about the intersection of tow rational infinitary languages.
An infinite sequence $\vec{c}_o, \ldots, \vec{c}_n, \ldots$ corresponds to some $\vec{u} \in L_1{}^{\omega} \times \ldots \times L_h{}^{\omega}$ iff there exists an increasing sequence of integers $n_1 < n_2 < \ldots < n_l < \ldots$ such that

$$\forall\, l \in \mathbb{N} \quad \pi_1(\vec{c}_{n_l}) \in C_{1\inf} \quad \text{and}$$

$$\exists\, m_{il} : n_l < m_{il} < n_{l+1} \quad \pi_i(\vec{c}_{m_{il}}) \in C_{i\inf}$$

One way this implication is clear.

Conversely if $\mathrm{card}\{n \mid \pi_i(\vec{c}_n) \in C_{i\mathrm{inf}}\} = \omega$ for all i one can take for n_1 any integer such that $\pi_1(c_{n_1}) \in C_{1\mathrm{inf}}$. Surely there exists $m_{21} > n_1$ such that $\pi_2(c_{m_{21}}) \in C_{2\mathrm{inf}}$ $m_{31} > m_{21}$ such that $\pi_3(c_{m_{31}}) \in C_{3\mathrm{inf}}$, and so on, up to $m_{h1} > m_{(h-1)1}$ such that $\pi_h(c_{m_{h1}}) \in C_{h\mathrm{inf}}$. Then there exists $n_2 > m_{k1}$ such that $\pi_1(c_{n_2}) \in C_{1\mathrm{inf}}$ and again $m_{22}, m_{32}, \ldots, m_{h2}$ such that $\pi_{j2}(c_{m_{j2}}) \in C_{j\mathrm{inf}}$.

Assume that S_i's are finite:

then one can write $L_1^{\omega} \times \ldots \times L_h^{\omega}$ as a finite union of products of the form $R_{(c_1,\ldots,c_h)}(R'_{(c_1,\ldots,c_h)})^{\omega}$ where

- $c_1 \in C_{1\mathrm{inf}}$
- $R_{(c_1,\ldots,c_h)} = \{\vec{f} \mid D \xrightarrow{\vec{f}} \langle c_1, \ldots, c_h\rangle \text{ in } S\}$
- $R'_{(c_1,\ldots,c_h)}$ is the set of all uniform finite vectors $\vec{g}$ such that

  $\forall\, i \in [h] \setminus \{1\}$ there exists g_{i1}, g_{i2} and $c'_i \in C_{i\mathrm{inf}}$ such that
  $\pi_i(\vec{g}) = g_{i1}g_{i2}$, $c_i \xrightarrow{g_{i1}} c'_i$ and $c'_i \xrightarrow{g_{i2}} c_i$ (in S_i).

Clearly each of the elements of this union can be realized by a finite uniform mts. So can then be their union.

We can state :

Property V. 2. 1. 2.

The cartesian product of purely infinitary rational languages $L_1 \subseteq A_1^{\omega}, \ldots, L_h \subseteq A_h^{\omega}$ is a purely infinitary uniformly rational relation.

As a consequence of Property V. 2. 1. 1. and V. 2. 1. 2.

Property V. 2. 1. 3.

The uniform product of infinitary rational languages is an infinitary uniformly rational relation.

Remark

If we look for a uniform mts S realizing

$L^{\infty}(S_1) \otimes \ldots \otimes L^{\infty}(S_h)$ we are lead to build it in two steps

$S = S_1 \otimes \ldots \otimes S_h$ will realize $L^{*}(S_1) \otimes \ldots \otimes L^{*}(S_h)$ and be such that $IR(p_1) \otimes \ldots \otimes IR(p_h) = R^{**}(S)$.

Then one builds S' such that $R^{\infty}(S') = L^{\omega}(S_1) \times \ldots \times L^{\omega}(S_h)$ and takes the disjoint union $S \cup S'$ (disjoint meaning that the sets of configurations of S and S' are disjoint).

$$\begin{aligned}
\text{Then} \quad R^{\infty}(S \cup S') &= R^{*}(S) \cup R^{\omega}(S') \\
&= L^{*}(S_1) \otimes \ldots \otimes L^{*}(S_h) \cup L^{\omega}(S_1) \times \ldots \times L^{\omega}(S_h) \\
&= L^{\infty}(S_1) \otimes \ldots \otimes L^{\infty}(S_h)
\end{aligned}$$

$$\begin{aligned}
\text{And since} \quad R^{**}(S') &\subseteq FG(L^{\omega}(S_1) \times \ldots \times L^{\omega}(S_h)) \\
&\subseteq FG(L^{\infty}(S_1)) \otimes \ldots \otimes FG(L^{\infty}(S_h)) = R^{**}(S)
\end{aligned}$$

one has $R^{**}(S \cup S') = IR(\vec{p})$ □

Example

Take $L_1^{fin} = \{f \in (a \cup e)^{*} \mid |f|_a \equiv 0 \pmod 2\}$

$L_1^{inf} = \{u \in (a \cup e)^{\omega} \mid |u|_a = \omega\}$

One can realize L_1 by the transition system S_1 :

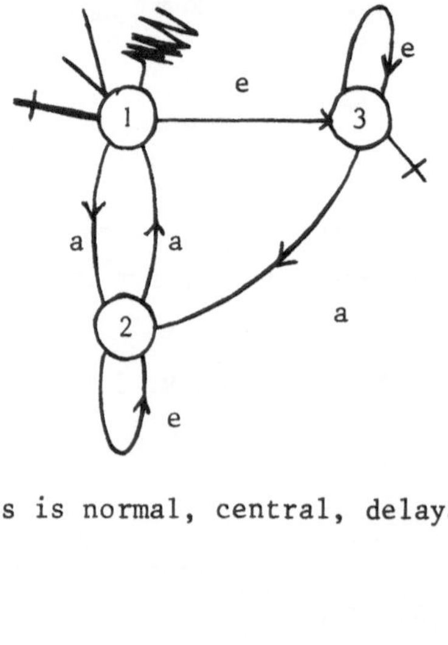

This process is normal, central, delayable but not closed nor ideal.

Take $L_2^{fin} = \{f \in (a \cup e)^* \mid |f|_b \quad 1 \pmod 2\}$
$L_2^{inf} = \{u \in (a \cup e)^\omega \mid |u|_b = \omega = L_1^{inf}$

This can be realized by S_2 :

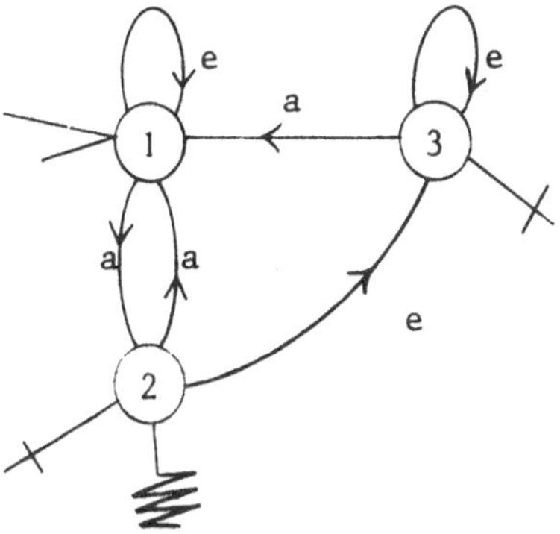

This process is also normal, central, delayable but not closed nor ideal.

In fact L_1 and L_2 are also realized by the two transition systems

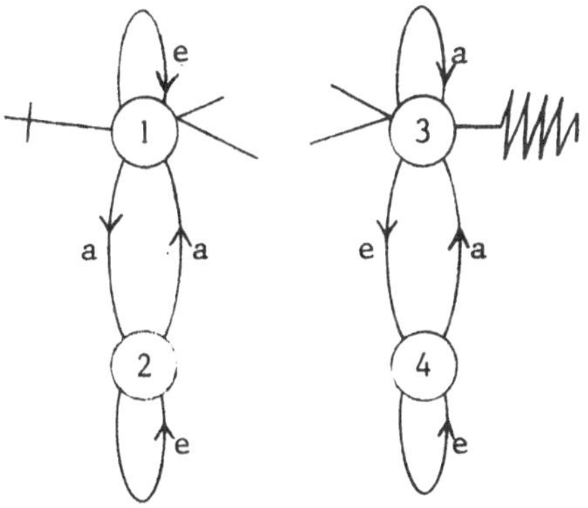

which is the disjoint union of S'_1 and S''_1 with obviously

$$L^*(S'_1) = L_1^{fin} \quad \text{and} \quad L^\omega(S''_1) = L_1^{inf} \quad .$$

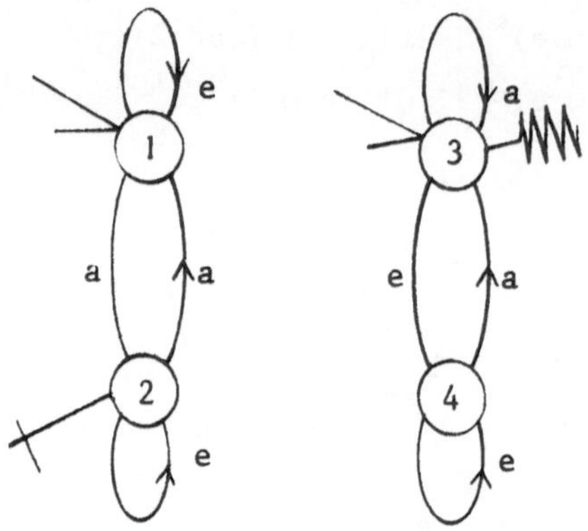

which is the disjoint union of S'_2 and S''_2 with obviously

$$L^*(S'_2) = L_2^{fin} \quad \text{and} \quad L^\omega(S''_2) = L_2^{inf} .$$

We can then build $S'_1 \circledast S'_2$ which is the uniform mts S' :

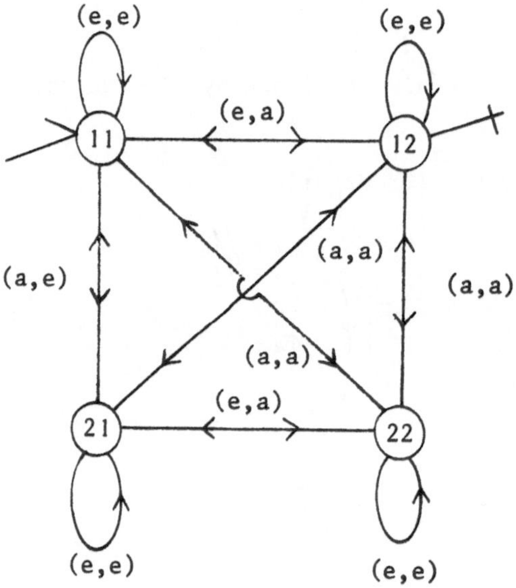

This recognizes clearly $L_1^{fin} \circledast L_2^{fin}$ and

$$\begin{aligned} R^{**}(S'_1 \circledast S'_2) &= (a \cup e)^* \circledast (a \cup e)^* \\ &= (\langle a,a\rangle \cup \langle a,e\rangle \cup \langle e,e\rangle \cup \langle e,e\rangle)^* \\ &= FG(L_1^{fin}) \circledast FG(L_2^{fin}) \end{aligned}$$

We build now S_3 to realize $L_1^{inf} \times L_2^{inf}$.

The product of S''_1 and S''_2 is the following:

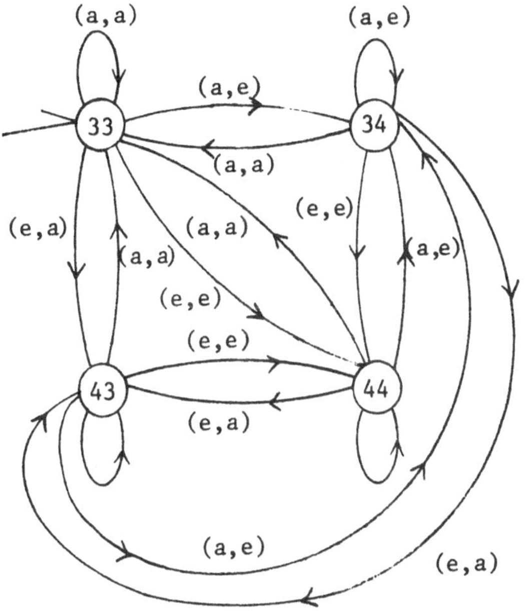

But this product is not of much use: one checks easily that $\{\vec{u} \mid (33) \overset{\vec{u}}{\twoheadrightarrow} (33)\} = \{\vec{u} \mid \vec{u}(n) = (a,a)$ for inifinitely many n's$\}$ is a subset of $L_1^{inf} \times L_2^{inf}$.

For all other configurations (34), (43), (44) the set $\{\vec{u} \mid (33) \overset{\vec{u}}{\twoheadrightarrow} (ij)\}$ is not in $L_1^{inf} \times L_2^{inf}$

Following the ideas in the proof we write then

$$L_1^{inf} \times L_2^{inf} = (e^* a \circledast (e \cup a)^*)(((e \cup a)^* \circledast c^* a)(e^* a \circledast (e \cup a)^*))^{\omega}$$

This allows us to build the mts S_3 :

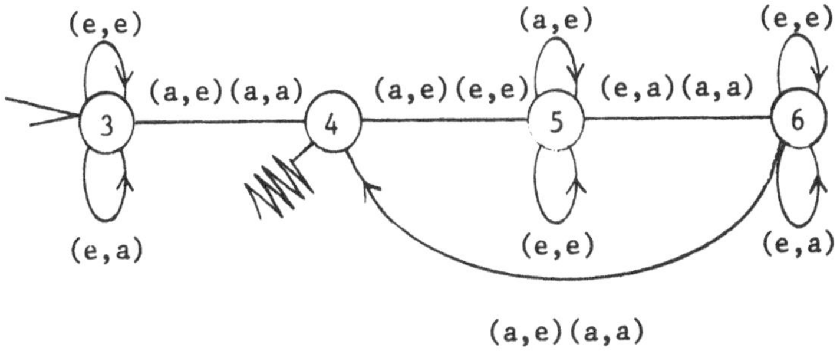

This mts S_3 is disjoint from $S'_1 \circledast S'_2$ and we can make then disjoint union $S_3 \cup S'_1 \circledast S'_2$: it realizes $L'_1 \circledast L'_2$ and

$$R^{**}(S_3 \cup S'_1 \circledast S'_2) = R^{**}(S_3) \cup R^{**}(S'_1 \circledast S'_2)$$
$$= (FG(L_1^{inf}) \circledast FG(L_2^{inf}) \cup FG(L_1^{fin}) \circledast FG(L_2^{fin})$$
$$= (a \cup e)^* \circledast (a \cup e)^*$$

In case of such a decomposition $S_1 = S'_1 \cup S''_1$ and $S_2 = S'_2 \cup S''_2$ with $L^\infty(S'_i) = L^*(S_i)$ and $L^\infty(S''_i) = L^\omega(S_i)$ for $i = 1, 2$

one usually has to build the product

$(S'_1 \cup \underline{S}''_1) \circledast (S'_2 \cup \underline{S}''_2)$ to get $R^{**}(S) = FG(L_1) \circledast FG(L_2)$

Our example was simplified by the fact that L_1 and L_2 were normal □

V. 2. 2. Synchronization

From the previous paragraph we know how to build (in the finite case) a mts S such that

$$R^\infty(S) = L^\infty(S_1) \circ \ldots \circ L^\infty(S_h) = HR^\infty(\vec{p})$$

and $$R^{**}(S) = IR(\vec{p}) .$$

We can restrict it to S by deleting all the transitions $(\vec{c}, \vec{a}, \vec{c}') \in T$ such that $\vec{a} \notin S$. Call this restriction $S|S$.

One has $$R^\infty(S|S) = HR^\infty(\vec{p}) \cap S^\infty$$
$$R^{**}(S|S) = IR(\vec{p}) \cap S^* .$$

In order to build a mts realizing $P = \langle \vec{p}, Syn \rangle$ we introduce the union $S_o +_\varphi S|S = S'$

the set of configurations of $S_o +_\varphi S|S$ is $C_o \times C = C_o \times C_1 \times \ldots \times C_h$

the alphabet is the same as the alphabet S of $S|S$

the set of transitions is the set

$$T' = \{(\langle c_o,\vec{c}\rangle, \vec{s}, \langle c'_o,\vec{c}'\rangle) \mid (c_o, \varphi(\vec{s}), c'_o) \in T \text{ and } (\vec{c},\vec{s},\vec{c}') \in T\}$$

the set of initial configurations is $D_o \times D$

the set of final configurations is $C_{ofin} \times C_{fin}$

One clearly has

$$R^*(S_o +_\varphi S|S) = HR^*(P)$$
$$R^{**}(S_o +_\varphi S|S) = IR(P) \quad \text{if } S_o \text{ is reduced}$$
$$R^\omega(S_o +_\varphi S|S) = HR^\omega(P) \quad \text{if } C_{oinf} = C .$$

We have the same difficulty as usual, if L_o is not closed, but as already remarked, it is reasonable to assume that a synchronization condition is closed: this means that if is satisfied by a system of processes from time 0 up to time n for all n then it is satisfied for ever.

We shall use also below the amalgamated product of two transition systems S_1, S_2 over the same alphabet A.

The amalgamated product

$S_1 \hat{\times} S_2$ is defined as the transition system S_3 where

$C_3 = C_1 \times C_2$

$A_3 = A_1 = A_2$

$T_3 = \{((c_1,c_2), a, (c_1',c_2')) \mid (c_1,a,c_1') \in T_1 \text{ and } (c_2,a,c_2') \in T_2\}$

$D_3 = D_1 \times D_2$

Let us assume the transition systems S_1 and S_2 are finite and reduced. One has for all $f \in A^*$, $c_1, c'_1 \in C$, and $c_2, c'_2 \in C_2$

$$(c_1,c_2) \xrightarrow{f} (c'_1,c'_2) \Leftrightarrow c_1 \xrightarrow{f} c'_1 \text{ in } S_1 \text{ and } c_2 \xrightarrow{f} c'_2 \text{ in } S_2$$

Whence $D_3 \xrightarrow{f} C_{1fin} \times C_{2fin} \Leftrightarrow f \in L^*(S_1) \cap L^\infty(S_2)$

$D_3 \xrightarrow{f} C_{1fin} \times C_2 \Leftrightarrow f \in FG(L^\infty(S_1)) \cap FG(L^\infty(S_2))$

If we now assume that $L^\infty(S_1) = L^*(S_1)$ and $L^\infty(S_2) = L^\omega(S_2)$ and write $L = L^*(S_1) \cup L^\omega(S_2) = L^\infty(S_1 \cup S_2)$

$$D_3 \xrightarrow{f} C_{1fin} \times C_2 \Leftrightarrow f \in L_1^{fin} \cap FG(L^{inf})$$

And thus $\{f \mid D_3 \xrightarrow{f} C_{1fin} \times C_2 \text{ in } S_1 \hat{\times} S_2\}$ is equal to the finitary part $\underline{\underline{L}}^{fin}$ of the center $\underline{\underline{L}}$ of L

$$D_3 \overset{u}{\twoheadrightarrow} C_1 \times C_{2inf} \Leftrightarrow f \in Adh(L_1^{fin}) \cap L^{inf}$$

And thus $\{u \mid D_3 \overset{u}{\twoheadrightarrow} C_1 \times C_{2inf}$ in $S_1 \hat{\times} S_2$ is the purely infinitary part L_n^{inf} of the largest normal language contained in L.

Example

We take L_1 and L_2 as above. And we impose the synchronization condition given by

$$S = \{\langle a,e\rangle, \langle e,a\rangle\}$$

$$\varphi(\langle a,e\rangle) = b$$

$$\varphi(\langle e,a\rangle) = \bar{b}$$

$$L_o = L(S_o) = (bb(\overline{bb} \cup \bar{b}))^{\infty}$$

We can then build $S'_1 \circledast S'_2 \mid S$:

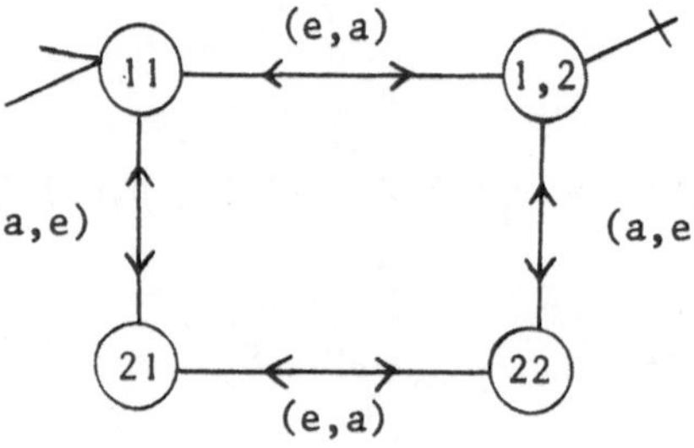

and $S_3 \mid S$:

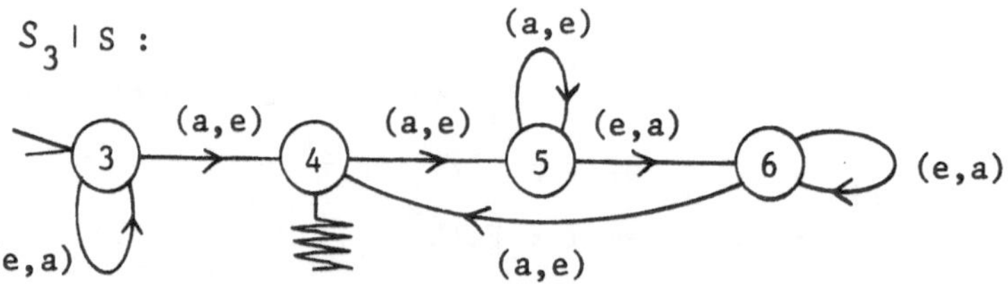

S_o can be taken as

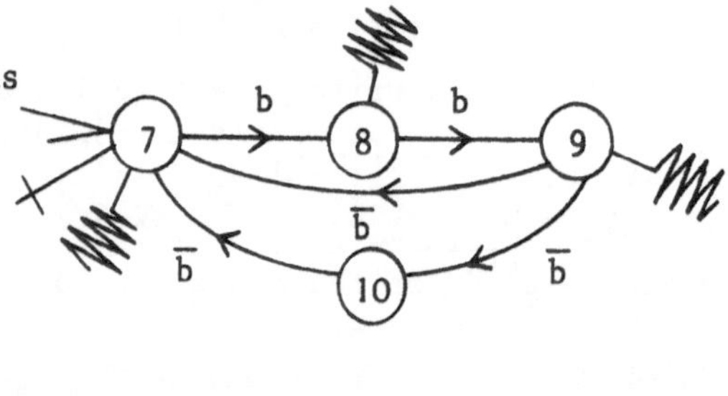

We build $S_o +_\varphi S'_1 \circledast S'_2$:

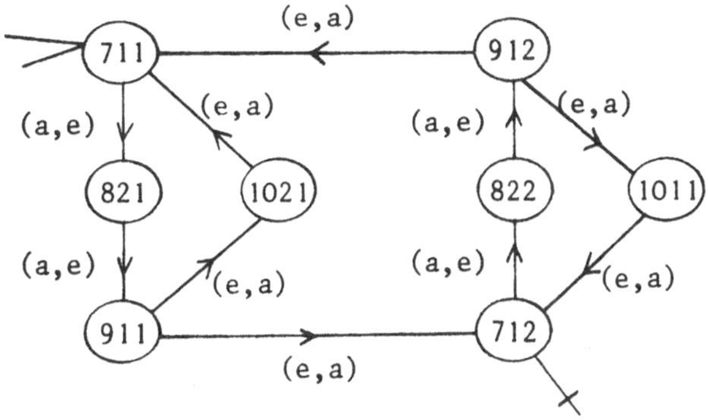

Clearly $R^*(S_o +_\varphi S'_1 \circledast S'_2) =$

$$((a^2,e^2)(e^2,a^2))^* \cup$$
$$((a^2,e^2)(e^2,a^2))^*(e,a)((a^2,e^2)(e^2,a^2))^* \cup$$
$$(((a^2,e^2)(e^2,a^2))^*(e,a)((a^2,e^2)(e^2,a^2))^*(a^2,e^2)(e,a))^*$$
$$((a^2,e^2)(e^2,a^2))^*(e,a)((a^2,e^2)(e^2a^2))^*$$

$S_o +_\varphi S_3$ is the following

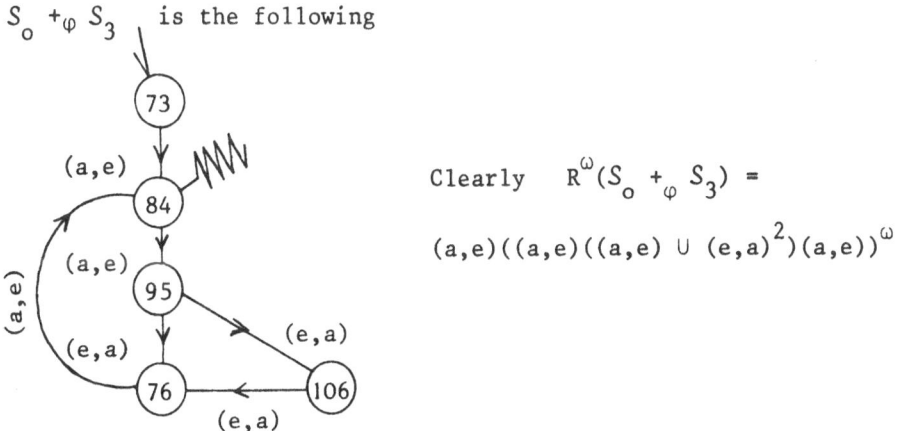

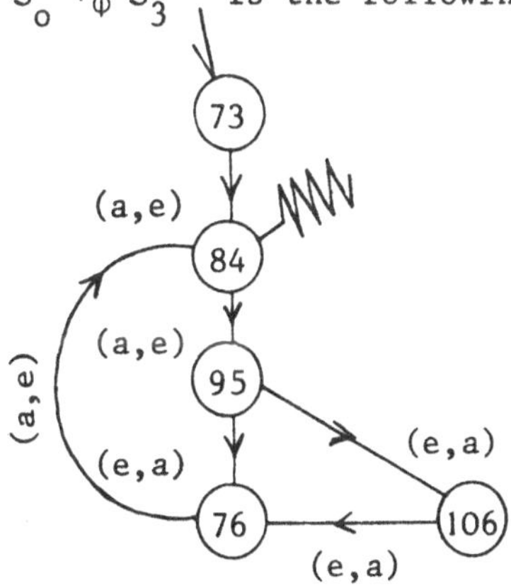

Clearly $R^\omega(S_o +_\varphi S_3) =$

$$(a,e)((a,e)((a,e) \cup (e,a)^2)(a,e))^\omega$$

The system P is realized by the disjoint union of these two mts. One can see immediately that P is safe since every configuration is coaccessible from $C_{fin} = \{712\}$ or $C_{inf} = \{84\}$.
On the contrary it is not entirely obvious that P is normal and central.

An alternative construction is the following:

the amalgamated product of $S'_1 \circledast S'_2$ and S_3 is

$$S_4 = (S'_1 \circledast S'_2) \hat{\times} S_3$$

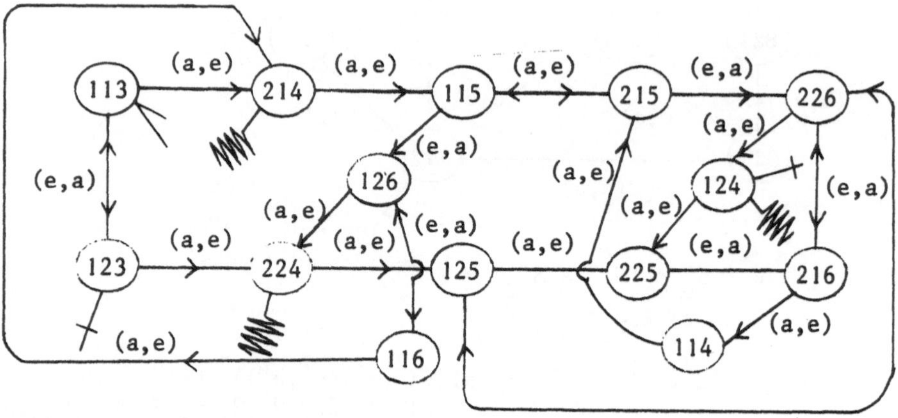

The mts S_4 is such that

$$R^{\infty}(S_4) = L_1 \circledast L_2 \quad \text{and}$$
$$R^{**}(S_4) = FG(L_1) \circledast FG(L_2)$$

As compared to $S'_1 \circledast S'_2 \cup S_3$ it has the advantage to be normal and permanent, i.e. each configuration is coaccessible from $C_{4fin} = \{123, 124, 125\}$ and from $C_{4inf} = \{214, 224, 124\}$

Then one builds $S_o +_\varphi S_4 = S_5$

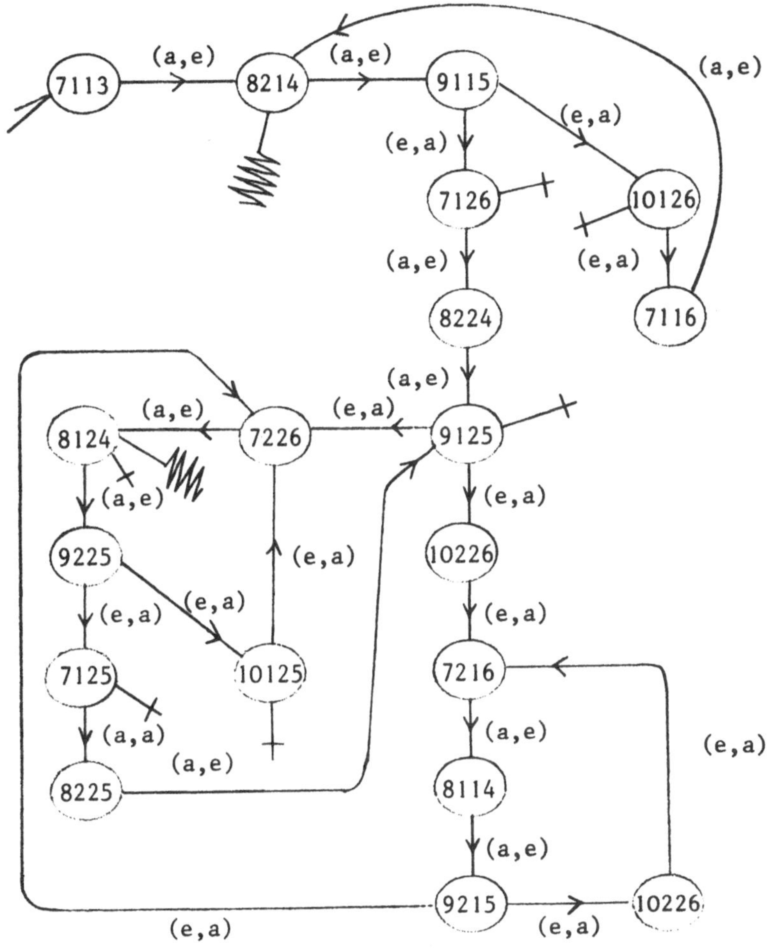

One checks immediately that S_5 is normal (since $C_5 \subseteq \text{Coacc}(C_{5fin})$) and central (since $C_5 = \{8214, 8124\}$ is alive and $C_5 \subseteq \text{Coacc}(C_{5inf})$) .

VI. CONCLUSION

These pages are just course notes, hastily written and imcomplete. The situation is not easy, in the first part we have left open the problem of computing the normal center of an infinitary language L . In the second part we said nothing or very little on computing the set of visible behaviors of a system of processes, even when all of them are delayable, and absolutely nothing about computing one real behavior corresponding to a visible one (this is the so-called serialization problem which is an important one tackled by various people, see [30]).

It is clear that results will come only from a careful examination of synchronization conditions (as in [20, 21]).

What we are most interested in, is the building, from S_0, S_1, ..., S_h, the subset S of A and the mapping φ of S into A, of a system S realizing

$$L = \{\vec{\alpha} \in S^{\infty} \mid \forall i \in [h] \quad \pi_i(\vec{\alpha}) \in L^{\infty}(S_i) \text{ and } \varphi(\vec{\alpha}) \in L^{\infty}(S_o)\}$$

We have attempted to build such an S using the operations of uniform product, sum mod φ, amalgamated product, etc.
As shown by our example, several constructions are possible and it is far from obvious which is best, i.e. which will result in the less costly algorithms to determine whether L has this or that property.
This should be the research area of a small team of researchers which is gathering in Paris around the author.

Remark added in proof:

The writing of these notes initiated in May 1981 was finished in October of the same year. Proof reading took place in January 1982. In 8 months the ideas of the author have developed, especially through oral exposition in various places and consequent discussions. Thus explains differences between the beginning and the end of these notes, especially between the general form of synchronization condition given in the introduction and the one described in paragraph IV. 2. (in fact they are equivalent, but this needs a proof not given here).

Bibliography

[1] A. Arnold and M. Nivat: "Metric interpretations of infinite trees and semantics of non deterministic recursive programs" Theor. Comp. Sci., Vol. 11, (1980), pp 181 - 205

[2] A. Arnold and M. Nivat: "Controlling behaviors of systems", in Proc. MFCS Conf., Lect. Notes Comp. Sci., Vol. 88, (1980)

[3] J. Beauquier and M. Nivat: "Applications of formal language theory to problems of security and synchronization", in Formal Language Theory, perspectives and open problems, (R. Book, ed.), Academic Press, New York, 1980

[4] L. Boasson and M. Nivat: "Adherences of languages", Jour. Comp. Syst. Sci., Vol. 20 (1980), pp 285 - 309

[5] N. Bourbaki: "Topologie Générale", Hermann, Paris (1967)
B. Trakhtenbrot and Y. Barzdin: "Finite automata: behavior and systems", North Holland, Amsterdam (1973)

[6] F. Boussinot: "Reseaux de processus avec suélange equitable: une approche du temps réel", Thèse d'Etat, Univ. Paris 7, (1981)

[7] M. Broy, R. Gnatz and M. Wirsing: "Semantics of non deterministic and non continuous constructs", in Lect. Notes Comp. Sci., No. 69, Springer-Verlag, Berlin (1979)

[8] J. Buchi: "On a decision method in restricted second order arithmetic", Int. Congress. Logic, Method Phil. Sci., Stanford Univ. Press (1962)

[9] R. Cohen and A. Gold: "Theory of ω -languages", Jour. Comp. Syst. Sci., Vol. 15 (1977), pp 169 - 184

[10] B. Courcelle: "Frontiers of infinite trees", RAIRO Info. Theor., Vol. 12 (1978), pp 319 - 337

[11] S. Eilenberg: "Automata, languages and machines", Vol. A, Academic Press, New York (1974)

[12] C. Elgot and J. Mezei: "On relations defined by generalized finite automata", IBM Tour. Res. Devel, Vol. 9 (1965), pp 47 - 68

[13] F. Gire: " Relations rationelles infinitaires", Thèse de 3ème cycle, Univ. Paris 7, 1981

[14] C.A.R. Hoare: "Communicating sequential processes", Com. Assoc. Comp. Mach., Vol. 21 (1978), pp 666 - 677

[15] L. Landweber: "Decision problems for ω -automata", Math. Syst. Th., Vol. 3 (1969), pp 376 - 384

[16] P. Lauer, P. Torrigiani, W. Shields: "COSY, a specification language based on pathes and processes", Acta Informatica, Vol. 12 (1979) pp 109 - 158

[17] M. Linna: "On ω -words and ω -computations", Annales Universitatis Turkuensis, Série A, Turku (1975)

[18] R. Mac-Naughton: "Testing and generating infinite sequences by a finite automaton", Inf. and Cont., Vol. 9 (1966), pp 521 - 530

[19] A. Mazurkiewicz: "Proving properties of processes", Report ICS-PAS 134, Polish Acad. Sci., Warsaw (1973)

[20] R. Milner: "A calculus of communicating systems", Lect. Notes Comp. Sci., No. 92 (1979)

[21] R. Milner: "On relating synchrony and asynchrony", Report No. CSR 75-80, Dep. Comp. Sci., Univ. Edinburgh (1980)

[22] M. Mycielsky and J. Taylor: "A compactification of the algebra of terms", Algebra Universalis, Vol. 6 (1976), pp 159 - 163

[23] M. Nivat: "Transductions des langages de Chomsky", Annales Inst. Fourier, Vol. 18 (1968), pp 339 - 456

[24] M. Nivat: "Mots infinis engendiés par une grammaire algebraique", RAIRO Info. Theor.,Vol. 11 (1977, pp 311 - 327

[25] M. Nivat: "Ensembles de mots infinis engendiés par une grammaire algébraique", RAIRO Info. Theor.,Vol. 12 (1978) pp 259 - 278

[26] M. Nivat: "Infinitary relations", in CAAP 81, Lect. notes Comp. Sci., No. 122 (E. Asteriano and C. Böhm, ed.), Springer Verlag, Berlin (1968)

[27] M. Nivat: "Systèmes de transition permanents et équitables" Revue technique Thomson CSF, Vol. 13 (1981), pp 55 - 79

[28] D. Park: " Concurrency and automata on infinite sequences", in Lect. Notes Comp. Sci., No. 104 (1981), pp. 167 - 183

[29] R. Redzicjowski: "The theory of general events and its application to parallel programming", Report TP 18-220, IBM Nordic Lab. (1972)

[30] G. Roucairol: "Contribution à l'étude des équivalences syntaxiques et transformations de programmes parallèles", Thèse d'Etat, Univ. Pierre et Marie-Curie, Paris 1978

[31] G. Vidal-Naquet: "Rationalité et déterminisme dans les réseaux de Petri", Thèse d'Etat, Univ. Pierre et Marie-Curie, Paris 1981

[32] E. Wiedmer: "Computing with infinite objects", Theor. Comp. Sci., Vol. 10 (1980), pp 133 - 155

Theme XI.

(M. Nivat)

A word of an infinitary language, (or possibly an infinitary lecture) on the same theme as IX

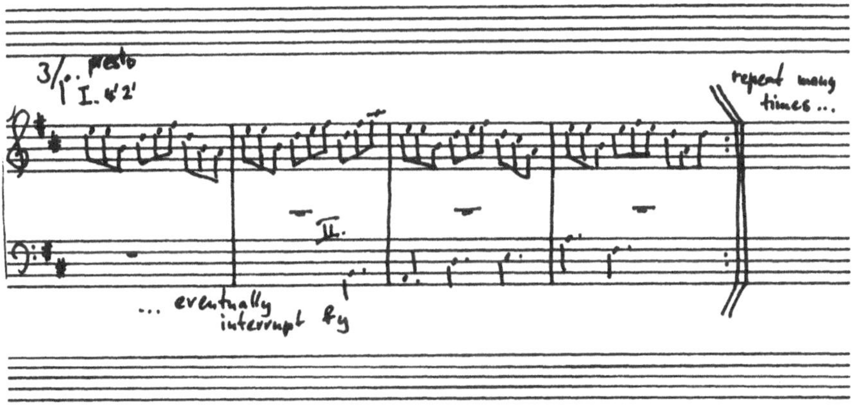

Part V

CONCURRENT PROGRAMS

As programmers began to write concurrent programs, they missed a considerable amount of the sound mathematical basis which they had for sequential programs and ran into additional methodological problems. Even today, the semantics of concurrent programs are presented in models of much less generality. The first of the following articles presents formal techniques for the development of multiprograms, while the second is an attempt to give a fixed point oriented approach to applicative multiprogramming. The final two papers are devoted to the application of communicating sequential processes, first to discrete event simulation and finally to the study of an operating system.

A TUTORIAL ON THE SPLIT BINARY SEMAPHORE

Edsger W.Dijkstra

Burroughs,
Plataanstraat 5,
5671 AL NUENEN,
The Netherlands

The purpose of this note is threefold. It has been written to introduce the student to

1) the technique of the so-called "split binary semaphore" --originally discovered, but at the same time not recommended by C.A.R.Hoare--

2) the use of formal techniques in the development of multiprograms

3) the now canonical example of "The readers and the writers", which the student must know anyhow; it will be used as a carrier for the other two purposes. The problem of the readers and the writers was designed by D.L. Parnas.

We consider two classes of cyclic processes, called "readers" and "writers" respectively. With "ncs" standing for "noncritical section", they can be described by the programs

reader: <u>do</u> true → ncs; READ <u>od</u> (1)

writer: <u>do</u> true → ncs; WRITE <u>od</u>

respectively. Here "READ" and "WRITE" denote their respective critical sections, critical in the sense that when a writer is engaged in its critical section, it must be <u>the only process</u> engaged in its critical section. This problem can be solved in many different ways --the problem is canonical in the sense that everybody proposing new synchronization primitives has solved it in his way-- ; we shall now solve it using a split binary semaphore.

M. Broy and G. Schmidt (eds.), Theoretical Foundations of Programming Methodology, 555–564.

Our first step is the introduction of variables, in terms of which we express our synchronization requirement formally; we call them "ar" and "aw" --short for "number of active readers" and "number of active writers" respectively-- and consider the following multiprogram:

```
initial state:    ar = 0, aw = 0, mutex = 1                  (2)

reader:  do true → ncs;
                   P(mutex); ar:= ar + 1; V(mutex);
                   READ;
                   P(mutex); ar:= ar - 1; V(mutex)
         od

writer:  do true → ncs;
                   P(mutex); aw:= aw + 1; V(mutex);
                   WRITE;
                   P(mutex); aw:= aw - 1; V(mutex)
         od
```

For multiprogram (2) the invariance of

P0: $ar \geq 0 \wedge aw \geq 0$

follows immediately in the usual way "from the topology of the program". (Associate with $reader_i$ an additional variable c_i, initially $= 0$ and satisfying $0 \leq c_i \leq 1$; replace $ar := ar + 1$ by $c_i, ar := 1, ar + 1$ and $ar := ar - 1$ by $c_i, ar := 0, ar - 1$ and observe the invariance of $ar = \Sigma c_i$, etc.)

Note. Looking at program (2) and interpreting it operationally, one might be tempted to say: clearly $ar \geq$ the number of readers engaged in READ, and as "a number of readers" is never negative, $ar \geq 0$ follows immediately, and similarly for $aw \geq 0$. I prefer not to do so, to prove the invariance of P0 from the uninterpreted text (2), and to conclude from P0 that it does not prohibit the interpretation of ar and aw as natural numbers. (End of Note.)

In terms of ar and aw I propose the required invariance of

P1: $aw = 0 \vee (aw = 1 \wedge ar = 0)$

as a suitable formal expression of our operational requirement "when a writer is engaged in its critical section, it must be _the only process_ engaged in its critical section". The term $aw = 0$ is intended to cover the situation when no writer is engaged in its critical section, the term $aw = 1 \wedge ar = 0$ should cover the case when a writer is engaged in its critical section.

Note. The attentive reader will already have decided that he should not expect more eloquence from me on the "suitability" of the proposal captured by P1 . (End of Note.)

In order to prevent ar:= ar + 1 from violating P1 we can make it into a guarded command of the form

$$wp(\text{"ar:= ar + 1"}, P1) \rightarrow ar:= ar + 1 \ .$$

The axiom of assignment gives for this guard

$$aw = 0 \vee (aw = 1 \wedge ar + 1 = 0)$$

but on account of P0 the second term is false, and we simplify

$$aw = 0 \rightarrow ar:= ar + 1 \quad .$$

In order to prevent aw:= aw + 1 from violating P1 we consider

$$wp(\text{"aw:= aw + 1"}, P1) \rightarrow aw:= aw + 1 \quad .$$

The axiom of assignment gives for this guard

$$aw + 1 = 0 \vee (aw + 1 = 1 \wedge ar = 0)$$

but, again, P0 admits simplification of this guard --its first term is false-- and we find

$$aw = 0 \wedge ar = 0 \rightarrow aw:= aw + 1 \quad .$$

The decrease ar:= ar - 1 is similarly guarded

$$wp(\text{"ar:= ar - 1"}, P1) \rightarrow ar:= ar - 1 \quad ,$$

and we get for the guard with the axiom of assignment

$$aw = 0 \vee (aw = 1 \wedge ar - 1 = 0) \quad ,$$

which, thanks to P1, can be simplified as

$$aw = 0 \rightarrow ar:= ar - 1 \quad .$$

The known invariance of P0 tells us that the precondition of ar:= ar - 1 implies wp("ar:= ar - 1", P0) , i.e. implies $ar - 1 \geq 0$; on account of P1 , this implies $aw = 0$, and, therefore, we can simplify to true → ar:= ar − 1

i.e. the decrease ar:= ar − 1 need not be guarded at all.

The verification that also aw:= aw − 1 need not be guarded is left to the reader.

Inserting the guards as derived we get

```
initial state:  ar = 0, aw = 0, mutex = 1                    (3)
reader:  do true → ncs;
§                  P(mutex); if aw = 0 → ar:= ar + 1 fi; V(mutex);
                   READ;
                   P(mutex); ar:= ar − 1; V(mutex)
         od
writer:  do true → ncs;
                   P(mutex);
§§                 if aw = 0 ∧ ar = 0 → aw:= aw + 1 fi; V(mutex);
                   WRITE;
                   P(mutex); aw:= aw − 1; V(mutex)
         od
```

Multiprogram (3) has been designed so as to leave P0 ∧ P1 invariant, and that is fine. It has, however, a major drawback: both alternative constructs, in the lines § and §§ respectively, may lead to abortion! The so-called split binary semaphore provides us with a technique for preventing the selection of the critical sections marked § and §§ respectively under those circumstances in which they would lead to abortion.

We replace the single binary semaphore mutex by three, also binary, semaphores -- m , r , and w say-- , related to the original mutex by

mutex = m + r + w ,

and replace in multiprogram (3) each P(mutex) by P(m) , P(r) , or P(w) --the three ways of decreasing mutex by 1-- and each V(mutex) by V(m) , V(r) , or V(w) --the three ways of increasing mutex by 1-- . (Because m , r , and w are semaphores, we thus guarantee $0 \le \text{mutex} \le 1$.) More precisely, we replace the P(mutex) marked by § by P(r) , replace the P(mutex) marked by §§ by P(w) , and the P(mutex) that opens a critical section that cannot lead to abortion by P(m) .

We can now avoid selection of an aborting critical section by guarding V(r) by $aw = 0$ and by guarding V(w) by $aw = 0 \wedge ar = 0$, because the precondition of a V-operation on a component of a split binary semaphore can be taken as the post-condition of the corresponding P-operation. Our analysis so far would suggest that it suffices to replace V(mutex) everytime by

```
Q:  if true → V(m)                                           (4)
     [] aw = 0 → V(r)
     [] aw = 0 ∧ ar = 0 → V(w)
    fi     .
```

This, however, is too naive. To start with: how do we translate the initialization mutex = 1 ? The initialization $m = 0,\ r = 1,\ w = 0$ is too restrictive: if all readers remain in their ncs , no writer could perform WRITE . The initialization $m = 0,\ r = 0,\ w = 1$ has to be rejected on similar grounds. In order to make the only remaining possible initialization $m = 1,\ r = 0,\ w = 0$ acceptable, readers and writers should encounter as first P-operation one that cannot lead to abortion. We can satisfy this requirement by inserting in both readers and writers of (3) after ncs

```
P(mutex); V(mutex);
```

before performing the substitution described above. This would lead to the following multiprogram.

```
initial state:  ar = 0, aw = 0, m = 1, r = 0, w = 0            (5)
reader:  do true → ncs; P(m); Q;
                   P(r) {aw = 0}; ar:= ar + 1; Q;
                   READ;
                   P(m); ar:= ar - 1; Q
         od
writer:  do true → ncs; P(m); Q;
                   P(w) {aw = 0 ∧ ar = 0}; aw:= aw + 1; Q;
                   WRITE;
                   P(m); aw:= aw - 1; Q
         od     .
```

Multiprogram (5) is, however, still too naive: the nondeterminacy that has been introduced by Q as given in (4) has lead to a system with the danger of deadlock. The recipe for its prevention is, however, universal.

1) At initialization and at each V-operation, the "type of the next P-operation" --i.e. the component of the split binary semaphore on which this program will perform its next P-operation-- must be uniquely defined. Our program (5) satisfies this condition, as do all programs without initial nondeterminacy nor nondeterminacy between a V-operation and the dynamically next P-operation. If a program does not satisfy this condition we can always make it satisfying it by introducing one or more extra components of the split binary semaphore, by replacing essentially

```
if true → P(component 1);...
 [] true → P(component 2);...
fi
```

by

```
if true → P(extra component); Q; P(component 1);...
 [] true → P(extra component); Q; P(component 2);...
fi
```

Task. Prove that for this substitution process only a finite number of different extra components is needed. (End of Task.)

2) With each component of the binary semaphore we associate a counter, initialized to the number of processes for which the first P-operation is of the corresponding type.

3) After each P-operation we insert a decrease by 1 of the counter associated with its type.

4) Before each Q we insert an increase by 1 of the counter associated with the type of the dynamically next P-operation. (Thanks to step 1 , this is a well-defined counter.)

Note. For the operationally minded: each counter can be interpreted as the number of processes "ready" or "headed" for a P-operation on the corresponding component. (End of Note.)

5) Strengthen in Q the guarding of each V(component) by the requirement that the corresponding counter is positive.

Associating with the semaphore components m , r , and w the counters bm , br , and bw respectively, carrying out the above transformations on program (5) leads to

```
initial state:                                                    (6)

     ar = 0, aw = 0, bm = number of processes, br = 0, bw = 0,

     m = 1, r = 0, w = 0

reader: do true → ncs;
                   P(m); bm:= bm - 1; br:= br + 1; Q;
                   P(r); br:= br - 1; ar:= ar + 1; bm:= bm + 1; Q;
                   READ;
                   P(m); bm:= bm - 1; ar:= ar - 1; bm:= bm + 1; Q
        od

writer: do true → ncs;
                   P(m); bm:= bm - 1; bw:= bw + 1; Q;
                   P(w); bw:= bw - 1; aw:= aw + 1; bm:= bm + 1; Q;
                   WRITE;
                   P(m); bm:= bm - 1; aw:= aw - 1; bm:= bm + 1; Q
        od
```

with Q short for

```
Q:      if bm > 0 → V(m)                                          (6')
         [] aw = 0 ∧ br > 0 → V(r)
         [] aw = 0 ∧ ar = 0 ∧ bw > 0 → V(w)
        fi
```

Note. The transformation of the introduction of the counters and of the strengthening of the guards in Q by the requirement that the corresponding counters be positive excludes the danger of deadlock. If the original requirement --in our case: the invariance of P1 -- entailed intrinsically the danger of deadlock, this danger is now made manifest by the danger of abortion in Q . The systematic procedure for dealing with that situation --i.e. how to strengthen P1 -- falls outside the scope of this tutorial. (End of Note.)

Our original system was free from deadlock, hence we must --see above note-- be able to prove the absence of the danger of abortion in Q . The precondition of Q implies everywhere in (6)

$$bm + br + bw = \text{number of readers} + \text{number of writers}$$

$$ar + br \leq \text{number of readers}$$

$$aw + bw \leq \text{number of writers} \quad .$$

<u>Task</u>. Verify the above three assertions. (End of Task.)

From the above we conclude

$$ar + aw \leq bm \quad , \text{ and hence}$$

$$bm > 0 \vee (ar = 0 \wedge aw = 0) \quad .$$

Assuming the number of processes to be larger than zero --otherwise the danger of abortion is absent anyhow!-- , we also have

$$bm > 0 \vee br > 0 \vee bw > 0$$

and from the last two relations it follows that at least one of the guards of Q as given in (6') is true.

* * *

The above form of Q as given in (6') is still very nondeterministic: we have the strategic freedom of strengthening the guards as long as we avoid the danger of abortion. As it stands, our solution does not exclude that readers or a writer are denied access to their critical section READ or WRITE without reason. Within Q we can give "priority" to V(r) or V(w) by strengthening the guard of V(m) to the conjunction of the complements of the other two guards:

Am: $(aw > 0 \vee br = 0) \wedge (aw > 0 \vee ar > 0 \vee bw = 0)$

Denoting by Ar and Aw respectively:

Ar: $aw = 0 \wedge br > 0$

Aw: $aw = 0 \wedge ar = 0 \wedge bw > 0$

we can now substitute for Q in (6)

Q: <u>if</u> Am → V(m) ▯ Ar → V(r) ▯ Aw → V(w) <u>fi</u> (7')

All (now superfluous) operations on bm can be omitted; using the postconditions of the P-operations, the substitution instances of Q as given by (7') can be simplified. Thus we derive from (6) the multiprogram

```
initial state:
    ar = 0, aw = 0, br = 0, bw = 0, m = 1, r = 0, w = 0          (7)

reader:
do true → ncs;
    P(m); br:= br + 1; if aw > 0 → V(m) ▯ aw = 0 → V(r) fi;
    P(r); br, ar := br - 1, ar + 1; if br = 0 → V(m)
                                    ▯ br > 0 → V(r)
                                    fi;
    READ;
    P(m); ar:= ar - 1; if ar > 0 ∨ bw = 0 → V(m)
                       ▯ ar = 0 ∧ bw > 0 → V(w)
                       fi
od

writer:
do true → ncs;
    P(m); bw:= bw + 1; if aw > 0 ∨ ar > 0 → V(m)
                       ▯ aw = 0 ∧ ar = 0 → V(w)
                       fi;
    P(w); bw, aw := bw - 1, aw + 1; V(m);
    WRITE;
    P(m); aw:= aw - 1; if br = 0 ∧ bw = 0 → V(m)
                       ▯ br > 0 → V(r)
                       ▯ bw > 0 → V(w)
                       fi
od
```

Remark. An inspection of the alternative constructs in (7) shows that only the very last one is nondeterministic: here we have, therefore, still a strategic freedom. Investigate the consequences of replacing the last guarded command bw > 0 → V(w) by bw > 0 ∧ br = 0 → V(w) . (End of Remark.)

* * *

In the above we have derived our final program as the end of a sequence of successive versions. We have done so for educational reasons, with the intention of introducing the different aspects of programs synchronized with the aid of a split binary semaphore one after the other. This is not meant as a suggestion that in the case of actual program design one should write down all those successive versions! The experienced programmer knows that "outside the critical sections" as delineated by the P-V pairs we have an invariant of the form

$$(m = 0 \vee Am) \wedge (r = 0 \vee Ar) \wedge (w = 0 \vee Aw)$$

and focusses his attention on the A's; having chosen them, he derives the final code. I make this remark because so-called "program transformations" are sometimes suggested as a practical way of program derivation --not by me, for as a rule it leads to an undue amount of writing-- .

A FIXED POINT APPROACH TO APPLICATIVE MULTIPROGRAMMING*)

Manfred Broy
Technische Universität München
Institut für Informatik
Postfach 20 24 20
D-8000 München 2

For a simple *nondeterministic applicative programming language* both *fixed point semantics* and *operational semantics* are defined. The operational semantics is given by a number of computation rules (term rewriting rules) representing a *data flow semantics* and the possibilities of inherent *parallelism* are studied. The language, which in the first instance implicitly uses the notions of multiprogramming, is extended by more explicit constructs, such that finally a general reduction language and data flow language with explicit *parallelism, communication* and *concurrency* is derived. For this extended language, too, a fixed point semantics is defined based on multidomains, and an operational semantics is given in the form of computation rules modelling the evaluation process of data driven reduction als well as data flow computations.

*) This research was carried out within the Sonderforschungsbereich 49 - Programmiertechnik - München

M. Broy and G. Schmidt (eds.), Theoretical Foundations of Programming Methodology, 565–623.

1. Introduction

The last two decades of computer science are characterised by an enormous progress in the formal foundations of programming languages and their semantics. In sequential, deterministic programming, most of the remaining questions are of quantitative (questions of programming in the large) rather than of qualitative nature. In the field of nondeterministic, parallel, communicating, concurrent programs the state of the art is less satisfactory. Although drastic efforts have been undertaken to investigate this field leading to a considerable amount of knowledge, we are far from having extensive, widely accepted theories for concurrent programming. Most existing theories do not cover all important aspects and/or they are too complex and complicated.

Nevertheless numerous papers have been published which suggest language constructs for concurrent programming. These papers have had a considerable impact on the field of concurrent programming, and in many cases helped in developing a better understanding. However, the lack of proper formal definitions of the semantics of such languages must be considered as a severe drawback. On the one hand, it seems impossible to enlarge on a programming methodology for the construction of concurrent software without having well-explored theoretical foundations. On the other hand, a properly designed programming language presumes a complete understanding of the underlying concepts, which is also impossible without having a formal theory. And last not least, mathematical foundations are an indispensable requirement for teaching concurrent programming. Therefore I strongly believe that it is necessary to investigate the concepts of concurrent programming in a joint consideration of both mathematical (denotational) semantics and its corresponding operational semantics. Here, fixed point theory seems to be the most adequate framework.

So in the sequel we try to develop a strictly fixed-point-oriented approach to the semantics of applicative multiprogramming. It should be noted, that the restriction to applicative languages is not a profound one. It only helps in concentrating on the central issues. The approach is based on a simple *non-deterministic programming language*. For this language both mathematical and operational semantics are given. The operational semantics consists of a set of computation rules which model the behavior of a simple *reduction machine*. Based on a thorough discussion, the language is stepwise extended to allow more general patterns of *communication* leading to *systems of communicating expressions*.

Before going deeper into the theory of applicative multiprogramming it seems useful to recall some of the most important notions, namely: nondeterminism, parallelism, communication, and concurrency.

Considering these notions isolated from each other (as far as this is possible) already causes some problems. However, combining these notions into the concepts of one language multiplies the difficulties. For example there are *different concepts* of nondeterminism (cf. /Kennaway, Hoare 80/, /Broy, Wirsing 81a/) which, taken for themselves, can be treated quite satisfactorily. In concurrent programming, however, some of these concepts are used side by side. For instance the scheduling of simple communication actions may be mapped onto straightforward-choice nondeterminism (*"erratic" nondeterminism*), while disjunctive ("multiple") waiting has to be mapped onto some kind of nondeterminism which delays the choice until one of the possibilities yields a defined way of resuming (local *"angelic" nondeterminism*).

Thus concurrency does not only require a free, straightforward choice between concurrent computations, but it requires a choice depending on certain termination properties of the concurrent computations. This brings about all of the problems of unbounded nondeterminism, its noncontinuity in the Egli-Milner ordering as found in the fairness-discussion (cf. /Park 80/, /Broy 81/, /Apt, Plotkin 81/) and even worse problems concerning monotonicity.

Yet another example for these increasing difficulties can be found when considering mutually communicating processes. If the possible sequences of communicated values are taken as defining the meaning of a process, then obviously a process cannot be considered as a nondeterministic function over flat domains. However, on nonflat domains the classical powerdomain construction does not work (cf. /Plotkin 76/, /Smyth 78/), since there the Egli-Milner "ordering" represents only a quasiordering.

In the following sections an attempt is undertaken to give a solution to these problems in one integrated approach completely based on fixed point theory.

Section 2 establishes the theory of *multidomains* constructed as the ideal completion of the set of finite multisets over the finite elements of an algebraic cpo. As an important example the multidomain of finite and infinite streams is considered.

Section 3 gives a simple *applicative nondeterministic language for multiprogramming* and its mathematical semantics based on multidomains.

Section 4 considers a nondeterministic computation rule for the *parallel evaluation* of nondeterministic recursive programs.

Section 5 extends the possibility of *communication* between processes evaluating expressions, which inherently can be found in the "communication" between the processes evaluating the arguments and the body of a function, to systems of expressions which mutually communicate by streams.

Section 6 studies the problems of *concurrency* by introduction of an ambiguity operator which is essentially more defined than the straightforward choice operator. This operator allows to "suppress" certain nondeterministic alternatives and in particular serves as a basis for defining for instance disjunctive (alternative, multiple, parallel) waiting. However, the introduction of this operator causes a radical change of our domain and our notion of computability.

2. Multidomains

For giving semantics to nondeterministic programs, /Plotkin 76/ suggests the use of *powerdomains*, which are particular subsets of the given domains representing the sets of possible values (cf. also /Smyth 78/). However, for nonflat domains the well-known Egli-Milner ordering does not work, since the induced ordering generally is only a quasiordering. Therefore in /Lehmann 76/ as an alternative to fixed point theory, category theory is suggested for giving meaning to nondeterministic programs.

In the sequel a different approach is given which uses multisets (cf. for instance /Dershowitz, Manna 79/) as suitable representations of the (multi-)set of possible results. Roughly speaking, a multiset is a "set, where multiple occurrences of elements are allowed". A nondeterministic program can be viewed as a finitary, possibly infinite tree. In the powerdomain construction the set of terminal nodes (including $\perp$ for infinite paths) is associated with the program. In our approach of multidomains, the multiset of terminal nodes is associated with the program (including $\perp$ for every infinite decision path).

For lack of space we omit all proofs.

2.1 Multisets

Given a set S , a <u>multiset</u> M over S is a total mapping

$$M: S \rightarrow \mathbb{N}^+$$

where $\mathbb{N}^+ =_{def} \mathbb{N} \cup \{\infty\}$. The set of multisets over a given set S is denoted by $\mathcal{M}(S)$. Every multiset M defines also a set

$$SET(M) = \{x \in S : M(x) \geq 1\}$$

The cardinality of a multiset M is denoted by $|M|$ where

$$|M| =_{def} \sum_{x \in SET(M)} M(x)$$

Trivially, every set $S1 \subseteq S$ denotes a multiset M with

$$M(x) = \begin{cases} 1 & x \in S1 \\ 0 & \text{otherwise} \end{cases}$$

Therefore we often use the set-notation for representing multisets.

For multisets M1, M2 we define the *multisetsum* $M1 \uplus M2$ and *multiset difference* $M1 \ominus M2$ by

$$(M1 \uplus M2)(x) = M1(x) + M2(x)$$

$$(M1 \ominus M2)\ (x) = \begin{cases} M1(x) - M2(x) & \text{if } M1(x) > M2(x) \\ 0 & \text{otherwise} \end{cases}$$

The set union and set intersection are extended to multisets by

$$(M1 \cup M2)\ (x) = \max\ \{M1(x),\ M2(x)\}$$

$$(M1 \cap M2)\ (x) = \min\ \{M1(x),\ M2(x)\}$$

Given a multiset M, we define a multiset $\biguplus_{x \in M} F(x)$ where $F: S \to \mathcal{M}(S)$

$$\Big(\biguplus_{x \in M} F(x)\Big)\ (y) =_{def} \sum_{x \in SET(M)} \sum_{i=1}^{M(x)} F(x)(y)$$

In analogy to the set notation, we write

$$M1 \subseteq M2 \quad \text{iff} \quad \forall\ x \in S : \quad M1(x) \leq M2(x)$$

Multisets can be viewed as a "natural" extension of set theory. Although it seems interesting to study the properties of multisets much more deeply, we restrict ourselves to the definition of multidomains as particular subsets of the set of multisets.

2.2 Multidomains

In this section multisets over a complete countable algebraic partially ordered set are considered.

A poset (DOM, $\sqsubseteq$) is a cpo (complete partially ordered set) iff

(i) DOM has a least element $\perp$

(ii) every directed subset X has a least upper bound lub(X)

An element $a \in DOM$ is called <u>finite</u> if for every directed subset $X \subseteq DOM : a \sqsubseteq lub(X) \Rightarrow \exists x \in X : a \sqsubseteq x$

DOM is called <u>countably algebraic</u> if

(i) the set of finite elements is countable ,
(ii) every element in DOM is the lub of a directed set of finite elements.

We shall refer to countably algebraic cpo's simply as <u>domains</u>.

In a domain DOM an element $x \in DOM$ is called <u>partial</u> if $\exists\, y \in DOM, y \neq x : x \sqsubseteq y$, <u>total</u> if x is not partial, i.e. x is maximal in DOM, <u>infinite</u> if x is not finite.

As a generalisation of the well-known Egli-Milner ordering on sets over DOM (cf. /Egli 75/, /Milner 73/) the relation $\sqsubseteq$ is defined on $M(DOM)$ by :

Let $M1, M2 \in M(DOM)$:

$$M1 \sqsubseteq M2 \quad \text{iff} \quad \exists\; G : SET(M1) \rightarrow M(DOM) :$$
$$(1) \quad \forall x \in SET(M1) : M1(x) \leq |G(x)| \;\wedge\; \forall y \in SET(G(x)) : x \sqsubseteq y$$
$$(2) \quad \biguplus_{y \in SET(M1)} G(y) = M2$$

Intuitively speaking, $M1 \sqsubseteq M2$ holds, i.e. M1 approximates M2, if M2 can be obtained from M1 by substituting every element x of M1 by nonempty multisets $G(x)$ which consist only of elements which can be approximated by x . This immediately gives:

If $M1 \sqsubseteq M2$, then $SET(M1) \sqsubseteq_{EM} SET(M2)$ where the Egli-Milner ordering (cf. /Egli 75/, /Milner 73/, /de Bakker 76/) is defined by

$$S1 \sqsubseteq_{EM} S2 \text{ iff } (\forall x \in S1\ \exists y \in S2 : x \sqsubseteq y) \wedge (\forall y \in S2\ \exists x \in S1 : x \sqsubseteq y).$$

Like the Egli-Milner ordering, the relation $\sqsubseteq$ in general does not define an ordering but only a quasiordering.

A multiset M can also be represented by a set S_M of pairs. S_M may be defined by

$$S_M = \{(n, x) \in \mathbb{N} \times DOM : 1 \leq n \leq M(x)\}$$

Lemma: $M1 \sqsubseteq M2$ iff $\exists H : S_{M2} \to S_{M1}$:

(1) H is surjective

(2) $\forall (i, y) \in S_{M2} : (j,x) = H(i,y) \Rightarrow x \sqsubseteq y$

(H is right-monotonic) □

Lemma: The relation "$\sqsubseteq$" defines a quasiordering on $M(DOM)$. □

However on a subset of $M(DOM)$ the relation "$\sqsubseteq$" even defines a cpo. To isolate this subset the cpo is constructed in the classical way as ideal completion of the finite elements.

Following the ideal-theory of /Scott 80 /, at first the finite elements of $M(DOM)$ are considered, i.e. the finite nonempty multisets of finite elements of DOM. The set of these multisets is denoted by $FM(FDOM)$ where $FDOM \subseteq DOM$ denotes the finite elements in DOM and for each set S $FM(S)$ denotes the multisets represented by the functions

$$f: S \to \mathbb{N} \text{ where } \{x \in DOM : f(x) > o\} \text{ is finite and nonempty.}$$

In contrast to the powerdomain, where the Egli-Milner ordering generally defines only a quasi-ordering even on finite sets of finite elements, the relation $\sqsubseteq$ denotes a partial ordering on $\mathcal{FM}$(FDOM).

Lemma: ($\mathcal{FM}$(FDOM), $\sqsubseteq$) forms a poset with minimal element $\{\bot\}$ □

Now $\mathcal{FM}$(FDOM) is extended to a complete partial ordering (cpo) by taking its ideal completion. A nonempty subset $I \subseteq \mathcal{FM}$(FDOM) is called an ideal iff

$$(1)\ M1 \sqsubseteq M2 \wedge M2 \in I \Rightarrow M1 \in I$$

$$(2)\ M1, M2 \in I \Rightarrow \exists M3 \in I: M1 \sqsubseteq M3 \wedge M2 \sqsubseteq M3$$

So I is an ideal if it is (1) downward closed and (2) directed. Let $\mathcal{ID}$(FDOM) denote the set of ideals. As well-known ($\mathcal{ID}$(FDOM), $\subseteq$) forms a countably algebraic cpo.

Since it seems more convenient to talk about multisets than about ideals, we isolate a subset $\mathcal{MD}$(DOM) of $\mathcal{M}$(DOM), such that ($\mathcal{MD}$(DOM), $\sqsubseteq$) is isomorphic to ($\mathcal{ID}$(FDOM), $\subseteq$). So every ideal is represented by one particular multiset. For every $M \in \mathcal{M}$(DOM) we define

$$I_M = \{M1 \in \mathcal{FM}(FDOM) : M1 \sqsubseteq M\}$$

If $M \in \mathcal{FM}$(FDOM) trivially I_M forms an ideal.

On the other hand we may associate a multiset M_I with every subset $I \subseteq \mathcal{FM}$(FDOM) by

$$M_I(x) = \operatorname*{glb}_{\substack{M1 \in I \\ y \sqsubseteq x \\ y \in FDOM}} \ \mathrm{lub}\ \{M2(z) : M1 \sqsubseteq M2 \in I \wedge y \sqsubseteq z \sqsubseteq x\}$$

Here lub and glb denote the least upper bound and greatest lower bound in IN^+.

Lemma: (1) For every ideal I with $M = M_I$ we have $I = I_M$.
(2) For ideals I1, I2 we have $I1 \subseteq I2 \Rightarrow M_{I1} \sqsubseteq M_{I2}$ □

Accordingly we define the multidomain $(MD(DOM), \sqsubseteq)$ by

$$MD(DOM) = \{M \in M(DOM): \exists\, I \in ID(FDOM): M_I = M \wedge I_M = I\}$$

Of course $FM(FDOM) \subseteq MD(DOM)$.

Since the multidomain is not closed under the usual multiset sum, we define a new sum operator for elements from the multidomain:

$$M_{I1} \uplus' M_{I2} = M_I \text{ where } I = \{M \in FM(FDOM): \exists M1 \in I1, M2 \in I2: M \sqsubseteq M1 \uplus M2\}$$

Obiviously on finite multisets:

$$M1 \sqsubseteq M1' \wedge M2 \sqsubseteq M2' \Rightarrow M1 + M2 \sqsubseteq M1' + M2'$$

Similary we define for $I \in ID(FDOM)$, $f: DOM \rightarrow ID(FDOM)$

$$\biguplus'_{y \in M_I} M_{f(y)} = M_{I'} \text{ where}$$

$$I' = \{M \in FM(FDOM) : \exists M' \in I, M_y \in f(y): M \sqsubseteq \sum_{y \in M'} M_y\}$$

For simplicity we write $\uplus$ for $\uplus'$ in the sequel and use $\uplus$ also for ideals.

Due to the results above the following proposition holds:

Lemma: The multidomain MD(DOM) forms a countably algebraic cpo, i.e. it forms a domain.

Note that the particular subset MD(DOM) of M(DOM) is determined by the fact, that only multisets are to be considered, which are approximable by finite multisets of finite elements, and the notion of approximability is determined by the particular choice of the ordering, which is defined such that the most important constructs which one wants to have in a programming language, such as function application, finite choice, and conditional are monotonic and continuous.

Now the continuity of some basic functions is established.

Lemma: Function extension is continuous:

Let the function

$$f : \mathrm{DOM}^n \rightarrow \mathrm{DOM}$$

be given and the function

$$F : \mathrm{DOM}^n \rightarrow MD(\mathrm{DOM})$$

be defined by

$$F(x) = \{f(x)\}$$

then if

(1) f is monotonic, then F is monotonic, too,

(2) f is continuous, then F is continuous, too. □

Lemma:

The multiset-sum is monotonic and continuous. □

As is well-known, the ordering on multidomains induces an ordering on multiple-valued functions. Let

$$f, g : \mathrm{DOM1} \rightarrow \mathcal{MD}(\mathrm{DOM2})$$

then we define

$$f \sqsubseteq g \quad \text{if} \quad \forall x \in \mathrm{DOM1} \;\; f(x) \sqsubseteq g(x)$$

Let Γ be a functional

$$\Gamma : (\mathrm{DOM1} \rightarrow \mathcal{MD}(\mathrm{DOM2})) \rightarrow (\mathrm{DOM1} \rightarrow \mathcal{MD}(\mathrm{DOM2}))$$

then Γ is called monotonic, if

$$f \sqsubseteq g \;\; \text{implies} \;\; \Gamma[f] \sqsubseteq \Gamma[g]$$

Γ is called continuous, if for every chain of continuous functions $\{f_i\}_{i \in \mathbb{N}}$:

$$\Gamma[\mathrm{lub}\{f_i\}] = \mathrm{lub}\{\Gamma[f_i]\}$$

Lemma:

Let $\Gamma : (\mathrm{DOM} \rightarrow \mathcal{MD}(\mathrm{DOM})) \rightarrow (\mathrm{DOM} \rightarrow \mathcal{MD}(\mathrm{DOM}))$

be defined by

$$\Gamma[f]\,(x) = \biguplus_{y \in T[f](x)} f(y)$$

where $T : (\mathrm{DOM} \rightarrow \mathcal{MD}(\mathrm{DOM})) \rightarrow \mathrm{DOM} \rightarrow \mathcal{MD}(\mathrm{DOM}))$

then Γ is

(1) monotonic, provided T is monotonic

(2) continuous, provided T is continuous.

To give an example how the multidomain construction works in a particular case, the multidomain of streams is considered in the following section.

2.3 The Multidomain of Streams

As an important example for a nonflat, algebraic domain for multiprogramming the domain of *streams* is considered (cf. /Landin 65/, /Burge 75/, /Dennis, Weng 79/). Let a countable set A of atoms be given, such that $\bot \notin A$. As usual $A^{\bot}$ denotes the corresponding flat domain.

The domain A^S of streams over A is defined by

$$A^S = A^* \cup (A^* \times \{\bot\}) \cup A^{\infty}$$

Here A^* denotes the set of finite streams, i.e. finite sequences of atoms from A , and includes ε , the *empty stream*. $A^* \times \{\bot\}$ denotes the set of *partial streams*, i.e. finite sequences of atoms ending with $\bot$, and includes $\bot$, the *totally undefined stream*. A^{∞} denotes the set of *infinite streams*, i.e. infinite sequences of atoms (which may also be represented by total functions $\mathbb{N} \to A$).

The following four functions are used on streams:

$$\begin{aligned} ap &: A^{\bot} \times A^S \to A^S \\ rest &: A^S \to A^S \\ first &: A^S \to A^{\bot} \\ isempty &: A^S \to \{tt, ff\}^{\bot} \end{aligned}$$

defined by

$$ap(a,s) = \begin{cases} \langle a \rangle \circ s & \text{if } a \in A,\ s \in A^S \\ \bot & \text{otherwise} \end{cases}$$

Let $a \in A$, $s \in A^S$, $s' = \langle a \rangle \circ s$, then

$rest(s') = s$, $rest(\varepsilon) = rest(\bot) = \bot$,

$first(s') = a$, $first(\varepsilon) = first(\bot) = \bot$,

$isempty(s') = ff$, $isempty(\varepsilon) = tt$, $isempty(\bot) = \bot$.

By $<a>$ the one-element sequence is denoted, and by $s \circ s'$ the concatenation of two sequences. Of course, $\varepsilon \circ s = s = s \circ \varepsilon$ and if s is infinite, i.e. $s \in A^{\infty}$ then $s \circ s' = s$ for all $s' \in A^S$. Note, however, that A^S is not closed with respect to concatenation since for $s \in A^* \times \{\bot\}$ and $s' \in A^* \setminus \{\varepsilon\}$: $s \circ s' \notin A^S$.

To make A^S into a domain, an ordering is needed. So we define for $s1, s2 \in A^S$:

$$s1 \sqsubseteq s2 \text{ iff } s1 = s2 \text{ or } \exists\, s3, s4 \in A^S \text{ such that}$$
$$s1 = s3 \circ <\bot> \text{ and } s2 = s3 \circ s4$$

Intuitively, $s1 \sqsubseteq s2$ holds, i.e. s1 "approximates" s2, if $s1 = s2$ or if s1 is a partial stream which is a prefix of s2 if $\bot$ is dropped at the end of s1. With this ordering A^S forms a countable algebraic cpo. Note that $A^{\bot}$ can be viewed as a proper subdomain of A^S.

<u>Lemma:</u> The functions ap, rest, top, isempty are monotonic and continuous.

<u>Proof</u>: omitted □

According to the results of section 2.2, the functions ap, rest, top, isempty can be trivially extended to multisets of streams. $MD(A^S)$ is called the *multidomain of streams*.

<u>Note</u>: One might also use mappings DOM → [0,1] instead of multisets and interprete this as the probability of a result (cf. /Francez, Rodeh 80/). Note, however, that this would lay severe restrictions on the implementations to assure that all probabilities are properly realized.

<u>end of note</u>

3. A Simple Language for Applicative Multiprogramming

In this section we define a simple nondeterministic programming language and its mathematical and operational semantics. By studying specific computation rules we analyse the possibilities of inherent parallelism leading to an (operational) data driven reduction semantics.

3.1 Syntax

The syntax of the language is close to λ-notation. However, only first-order functions are considered and the fixed point operator is replaced by the possibility of defining a system of mutually recursive functions.

```
< program >::=⌈{funct < function identifier > ■ < funct abstract>,}*
               < expr > ⌋
< expr > ::= < funct appl > | < cond > | < choice> |
             < object >
< funct appl > ::= < function > ({< expr > {, < expr > }* })
< cond > ::=  if  < expr > then  < expr > else < expr > fi
< choice > ::=  < expr > ▯  < expr >
< object > ::=  < primitive object > | < identifier >
< funct abstract > ::= λ {< identifier > {, < identifier >}* } :
                        < expr >
< function > ::= < funct abstract > | < function identifier > |
                 < primitive function >
```

Here we assume a domain **DOM** of semantic values called primitive objects including ff and tt for the boolean values and the natural numbers. Furthermore we assume a set P of primitive function symbols where for every $g \in P$ an arity n is given and an n-ary partial function

$$\tilde{g} : DOM^n \rightarrow DOM$$

where $\tilde{g}(x_1, \ldots, x_n) = \bot$, if one of the x_i is partial, and g is strict, monotonic, and continuous.

An expression is closed, if no free identifiers occur in it. An expression is called primitive, if it is a term built from primitive functions and objects only.

3.2 Mathematical Semantics

In this section we define a mathematical semantics by giving a function (cf. /Broy et al. 78/):

$$B : EXPR \rightarrow \mathcal{MD}(DOM)$$

where **EXPR** denotes the set of closed expressions, i.e. the set of expressions in which no free identifiers or nonprimitive function symbols occur.

$$B[\underline{if}\ C\ \underline{then}\ E1\ \underline{else}\ E2\ \underline{fi}] = \left(\biguplus_{tt \in B[C]} B[E1]\right) \uplus \left(\biguplus_{ff \in B[C]} B[E2]\right) \uplus \left(\biguplus_{\bot \in B[C]} \{\bot\}\right),$$

$$B[E1 \mathbin{[\!]} E2] = B[E1] \uplus B[E2],$$

$$B[(\lambda\ x_1, \ldots, x_n : E_{n+1})\ (E_1, \ldots, E_n)] =$$

$$\biguplus_{e_1 \in B[E_1]} \cdots \biguplus_{e_n \in B[E_n]} B[E_{n+1}\ [e_1/x_1, \ldots, e_n/x_n]]$$

$$B[e] = \{e\} \qquad \text{for } e \in DOM$$

$$B[g(E_1, \ldots, E_n)] = \biguplus_{e_1 \in B[E_1]} \ldots \biguplus_{e_n \in B[E_n]} \{\tilde{g}(e_1, \ldots, e_n)\}$$

for n-ary $g \in P$

Given an expression E in which the nonprimitive function symbols $f_1, \ldots, f_n$ occur in n_i-ary function applications, then for each set

$\tilde{f}_1, \ldots, \tilde{f}_n$ of n_i-ary functions $\tilde{f}_i : DOM^{n_i} \to \mathcal{MD}(DOM)$ by

$$B[f_i(E_1, \ldots, E_n)] = \biguplus_{e_1 \in B[E_1]} \ldots \biguplus_{e_n \in B[E_n]} \tilde{f}_i(e_1, \ldots, e_n)$$

the semantics $B[E]$ is fixed. To express that the function $\tilde{f}_i$ is to be taken for the symbol f_i we often write

$$B[E[\tilde{f}_1/f_1, \ldots, \tilde{f}_n/f_n]]$$

Given an expression E in which the free identifiers $x_1, \ldots, x_n$ occur, in principle one might consider arbitrary closed expressions $E_1, \ldots, E_n$ and the multiset $B[E[E_1/x_1, \ldots, E_n/x_n]]$. However, since we would like to consider free identifiers as identifiers for values rather than as identifiers for expressions, we restrict ourselves to the substitution of semantic values for free identifiers (cf. the remark on call-by-value versus call-by-name in section 4).

According to the propositions in section 2, all the functions defined so far are monotonic and continuous and therefore we may define the semantics of programs:

$$B[\lceil \underline{\text{funct}}\ f_1 \equiv F_1, \ldots, \underline{\text{funct}}\ f_n \equiv F_n, E \rfloor] = B[E[\tilde{f}_1/f_1, \ldots, \tilde{f}_n/f_n]]$$

where
$$\tilde{f}_i,\ f_i^{(j)} : \mathrm{DOM}^{n_i} \to \mathcal{MD}(\mathrm{DOM})$$

are defined by

$$f_i^{(o)}(e_1,\dots,e_{n_i}) = \{\bot\}$$

$$f_i^{(j+1)}(e_1,\dots,e_{n_i}) = B[(F_i[f_1^{(j)}/f_1,\dots,f_n^{(j)}/f_n])(e_1,\dots,e_{n_i})]$$

$$\tilde{f}_i = \underset{j\in\mathbb{N}}{\mathrm{lub}}\ \{f_i^{(j)}\}$$

The continuity of the operations of our language guarantees

$$(*)\quad \tilde{f}_i(e_1,\dots,e_{n_i}) = B[F_i[\tilde{f}_1/f_1,\dots,\tilde{f}_n/f_n]\ (e_1,\dots,e_{n_i})]$$

and that $\tilde{f}_1,\dots,\tilde{f}_n$ are the least fixed points of the equation (*).

Note, that for the language considered so far it is not necessary to work with multidomains, if DOM is flat, since powerdomains are sufficient for flat domains (cf. /Broy et al. 78/). However, working with multidomains, we need not necessarily restrict ourselves to flat domains.

4. Operational Semantics: A Nondeterministic Computation Rule for Parallel Evaluation

Now we consider the program

$$\lceil \underline{\text{funct}}\ f_1 \equiv F_1,\dots,\ \underline{\text{funct}}\ f_n \equiv F_n,\ E \rfloor$$

and give a number of rewrite rules which serve as basic computation steps.

To give a proper definition, at first the predicate **ismaximal** is defined. Intuitively, an expression is maximal if none of the rewrite rules can be immediately applied, i.e. it is maximal, iff one or more free identifier occur in all decisive positions. The maximality is specified by the following axioms, i.e. ismaximal is the least (weakest) predicate fullfilling the following axioms:

ismaximal (x) for every identifier x.

$\forall i,\ 1 \le i \le n$: (ismaximal($E_i$) ∨ E_i ∈ DOM) ∧

$\exists i,\ 1 \le i \le n$: ismaximal($E_i$) ⇒ ismaximal($g(E_1, \ldots E_n)$)

ismaximal (C) ⇒ ismaximal (<u>if</u> C <u>then</u> E1 <u>else</u> E2 <u>fi</u>)

ismaximal (E_1) ∧...∧ ismaximal(E_{n+1})⇒ismaximal(($\lambda x_1, \ldots x_n : E_{n+1}$)($E_1, \ldots E_n$))

Note, that one might be less restrictive in defining the maximality of the conditional.

The computation rule <u>call-in-parallel</u> (<u>parallel evaluation</u>) is specified by the following rewrite rules:

(1) <u>Evaluation of conditional expressions</u>

<u>if</u> tt <u>then</u> E1 <u>else</u> E2 <u>fi</u> → E1,

<u>if</u> ff <u>then</u> E1 <u>else</u> E2 <u>fi</u> → E2,

C → C' ⇒ <u>if</u> C <u>then</u> E1 <u>else</u> E2 <u>fi</u> → <u>if</u> C' <u>then</u> E1 <u>else</u> E2 <u>fi</u>,

(2) <u>Evaluation of choice</u>

(E1 ▯ E2) → E1,

(E1 ▯ E2) → E2,

(3) <u>Unfold of recursively defined functions</u>

$f_i(E_1, \ldots, E_n) \rightarrow F_i(E_1, \ldots, E_n)$,

(4) <u>Evaluation of function applications</u>

$\forall i, 1 \leq i \leq n : (E_i \to E_i' \lor ((\text{ismaximal}(E_i) \lor E_i \in \text{DOM}) \land E_i = E_i'))) \land$
$(\exists i, 1 \leq i \leq n : E_i \to E_i') \Rightarrow g(E_1, \ldots, E_n) \to g(E_1', \ldots, E_n'),$

$(\forall i, 1 \leq i \leq n + 1 : (E_i \to E_i' \lor (\text{ismaximal}(E_i) \land E_i = E_i'))) \land$
$(\exists i, 1 \leq i \leq n + 1 : E_i \to E_i') \Rightarrow$

$$(\lambda x_1, \ldots, x_n : E_{n+1})(E_1, \ldots, E_n) \to (\lambda x_1, \ldots, x_n : E_{n+1}')(E_1', \ldots, E_n')$$

(5) <u>Communication of computed arguments</u>

$E_i \in \text{DOM} \Rightarrow (\lambda x_1, \ldots, x_n : E_{n+1}) (E_1, \ldots, E_n) \to$

$$(\lambda x_1, \ldots, x_{1-1}, x_{i+1}, \ldots, x_n : E_{n+1} [E_i / x_i])$$
$$(E_1, \ldots, E_{i-1}, E_{i+1}, \ldots, E_n).$$

(6) <u>Termination of the evaluation of function applications</u>

$E_{n+1} \in \text{DOM} \Rightarrow (\lambda x_1, \ldots, x_n : E_{n+1})(E_1, \ldots, E_n) \to E_{n+1},$
$(\lambda : E_{n+1}) () \to E_{n+1},$

(7) <u>Simplification of primitive expressions</u>

E primitive, $E \in \text{DOM}$, $B[E] = \{e\} \Rightarrow E \to e.$

Note that this rule can even be applied to expressions containing
free identifiers, as far as these identifiers do not become
decisive, i.e. ismaximal (E) does not hold (cf. <u>partial evaluation</u>,
<u>mixed computation</u> in /Ershov 78/).

A computation rule R is called

- <u>consistent</u>, if for every pair of programs

$P = \lceil \underline{\text{funct}} \ f_1 \equiv F_1, \ldots, \underline{\text{funct}} \ f_n \equiv F_n, E \rfloor ,$

$P' = \lceil \underline{\text{funct}} \ f_1 \equiv F_1, \ldots, \underline{\text{funct}} \ f_n \equiv F_n, E' \rfloor$

we have

$$E \ R \ E' \Rightarrow B[P'] \subseteq B[P],$$

i.e. if P' is descendant (cf. /McCarthy 63/, or implementation cf. /Broy et al. 78/, /Broy et al. 80/) of P.

- complete, if for every program P with $e \in$ SET(B[P]) there exists a finite sequence of expressions $\{E_i\}_{0 \leq i \leq n}$ such that

$$E = E_o \; R \; E_1 \; \dots \; R \; E_n = e$$

holds, provided e is total and finite.

- finitary, iff for every program P there exists only a finite number of expressions E' such that E R E' ,

- effective, iff for every program P there exists an infinite sequence $\{E_i\}_{i \in \mathbb{N}}$ of expressions with $E \, R \, E_o$ and $E_i \; R \; E_{i+1}$, at most if $e \in B[P]$ with e infinite or partial.

Due to König's Lemma, a finitary, effective rule can only compute programs P with (provided DOM is flat):

$$| \, B[P] \, | < \infty \quad \text{or} \quad \bot \in B[P]$$

This notion is directly related to the notion of finite approximability (cf. section 2) of multidomains over flat domains.

Lemma: The rule "$\rightarrow$" is

(1) consistent
(2) complete
(3) finitary
(4) effective

Sketch of proof:

Let f_j be defined by $f_j^o = \lambda \; x_1, \dots, x_{n_j} : \bot$, $f_j^{i+1} =$

$$F_j[f_1^i / f_1, \dots, f_n^i / f_n].$$

(1) According to the definition of our language all rules are consistent.

(2) According to the definition of fixed points, there exists $i \in \mathbb{N}$ such that for $e \in B[E[f_1^i/f_1,\dots,f_n^i/f_n]]$ appropriate application of the rules leads from

$$E[f_1^i/f_1,\dots, f_n^i/f_n] \quad \text{to} \quad e .$$

(3) Proof by structural induction on E .

(4) Assume all $e \in B[P]$ are finite and total, then there exists an $i \in \mathbb{N}$ such that

$$B[P] = B[E[f_1^i/f_1,\dots f_n^i/f_n]].$$

Every computation step reduces

- the number of occurring "λ"-symbols or
- the number of <u>if</u>-<u>fi</u> constructs or
- the number of formal parameters or
- the number of primitive function symbols or
- the number of λ-symbols

and only a finite number of them may occur.

Since E itself cannot become maximal (since all $e \in B[E]$ are total), after a finite number of computation steps the computation must terminate with $e \in B[E]$, at least if all $e \in B[E]$ are finite and total. □

<u>Example:</u>

Let us consider the recursive program:

⌈<u>funct</u> f ≡ F, f(1,1)⌋

with the abbreviation:

$F \mathrel{\hat{=}} \lambda\, x, y : R[f, x, y]$

$R[f,x,y] \mathrel{\hat{=}}$ <u>if</u> x=o <u>then</u> 2 ∗ y <u>else</u> f(x-1, f(x-1, y+1))<u>fi</u>

For the function application f(1,1) we obtain the computation sequence:

f(1,1) →
F(1,1) →
(λy : R[f, 1, y]) (1) →
R[f, 1, 1] →
if ff **then** 2 * 1 **else** f(1-1, f(1-1, 1+1)) **fi** →
f(1-1, f(1-1, 1+1)) → F(1-1, f(1-1, 1+1)) →
F(0, F(1-1, 1+1)) →
(λ y : R[f, 0, y]) (F(1-1, 1+1)) →
(λ y : **if** tt **then** 2 * y **else** f(0-1, f(0-1, y+1)) **fi**) (F(0,2)) →
(λ y : 2 * y) (**if** tt **then** 2*2 **else** f(0-1, f(0-1, 2+1)) **fi**) → ...
(λ y : 2 * y) (4) →
2 × 4 →
8

<u>end of example</u>

The main difference between the computation rule "→" and the rules given in /Manna et al. 73/ is found in the different substitution mechanism. In /Manna et al. 73/ all substitutions are UNFOLDINGs replacing an identifier f_i for a recursively defined function in one *indivisible* action (where $F_i = \lambda x_1, \ldots, x_n : E$)

$$f_i(E_1, \ldots, E_n) \rightarrow E\,[E_1/x_1, \ldots, E_n/x_n]$$

where the $E_1, \ldots, E_n$ all are deterministic (since in /Manna et al. 73/ only deterministic expressions are considered), while in the rule "→" this indivisible action is divided in up to n+1 independent substitution steps.

Note, that guarded expressions (as a counterpart to guarded commands in /Dijkstra 76/) can be expressed as a notational extension by the rules (cf. also /Bauer, Wössner 81/, page 69):

$\underline{\text{if}}$ C $\underline{\text{then}}$ E $\underline{\text{fi}}$ = $\underline{\text{if}}$ C $\underline{\text{then}}$ E $\underline{\text{else}}$ $\perp$ $\underline{\text{fi}}$

$\underline{\text{if}}$ C_1 $\underline{\text{then}}$ E_1 $[\!]$ ... $[\!]$ C_n $\underline{\text{then}}$ E_n $\underline{\text{fi}}$ = $(G_1$ $[\!]$ ... $[\!]$ $G_n)$

where

G_i = $\underline{\text{if}}$ C_i $\underline{\text{then}}$ E_i $\underline{\text{else}}$ $\underline{\text{if}}$ C_1 $\underline{\text{then}}$ E_1 $[\!]...[\!]$ C_{i-1} $\underline{\text{then}}$ E_{i-1}
$[\!]$ C_{i+1} $\underline{\text{then}}$ E_{i+1} $[\!]...[\!]$ C_n $\underline{\text{then}}$ E_n $\underline{\text{fi}}$ $\underline{\text{fi}}$

Thus guarded expressions can be seen as a notational combination of choice and conditional expression.

Remark: Call-by-Value versus Call-by-Name Revisited

As is well-known, for recursively defined functions call-by-value and call-by-name rules may produce different results, if expressions with undefined values occur as arguments. Then the function corresponding to call-by-value may be strictly less defined than the function defined by call-by-name . In particular one can give proper fixed point theories for each of these rules ("smash product" of domain $(S^n)^\perp$ versus "cartesian product" $(S^\perp)^n$) (cf. /Bauer, Wössner 81/).

For nondeterministic functions still another difference between call-by-value and call-by-name becomes apparent. Consider the examples:

(P1) $\lceil$ $\underline{\text{funct}}$ $f_1 \equiv \lambda\, x : x+x,\ \ f_1(0\ [\!]\ 1)$ $\rfloor$

(P2) $\lceil$ $\underline{\text{funct}}$ $f_2 \equiv \lambda\, x : 2*x,\ \ f_2(0\ [\!]\ 1)$ $\rfloor$

In strict call-by-value we obtain B[P1] = B[P2] = {0,2}, while (as pointed out in /Hennessy, Ashcroft 76/) in straightforward call-by-name semantics we obtain B[P1] = {0, 1, 1, 2} and B[P2] = {0, 2}, although from the mathematical point of view the functions f_1 and f_2 are (in a deterministic environment) equivalent. So in straightforward call-by-name semantics one is forced to consider functions as mappings $MD(S^\perp)^n \to MD(S^\perp)$

whereas in call-by-value it suffices to take $(S^n)^\perp \to \mathcal{MD}(S^\perp)$ (cf. /Astesiano, Costa 79/, /Benson 79/, /Hennessy 80/).

In our definition of mathematical semantics a mixture of call-by-value and call-by-name is used which is called call-time choice in /Hennessy, Ashcroft 77/. It can be evaluated by an extension of delayed evaluation (cf. /Vuillemin 74/) or call-by-need (cf. /Wadsworth 71/) to nondeterministic functions. It allows one to consider nondeterministic functions as elements from $(S^\perp)^n \to \mathcal{MD}(S^\perp)$. The parallel evaluation rule (call-in-parallel) as defined in this section contrasts the implementation of call-time-choice (and thus of call-by-name in the deterministic case) by delayed evaluation (call-by-need) by a method which does not *delay* the evaluation of the arguments until they become decisive and thus are *needed*, but starts the evaluation of the body of the function and of the arguments in parallel, simply eliminating computations of arguments which apparently are no longer needed. So one might, in analogy, talk of enforced evaluation.

With the computation rule "$\to$" a function application

$$(\lambda\ x_1,\ldots,x_n\ :\ E_{n+1})\ (E_1,\ldots,E_n)$$

is evaluated by n independent processes evaluating E_i which communicate their results under the identifier x_i to the process E_{n+1}. If the evaluation of E_{n+1} needs a value for some identifer x_i, then its evaluation stops and waits until the value is communicated. If the value is never communicated, i.e. if the evaluation of E_i fails or does not terminate, than the process waits forever.

Here an important difference between call-by-value and call-by-name (or more precisely call-time-choice) can be seen. In the case of call-by-name the whole system of processes (of evaluations) terminates iff the process terminates which evaluates E_{n+1} (which needs the termination of all processes evaluating the expressions E_i for which x_i is actually needed in E_{n+1}), while in the case of call-by-value the whole system of processes terminates, iff all processes terminate themselves. For parallel evaluation using call-by-value see /Broy 80a/.

end of remark

The evaluation rule described in this section can be seen as operational semantics for a kind of a data flow language where, however, in contrast to the straightforward concept of demand driven evaluation (cf. /Dennis 74/, /Kosinski 73, 77/) much more is done in parallel. Of course, the programs of our languages can also be represented as graphs (cf. function graphs in /Keller 80/), such that the evaluation rule describes a graph reduction process.

One may also think of a reduction machine for an efficient implementation of the evaluation rule. Note, that the classical straightforward transformation of applicative programs (see /Bauer, Wössner 81/) generally means a transition to strict innermost evaluation and destroys the possibilities of inherent parallelism (cf. /Broy 80a/).

5. Applicative Communicating Systems

Parallel evaluations of sub-expressions (such as parameters of a function) which do not interfere do not cause any problems. They can be specified, considered, computed and analysed separately. If, however, there is some possibility of communication, i.e. the possibility of transmitting (intermediate) results from one expression to the other, and if the "behavior" of the expressions is influenced by these communicated results, then it is more difficult to consider the meaning of such expressions.

5.1 Specific Rewrite Rules for Communications

The phenomenon of communication can even be observed in the computation rules of the preceding section. A function application

$$(\lambda\ x_1, \ldots, x_n : E_{n+1})\ (E_1, \ldots, E_n)$$

can be considered as a system of n+1 expressions, the evaluations of which can be viewed as *communicating processes*. If one of the processes evaluating E_i has successfully terminated with result e_i, then this result is communicated to the process evaluating E_{n+1} and we obtain

$$(\lambda\, x_1, \ldots, x_{i-1}, x_{i+1}, \ldots, x_n : E_{n+1}[e_i/x_i])\ (E_1, \ldots, E_{i-1}, E_{i+1}, \ldots, E_n)$$

Obviously this is only a very restricted type of communication, which only occurs if the sending process has already terminated and the communication is its last action.

To obtain more general types of communication, it seems adequate to consider a particular domain

$$\text{DOM} = (\text{ATOM} \cup \text{SEQ})^{\perp}$$

where $\text{SEQ} = \text{ATOM}^*$. We use the functions isempty, first, rest as defined in section 2 and the function

$$\text{app} : \text{ATOM}^{\perp} \times \text{SEQ}^{\perp} \rightarrow \text{SEQ}^{\perp}$$

defined by

$$\text{app}(e, s) = \begin{cases} \text{ap}(e,s) & \text{if } e \neq \perp,\ s \neq \perp \\ \perp & \text{otherwise} \end{cases}$$

where ap is defined as in section 2.

With this definition, DOM is a flat domain, provided $ATOM^{\perp}$ is flat. It is a proper subdomain of the domain of streams. In contrast to the usual primitive functions, we define an application app(e, s) as maximal iff ismaximal(e) holds.

Now specific computation rules for evaluating stream processing functions can be given:

$$\forall i,\ 1 \leq i \leq n+1 : (E_i \rightarrow E'_i \lor (\text{ismaximal}(E_i) \land E_i = E'_i)) \land e \in ATOM \Rightarrow$$

$$(\lambda\ x_1,\ldots,x_n : E_{n+1})(E_1,\ldots,E_{i-1},\text{app}(e,E_i),E_{i+1},\ldots,E_n) \rightarrow$$
$$(\lambda\ x_1,\ldots,x_n : E'_{n+1}[\text{app}(e,x_i)/x_i])(E'_1,\ldots,E'_n)$$

and for $e \in ATOM$

$$(\lambda\ x_1,\ldots,x_n : \text{app}(e,E_{n+1}))(E_1,\ldots,E_n) \rightarrow \text{app}(e,(\lambda x_1,\ldots,x_n : E_{n+1})$$
$$(E_1,\ldots,E_n))$$

Lemma: If the computation rules of the preceding sections are complemented by the rules above, the resulting rule is also consistent, complete, finitary and effective.

Proof: The consistency of the rules immediately follows from the language definition and the definition of the function app. Finitarity and completeness is trivial. The effectiveness follows from the fact that all elements $s \in SEQ$ are of finite length.

□

Note that the rule above allows to shorten the computations because if E_{n+1} is maximal because x_i is decisive, a substitution of x_i by app(e, x_i) should allow E_{n+1} to proceed.

Example: The producer/ consumer problem

```
⌈funct produce  ≡ λ x : if x = o then empty
                               else app(product(x),produce(x-1)) fi

 funct consume  ≡ λ s : if isempty(s) then t
                                   else g(first(s),consume(rest(s))
                        fi

 consume(produce(n))
```

for some $n \in \mathbb{N}$, arbitrary functions product and g, $t \in$ ATOM.

In this example we obtain, assuming product(i) = $p_i \in$ ATOM

consume(produce(10)) →

(λ s : ...) ((λ x : ...) (10)) → ...

(λ s : ...) (app(product(10), produce(9))) → ...

(λ s : ...) (app(p_{10}, ap(p_9, (λ x : ...) (8)))) →

(λ s : if isempty(app(p_{10}, s)) then ... fi) (ap(p_9,...)) → ...

However we cannot conclude in this step

isempty(ap(p_{10}, s)) → ff

since s may be ⊥. Therefore in our example the consumer process has to wait until the producer process has successfully terminated. So the benefits of the two rules of communication cannot be fully exploited.

end of example

So one really has to consider the non flat domain of streams instead of the flat domain used so far.

5.2 Streams

To cope with this problem we would like to have rules like

$$\text{isempty}(\text{ap}(e, s)) = \text{ff},$$
$$\text{first}\ (\text{ap}(e, s)) = e\ ,$$
$$\text{rest}\ (\text{ap}(e, s)) = s\ ,$$

for all $s \in$ STREAM , $e \in$ ATOM i.e. we need a *nonstrict* constructor function like the one used in *lazy evaluation* (cf. /Henderson , Morris 76/, /Friedmann, Wise 76/. So we define now for DOM = ATOM $\cup$ STREAM where STREAM = SEQ $\cup$ (ATOM$^*\times\{\bot\}$) $\cup$ ATOM$^\infty$

$$.\&. \ : \text{ATOM}^\bot \times \text{STREAM} \rightarrow \text{STREAM}$$
$$\underline{\text{first}}. : \text{STREAM} \rightarrow \text{ATOM}^\bot$$
$$\underline{\text{rest}}. : \text{STREAM} \rightarrow \text{STREAM}$$
$$\underline{\text{isempty}}. : \text{STREAM} \rightarrow \{\text{tt}, \text{ff}\}^\bot$$
$$\text{empty} : \rightarrow \text{STREAM}$$

$$B[E \ \& \ S] = \biguplus_{e \in B[E]} \ \biguplus_{s \in B[S]} \{\text{ap}(e, s)\}$$

$$B[\underline{\text{first}}\ S] = \biguplus_{s \in B[S]} \{\text{first}(s)\}$$

$$B[\underline{\text{rest}}\ S] = \biguplus_{s \in B[S]} \{\text{rest}(s)\}$$

$$B[\underline{\text{isempty}}\ S] = \biguplus_{s \in B[S]} \{\text{isempty}(s)\}$$

$$B[\text{empty}] = \{\ \varepsilon\ \}$$

where ap, first, rest, isempty are defined as in section 2.

From the results of section 2 the continuity of the contructs &, first, rest and isempty immediataly follows. Now our computation rules can be complemented by the rules:

$\underline{\text{first}}\ (E \,\&\, S) \rightarrow E,$

$\underline{\text{rest}}\ (e \,\&\, S) \rightarrow S \qquad \text{for} \quad e \in \text{DOM}, \ e \neq \bot$

$\underline{\text{isempty}}\ (e \,\&\, S) \rightarrow ff \qquad \text{for} \quad e \in \text{DOM}, \ e \neq \bot$

$E \rightarrow E' \;\Rightarrow\; E \,\&\, S \;\rightarrow\; E' \,\&\, S\ ,$

$E \rightarrow E' \;\Rightarrow\; \underline{\text{first}}\ E \;\rightarrow\; \underline{\text{first}}\ E'\ ,$

$E \rightarrow E' \;\Rightarrow\; \underline{\text{rest}}\ E \;\rightarrow\; \underline{\text{rest}}\ E'\ ,$

$E \rightarrow E' \;\Rightarrow\; \underline{\text{isempty}}\ E \;\rightarrow\; \underline{\text{isempty}}\ E'\ ,$

$\underline{\text{isempty}}\ \text{empty} \rightarrow tt,$

ismaximal (E & S) = ismaximal (E),

ismaximal (<u>first</u> S) = ismaximal (<u>rest</u> S) = ismaximal (<u>isempty</u> S)= ismaximal (S).

Using these rules one obtains really systems of communicating expressions, however, the communication paths are still noncyclic. For expressing real "feed back" we introduce systems of mutually recursively defined streams.

5.3 Systems of Expressions Communicating by Streams

For allowing cyclic communication paths, systems of expressions which mutually communicate via streams are considered using the following syntax:

< c-program > ≡ ⌈{<u>funct</u> <function identifier>≡<funct abstract>,}*
{<u>stream</u><stream identifier>≡<expr>,}* <expr>⌋

Throughout this section we consider the scheme R of the from:

$$\lceil \underline{\text{funct}}\ f_1 \equiv F_1, \ldots, \underline{\text{funct}}\ f_n \equiv F_n\ ,$$
$$\underline{\text{stream}}\ s_1 \equiv S_1, \ldots, \underline{\text{stream}}\ s_m \equiv S_m,\ E \rfloor$$

where s_i does not occur in F_j.
To define a semantics for R again fixed point theory is applied.

A system of expressions communicating by streams corresponds to a mutually recursive definition of streams. This coincides with the result of /Broy 80b/, where it is shown that every tail-recursive system of concurrent processes working on shared variables within conditional critical regions can be transformed into a system of mutually recursive nondeterministic procedures. If the right-hand side E of a stream equation consists of a deterministic expression functionally no difference can be found between the stream equation <u>stream</u> s ≡ E and the recursive definition of a nullary function <u>funct</u> s ≡ λ : E (cf. the proposal of /Kahn 74/, /Kahn, MacQueen 77/).

Note, however, that for instance the system S1

⌈<u>funct</u> f ≡ λ : (1 ▯ 2) & empty, <u>first</u> f()=<u>first</u> f()⌋

is different from the system S2

⌈<u>stream</u> s ≡ (1 ▯ 2) & empty, <u>first</u> s = <u>first</u> s⌋

since functions are treated by simple substitution ("call-by-name"-like with respect to the body of the function), while streams are to be substituted (elementwise) only after evaluation ("call-by-value" or more precisely "call time choice"), such that SET(B[S1]) = {tt, ff} while SET(B[S2]) = {tt}.

So to avoid mixing alternative possibilities of behavior (cf. the "merge anomaly" described in /Keller 78/, /Brock, Ackermann 81/) we have to consider the stream equations as <u>multisets of fixed point equations</u> rather than one fixed point equation for a multiset-valued ("nondeterministic") function. To explain these problems let us consider the following example:

Example: Consider the two programs

P1 : $\lceil$ funct f ≡ λ : (O & f()) ▯ ((1 + first f())&f()), first f() $\rfloor$

P2 : $\lceil$ stream s ≡ (O & s) ▯ ((1 + first s) & s) , first s $\rfloor$

Our definition gives SET(B[P1]) = $\mathbb{N} \cup \{\bot\}$, however for the stream we would like to obtain SET(B[P2]) = $\{0\} \cup \{\bot\}$

end of example

Intuitively speaking, a stream equation defines a process with one determinate identity for each of its applications in one particular instance of evaluation, while a function equation defines a nondeterminate function where for every application a new individual choice can be taken.

Therefore we introduce a new semantic function BF , associating with every expression E in which at most the free stream identifiers $s = (s_1, \ldots, s_m)$ occur a functional which gives for every m-tuple of streams a multiset of deterministic partial functions, which can be taken as functions for defining streams as their least fixed points.

$$BF : EXP[s] \rightarrow (STREAM^m \rightarrow \mathcal{MD}(STREAM^m \rightarrow STREAM))$$

Now we define:

BF [if C then E1 else E2 fi] ($\tilde{s}$) =

$$\biguplus_{p \in BF[C](\tilde{s})} \begin{cases} \biguplus_{f \in BF[E1](\tilde{s})} \{cond(p,f)\} & \text{if } p(\tilde{s}) = tt \\ \biguplus_{f \in BF[E2](\tilde{s})} \{cond(\neg p,f)\} & \text{if } p(\tilde{s}) = ff \\ \{\Omega\} & \text{otherwise} \end{cases}$$

$BF[s_i](\tilde{s}) = \{\lambda\ s_1,\ldots,s_n : s_i\}$

$BF[E1 \,[]\, E2](\tilde{s}) = B[E1](\tilde{s}) \uplus B[E2](\tilde{s})$

$BF[e](\tilde{s}) = \{\lambda\ s_1,\ldots,s_m : e\}$ for $e \in DOM$

$BF[(\lambda\ x_1,\ldots,x_n : E_{n+1})\ (E_1,\ldots,E_n)]\ (\tilde{s}) =$

$$\biguplus_{f_1 \in BF[E_1](\tilde{s})} \cdots \biguplus_{f_n \in BF[E_n](\tilde{s})} BF[E_{n+1}\,[f_1(s)/x_1 \ldots f_n(s)/x_n]](\tilde{s})$$

$BF[g(E_1,\ldots,E_n)]\ (\tilde{s}) =$

$$\biguplus_{f_1 \in BF[E_1](\tilde{s})} \cdots \biguplus_{f_n \in BF[E_n](\tilde{s})} \{\lambda\ s : g(f_1(s),\ldots,f_n(s))\}$$

where Ω denotes the totally undefined function and cond:
$(STREAM^m \to \{tt, ff, \bot\}) \times (STREAM^m \to DOM) \to (STREAM^m \to DOM)$
is defined by cond $(p, f)(\tilde{s})$ = if $p(\tilde{s})$ then $f(\tilde{s})$ else $\bot$.
So given the system R, we associate with f_i the least fixedpoint

$BF[f_i] = BF[E_i]$ where $F_i = \lambda x_1,\ldots,x_{n_i} : E_i$.

So we may define

$$BF[f_i(E_1,\ldots,E_{n_i})]\ (\tilde{s}) = \biguplus_{h_1 \in BF[E_1](\tilde{s})} \cdots \biguplus_{h_{n_i} \in BF[E_{n_i}](\tilde{s})}\ \biguplus_{f \in BF[f_i](h_i(\tilde{s}_1),\ldots,h_{n_i}(\tilde{s}))} \{\lambda s : f(h_1(s),\ldots,h_{n_i}(s))\}.$$

We associate with $s = s_1,\ldots s_n$ the multiset S:

$$S = \biguplus_{s \in STREAM^m}\ \biguplus_{h_1 \in BF[S_1](s)} \cdots \biguplus_{h_m \in BF[S_m](s)} \begin{cases} \{s\} & \text{if } s = Y\tilde{s} : (h_1(\tilde{s}),\ldots,h_m(\tilde{s})) \\ \emptyset & \text{otherwise} \end{cases}$$

where Y denotes the least-fixed-point operator for streams.

In particular systems of expressions communicating by streams can be used to represent networks of processes as considered for instance in /Arnold 79/.

Again a number of computation rules can be given to evaluate such systems of communicating expressions.

Now we define 4 rewrite rules. We use the abbreviations:

$$R1 = (\underline{\text{stream}}\ s_1 = S_1, \ldots, \underline{\text{stream}}\ s_m = S_m, S_{m+1})$$

$$R2 = (\underline{\text{stream}}\ s_1 = S'_1, \ldots, \underline{\text{stream}}\ s_m = S'_m, S'_{m+1})$$

(1) Parallel computation

$$\forall i, 1 \leq i \leq m+1 : (S_i \rightarrow S'_i \vee (\text{ismaximal}(S_i) \wedge S_i = S'_i)) \Rightarrow R1 \rightarrow R2$$

(2) Communication

$$(S_j = e \,\&\, \tilde{S}_j \wedge \forall i, 1 \leq i \leq m+1 : (i \neq j \Rightarrow S'_i = S_i[e \,\&\, s_j/s_j]) \wedge S'_j = \tilde{S}_j[e \,\&\, s_j/s_j]) \Rightarrow R1 \rightarrow R2$$

(3) Termination of single processes

$$(S_j = \text{empty} \wedge \forall i, 1 \leq i \leq m+1 : (i \neq j \Rightarrow S'_i = S_i[\text{empty}/s_j]) \Rightarrow R1 \rightarrow (\underline{\text{stream}}\ s_1 = S'_1, \ldots, \underline{\text{stream}}\ s_{j-1} = S'_{j-1}, \underline{\text{stream}}\ s_{j+1} = S'_{j+1}, \ldots, \underline{\text{stream}}\ s_m = S'_m, S'_{m+1})$$

(4) Termination of the system

$$S_{m+1} = e \Rightarrow R1 \rightarrow e \qquad \text{where } e \in \text{DOM}$$

The first 2 of these rules can be combined into one rule which guarantees the simultaneous progress of the evaluations of the single processes (note that this is just one possibility for a computation rule assuring the continuous computation and communication progress for all processes).

$\forall i,\ 1 \leq i \leq m+1$:

$$(((S_i \to \tilde{S}_i \vee (\text{ismaximal}(S_i) \wedge S_i = \tilde{S}_i)) \wedge C_i = s_i) \vee$$
$$(S_i = e_i \,\&\, \tilde{S}_i \wedge C_i = e_i \,\&\, s_i \wedge i \neq m+1)) \wedge$$
$$S_i' = \tilde{S}_i\,[C_1/s_1,\ldots,C_m/s_m] \Rightarrow R1 \to R2 .$$

For communicating intermediate results to the outside world one might add rules like (cf. rule (4)):

$$S_{m+1} = e \,\&\, S'_{m+1} \quad \Rightarrow \quad R1 \to e \,\&\, R2$$

Again the computation rule resulting from adding this rewrite rule to the previous rules is effective, finitary, consistent and complete.

Similar rules may be considered as an extension of the technique of call-by-need (cf. /Wadsworth 72/) or delayed evaluation (cf. /Vuillemin 74/) to primitive functions which previously have always been assumed to be strict (cf. /Manna et al. 73/) leading to the concept of lazy evaluation (cf. /Friedman, Wise 76/ , /Henderson, Morris 76/, /Bauer 79/). This allows a treatment of infinite objects such as infinite streams by particular computation rules. The rules given above, however, then could be called "speedy evaluation", or also "busy evaluation" or, extending the notion of section 4 to primitive functions, parallel evaluation, since the evaluation of an infinite stream is not delayed until a selector function is applied as in lazy evaluation, but the evaluation is continuously enforced such that an infinite stream is consecutively generated.

Example:

Let the function f be defined by :

$$\underline{\text{funct}}\ f \equiv \lambda\ t,\ n : \underline{\text{if}}\ n = 0\ \underline{\text{then}}\ \text{empty}\ \underline{\text{else}}\ (\underline{\text{first}}\ x+1)\ \&\ f(\underline{\text{rest}}\ x,\ n-1)\ \underline{\text{fi}}\ ,$$

and the interactive system E by :

$$E \mathrel{\hat{=}} \lceil \underline{\text{stream}}\ t \equiv 0\ \&\ f(t,\ 2),\ t \rfloor .$$

We obtain a computation sequence for E by our computation rule. UNFOLDING f and simplifying the result we get:

$$\lceil \underline{\text{stream}}\ t \equiv 1\ \&\ f(t,\ 1),\ 0\ \&\ t \rfloor$$

Applying the rule again we get :

$$\lceil \underline{\text{stream}}\ t \equiv f(1\ \&\ t,\ 1),\ 0\ \&\ 1\ \&\ t \rfloor$$

and: $\lceil \underline{\text{stream}}\ t \equiv 2\ \&\ f(t,\ 0),\ 0\ \&\ 1\ \&\ t \rfloor$

and: $\lceil \underline{\text{stream}}\ t \equiv f(2\ \&\ t,\ 0),\ 0\ \&\ 1\ \&\ 2\ \&\ t \rfloor$

and: $\lceil \underline{\text{stream}}\ t \equiv \text{empty},\ 0\ \&\ 1\ \&\ 2\ \&\ t \rfloor$

and finally : 0 & 1 & 2 & empty

Since no nondeterminism is involved, we can use simple iteration to compute the fixed point of the stream function: The iteration gives

$$K^{(0)} = \{\bot\}$$
$$K^{(1)} = \{0\ \&\ \bot\}$$
$$K^{(2)} = \{0\ \&\ 1\ \&\ \bot\}$$
$$K^{(3)} = \{0\ \&\ 1\ \&\ 2\ \&\ \bot\}$$
$$K^{(4)} = \{0\ \&\ 1\ \&\ 2\ \&\ \text{empty}\}$$
$$K^{\infty} = K^{(4+n)} = K^{(4)}$$

<u>end of example</u>

Example: Applicative Loops

"Applicative loops"(the name is due to Keller) can be seen as specific "iterative" programs producing an (eventually infinite) stream. They can conveniently be used to compute sequences of results of a function given in the form of course-of-value recursion. Consider for instance

```
funct f ≡ λ n : if n=0 then E0
            else if n=1 then E1
                         else  g(f(n-2), f(n-1), n)  fi
```

then with

```
stream s ≡  E0 & (E1 & h(s, 2))
```

where

```
funct h ≡ λ s, n : g(first s, first rest s, n) & h(rest s, n+1)
```

One simply proves, that

$$\underline{\text{first}}\ \underline{\text{rest}}^{\,i}\ s = f(i)$$

This is a simple example for a formally justified transformation from classical applicative programs to stream processing programs. Note that the program above consists of an "inner most" recursive definition of the stream s (in contrast to "outermost"-recursion or tail-recursion for recursive functions representing loops).

end of example

<u>Example</u>

If a program is required which generates the infinite stream of all numbers >1 of the form $2^i \times 3^j \times 5^k$ (cf./Dijkstra 76/), in ascending order one may use three communicating streams:

```
⌈funct streammult ≡ λ n, s : (n × first s) & streammult(n,rest s),
 funct merge ≡ λ s1, s2 : if first s1 < first s2
                             then first s1 & merge(rest s1, s2)
                             else first s2 & merge(s1, rest s2) fi,
 stream s1 ≡ streammult(5, 1 & s1),
 stream s2 ≡ merge(streammult(3, 1 & s2), s1),
 stream s3 ≡ merge(streammult(2, 1 & s3), s2),        s3          ⌋
```

The correctness of this program may quite straightforwardly be proved using induction.

<u>end of example</u>

6. Concurrency

One of the most intricate issues in multiprogramming is that of concurrency. Analogously to everyday life one may talk of two (or more) concurrent candidates (processes, expressions, programs), if these two candidates both compete for something (for instance to be served or to be elected). For resolving a competition a choice has to be made.

One may consider an expression (E1 [] E2) as a competition of the expressions E1 and E2 for being chosen. However, in contrast to everyday life, this choice is performed in a totally arbitrary way without taking into account any of the particular properties of E1 or E2. Therefore in the expression

⌈**stream** s1 ≡ S1, **stream** s2 ≡ S2, **if** C(s1,s2)
then **first** s1 **else** **first** s2 **fi**⌋

there is no way to formulate the predicate C such that the first alternative is chosen if **first** s1 $\neq \perp$ and the second one is chosen if **first** s2 $\neq \perp$ (and ambiguously one of them is chosen if both are $\neq \perp$). If such a predicate would be definable, then functions g would be definable, such as the parallel or

$$g(tt, \perp) = tt, \quad g(\perp, tt) = tt, \quad g(ff, ff) = ff$$

According to /Hennessy, Ashcroft 80/ such a function is not definable in a nondeterministic language like the one defined in section 3. Note that all definable functions are either constant or strict in at least one argument. But even the "parallel" or would not solve the above problem.
However, one may use a more strongly defined choice operator, such as McCarthy's ambiguity operator, the meaning of which is specified in /McCarthy 63/ as follows:

> "We define a basic ambiguity operator amb(x, y), whose possible values are x or y when both are defined, otherwise, whichever is defined."

Formally the definition may be written:

$$\text{amb}(x, y) = \begin{cases} x \,[\!]\, y & \text{if } x \neq \perp,\ y \neq \perp \\ x & \text{if } x \neq \perp,\ y = \perp \\ y & \text{if } x = \perp,\ y \neq \perp \\ \perp & \text{if } x = \perp,\ y = \perp \end{cases}$$

and C(s1, s2) in the program above may be expressed by
amb(¬ **isempty** s1 , **isempty** s2).

This is a nonstrict extension (and so not a natural one) of the choice operator.

However, the ambiguity operator causes several problems:

- it allows writing noncontinuous functions such as

  <u>funct</u> f ≡ λ x : amb(x, f(x+1))

  where f(0) represents a multiset which is not finitely approximable, i.e. it is not a member of the multidomain (cf. also the discussion on *fairness* in /Park 80/, /Broy 81/, /Apt, Plotkin 81/, /Broy, Wirsing 81b/)

- it is not even monotonic, neither in the Egli-Milner ordering nor the multiset-ordering, since $\{1\} \sqsubseteq \{1\}$, $\{\bot\} \sqsubseteq \{2\}$, but $amb(1, \bot) = \{1\} \not\sqsubseteq \{1,2\} = amb(1,2)$.
  This is even worse since in the case of noncontinuity but monotonicity (like in the fairness-discussion) one can still work with least fixed points (cf. /Apt, Plotkin 81/).

Moreover one cannot hope to obtain finitary, consistent, complete, and effective computation rules which compute f(o), since, due to König's Lemma, a finitary tree (i.e. a tree with a finite number of branches at each node) may only contain an infinite number of nodes if there exists an infinite path (i.e. the tree is infinite) and therefore a nonterminating computation for f(o) can not be excluded by a complete, finitary rule.

So one has to choose between dropping completeness or finitarity. Since infinitary rules do not seem very realistic, because no real machine can be assumed which makes an infinite number of choices within finite time, we drop the requirement of completeness and rather consider an infinite set of finitary, effective and consistent computation rules, such that for each feasible value x there exists at least one rule which computes x .

This very clearly reflects the needs of software for parallel processsing. For every program P any implementation or system S does not guarantee that every feasible value x of P can be the result of running P on S, but every result x computed by S when applied to P is a feasible value. Thus one might talk of "loose" nondeterminism here (as an interpretation of this notion as found in /Park 80/).

In accordance with McCarthy's ambiguity operator a choice operator ∇ is used in infix notation specified by

$$B[E1 \nabla E2] = \biguplus_{e1 \in B[E1]} \; \biguplus_{e2 \in B[E2]} B[amb(e1, e2)]$$

<u>Note</u>: The use of a special element <u>none</u> as advocated in /Henderson 80/ corresponds to backtracking on <u>finite</u> errors and reflects a much simpler concept, causing no problems with continuity.

<u>end of note</u>

As a consequence, "∇" is not continuous and not even monotonic.

Therefore we use another approach, exploiting the following two facts:

<u>Lemma:</u> All language constructs are monotonic with respect to the descendent relation, i.e. multiset-inclusion:

$$B[E] \subseteq B[E'] \text{ implies } B[CN[E]] \subseteq B[CN[E']] \quad \text{for all}$$

contexts CN .

<u>Proof</u>: Structural induction (cf. /Broy et al. 78/) □

This property is in particular of interest for program development by refinement. Now we define a second choice operator Δ :

$$B[E1 \; \Delta \; E2] = \biguplus_{e1 \in B[E1]} \; \biguplus_{e2 \in B[E2]} \{e1, e2\}$$

Trivially Δ is monotonic and continuous.

<u>Lemma:</u> Let E be an expression and $\tilde{E}$ be the expressiong resulting from replacing all occurrences of ∇ in E by Δ then
$B[E] \subseteq B[\tilde{E}]$

<u>Proof</u>: $B[E1 \, \nabla \, E2] \subseteq [E1 \, \Delta \, E2]$ for all E1, E2.

□

Given a program P:

$\lceil$ <u>funct</u> $f_1 \equiv F_1, \ldots,$ <u>funct</u> $f_n \equiv F_n$,
<u>stream</u> $s_1 \equiv S_1, \ldots,$ <u>stream</u> $s_m \equiv S_m$, E$\rfloor$

and the program $\tilde{P}$ resulting from substituting all occurrences of "∇" in P by "Δ". The semantics of $\tilde{P}$ is well-defined. So functions $\tilde{f}_i$ can be associated with the recursive definitions in $\tilde{P}$ by taking least fixed points

$$\tilde{f}_i : DOM^{n_i} \rightarrow \mathcal{MD}\ (DOM)$$

of the equations $\tilde{f}_i \;(x_1, \ldots, x_{n_i}) = B[(\tilde{F}_i[\tilde{f}_1/f_1, \ldots, \tilde{f}_n/f_n])$ $(x_1, \ldots, x_{n_i})]$. Now we associate functions f_i' with the functions in P by taking the $\supseteq$-least (i.e. $\subseteq$-greatest) fixed points

$$f_i' : DOM^{n_i} \rightarrow \mathcal{M}\ (DOM)$$

of the equations

$$f_i' \;(x_1, \ldots x_{n_i}) = B[\,(F_i[f_1'/f_1, \ldots, f_n'/f_n])(x_1, \ldots x_n)]$$

where $f_i' \;(x_1, \ldots x_{n_i}) \subseteq \tilde{f}\;(x_1, \ldots, x_{n_i})$.

Due to the $\subseteq$ - monotonicity and Tarski's fixed point theorem such fixed points exist and are uniquely defined.
Note, that the definition of the constructs of our language (apart from fixed point definitions) also works for arbitrary multisets not contained in $\mathcal{MD}$(DOM).

Similarly we define

$$BF[E1 \nabla E2](\tilde{s}) = \biguplus_{h1 \in BF[E1](\tilde{s})} \biguplus_{h2 \in BF[E2](\tilde{s})} \begin{cases} \{h1,h2\} & \text{if } h1(\tilde{s}) \neq \bot,\ h2(\tilde{s}) \neq \bot \\ \{h1\} & \text{if } h1(\tilde{s}) \neq \bot,\ h2(\tilde{s}) = \bot \\ \{h2\} & \text{if } h1(\tilde{s}) = \bot,\ h2(\tilde{s}) \neq \bot \\ \{\Omega\} & \text{if } h1(\tilde{s}) = \bot,\ h2(\tilde{s}) = \bot \end{cases}$$

Now in systems of recursively defined streams we may associate with the f_i a multiset of functions $BF[f_i]$ as the $\subseteq$ - maximal fixed point of

$$BF[f_i] = BF[E_i] \qquad \text{where } F_i = \lambda x_1,\ldots,x_{n_i} : E_i \text{ and}$$

$$BF[f_i] \subseteq BF[\tilde{f}_i]$$

and the $BF[\tilde{f}_i]$ are the fixed points associated with $\tilde{P}$.
So we may define

$$S = \biguplus_{s \in STREAM^m} \biguplus_{h_1 \in BF[S_1](s)} \cdots \biguplus_{h_m \in BF[S_m](s)} \begin{cases} \{s\} & \text{iff } s = Y\tilde{s}:h_1(\tilde{s}),\ldots,h_m(\tilde{s}) \\ \emptyset & \text{otherwise} \end{cases}$$

Note, that the functions in $BF[f_i](\tilde{s})$ are still monotonic and continuous in spite of using the ambiguity operator.

Finally

$$B[P] = \biguplus_{(s'_1,\ldots,s'_m) \in S} B[E[s'_1/s_1,\ldots, s'_m/s_m, f'_1/f_1,\ldots,f'_n/f_n]]$$

where the f'_i in E are defined by the fixed points above.

For lack of space we restrict ourselves to one particular rewrite rule for the operator ∇ . At first a condition for the maximality is given (note that the expression (E1 $[]$ E2) is never maximal)

$$\text{ismaximal}(E1) \wedge \text{ismaximal}(E2) \Rightarrow \text{ismaximal}(E1 \nabla E2)$$

As a rewrite rule one may use

$$\neg \text{ismaximal}(E1 \nabla E2) \wedge$$
$$(E1 \to E1' \vee (\text{ismaximal}(E1) \wedge E1 = E1')) \wedge$$
$$(E2 \to E2' \vee (\text{ismaximal}(E2) \wedge E2 = E2')) \Rightarrow (E1 \nabla E2) \to (E1' \nabla E2') ,$$

$$e \in DOM \setminus \{\bot\} \Rightarrow (e \nabla E) \to e, \quad (E \nabla e) \to e ,$$

and in addition for streams

$$e \in ATOM \setminus \{\bot\} \Rightarrow ((e \,\&\, s1) \nabla s2) \to (e \,\&\, s1), (s1 \nabla (e \,\&\, s2)) \to (e \,\&\, s2).$$

Complementing the computation rule of the preceding section by these rewrite rules leads to an effective, consistent and finitary rule, which is no longer complete, however. Note that replacing $E1 \to E1'$ and $E2 \to E2'$ by $E1 \overset{n}{\to} E1'$ and $E1 \overset{m}{\to} E1'$ resp., $n, m \in \mathbb{N} \setminus \{o\}$, where $E \overset{n}{\to} E'$ means: there exist expressions $E_1, \dots, E_{n-1}$ such that $E \to E_1 \to \dots \to E_{n-1} \to E'$ holds, immediately leads to an infinite family of effective, consistent and finitary rules, such that for each feasible value x there is a rule which possibly computes x . We like to talk about "loose implementations of nondeterminism" interpreting the notion "loose nondeterminism" in /Park 80/ (cf. also /Broy, Wirsing 81a/).

The introduction of the ∇-operator into the nondeterministic programming language essentially changes our notion of computability. According to the results of /Chandra 78/ now all sets in Σ_1^1 may occur as domains of nonterminating computations. This fact is no longer surprising, if one adapts the notion of *loose nondeterminism* (coined in /Park 80/) to computation rules for nondeterministic functions. Let CR be the set of finitary, consistent and effective computation rules. A computation rule $r \in$ CR may be considered as a recursive multiset function

$$r : \text{PROGRAM} \rightarrow M(\text{DOM})$$

A function

$$f : \text{DOM}^n \rightarrow MD(\text{DOM})$$

is called *defineable* iff there exists a program $P(x_1,\ldots,x_n)$

$\forall\, x_1,\ldots,x_n \in \text{DOM}$:

$$\text{SET}(B[f(x_1,\ldots,x_n)]) = \{y \in \text{DOM} : \exists\, r \in \text{CR} : P(x_1,\ldots,x_n) \overset{r}{\rightarrow} y\}$$

Accordingly not *one* complete computation rule is considered, but an infinite number of computation rules, each of which is not complete, but which together are "complete". This properly models the (in principle) infinite number of implementations of a concurrent programming language, each of which is not required to be complete, but consistent, effective and (according to the impossibility to do an infinite number of decisions in finite time) finitary.

Note, that the introduction of (and)- and (or)- nodes into procedural programming languages as used in /Manna 70/ has some similarities to our ∇ - and $[]$-operator resp. However, in our nondeterministic programming language, we can prove $C[E1 \,[]\, E2] = C[E1] \,[]\, C[E2]$ for each context C, however $C[E1 \,\nabla\, E2] = C[E1] \,\nabla\, C[E1]$ does not hold, while in /Manna 70/ one always has $C[S1$ (or) $S2] = C[S1]$ (or) $C[S2]$ as well as $C[S1$ (and) $S2] = C[S1]$ (and) $C[S2]$. For this simpler case with more convenient algebraic properties, /Chandra 74/ proves that both concepts are independent, i.e. that (or) cannot be expressed by (and) and vice versa. In our language we have: if $tt \,\nabla\, ff$ then E1 else E2 fi is equivalent to $(E1 \,[]\, E2)$

and not to E1 ∇ E2 , which would be the case if one systematically translated the concept of alternation (as the other approach is called in /Chandra et al. 81/) into our language.

Note, that the ambiguity operator ∇ can also be used to specify the logical "parallel or" (and "parallel and"):

funct paror ≡ λ x, y :
if x then tt else y fi ∇ if y then tt else x fi

Interestingly the function paror can be viewed as a determinate function (in the powerdomain approach) with paror(tt,x) = paror(x, tt) = tt, paror(ff,ff) = ff, paror(⊥, ⊥) = ⊥ for all x ∈ {tt, ff, ⊥}, while in the multidomain approach it is non-deterministic.

In a similar way the ambiguity operator ∇ can be used to specify disjunctive (multiple) waiting. Given a guarded wait command:

await B_1 then E_1 ∇ ... ∇ B_n then E_n endwait

which can be defined by

(λ $b_1, \ldots, b_n$:
if b_1 ∇ $\tilde{B}_1$ then E_1
else await b_2 then E_2 ∇ ...
∇ b_n then E_n endwait fi) ($B_1, \ldots, B_n$)

where

$\tilde{B}_1$ = if b_2 then ff else ⊥ fi ∇ ... ∇ if b_n then ff else ⊥ fi

One proves easily, that in the await - construct the pairs (B_i then E_i) can be arbitrarily permuted without changing the semantics, i.e. that in spite of the unsymmetric expression above the await - construct is symmetric in its guards.

Example: Disjunctive Waiting

```
⌈funct table ≡ λ s : g(first s) & table (rest s),
 stream  s1  ≡ table(1 & s1), stream s2 ≡ table(2 & s2), merge(s1,s2)⌋

 where

  funct merge ≡ λ s1, s2 :
     await ¬ isempty s1 then (first s1)& merge(rest s1,s2)
      ∇    ¬ isempty s2 then (first s2) & merge(s1, rest s2)
      ∇      isempty s1 ∧  isempty s2 then empty          endwait
```

and g is some recursive or primitive function. The result of this program is an infinite stream, iff at least one of the streams s1 or s2 is infinite. Note that it is not a fair merging in the sense of /Keller 78/, since if both s1 and s2 are infinite, s1 as well as s2 may be a possible result, where s2 (or s1 resp.) do not contribute anything. Fair merging seems only to be a fair concept (cf. /Broy 81/), if we switch to real time processing.

end of example

Our system is completely free of *global nondeterminism* , i.e. all decisions can be made by the single processes without any feedback of other processes. Of course, certain decisions can only be recognized as feasible after a number of communication steps, however, when the communication has taken place, the decision can be made *locally*.

In Milner's CCS (cf. /Milner 80/) or Hoare's CSP (cf. /Hoare 78/) the communication is generally coupled with a nondeterministic choice. However, a single process cannot decide by itself which alternative for communication is to be chosen. The "rendezvouz"-concept needs some coordination *before* an actual communication occurs. In particular, if no priorities between the processes are given, a global instance is needed to resolve conflicts. As outlined in /Francez, Rodeh 80/ there is no way of getting a fully distributed, symmetric implementation without using probabilistic computation techniques. These techniques, however, may not be satisfactory, since they may restrict possible implementations on real computing systems in an inadequate way.

7. Concluding Remarks

Originally the issues of concurrent programming have been motivated by particular properties of multi-processor machines. Due to the fact that these machines are of the von Neumann type, early proposals and investigations were strictly procedure-oriented, centering around the problem of how to protect and synchronize the access to shared memory, which was considered as the only way of communication between programs executed in parallel. First attemps to overcome these difficulties can be found in the single assignment approach (cf. /Tesler, Enea 68/). This proposal was completed to data flow language concepts (cf. /Dennis 74/), based on the demand driven evaluation (cf. also /Backus 78/). Numerous papers have been published on this issue, few of them, however, containing much formal foundation. A related approach was developed in /Friedman, Wise 78/, where already the concepts of lazy evaluation are incorporated to obtain a LISP-extension which is suitable for applicative multiprogramming. There "ferns" are used instead of streams.

Other approaches use "tagged" values (cf. /Kosinski 77/, /Kosinski 79/) or "scenarios" (cf. /Brock, Ackermann 81/).

A far developed approach is Milner's Calculus of Communicating Systems (CCS, cf. /Milner 73/, /Milne , Milner 77/, /Milner 80/). In CCS a communication mechanism is integrated into an applicative programming language. The communication in CCS follows the rendevouz principle (like CSP in /Hoare 78/) and therefore can be seen as the applicative counterpart of CSP.

A paper with much impact on the field of concurrent programming is /Kahn, MacQueen 77/ , where an approach is outlined which is very similar to the streams described in section 2.3 and section 5 (cf. also /MacQueen 79/). However Kahn's approach does not include nondeterminism nor concurrency in the sense of section 6 (cf. the discussion at the end of /Keller 78/).

In the field of nondeterministic applicative programming languages several papers on fixed point theory have been published (cf. for instance /Hennessy, Ashcroft 76, 77, 80/, /de Bakker 76/, /Arnold, Nivat 77/, /Broy et al. 78/, /Nivat 80/). For surprisingly few approaches to multiprogramming, however, attempts have been undertaken to give a fixed point semantics (cf. for instance /Arvind, Gostelow 78/, /Hewitt , Baker 78/, /Kosinski 79/).

We prefer a fixed point oriented approach for the following reasons: First, the joint consideration of operational and mathematical semantics gives valuable insights into the structure of the concepts. Second, we can always check whether our intuition actually leads to computable, formally sound semantics. Third, the technical difficulty and complexity of particular concepts gives hints on their comprehensive complexity and also on the difficulties to find appropriate methodologies for the design and verification of such programs.

Of course the semantics of the language for applicative multiprogramming could be described also by algebraic means along the lines of /Broy, Wirsing 80/. However, in addition to the reasons cited above it seems worthwhile to develop the concepts of multiprogramming from the nowadays well-understood concepts of sequential programming.

Hence in the preceding sections a strictly fixed point oriented approach to the concepts of nondeterminism, parallelism, communication and concurrency has been undertaken. Not all results are satisfactory yet although most of the important notions can be described properly and, to some extent, fit in naturally with the framework of fixed point theory. The extension and application of this approach to more explicitly communication-oriented languages like CCS or to procedural languages challenges further investigations. Moreover, the consequences of the identifications of certain infinite multisets in multidomains are not completely understood by the author and deserve further attention.

Acknowledgement

I gratefully acknowledge stimulating discussions with Prof. F. L. Bauer and Prof. G. Seegmüller. I am indebted to B. Möller and M. Wirsing for a number of valuable remarks. Thanks go to T.A. Matzner ("TAM") for carefully reading drafts.

References

/Apt, Plotkin 81/
K.R. Apt, G. Plotkin: A Cook's Tour of Countable Nondeterminism. In: S. Even, O. Kariv (eds.): 8th International Colloquium on Automata, Languages and Programming, Haifa 1981, Lecture Notes in Computer Science 115, Berlin - Heidelberg - New York: Springer 1981, 479-494

/Arnold 79/
A. Arnold: Operational and Denotational Semantics of Nets of Processes. Universite P. et M. Curie, Universite Paris 7, Laboratoire Informatique Theoretique et Programmation, LITP Report No. 79-35, June 1979

/Arnold, Nivat 77/
A. Arnold, M. Nivat: Nondeterministic Recursive Program Schemes In: M. Karpinski (ed): Fundamentals of Computation Theory. Lecture Notes in Computer Science 56, Berlin - Heidelberg - New York: Springer 1977, 12-21

/Arvind, Gostelow 78/
Arvind, K.P. Gostelow: Some Relationships between Asynchronous Interpreters of a Dataflow Language. In: /Neuhold 78/, 96-119

/Astesiano, Costa 79/
E. Astesiano, G. Costa: Sharing in Nondeterminism. In: H.A. Maurer (ed.): 6th Int. Coll. on Algorithms, Languages and Programming. Lecture Notes in Computer Sciences 71, Berlin - Heidelberg - New York: Springer 1979, 1-13

/Backus 78/
J. Backus: Can Programming be Liberated from the von Neumann Style? A Functional Style and its Algebra of Programs. Comm. ACM 21:8, August 1978, 613-641

/de Bakker 76/
J.W. de Bakker: Semantics and Termination of Nondeterministic Recursive Programs. In: S. Michaelson, R. Milner (eds.): Proc. of the 3rd International Colloquium on Automata, Languages and Programming, Edinburgh: Edinburgh University Press 1976, 435-477

/Bauer 79/
F.L. Bauer: Detailization and Lazy Evaluation, Infinite Objects and Pointer Representation. In: /Bauer, Broy 79/, 235-236

/Bauer, Broy 79/
F.L. Bauer, M. Broy (eds.): Program Construction. Lecture Notes of the International Summer School on Program Construction, Marktoberdorf 1978. Lecture Notes in Computer Science 69, Berlin - Heidelberg - New York: Springer 1979

/Bauer, Wössner 81/
F.L. Bauer, H. Wössner: Algorithmische Sprache und Programmentwicklung. Berlin - Heidelberg - New York: Springer 1981

/Benson 79/
D.B. Benson: Parameter Passing in Nondeterministic Recursive Programs. Journal of Computer and System Sciences 19, 1979, 50-62

/Brock, Ackermann 81/
J.D. Brock, W.B. Ackermann: Scenarios: A Model of Non-determinate Computation. In: J. Diaz, I. Ramos (eds.): Formalization of Programming Concepts, Peniscola 1981, Lecture Notes in Computer Science 107, Berlin - Heidelberg - New York: Springer 1981, 252-267

/Broy 80a/
M. Broy: Transformation parallel ablaufender Programme. Technische Universität München, Dissertation an der Fakultät für Mathematik, Februar 1980

/Broy 80b/
M. Broy: Transformational Semantics for Concurrent Programs. IPL 11:2, October 1980, 87-91

/Broy 81/
M. Broy: Are Fairness-assumptions Fair? In: Proc. of the Second International Conference on Districuted Computing Systems, Paris April 8-10, 1981, IEEE 1981, 116-125

/Broy, Wirsing 80/
M. Broy, M. Wirsing: Initial versus Terminal Algebra Semantics for Partially Defined Abstract Types. Technische Universität München, Institut für Informatik, TUM-I8018, December 1980

/Broy, Wirsing 81a/
M. Broy, M. Wirsing: On the Algebraic Specification of Nondeterministic Programming Languages. In: E. Astesiano, C. Böhm (eds.): 6th Colloquium on Trees in Algebra and Programming, Genua 1981, Lecture Notes in Computer Science 112, Berlin - Heidelberg - New York: Springer 1981, 162-179

/Broy, Wirsing 81b/
M. Broy, M. Wirsing: Unbounded Nondeterminism - An exercise in Abstract Data Types. INRIA 1981

/Broy et al. 78/
M. Broy, R. Gnatz, M. Wirsing: Semantics of Nondeterministic and Noncontinous Constructs. In: /Bauer, Broy 79/, 553-592

/Broy et al. 80/
M. Broy, H. Partsch, P. Pepper, M. Wirsing: Semantic Relations in Programming Languages. In: S.H. Lavington (ed.): Information Processing 80, Proceedings of the IFIP Congress 80, Amsterdam - New York - Oxford: North-Holland Publ. Comp. 1980, 101-106

/Burge 75/
W.H. Burge: Stream Processing Functions. IBM Journal of Research and Development 19, January 1975, 12-25

/Chandra 74/
A.K. Chandra: The Power of Parallelism and Nondeterminism in Programming. In: J.L. Rosenfeld (ed.): Information Processing 74, Proc. of the IFIP Congress 74, Amsterdam: North Holland Publ. 1974, 461-465

/Chandra 78/
A.K. Chandra: Computable Nondeterministic Functions. Proc. of the 19th Annual Symposium on Foundations of Computer Science, October 1978, 127-131

/Chandra et al. 81/
A.K. Chandra, D.C. Kozen, L.J. Stockmeyer: Alternation. J. ACM 28:1, January 1981, 114-133

/Dennis 74/
J.B. Dennis: First Version of a Data Flow Procedure Language. In: B. Robinet (ed.): Colloque sur la Programmation, Lecture Notes in Computer Science 19, Berlin - Heidelberg - New York: Springer 1974, 362-367

/Dennis, Weng 79/
J.B. Dennis, K. K.-S. Weng: An Abstract Implementation for Concurrent Computation with Streams. In: Proc. of the 1979 International Conference on Parallel Processing, August 1979, 35-45

/Dershowitz, Manna 79/
N. Dershowitz, Z. Manna: Proving termination with Multiset Orderings. Comm. ACM 22:8, August 1979, 465-476

/Dijkstra 76/
E.W. Dijkstra: A Discipline of Programming. Prentice Hall, Englewood Cliffs N.J. 1976

/Egli 75/
H. Egli: A Mathematical Model for Nondeterministic Computations. Unpublished report ETH Zürich 1975

/Ershov 78/
A.P. Ershov: On the Essence of Compilation. In: /Neuhold 78/, 391-418

/Francez, Rodeh 80/
N. Francez, M. Rodeh: A Distributed Abstract Data Type Implemented by a Probabilistic Communication Scheme. 21st Annual Symposium on Foundations of Computer Science, October 1980

/Friedmann, Wise 76/
D.P. Friedmann, D.S. Wise: CONS Should not Evaluate its Arguments. In: S. Michaelson, R. Milner (eds.): Proc. of the 3rd International Colloquium on Automata, Languages and Programming. Edinburgh: Edinburgh Univ. Press, 1976, 257-284

/Friedmann, Wise 78/
D.P. Friedmann, D.S. Wise: Applicative Multiprogramming. Indiana University, Computer Science Department, Technical Report 72, Januar 1978, revised December 1978

/Henderson 80/
P. Henderson: Functional Programming: Application and Implementation. Englewood Cliffs, NJ: Prentice Hall International 1980

/Henderson, Morris 76/
P. Henderson, J.H. Morris: A Lazy Evaluator. University of Newcastle upon Tyne, Computing Laboratory, Techn. Report Series No. 85

/Hennessy 80/
M.C.B. Hennessy: The Semantics of Call-by-Value and Call-by-Name in a Nondeterministic Environment. SIAM J. Comput. 9:1, February 1980, 67-84

/Hennessy, Ashcroft 76/
M. Hennessy, E.A. Ashcroft: The Semantics of Nondeterminism. In: S. Michaelson, R. Milner (eds.): Proc. of the 3rd International Colloquium on Automata, Languages and Programming, Edinburgh: Edinburgh University Press 1976, 479-493

/Hennessy, Ashcroft 77/
M. Hennessy, E.A. Ashcroft: Parameter Passing mechanisms and Nondeterminism. In: Proceedings of the 9th Annual ACM Symposium on Theory of Computing, May 1977, 306-311

/Hennessy, Ashcroft 80/
M. Hennessy, E.A. Ashcroft: A Mathematical Semantics for Typed λ-calculus. Theoretical Computer Science 10, 1980, 227-245

/Hewitt, Baker 78/
C. Hewitt, H. Baker: Actors and Continuous Functionals. In: /Neuhold 78/, 367-390

/Hoare 78/
C.A.R. Hoare: Communicating Sequential Processes. Comm. ACM 21:8, August 1978, 666-677

/Kahn 74/
G. Kahn: The Semantics of a Simple Language for Parallel Processing. In: J.L. Rosenfeld (ed.): Information Processing 74, Proc. of the IFIP Congress 74, Amsterdam: North-Holland 1974, 471-475

/Kahn, MacQueen 77/
G. Kahn, D. MacQueen: Coroutines and Networks of Parallel Processes. In: B. Gilchrist (ed.): Information Processing 77, Proc. of the IFIP Congress 77, Amsterdam: North-Holland 1977, 994-998

/Keller 78/
R.M. Keller: Denotational Models for Parallel Programs with Indeterminate Operators. In: /Neuhold 78/, 337-366

/Keller 80/
R.M. Keller: Semantics and Applications of Function Graphs. University of Utah, Department of Computer Science, Technical Report UUCS-80-112, October 1980

/Kennaway, Hoare 80/
J.R. Kennaway, C.A.R. Hoare: A Theory of Nondeterminism. In: J. de Bakker, J.v.d. Leuwen (eds.): Proc. of the 7th International Colloquium on Algorithms, Languages and Programming. Lecture Notes in Computer Science 85, Berlin - Heidelberg - New York: Springer 1980, 338-350

/Kosinski 73/
P.R. Kosinski: A Data Flow Language for Operating Systems Programming. SIGPLAN Notices 8:9, September 1973, 89-94

/Kosinski 77/
P.R. Kosinski: A Straightforward Denotational Semantics for Nondeterminate Data Flow Programs. Proc. of the 5th Annual Symposium on Principles of Programming Languages, 1977

/Kosinski 79/
P.R. Kosinski: Denotational Semantics of Determinate and Nondeterminate Data Flow Programs. MIT, Laboratory for Computer Science, TR-220, May 1979

/Landin 65/
P.J. Landin: A Correspondence Between ALGOL 60 and Church's Lambda-Notation: Part I. Comm. ACM 8:2, February 1965, 89-101

/Lehmann 76/
D.J. Lehmann: Categories for Fixpoint-Semantics. Proc. of the 17th Annual Symposium on Foundations of Computer Science 1976, 122-126

/MacQueen 79/
D.B. MacQueen: Models for Distributed Computing. IRIA Rapport de Recherche No 351, April 1979

/Manna 70/
Z. Manna: The Correctness of Nondeterministic Programs. Artificial Intelligence 1, 1970, 1-26

/Manna et al. 73/
Z. Manna, S. Ness, J. Vuillemin: Inductive Methods for Proving Properties of Programs. Comm. ACM 16:8, August 1973, 491-502

/McCarthy 63/
J. McCarthy: A Basis for a Mathematical Theory of Computation. In: P. Braffort, D. Hirschberg (eds.): Computer Programming and Formal Systems, Amsterdam: North-Holland 1963

/Milne, Milner 77/
G. Milne, R. Milner: Concurrent Processes and their Syntax. University of Edinburgh, Department of Computer Science, CSR-2-77, May 77

/Milner 73/
R. Milner: Processes: A Mathematical Model of Computing Agents. Proc. Logic Colloquium, Bristol, Amsterdam: North-Holland 1973, 157-173

/Milner 80/
R. Milner: A Calculus of Communicating Systems. Lecture Notes in Computer Science 92, Berlin - Heidelberg - New York: Springer 1980

/Neuhold 78/
E.J. Neuhold (ed.): Formal Descriptions of Programming Concepts. Amsterdam: North-Holland 1978

/Nivat 80/
M. Nivat: Nondeterministic Programs: An Algebraic Overview. In: S.H. Lavington (ed.): Information Processing 80, Proc. of the IFIP Congress 80, Amsterdam - New York - Oxford: North-Holland Publ. Comp. 1980, 17-28

/Park 80/
D. Park: On the Semantics of Fair Parallelism. In: D. Björner (ed.): Abstract Software Specification. Lecture Notes in Computer Science 86, Berlin - Heidelberg - New York: Springer 1980, 504-526

/Plotkin 76/
G. Plotkin: A Powerdomain Construction. SIAM J. on Computing 5, 1976, 452-486

/Scott 80/
D. Scott: Lectures on a Mathematical Theory of Computation. University of Oxford, Mathematical Institute, Preliminary Version, completed November 1980

/Smyth 78/
M.B. Smyth: Power Domains. J. CSS 16, 1978, 23-36

/Tesler, Enea 68/
L.G. Tesler, H.J. Enea: A Language Design for Concurrent Processes. Spring Joint Computer Conference 1968, 403-408

/Vuillemin 74/
J. Vuillemin: Correct and Optimal Implementation of Recursion in a Simple Programming Language. J. Comp. Sci. 9:3, June 1974, 332-354

/Wadsworth 71/
C. Wadsworth: Semantics and Pragmatics of Lambda Calculus. Oxford, Ph. D. Dissertation 1971

Theme XII.

(M. Broy)

His way of reminding other lecturers
of the time limit is best represented
by Mussorgskij's "Promenade"

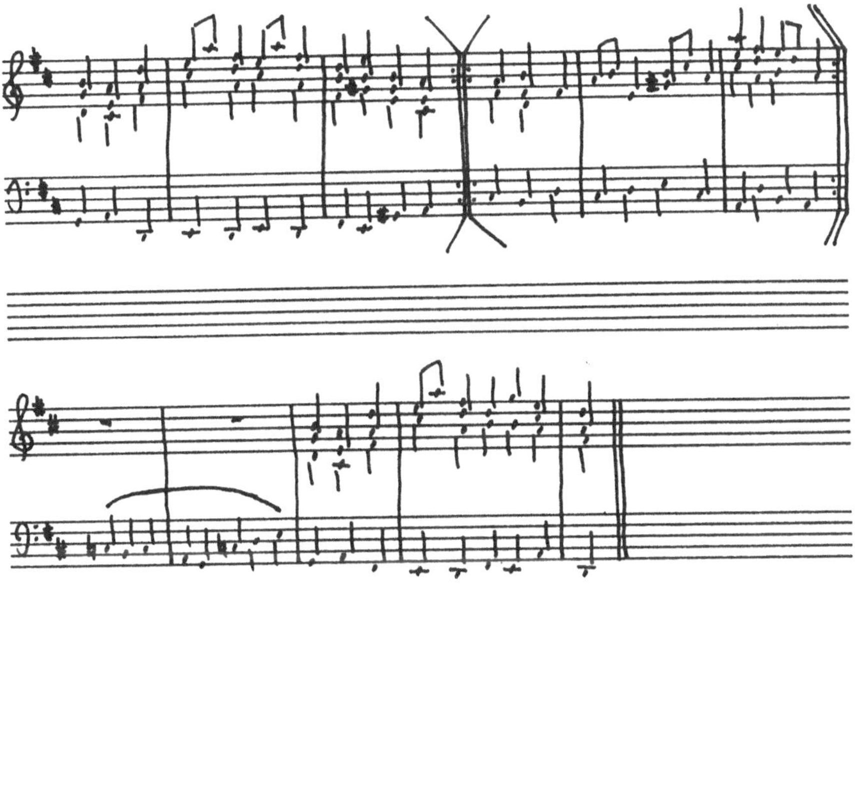

DISCRETE EVENT SIMULATION BASED ON COMMUNICATING SEQUENTIAL PROCESSES

W.H. Kaubisch and C.A.R. Hoare

Oxford University Computing Laboratory
Programming Research Group
45 Banbury Road, Oxford

This paper suggests a minimal set of primitive concepts required in the construction of algorithms for discrete event simulation. The basic concept is the communicating sequential process [CSP]; however, this is reinterpreted as a quasiparallel process, operating in simulated time. The most important features of simulation are shown to be implementable as communicating processes, and a nontrivial example of a simulation algorithm is given.

Key words and phrases: Programming Languages, Primitive Concepts, Discrete Event Simulation, Quasiparallel Processing, Communicating Sequential Processes.

CR Categories: 4.22, 3.65

*This research was supported by a grant and a senior fellowship from the Science Research Council of Great Britain.

INTRODUCTION

Research into programming languages has produced a wide variety of proposed designs. Each design attempts to improve upon its rivals, often by introducing additional "powerful" features, perhaps oriented towards a particular application area. As a result, some of the languages have been quite complicated to implement and even to understand; but many programmers have taken this as a challenge to their skill and ingenuity.

M. Broy and G. Schmidt (eds.), Theoretical Foundations of Programming Methodology, 625–642.

This paper takes exactly the opposite approach; it attempts to remove as many features as possible from previously proposed languages, and replace them by the barest minimum set of primitive structures, which are adequate for the intended purposes. It suggests that, even in a specialised application area such as discrete event simulation, most requirements can be met by a few general-purpose features.

The paper starts with a brief survey of the main requirements of a programming language designed for discrete event simulation, namely, resources, processes, simulated time, queues, statistics, and random numbers. These are illustrated by features and examples of the use of existing simulation languages, GPSS [5], SIMULA 67 [2], and SIMONE [4].

The next following section introduces the concept of a communicating sequential process [1], which is interpreted as operating in simulated time instead of real time. The proposed notation is described by means of annotated examples, since a fuller and more formal description is already available [1]; however, as an experiment, this earlier language has been extended by permitting output commands to appear in guards.

In section 4, a series of examples illustrate how this language can be used to implement all the basic requirements of simulation, as surveyed in section 2. It culminates in a simple but complete simulation algorithm, the machine shop [3].

2. THE REQUIREMENTS OF DISCRETE EVENT SIMULATION.

This section surveys the requirements of discrete event simulation, and the features which have been added to programming languages to adapt them for this purpose.

2.1. Resources.

Among its definitions for "resources", the Oxford English Dictionary gives:

1. A means of supplying some want or deficiency; a stock or reserve one may draw upon when necessary.

2. An action or procedure to which one may have recourse.

3. The capability of adapting means to ends ...

The first definition refers to stocks and reserves, which we may interpret as being supplies of inanimate objects or materials, for example, the components or metal used in a machine shop. The second definition refers to actions and procedures executed by agents, which are of interest not because of their physical substance but because they accomplish some desired result; the machines and machine operators in the machine shop are examples of this type of agent. According to the third definition, we may say that the machine shop itself is a resource, because of its ability to adapt the means at its disposal (i.e. the metal, the machines and the operators) to meet a given set of orders.

Simulation models and languages must be capable of representing such widely varying types of resources. In the simplest cases, resources may be represented by simple variables; but more generally their representation will require the use of structures, e.g. the PASCAL [7] Record structures, or even structures with associated rules of access, such as the SIMULA 67 Class [2], or module of MODULA [6], or by the Monitor of SIMONE [4].

As an example, in the case of the machine shop, the stock of metal may be represented as an integer giving the amount of stock on hand e.g.

```
METALONHAND : INTEGER;
```

The simplest representation of a (single resource) machine is as a boolean variable, indicating whether that machine is in use:

```
MACHINEINUSE : BOOLEAN;
MACHINEINUSE := FALSE;
```

Going on to a more elaborate example, consider a group of machines, each of which is capable of doing the same job, but each of which has a different running cost. Each machine could now be represented as a RECORD:

```
MACHINE = RECORD
            RUNNING COST : INTEGER;
            INUSE : BOOLEAN
          END;
```

and a group of ten of these machines can be represented as an array:

```
MACHINEGROUP : ARRAY [1..10] OF MACHINE;
```

Finally, assume that the machine group has a foreman whose job it is to decide about the usage of his machines. When a customer wishes to use a machine he must first ask the foreman for permission, and when he has finished, he must inform the foreman of this fact. In the language GPSS, a resource like the foreman would be represented by the built-in STORAGE feature. In other languages, it must be programmed explicitly - for example, as a monitor in SIMONE:

```
MONITOR FOREMAN;
  ... declaration of local variables ...;
    PROCEDURE REQUEST;
          ... body of request ...;
    PROCEDURE RELEASE;
          ... body of release ...;
    initialisation of local variables ...
END
```

Here, the procedures REQUEST and RELEASE may be called from outside the monitor by qualified calls:

(1) FOREMAN.REQUEST;
which acquires a machine from the foreman, possibly after some delay.

and (2) FOREMAN.RELEASE;
which returns a machine to the foreman for reallocation.

There is a qualitative difference in complexity between the foreman and the previous examples. One could write ever more complex records to represent ever more complex resources; but they remain inanimate, and the manner and sequence of access to them is determined solely or mainly by the accessing program. However, a monitor or a class has the capability of apparently autonomous behaviour. The foreman, in his efforts to grant a request, may himself initiate requests for other resources; for example, he may make choices about maintenance schedules, may call for a mechanic, etc. Hence, one request may set off a chain of other events, and the interaction between the resources may become quite complex. We shall see later that the representation of such a complex resource can take advantage of the full generality of a communicating sequential process.

2.2. Processes.

A process is an independent action or series of actions leading to the realisation of some result. Processes can interact with each other when they compete for resources or communicate. Apart from these interactions, processes are independent of each other; and in particular they make independent progress in time, in that a number of processes will be executed concurrently (in parallel). In a simulation, this concurrency is usually implemented by interleaving actions from all the active processes (in quasiparallel).

In a machine shop simulation, an example of a process would be an order which flows through the shop. Each order specifies the series of steps required to produce the desired product; each order is independent of the other orders except insofar as it uses a common set of resources (machines). In SIMONE [4] an order may be represented as a process:

```
PROCESS ORDER;
BEGIN ... declaration and initialisation of local
          variables ...;
  FOR I :=1 TO NUMBEROFSTEPS DO
 BEGIN request, use and release the machine required
       for this step ...
 END
END;
```

The process-like quality of the ORDER is obvious, because it is not "used" by any other process. We have already seen that type 2 and 3 resources also exhibit process-like characteristics relative to the resources they manipulate, e.g., from the point of view of ORDER, the FOREMAN is a resource, yet the FOREMAN itself behaves as a process relative to the machines, materials and mechanics which it schedules. Thus it appears that the processes and resources of a simulation algorithm display a multilevel tree organisation. At the bottom level are the type 1 inanimate resources, at the top level are the pure processes; and in between are the type 2 and 3 resources. Looking from the top down, every structure looks like a resource; looking from the bottom upward, they look like processes. The process and resource concepts are relative rather than absolute. Even the processes (orders) at the top level would appear to be resources if we were to add another level to the tree, for example, customers who originate and cancel orders. It is this insight which will enable us to represent both processes and resources of type 2 and 3 by a single primitive program structure, the communicating sequential process.

2.3. Time.

A simulation algorithm describes not only the static elements and relations of the system being modelled, but also the dynamic behaviour and interactions of the processes and resources as they evolve in time. But the passage of time must itself be simulated as the algorithm is executed; and each process which engages in an activity which is intended to take an appreciable amount of time must specify its duration explicitly.

In SIMULA 67 and SIMONE, the current value of simulated time may be discovered by a call on the parameterless function

TIME, for example IF TIME<FIVEOCLOCK THEN ....

When a process is to engage in an independent activity which will last D units of time, this is indicated in the program by a call on the standard procedure HOLD(D). If the value of TIME before this call is T, then the value of TIME after the call will be T+D. Thus the effect of HOLD is simply to suspend the calling process until the elapse of the specified duration of simulated time. For example, the loop of the ORDER process (of the previous section) may be given in greater detail:

```
FOR I := 1 TO NUMBEROFSTEPS DO
BEGIN FOREMAN.REQUEST;
      HOLD(USAGETIME);
      FOREMAN.RELEASE;
END.
```

Here, we are not interested in the details of what the order does with the machine; we are interested only in the fact that its usage of the machine continues during the specified interval in model time, during which the process engages in no other interaction or change of state.

Simulated time must not be confused with the real time taken by a computer in execution of the commands of a simulation program. If a program does not contain any HOLD operation, the entire program would be executed at the same instant of simulated time, though it would certainly take some real execution time on a computer. Conversely, when every process of a program is engaged in a HOLD, they are using no computer time; but on each such occasion, the implementation of the simulation language steps on the value of simulated time to the earliest value which would permit a process to resume execution after its HOLD. Thus it may be said that movement in simulated time takes no execution time, and vice versa.

2.4. Queues.

A process which requires use of a resource will usually have to wait if that resource is busy. If several processes have to wait for the same resource, they will have to form some kind of queue. When the resource becomes free, a choice must be made between the waiting processes, on the basis of some specified scheduling discipline. A simulation is often concerned with the relation between scheduling discipline and the acceptability of response times.

In GPSS, queues are not represented explicitly, but there is an implicit queue associated with each FACILITY or STORAGE. In SIMULA 67, a queue is represented by the built-in SIMSET class. In SIMONE, where a resource is represented as a monitor, a queue of processes waiting for the resource can be represented as a condition variable local to the monitor. For example, local to the FOREMAN monitor of section 2.1, there might be declared

```
FREE : INTEGER;
```

which contains the number of free machines, or (if none) the negative of the number of waiting orders, and

```
Q : CONDITION;
```

representing the queue on which the orders wait. Now the procedures of the foreman could be written:

```
PROCEDURE REQUEST;
   BEGIN FREE := FREE-1;
         IF FREE < 0 THEN Q.WAIT
   END;

PROCEDURE RELEASE;
   BEGIN FREE := FREE+1;
         IF FREE ≤ 0 THEN Q.SIGNAL
   END;
```

The command Q.WAIT suspends the process which called REQUEST; the command Q.SIGNAL causes resumption of the process (if any) which earliest executed Q.WAIT. Thus a condition variable Q implements a policy of "first in first out" (fifo) scheduling. Other scheduling disciplines can be specified by a "scheduled condition".

2.5. Statistics.

A simulation program is often in principle non-terminating, in the sense that there is no well-defined state in which it can be said to have arrived at "the answer". Instead, one generally allows a simulation to cycle through a given number of operations; or, alternatively, to execute for a given duration of simulated time. Hence, its state when it terminates is unpredictable; and even if it were, it would be of no real interest. Instead, one is interested in the history of the states through which the simulation has passed. Two runs of the same model ending in the same state but having different histories are not considered as equivalent.

The history of the execution of a simulation is simply the set of "values" of all the components of the simulation at each moment of simulated time. In general, however, the entire set is not of interest and some subset must be selected and summarised to produce the required set of statistics. One method of doing this is to accumulate a histogram of relevant observations. Some special purpose languages would contain such a facility built-in; but in SIMULA 67 it must be programmed in the language itself, using the general-purpose structure provided by the class. For example, suppose that a histogram requires three parameters.

(1) N, the number of intervals.
(2) LOW, the lower bound of the lowest interval; it should be non-negative.
(3) HIGH, the upper bound of the highest interval; it should be greater than LOW.

The class provides two procedures:

(4) RECORD(X), which records the observation X in the histogram;
X should be between LOW and HIGH. ($LOW \leq X < HIGH$)
(5) PRINT, which prints the histogram in some suitable graphic representation (which we will not specify here).

The entire class can be constructed:

```
CLASS HISTOGRAM (N,LOW,HIGH);
  INTEGER N, LOW,HIGH;

BEGIN INTEGER COUNT;
  INTEGER ARRAY HISTO [0:N-1];

PROCEDURE RECORD (X); INTEGER X;
   BEGIN INTEGER I;
     I :=N*(X-LOW)÷(HIGH-LOW);
     HISTO[I] :=HISTO[I]+1;
     COUNT :=COUNT+1
   END;

PROCEDURE PRINT
   ... body of print ...;

FOR COUNT :=0 STEP 1 UNTIL N-1 DO
       HISTO[COUNT] :=0;
COUNT :=0
END;
```

However, extensive statistics gathering written by the programmer tends to clutter the program and obscure the model. Hence, there is a case to be made for certain semi-automatic facilities (as in GPSS), though this option carries with it the problem that the volume of (perhaps unwanted) statistics can become quite large.

2.6. Random Numbers.

In a simulation program, the parameters of a process (e.g. its start time, service times) may be specified in the normal way by the program itself or by its input data; but it is often more convenient to select them at random in accordance with some known or conjectured distribution (e.g. a uniform distribution between given limits, or a negative exponential with a given mean). A language like GPSS provides a range of random number drawing facilities to assist in the construction of probabilistic models; but in a general purpose language these facilities need not be built-in, since they can be programmed by a pseudorandom multiplicative technique, using some suitable MULTIPLIER and LIMIT. For example, a generator of a random number between zero and one can be implemented as a SIMULA 67 class, and used by repeated calls on the SAMPLE procedure:

```
CLASS RANDOM (SEED); INTEGER SEED;
  BEGIN REAL PROCEDURE SAMPLE;
       BEGIN SAMPLE  :=SEED/LIMIT;
             SEED :=SEEDxMULTIPLIER;
             SEED :=SEED-(SEED÷LIMIT)xLIMIT;
       END
END
```

This simple multiplicative algorithm is chosen merely for the sake of the example.

2.7. Summary.

Table 1 shows how the six essential features of discrete simulation are represented in three languages designed for the purpose. The fourth column provides a comparison with the language described later in this paper. It can be seen that GPSS provides the widest range of built-in special-purpose features, and CSP leaves the most to be programmed in the language itself, possibly with some loss of convenience and efficiency.

Table 1.

	GPSS	SIMULA 67	SIMONE	CSP
Resources	FACILITY, STORAGE	ALGOL data structures, CLASS	PASCAL data structures, MONITOR	PASCAL data structures, PROCESS
Processes	TRANSACTION	PROCESS CLASS	PROCESS	PROCESS
Time	ADVANCE	Built-in HOLD	Built-in HOLD	Built-in HOLD
Queues	implicit	SET	CONDITION variable	Programmer responsibility
Statistics	Built-in feature	Programmer responsibility	Programmer responsibility	Programmer responsibility
Random Numbers	Built-in generators	Built-in generators	Built-in generators	Programmer responsibility

3. COMMUNICATING SEQUENTIAL PROCESSES.

A complete description of communicating sequential processes has been given in a previous paper [1]. This section contains a series of annotated examples, selected from the area of discrete event simulation. They may serve as revision for a reader who is already familiar with the previous paper; otherwise, the reader is recommended to study the previous paper, supplementing or replacing its examples by those of this section.

The main difference between the general-purpose language described previously, and the special-purpose language described here is that here the processes are interpreted as being executed in simulated time (in quasiparallel) instead of in real time (genuine concurrency). The mechanism of simulated time was described in 2.3. The question arises whether simulated time could have been implemented as a TIMER process, using only the general-purpose features of the language.

This could in fact be done, on condition that:

(1) the TIMER maintains a count of all quasiparallel processes in the system; i.e., all those which make calls of HOLD.

(2) No quasiparallel process communicates with any other process except the TIMER.

These restrictions are required to ensure that the TIMER can detect when the number of processes which have executed a HOLD is equal to the total number of quasiparallel processes, so that it can advance simulated time and resume the process which is due to be resumed the earliest. However, the restrictions are unacceptably severe; and unless some general-purpose method can be found of triggering the TIMER process when there is nothing else left to do, it would seem necessary to include an automatic timer as a built-in feature of a special-purpose language intended for discrete event simulation.

One minor extension has been made to the language described previously, in that output commands are permitted to appear as guards in alternative and repetitive commands. This gives a useful increase in the convenience of use of the language, although it may lead to implementation problems on multiple processors with disjoint stores.

3.1. Assignment Commands.

(1) IN := IN+1 adds one to IN

(2) (N,LOW,HIGH) := (10,0,100) a multiple assignment, assigning to each target variable on the left the value of the corresponding element on the right.

(3) REQUEST(n) := REQUEST(3) the tags "REQUEST" on the left and right are matching, so the effect is the same as n :=3.

(4) GRANTED( ) :=GRANTED( ) the assignment of matching signals has no effect.

(5) ACQUIRE( ) :=RELEASE( ) fails, owing to mismatch of tags

3.2. Parallel Commands.

(1) [Q:: queue||U:: user]

Here, "queue" and "user" stand for command lists which are to be executed concurrently, and Q and U are identifiers which name these processes. The processes start simultaneously, and the parallel command ends successfully only if and when both of them have successfully terminated.

(2) [FOREMAN (J:1..10):: foreman]

Here "foreman" stands for a command list, possibly containing the bound variable J. This example specifies ten processes, with names FOREMAN (1), FOREMAN (2), ... FOREMAN (10). The actions of each are specified by the identical text "foreman", except that the value of J in each process gives the index of its name.

3.3. Input and Output Commands.

(1) READER?STARTTIME - from the READER input an integer value, and assign it to STARTTIME.

(2) PUNCH!"*" - to the punch, output the character "*".

(3) U?(N,LOW,HIGH) - from process named U, input a group of three values, and assign them to variables, N, LOW, and HIGH.

(4) HISTOGRAM!(10,0,100) - to process HISTOGRAM, output the three values 10,0,100.

Note: if a process named HISTOGRAM issues command (3), and a process named U issues command (4), these are executed simultaneously, and have the same effect as the structured assignment:

```
    (N,LOW,HIGH) := (10,0,100),
i.e. N := 10; LOW := 0; HIGH := 100
```

(5) ALLOC!ACQUIRE( ) - to process ALLOC send a signal ACQUIRE( )

(6) U?ACQUIRE( ) - from process named U, accept a signal ACQUIRE( ).

(7) ORDER(I)?REQUEST( ) - from the ith element of an array of ORDER processes accept a signal REQUEST( ).

3.4. Alternative and Repetitive Commands.

```
(1)  [FREE<0 → Q!I
     ▯FREE≥0 → ORDER(I)!GRANTED( )
     ]
```

If FREE is negative, I is output to Q; otherwise a GRANTED( ) signal is sent to ORDER(I).

```
(2)  I :=1;
     *[I≤NOFSTEPS → ...; I :=I+1]
```

The body ... is repeated NOFSTEPS times, once for each value of I between 1 and NOFSTEPS inclusive.

```
(3)  *[U!(SEED/LIMIT) →
          SEED := (SEEDxMULTIPLIER) MOD LIMIT;
      ]
```

Repeatedly outputs a number to U, and then computes a new value of SEED. Terminates when U terminates.

```
(4)  *[U?RELEASE( ) → FREE := FREE+1
      ▯FREE>0; U?ACQUIRE( ) → FREE := FREE-1
      ]
```

Each repetition either accepts a RELEASE( ) signal from U and then adds one to FREE, or it accepts an ACQUIRE( ) signal from

U and then subtracts one from FREE; but this second alternative can occur only if FREE is originally greater than zero. Thus FREE can never go negative.

```
(5)  *[(I:1..100)ORDER(I)?ACQUIRE( )
               → ORDER(I)?RELEASE( )
      ]
```

Repeatedly accepts ACQUIRE( ) signals from any one of 100 ORDER processes. The bound variable I gives the index of the acquiring ORDER on each occasion. The body of the loop accepts a RELEASE( ) signal from the same I^{th} ORDER. The repetitive command terminates when all hundred ORDER processes have terminated.

```
(6)  *[IN<OUT+100; U?BUFFER(IN MOD 100) →
                    IN := IN+1
      □OUT<IN; U!BUFFER(OUT MOD 100) →
                    OUT := OUT+1
      ]
```

Repeatedly, on request from U,

either (1) (Provided that IN<OUT+100) inputs a value from U, and stores it in the appropriate element of an array BUFFER

or (2) (Provided that OUT< IN) outputs the value of the appropriate element of BUFFER to U.

The repetitive command terminates when U does.

4. EXAMPLES.

In this section, we present a series of examples to show how communicating sequential processes can be used to implement the basic requirements of discrete event simulation, as described in section 2; but the topics are treated in the opposite order.

4.1. Random Numbers.

Problem: Write a process RANDOM to represent a stream of random numbers, as described in section 2.6. The name of the using process is U. The process first inputs from U the value of its seed, and then it outputs a series of random numbers starting with one derived directly from the seed.

Solution:

```
RANDOM::

[SEED: INTEGER; U?SEED; SEED:=SEED MOD LIMIT;
  *[U!(SEED/LIMIT)→
        SEED :=(SEED×MULTIPLIER) MOD LIMIT
]]
```

4.2 Statistics: Histogram

Problem: Write a process to represent a histogram, as described is section 2.5. The name of the using process is U. The histogram first inputs its parameters N, LOW, and HIGH; it then inputs and records a series of integers from U. When U terminates, the histogram is automatically printed in some suitable graphic notation. If any input value is invalid, the process aborts.

Solution: HISTOGRAM::

```
[N,LOW,HIGH, COUNT: INTEGER;
 U,(N,LOW,HIGH); COUNT :=0; [N>0&0≤LOW&LOW<HIGH → SKIP];
 HISTO: ARRAY(0..N-1) OF INTEGER;
 FOR I =0..N-1 DO HISTO(I) :=0; X:INTEGER;
 *[U?X → I:INTEGER; [LOW≤X&X<HIGH → SKIP];
        I :=Nx(X-LOW)÷(HIGH-LOW);
        HISTO(I) :=HISTO(I)+1;
        COUNT :=COUNT+1
  ];.. print the value of HISTO ...
]
```

4.3 Queues: A Fifo Discipline.

Problem: Write a process Q to implement a fifo queue of integers for a user U. The user appends an integer I to the queue by an output command Q!I. It removes the first member of the queue by the input command Q?F which assigns to F the value removed; this will be delayed if the queue is empty. The maximum length of the queue is 100.

Solution:

```
Q::  IN,OUT: INTEGER; IN :=0; OUT :=0;
BUFFER: ARRAY (0..99) OF INTEGER;
*[IN<OUT+100; U?BUFFER(IN MOD 100) →
                  IN := IN+1
[] IN>OUT; U!BUFFER(OUT MOD 100) →
                  OUT := OUT+1
]
```

4.4. Single Resource Allocator.

Problem: A single input device is to be shared among an array of processes

ORDER(I: 1 .. 100):: ...

Each order acquires the device by a command

ALLOC!ACQUIRE( );

it then uses the device, and finally releases it by;

ALLOC!RELEASE( );

Write the process ALLOC, which ensures that at most one ORDER at a time can use the device.

Solution:

```
ALLOC ::
*[(I:1 .. 100) ORDER(I)?ACQUIRE( ) →
               ORDER(I)?RELEASE( )
 ]
```

4.5. Multiple Resource Allocation.

Problem: Write a process FOREMAN to allocate 10 machines among an array of 100 processes:

```
ORDER(I:1) .. 100):: ...
```

An order acquires a machine by a pair of commands:

```
FOREMAN!REQUEST( ); FOREMAN?GRANTED( );
```

and it releases a machine by

```
FOREMAN!RELEASE( )
```

When there are no free machines, the foreman uses a fifo queue to store the identity of the orders whose requests cannot yet be granted.

Solution:

```
FOREMAN::
[Q:: see example (4.3) ...
||U:: FREE: INTEGER; FREE :=10;
  *[(I: 1..100) ORDER(I)?REQUEST( ) →
          FREE := FREE-1;
          [FREE<0 → Q!I
          □FREE≥0 → ORDER(I)!GRANTED( )
          ]

   □(I: 1..100) ORDER(I)?RELEASE( ) →
          FREE := FREE+1;
          [FREE>0 → SKIP
          □FREE≤0 → F:INTEGER;
                    Q?F; ORDER(F)!GRANTED( )
          ]
   ]
]
```

4.6. Machine Shop. [3]

A machine shop contains ten groups of ten machines each. Each group of machines is scheduled by a foreman using a fifo discipline. The machine shop must process a hundred orders. Each order has the following parameters:

1. STARTTIME - the simulated time at which the order enters the shop for processing.

2. NOFSTEPS - the number of steps required for processing the order.

3. For each step, numbered between 1 and NOFSTEPS, there are two parameters:
   (3.1) MACHGROUP - the number of the machine group required to carry out this step

   (3.2) SERVICETIME - the amount of time required to process this step.

These parameters for each order may be read from the input device READER. Each order must acquire exclusive access to the reader before reading its parameters.

Solution: The overall structure of the solution is:

```
[ALLOC:: ... see example (4.4) ...
||U::[FOREMAN(J: 1..10):: ... see example (4.5) ...
     ||ORDER(K: 1..100):: ... see below ...
     ]
]
```

The order process array is:

```
ORDER(K: 1..100)::
```

<u>comment</u> read the parameters for this order;

```
ALLOC!ACQUIRE( ); STARTTIME,NOFSTEPS: INTEGER;
READER?STARTTIME; READER?NOFSTEPS;
MACHGROUP,SERVICETIME: ARRAY(1..NOFSTEPS) OF INTEGER;
I: INTEGER; I := 1;
*[I≤NOFSTEPS → READER?MACHGROUP(I);
               READER?SERVICETIME(I);
               I := I+1
]; ALLOC!RELEASE( );
```

<u>comment</u> start the simulation proper;

```
HOLD(STARTTIME); I := 1;
*[I≤NOFSTEPS → J: INTEGER; J :=MACHGROUP(I);
               FOREMAN(J)! REQUEST( );
               FOREMAN(J)?GRANTED( );
               HOLD(SERVICETIME(I));
               FOREMAN(J)!RELEASE( );
               I := I+1
];
```

5. CONCLUSION.

This paper has shown by example that the general purpose concept of a communicating sequential process is adequate for many of the requirements of a special purpose discrete event simulation language, provided that the concept of simulated (quasiparallel) time is also built into the language. Whether this too can be implemented by some reasonable general-purpose feature is an open question.

The notations described and used in this paper are not recommended for general use as a programming language, since they still suffer from many of the defects summarised in [1], namely,

(1) The static upper bound on the size of an array, including an array of processes. This defect has been masked in the example problems by artificial simplification.

(2) The absence of aids to the construction and use of libraries of standard processes.

(3) The non-existence of an efficient implementation.

Suggestions for the solution of these problems have not been given in this paper.

References.

[1] Hoare, C.A.R. Communicating Sequential Processes Commun. ACM 21,8. (Aug. 1978) 666-677

[2] Dahl, O.-J., Myhrhang, B., Nygaard, K., The Simula 67 Common Base Language NCC. Forskningsreien 1B Oslo (1968).

[3] Dahl, O.-j., Dijkstra, E.W., Hoare, C.A.R., Structured Programming. Academic Press. (1972).

[4] Kaubisch, W.-H., Perrott, R.H., Hoare, C.A.R., Quasiparallel Programming. Software Practice and Experience 6. (1976) 341-356.

[5] Gordon, G., A General Purpose System Simulation Program Eastern Joint Computer Conference Dec. 1961.

[6] Wirth, N., Modula: a language for modular multiprogramming. Software Practice and Experience. 7.3. (1977).

[7] Wirth, N., The Programming Language PASCAL. Acta Informatica 1,1 (1971) 35-63.

STRUCTURE OF AN OPERATING SYSTEM

C.A.R. Hoare*, R.M. McKeag **.

*Programming Research Group, Oxford University Computing Laboratory, Oxford OX2 6PE. England
**Department of Computer Science, The Queen's University, Belfast, BT7 1NN, N. Ireland.

This paper suggests that the structure of an operating system can be clearly expressed as a hierarchy of communicating sequential processes. The suggestion is illustrated by the development of an absurdly simple multiprogrammed batch processing system. It is hoped that the structuring methods and notations may be more widely useful.

Key words and phrases: programming languages, operating systems, program structure, communicating sequential processes.

C.R. Categories. 4.22, 4.32, 4.35.

1. INTRODUCTION.

Some of the reasons for planning and maintaining a clear structure of a large computer program are

(1) To enable the design to proceed in an orderly and intellectually manageable fashion.

(2) To enable different parts of the design to be implemented reliably by different programmers at different times.

(3) To enable the program to be tested systematically in a way that contributes to confidence in its overall correctness.

M. Broy and G. Schmidt (eds.), Theoretical Foundations of Programming Methodology, 643–658.

(4) To enable the program to be readily modified in its general configuration or in the detail of its parts, without risk of unexpected interactions.

(5) To enable the programming conventions which guarantee soundness of the whole structure to be enforced as far as possible by "compile time" checks.

Edsger W. Dijkstra [2] has suggested that an operating system should be structured as a series of levels, each of which uses the lower levels to transform some of the bare hardware into a more desirable family of virtual resources for the benefit of the higher levels, and the highest level consists of the virtual machines in which the user programs run. This paper suggests that the concept of the communicating sequential process is a suitable one for expressing such a structure, and illustrates the suggestion by the stepwise development of an absurdly simple batch processing system.

Of course, when an operating system is expressed in a higher level language, the lowest level implemented by software (or microcode, or even hardware) will be the necessary "run-time support" for that language, and cannot reasonably be implemented as part of a program expressed in that language. In the case of a language incorporating communicating sequential processes, the run-time support must include the allocation of local storage and processor(s) to the processes, and administration of the communication between them, as well as any simulation required to make peripheral devices with interrupts look like communicating sequential processes. This will usually require several hundred machine code or microcode instructions, depending on the vagaries of hardware interface designs.

The language of communicating sequential processes has been described in a previous paper [4]. However, for the elucidation of the structure of an operating system, several extensions are desirable. Section 2 describes a method for dynamic establishment and disestablishment of communication channels between processes. Section 3 describes a scope rule which assists in the multilevel structuring of programs. Section 4 introduces the parallel repetitive command in which the number of activations of a process is not bounded *a prioti*, but is determined by the needs of the rest of the program. Finally, for convenience and symmetry, we have used output commands as guards, in the same way as input commands. This may cause some difficulty for a distributed implementation.

These extensions are not so well suited to implementation on arrays of processors with disjoint main storage; and even when implemented on a single processor (or multiprocessor with shared main store), considerable optimisation of processor allocation, storage allocation, and message passing may be required to achieve reasonable efficiency. It is left as an open question how far such optimisations can be accomplished automatically by a translator, and how far the programmer can guide the optimisation.

The general structuring methods described in this paper reproduce some of the facility of the class and inner concepts of SIMULA 67-1 indeed, an operating system structure based on these SIMULA concepts was presented in [5]. The major advance of the present paper is the introduction of forms of parallelism, input, and output which seems to bring a great conceptual simplification and unification, though possibly at the expense of postponing problems of efficient implementation. If these problems can be solved, it is hoped that the usefulness of these concepts will extend more widely than operating systems.

2. DECLARATION OF COMMUNICATION CHANNELS.

We allow a process name to feature in place of a type in the declaration of a variable, e.g.

ℓp: lineprinter   x:X

If the declaration of "ℓp" (on the left) occurs in a process named "X", and the declaration of "x" (on the right) occurs in a process named "lineprinter", then these two declarations are executed simultaneously, and their effect is to set up a new communication channel between "lineprinter" and "X". The channel is broken down again as soon as either process exits from the scope of the declaration.

Within the scope of "ℓp", there may occur output (or input) instructions, e.g.

ℓp!heading; ℓp!concatenate ("COST IS", decimal (cost));

which communicate through the dynamically established channel to a corresponding input (or output) command within the scope of "x", e.g.

ℓ: line; x?ℓ;

An attempt to communicate using a channel which has been broken down (by exit from scope) will fail, in the same way as communication with a terminated process; and this failure can cause termination of a repetitive command in which the input command appears as a guard.

The purpose of this extension is to permit processes internal to one process to set up communication channels with another process or even the latter's internal subprocesses. Since names of internal subprocesses are local, it is not possible for another process to use these names directly. That is why it is necessary for each subprocess to declare a new local name (e.g. "ℓp","x") by which it refers to its communicant. The declaration of a communication channel can appear as a guard, and will fail if the named process has terminated.

Example 1. Allocation of a single resource.

Problem: A single lineprinter is to be shared among the processes local to a process "X". The lineprinter is to be acquired by declaration (as described above); and the user process may then repeatedly output to it values of type "line". On exit from the scope of the declaration, the lineprinter is released, and is able to respond to further declarations from the same or another process in "X". But only one process at a time should use the lineprinter.

Use English instead of machine code for instructions to the hardware of the lineprinter.

Solution:

lineprinter:: *[x: X→*[ℓ: line; x?ℓ → ...print ℓ ...]]

Notes:

(1) The "lineprinter" process consists of a repetitive command, which terminates if and when the process "X" is terminated; this can happen only when all internal processes of "X" have terminated.

(2) Each repetition first "acquires" a client process from "X", and gives it a local name "x".

(3) It then embarks on an inner repetitive command, each repetition of which prints one line sent by "x".

(4) Each output line is single-buffered in "ℓ", so that "x" can proceed while "lineprinter" is waiting for the hardware to accept the line.

(5) The inner repetitive command terminates when the client process "x" leaves the block in which the given activation of a "lineprinter" was declared.

(6) The outer repetition is then ready again to respond to another output declaration of a "lineprinter" from within "X", either from the same or from a different process.

(7) If "X" contains a nested declaration of a "lineprinter", the "lineprinter" process can never respond to it, and the two processes will be deadlocked.

(8) However, normal scope rules provide a compile-time check against a process in "X" using the actual lineprinter before "acquiring" it, or after "releasing" it; and it is impossible to "forget" to release it (provided, of course, that "X" contains no machine code).

Often there will be not just one resource of a particular type but several resources; and it does not matter which of them is used on a particular occasion. This case is readily treated by a parallel command, with one process per resource; e.g.:

[resource 1 || resource 2 || resource 3]

If the code for the three resources is quite similar, it is more convenient to use the notation of the parallel array, e.g.

[Y(i: 1 .. 3) :: resource].

Here "resource" stands for the code which represents the resource. It may contain (but not assign to) the bound variable "i". The parallel array is equivalent to writing out the code for the resource three times, each time with a different value for "i", ranging between one and three:

[Y(1) :: $resource_1$ || Y(2) :: $resource_2$ || Y(3) :: $resource_3$]

Example 2.

Problem: Same as example 1 but with two lineprinters instead of one.

Solution. lineprinter :: [Y(i: 1 .. 2) :: one lineprinter]

where "one lineprinter" stands for

[x: X→[ℓ: line; x?ℓ→ ... print ℓ on printer i ...]]

Notes:

(1) The "lineprinter" process now contains an array of two internal subprocesses, each of which deals with one lineprinter, in the same way as in the solution to the previous problem.

(2) From within "X", or from within a subprocess of "X", a lineprinter may be acquired and used in exactly the same

way as before: "ℓp: lineprinter". One of the two processes of the "lineprinter" array will respond to this declaration when it is ready to execute its input declaration. There is no way in which "x" may find out which lineprinter it is using.

(3) If "X" contains doubly nested lineprinter declarations, deadlock will result; or if each of two concurrent processes contains a singly nested declaration, there is a risk of deadly embrace [3].

(4) The name "Y" of the local array is never used, since the lineprinters do not need to communicate with each other.

Example 3. A simple multiprogramming system.

A simple multiprogramming system consists of a fixed number of processes (say three), each of which executes a batch of jobs submitted on one of (say) two cardreaders, and prints their output on one of (say) two lineprinters. Each user program is executed in one of three virtual machines. Thus the overall structure of the system is:

```
[X :: [Y(i: 1 .. 3) :: batch processor]
   ||lineprinter :: ... see example 2 ...
   ||cardreader ::  ... left as an exercise ...
   ||virtualmachine :: ... explained below ...
]
```

Each of the three batch processors consists of a repetitive command, which terminates when a switch is off; i.e. "batch processor" stands for:

```
*[switch? on ( ) → execute one job]
```

In order to execute a job, it is necessary to acquire a card reader, a lineprinter, and a "virtual machine", which provides the main storage within which the user's job will be executed. We specify that this store is initialised to contain a standard user program, say a load-and-go compiler, or a control language interpreter. This program is triggered and proceeds in parallel with the batch processor. But since the virtual machine has no input or output devices, it must communicate with the operating system to perform all required reading and printing. It also informs the operating system on completion of each timeslice of (say) ten thousand instructions. This enables the operating system to maintain an account of the cost of each job, and print it out afterwards.

(In practice, the concept of a virtual machine will be implemented by setting base and limit registers, and using supervisor entry and exit instructions; the details are not relevant to the conceptual structure of the operating system. In fact, the relationship between a virtual machine and a batch processor is not necessary for an understanding of the remainder of this paper).

The following program assumes that the user's job always terminates after a reasonable time, and always reads the right number of cards.

"execute one job" stands for:

```
cost: integer; cost := 0;
cr: cardreader; ℓ: line;
ℓp: lineprinter;
job: virtualmachine;
*[job? timeslice ( )→ cost := cost+timecharge
 □job?ℓ →ℓp!ℓ; cost := cost+linecharge
 □job? input ( )→ card: line; cr? card; job!card;
                  cost := cost+cardcharge
]; comment terminates when job exits from scope;
ℓp! concatenate ("COST IS", decimal (cost))
```

Notes:

(1) The operating system is not subject to deadly embrace, because the resources are always acquired in the same order by each batch processor, and they are all released before any of them is acquired again.

(2) We have assumed that each job behaves correctly, in that it reads exactly the right cards from the batch, and terminates after a reasonable time.
The operating system described above is very simple, but that is its only merit - it would be dreadful to use! Among other defects:
  (a) It contains no provision for breakdown of hardware components.
  (b) The assumption of note (2) is wholly unrealistic.
  (c) In practice, shortage of peripheral equipment limits the actual degree of multiprogramming to two, because one of the batch processors will always be waiting for peripherals.

Defects (b) and (c) will be mitigated in the development of later examples.

3. Hole in scope.

In ALGOL 60, when the name of a procedure occurs inside the body of that same procedure, it refers to a recursive activation of that procedure. This is obviously a useful feature of the language, but it creates a slight difficulty in the multilevel structuring of a program. Suppose, for example, that a high level program uses the function "cos", without caring how it is implemented. At a lower level, it is decided that the standard function "cos" is not suitable, and it should be replaced by a programmed procedure. The scope rules of ALGOL 60 provide an excellent method of doing this, without changing the text of the high level program; simply replace the "high level program" by:

```
begin real procedure cos(x); ... new cos procedure body ...;
      high level program
end
```

All occurrences of the identifier "cos" in the high level program are "captured" by the local declaration, and do not reach the more global standard function. But the difficulty occurs when the new cos procedure body needs to call the <u>standard</u> function "cos"; since in ALGOL 60 a use of this identifier would make a recursive call on the <u>new</u> "cos", which is certainly not wanted!

For this reason we adopt a different scope rule for names of communicating sequential processes. We permit the body of a process to mention its own name (or that of a textually enclosing process) in an input or output command, but specify that this denotes a <u>non</u> enclosing process with that same name. Thus the scope of a process name extends over all other processes in the same parallel command, but it does <u>not</u> include the body of the named process.* In all other respects, the normal ALGOL scope rules still apply - a name denotes the process to which that name is prefixed in the smallest enclosing parallel command. (In fact this is the only reasonable interpretation of a process which mentions its own name; since an attempt to communicate with itself (or an enclosing process) would be always unsuccessful and an attempt at recursive communication would seem meaningless. But further discussion of recursion is beyond the scope of this paper).

* This rule conflicts with the use of process array names in [4].

Less formally, if we regard a process as the "ancestor" of all its internal subprocesses, then a process name occurring in an input or output command always refers to its brother, or its uncle, or its great uncle, or its great great uncle, etc; and it always refers to the closest possible member of this series. It never refers to a direct ancestor.

Example 4.

Problem: The multiprogramming system of example 3 assumes that each job will read all the cards relevant to that job, and no more. This is an unrealistic assumption. We shall therefore stipulate that the cards of each job are followed by a special separator card, and we need to rewrite the system to ensure that if any job attempts to read beyond the separator its input requests will be met by the simple trick of replicating the separator card; and if the job terminates before the separator card is read, the remaining cards of the job (including the separator) are read and ignored, so that the next job will start properly at the beginning of its card deck.

An additional advantage of the separator card is that the cardreader can be deallocated as soon as the separator is read. Thus, if the jobs tend to finish their input early, it will often be possible for more than two jobs to run concurrently.

Solution:

replace "execute one job" in example 3 by

[X :: execute one job || cardreader :: separate input]

where "separate input" is

```
x:X;
[cr: cardreader;
 c:line; cr?c; comment read one card ahead;
*[c≠separator; x!c→cr?c];
comment either c = separator or x has terminated;
*[c≠separator→cr?c]; comment skip unread cards (if any);
]; comment the real cardreader is released here;
*[x!separator→skip]
```

Notes:

(1) The declaration "x:X" responds to the declaration "cr:cardreader" in "execute one job", which is now the closest process with name "X".

(2) The declaration "cr: cardreader" acquires a real cardreader from the more global process with the name "cardreader",

even though it occurs within a process which itself is named "cardreader".

(3) The last repetitive command sends separator cards to satisfy any additonal input commands from "x".

Example 5.

Problem: The output produced by a batch also requires separator lines, so that material output by each job can be conveniently detached and returned to its owner. Adapt the system of example 4 to ensure that output from each job is followed by a separator; and that any attempt to output further lines after a separator is ignored

It is advantageous also to delay acquisition of the "real" lineprinter until the first line has to be output. Thus if the jobs tend to engage in significant computation before their first output, it will often be possible for more than two jobs to run concurrently.

Solution: replace "execute one job" by

```
[X :: execute one job || lineprinter :: separate output]
```

where "separate output" is

```
x:X; ℓ: line;
*[x?ℓ → [ℓp:lineprinter; ℓp!ℓ;
       *[ℓ≠separator; x?ℓ → ℓp!ℓ];
        [ℓ=separator → skip [] ℓ≠separator → ℓp!separator]
      ] ; comment real lineprinter released here;
      *[x?ℓ → skip] comment ignore lines after separator
]
```

Notes:

(1) The first loop is iterated at most once (and not at all, if a job has no output).

(2) In practice it would be a good idea to ensure that all separator lines are printed on double-page boundaries, to facilitate bursting by an operator.

(3) Removing the unnecessary nesting, the overall structure of the operating system is now:

```
[X :: [Y(i: 1 .. 3) :: *[switch?on ( ) →
                            [X:: execute one job
                            ||cardreader :: separate input
                            ||lineprinter :: separate output
                            ]
                         ]
      ]
||cardreader: ... left as an exercise ...
||lineprinter: ... see example 2 ...
||virtualmachine: ... explained above ...
]
```

4. The Parallel repetitive command.

A parallel repetitive command is like the normal sequential repetitive command, in that it involves a dynamically determined number of activations of its body; it differs only in that each activation proceeds in parallel with all those that started earlier. The notation for a parallel repetitive command will be the same as that of the sequential repetition, except that a double star ** will be used in place of the single star *.

To ensure disjointness, we must stipulate that the body of a parallel repetitive command must not update any global variables at all. Consequently, the normal method of termination of repetitive commands (when their Boolean guards become false) is not applicable; so the guards on a parallel repetitive command must be input or output guards or declarations, which cause termination when all their sources and destinations have terminated.

Example 6.

Problem: The efficiency of a multiprogramming system can be greatly increased by the technique of spooled (pseudoofflined) output. The lines, when output by a job, are not transmitted to a real lineprinter; instead, they are copied to a file on backing store, and actual printing is started only when the job is complete. Adapt the system of example 5 by including spooled output; assume an implementation of the concept of a file, to which lines may be output, and which must be rewound before they can be input again.

Solution: The only change required is to replace the code of example 2 by

```
lineprinter :: [X :: output spooling || ... example 2 ...]
```

where "output spooling" stands for

```
**[x:X → f:file; ℓ:line;
   *[x?ℓ → f!ℓ]; comment x has terminated;
   f!rewind ( );
   ℓp: lineprinter; comment only now, acquire a real
                                              lineprinter;
   *[f?ℓ → ℓp!ℓ]; comment the file has terminated;
  ]
```

Notes:

(1) There is no *a priori* limit to the number of activations of this parallel repetitive command that may be in concurrent execution.
(2) Each activation is initiated by a declaration "ℓp: lineprinter" from within the more global "X", i.e., the batch processors.
(3) When the more global "X" terminates, no further activations of the parallel repetitive command are initiated. The command then terminates after all outstanding activations have terminated.
(4) Of course, since there are only two real lineprinters, only two activations of the command can be executing their second loop simultaneously; and since there are only three jobs in concurrent execution, at most three of them can be executing their first loop. All the rest of them will be rewinding their files or waiting for a lineprinter or a file.

Example 7.

Problem: Efficiency of operation may also be increased by spooling of input.

Solution: replace the original "cardreader" process of example 3 (exercise) by cardreader

```
::[X :: input spooling || cardreader :: ...]
```

where "input spooling" stands for

```
**[x:X → f: file; c: line;
   [cr: cardreader; comment real card reader;
    cr?c;
    *[c≠separator → f!c; cr?c]; comment c = separator;
   ]; comment real cardreader released here;
   f!separator; f!rewind ( );
   comment now the user x can begin input;
   *[f?c → x!c];
   comment we have already ensured that the job will read
   all cards up to the separator, and no more;
  ]
```

The maximum number of concurrent instances of the input spooling is equal to the number of concurrent batch processors; it would therefore be advantageous to increase the number of batch processors to (say) ten. Of course, it is likely that most of them will spend most of their time waiting for a virtual machine in which to run a job; but this is good, because it ensures that there will usually be a load of work waiting for the central processor(s).

5. Summary.

After all the developments of the previous sections, it is helpful to display the overall structure of the complete spooled multiprogramming system (see figure 1).

```
[X :: [Y(i: 1 .. 10) :: *[switch?on ( ) →
                           X :: execute one job
                           || cardreader :: separate input
                           || lineprinter :: separate output
     ]                  ]
||cardreader :: [X :: input spooling
              || cardreader:: ...     exercise ...]
||lineprinter :: [X :: output spooling
              || lineprinter:: ...    example 2 ...]
||virtual machine :: ... explained above ...
]
```

Figure 1.

The persistent (even perverse) reuse of the same identifiers at every branch and every level of the structure is the result of the way we have chosen to present its development by stepwise enrichment. The structure can also be displayed pictorially without redundant names. Figure 2 omits the virtual machines and the filing system: it indicates physical containment of processes and subprocesses by solid lines, and communication channels by dotted lines. A process with more than one instance is indicated by a double box.

For practical use, this multiprogramming system still suffers from many defects, including

(1) There should be a cost limit imposed on each job,
(2) Cards unread by a job should be printed out, to assist in diagnosis.
(3) Better methods are required for job identification and accounting.
(4) No provision is made for rerunning jobs which have already been input when the hardware breaks down.
(5) It is not possible to ensure that more urgent jobs overtake less urgent ones.
(6) There is no way of avoiding the spooling of exceptionally long files.
(7) No job can use more than one input and one output file.
(8) No job can use files as temporary working storage, ..
(9) ... or for long term storage of information to be processes by succeeding jobs.

A remedy of these defects requires introduction of a "job description card" and a major revision of "execute one job"; and the overall system would be much more complicated than that described in this paper. It is to be hoped that the same structuring methods may be helpful in controlling the additional complexity by assisting in the achievement of some of the objectives of program structure listed at the beginning of this paper.

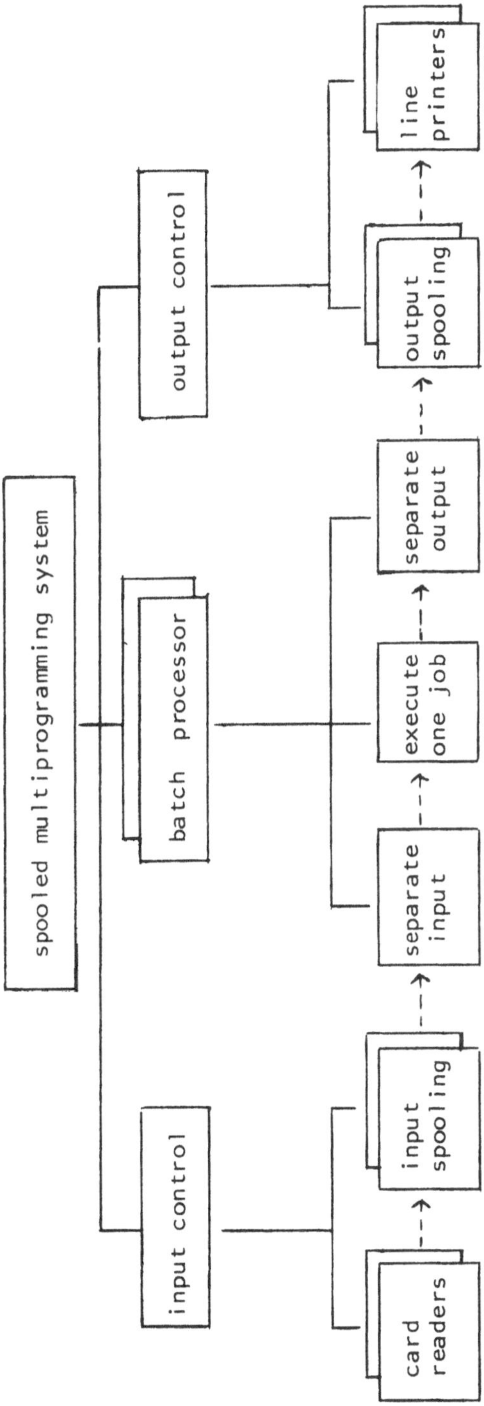
spooled multiprogramming system
input control
batch processor
output control
card readers
input spooling
separate input
execute one job
separate output
output spooling
line printers

Figure 2.

Acknowledgements.

The work of the first author was supported by a Senior Fellowship of the Science Research Council of Great Britain.

The ideas reported arose from a research project into "A Model Operating System", also supported by the SRC; and have benefited from an implementation of a similar approach by D.W. Bustard and S.A.J. Clarke.

They have also been improved by the useful advice of M.V. Wilkes, E.W. Dijkstra, C. Hewitt, M.K. Harper and D.W. Bustard.

References.

1. Birtwhistle, G.M. Simula Begin. Auerbach, London, 1973.

2. Dijkstra, E.W. The Structure of the T.H.E. multiprogramming system. C.A.C.M. 11, 5 (May 1968), 341-346.

3. Dijkstra, E.W. Co-operating sequential processes. In Programming Languages (F. Genuys, ed.), 43-112. Academic Press, London, 1968.

4. Hoare, C.A.R. Communicating sequential processes. C.A.C.M. 21, 8 (August 1978), 666-677.

5. Hoare, C.A.R. The Structure of an Operating System. Draft May 1975 in "Language Hierarchies and Interfaces" Lecture Notes in Computer Science No. 46 Springer 1976.